2nd International Symposium on Frontier of Applied Physics (ISFAP 2016)

Journal of Physics: Conference Series Volume 817

Jakarta, Indonesia
3 – 5 October 2016

Editors:

Febty Febriani

ISBN: 978-1-5108-3902-1
ISSN: 1742-6588

Printed from e-media with permission by:

Curran Associates, Inc.
57 Morehouse Lane
Red Hook, NY 12571

Some format issues inherent in the e-media version may also appear in this print version.

Printed by Curran Associates, Inc. (2017)

Institute of Physics
Dirac House, Temple Back
Bristol BS1 6BE UK

Phone: 44 1 17 929 7481
Fax: 44 1 17 920 0979

techtracking@iop.org

Additional copies of this publication are available from:

Curran Associates, Inc.
57 Morehouse Lane
Red Hook, NY 12571 USA
Phone: 845-758-0400
Fax: 845-758-2633
Email: curran@proceedings.com
Web: www.proceedings.com

Table of contents

Volume 817

2nd International Symposium on Frontier of Applied Physics (ISFAP 2016)
3–5 October 2016, Jakarta, Indonesia

Accepted papers received: 28 February 2017
Published online: 7 April 2017

Preface

Papers

Advanced Materials, Materials for Energy and Nanotechnology (AM)

Laser and Optoelectronics (LO)

Extractive Metallurgy and Materials Process Engineering (EM)

Surface and Corrosion Science (SC)

IOP Conf. Series: Journal of Physics: Conf. Series **817** (2017) 011001 doi:10.1088/1742-6596/817/1/011001

Preface: Proceedings of the International Symposium on Frontier of Applied Physics (ISFAP) 2016

The International Symposium of Frontier of Applied Physics (ISFAP) 2016 was a scientific meeting that is technically organized by Research Center for Physics in conjunction with Indonesian Science Technology Festival organized by Deputy of Engineering Sciences, Indonesian Institute of Sciences (LIPI). ISFAP was held on 3 - 5 October, 2016 at Indonesia Convention Exhibition (ICE), Serpong, Indonesia. The event was supported by Indonesian Physical Society (HFI), Indonesian Institute of Sciences (LIPI), and Indonesian Society for Microscopy (Ina-SM).

This symposium aimed to bring together researchers, academicians, and scholars all around the world to share information and their research results. This event was also expected to become a beneficial media in exchanging their innovative ideas and build research collaborations in the future. The symposium covered all scientific and technical activities in physics – from basic to the advanced applications – especially those related to advanced material sciences for energy, nanotechnology, geophysics, lasers and optoelectronics, magnetism and superconductivity, hybrid, ceramics, and composite materials, material process engineering, surface and corrosion science.

We specially invited 4 scientists as keynote speakers, 4 researchers as invited speakers, and 4 researchers as selected speakers from various countries to share their knowledge to all participants. The total participants attending this symposium were 133 persons. This symposium has successfully attracted 109 submitted papers. Papers that have been presented in the ISFAP 2016 were eligible for publishing in IOP Journal of Physics: Conference Series. The editorial board has decided to accept 75 papers from Indonesia, Philippines, Nigeria, South Africa, Saudi Arabia, Egypt, Malaysia, Germany, and Taiwan. The accepted papers have passed through strict reviewing process by scientific committee of this symposium. EDAS system has been used to facilitate the submission, reviewing, and revision processes.

We would like to appreciate all of our co-organizers, supporting societies, sponsors, and Research Center for Physics staffs who have contributed generous amount of grants and effort to support us during the symposium. We hope this symposium has successfully enabled us to share and communicate our knowledge and skills and to build collaborations among the participants. We are looking forward to welcoming you in Indonesia again. See you in ISFAP 2017!

IOP Conf. Series: Journal of Physics: Conf. Series **817** (2017) 011002 doi:10.1088/1742-6596/817/1/011002

Editorial Board

Eni Sugiarti (eni.sugiarti@lipi.go.id)
Indonesian Institute of Sciences (LIPI)
Research Center for Physics
Komplek PUSPIPTEK Serpong,
Tangerang Selatan 15314,
Indonesia

Titi Anggono (titi.anggono@lipi.go.id)
Indonesian Institute of Sciences (LIPI)
Research Center for Physics
Komplek PUSPIPTEK Serpong,
Tangerang Selatan 15314,
Indonesia

Isnaeni (isnaeni@lipi.go.id)
Indonesian Institute of Sciences (LIPI)
Research Center for Physics
Komplek PUSPIPTEK Serpong,
Tangerang Selatan 15314,
Indonesia

Yuliati Herbani (yuliati.herbani@lipi.go.id)
Indonesian Institute of Sciences (LIPI)
Research Center for Physics
Komplek PUSPIPTEK Serpong,
Tangerang Selatan 15314,
Indonesia

Ferensa Oemry (ferensa.oemry@lipi.go.id)
Indonesian Institute of Sciences (LIPI)
Research Center for Physics
Komplek PUSPIPTEK Serpong,
Tangerang Selatan 15314,
Indonesia

Hubby Izzuddin (hubby.izzuddin@lipi.go.id)
Indonesian Institute of Sciences (LIPI)
Research Center for Physics
Komplek PUSPIPTEK Serpong,
Tangerang Selatan 15314,
Indonesia

Widi Astuti (widi.astuti@lipi.go.id)
Indonesian Institute of Sciences (LIPI)
Research and Development Division for Mineral Technology
Jl. Ir Sutami Km.15 Tanjung Bintang, Lampung Selatan
Lampung. 35361
Indonesia

Wahyu Bambang Widayatno (wahyu.bambang.widayatno@lipi.go.id)
Indonesian Institute of Sciences (LIPI)

Research Center for Physics
Komplek PUSPIPTEK Serpong,
Tangerang Selatan 15314,
Indonesia

Deni Shidqi Khaerudini (deni.shidqi.khaerudini@lipi.go.id)
Indonesian Institute of Sciences (LIPI)
Research Center for Physics
Komplek PUSPIPTEK Serpong,
Tangerang Selatan 15314,
Indonesia

Kirana Yuniati Putri (kirana.yuniati.putri@lipi.go.id)
Indonesian Institute of Sciences (LIPI)
Research Center for Physics
Komplek PUSPIPTEK Serpong,
Tangerang Selatan 15314,
Indonesia

Fredina Destyorini (fredina.destyorini@lipi.go.id)
Indonesian Institute of Sciences (LIPI)
Research Center for Physics
Komplek PUSPIPTEK Serpong,
Tangerang Selatan 15314,
Indonesia

Nurfina Yudasari (nurfina.yudasari@lipi.go.id)
Indonesian Institute of Sciences (LIPI)
Research Center for Physics
Komplek PUSPIPTEK Serpong,
Tangerang Selatan 15314,
Indonesia

Qolby Sabrina (qolby.sabrina@lipi.go.id)
Indonesian Institute of Sciences (LIPI)
Research Center for Physics
Komplek PUSPIPTEK Serpong,
Tangerang Selatan 15314,
Indonesia

Risma Sundawa (risma.sundawa@gmail.com)
Indonesian Institute of Sciences (LIPI)
Research Center for Physics
Komplek PUSPIPTEK Serpong,
Tangerang Selatan 15314,
Indonesia

IOP Conf. Series: Journal of Physics: Conf. Series **817** (2017) 011003 doi:10.1088/1742-6596/817/1/011003

Peer review statement

All papers published in this volume of *Journal of Physics: Conference Series* have been peer reviewed through processes administered by the proceedings Editors. Reviews were conducted by expert referees to the professional and scientific standards expected of a proceedings journal published by IOP Publishing.

Synthesis and characterization of 8YSZ/5GDC composite and dissimiliar-bulk by sonochemistry assisted sol-gel method

A K Rivai[1], N Shabrina[2] and A R Setiawan[2]

[1]Center for Science and Technology of Advanced Materials, BATAN, Kawasan PUSPIPTEK Gedung 71, Tangerang Selatan, Banten 15314
[2]Material Science Department, FTMD, ITB, Ganesha No.10 Bandung 40132

Email: rivai.abukhalid@batan.go.id

Abstract. Synthesis and characterization of 8YSZ (8%mol Y_2O_3 – ZrO_2) and 5GDC (5%mol Gd_2O_3–CeO_2) combination in the form of composite and dissimilar bulk using sonochemistry assisted sol-gel method have been done. 8YSZ and 5GDC powders were produced from sol-gel route method. Each of the powder was then ultrasonicated for 30 hours with 25kHz±50Hz of frequency and 39% of amplitude. Toluene was used as a solution for the ultrasonication process. 8YSZ/5GDC composite was made by mixing the powders of 50%:50% in weight, and afterward compacted become a pellet. 8YSZ/5GDC dissimilar bulk was made by connected between 8YSZ pellet and 5GDC pellet at a flat part, and afterward re-compacted become a single pellet. Then, the pellets were sintered at 1400°C of temperature for 3 hours. The results showed that the ionic conductivity of the composite is higher than the dissimilar bulk.

1. Introduction

Yttria-Stabilized Zirconia (YSZ: Y_2O_3-ZrO_2) ceramic is widely used in various fields, such as thermal barrier coating (TBC) for high temperature materials [1], oxygen sensor for nuclear reactors[2, 3], and Solid Oxide Fuel Cell (SOFC) [4] because of its superior characteristics. SOFC is an electrochemical cell that converts the chemical energy of the fuel into the electrical energy without any combustion process [4]. Development of SOFC is directed for lowering their operating temperature in order to reduce the fabrication costs and to expand the area of application of this technology. Lowering the operating temperatures of solid oxide fuel cells (SOFCs) from conventional 900–1000◦C to the intermediate temperature range of 600–700◦C allows a wider selection of materials for SOFC components, a better thermal integrity of the cells, and thus a longer life time of the cells [4]. Electrolyte materials with higher ionic conductivity are needed in order to improve the SOFC efficiency at lower operating temperatures. YSZ is a conventional electrolyte material with high ion conductivity of 0.1 S/cm at 1000°C; however, its high operation temperature significantly limits its applications due to the materials' thermomechanical degradation and raises the overall cell costs because of the ceramic interconnects required at high operation temperatures [4]. This material reacts with cathode LSM (lantahum strontium manganite) at higher temperature producing insulating phases such as $La_2Zr_2O_7$ and $SrZrO_3$ at the cathode-electrolyte interface [5]. It blocks the ion migration. In addition, the ionic conductivity of YSZ in the intermediate operating temperature (600°C) is low.

GDC (Gadolinia-Doped Ceria: Gd_2O_3–CeO_2) has been considered as one of the most promising electrolyte materials for the development of intermediate temperature SOFCs because of its high ionic conductivity [6, 7]. Furthermore, the reduction of ceria also causes lattice expansion of the ceria electrolyte at the anode side. However, Ce^{4+} reduces to Ce^{3+} under reducing anode atmosphere at elevated temperatures. This induces an electronic conductivity which decreases an ionic conductivity and destabilizes the chemical reactivity. One proposed approach to prevent the OCV (open circuit voltage) drop and to improve the chemical and mechanical stability of doped ceria electrolytes in reducing atmospheres is to coat a thin YSZ film on the anode side of a doped ceria electrolyte [4, 7-9]. Since the ionic conductivity of YSZ is lower than GDC at an intermediate temperature, the YSZ film as an electronic and hydrogen blocking layer should be adequately thin to minimize its contribution to the total electrolyte resistance.

In this paper, we develop a simple, rapid and cost-effective method based on the sonochemistry assisted sol-gel method processing. The sol-gel technique has been widely used in recent years to form fuel cell components with enhanced microstructures for SOFCs [10]. However, the sol-gel method also has some disadvantages such as a complex synthesis route; the quality of the coating was influenced much by many process parameters. Furthermore, the typical gels show inherently large shrinkage, usually 20–25 vol%, which will introduce huge shrinkage and stresses during drying/heating process. Therefore, the sol-gel process also needs further development to deposit micro-structured materials in a selective area such as the triple-phase boundary of a fuel cell [10].

The alternative potential method that does not required mixing with other substance is the sonochemistry method. This method offers some advantages, such as simple and rapid preparation, direct and precise control of stoichiometry, homogeneity and high purity [11]. As an assisted method, the sonochemistry method is very useful to support with other synthesis methods [11-13].

In this work, we have synthesized and investigated the combination of 8YSZ and 5GDC in the form of composite and dissimilar bulk by sonochemistry-assisted sol-gel method. In order to investigate the performance of the synthesized materials, the conductivity performance tests of 8YSZ/5GDC in the form of composite and dissimilar bulk have been observed.

2. Experiment

The powders of 8YSZ (8%mol $Y_2O_3 - ZrO_2$) and 5GDC (5%mol $Gd_2O_3 - CeO_2$) were synthesized by sonochemistry-assisted sol-gel method using Zirkonium(IV) oxide chloride-octahidrate, Yttrium(III) nitrate hexa-hydrate and cerium nitrate hexa-hydrate ($Ce(NO_3)3.6H_2O$ E.Merck), gadolinium(III) nitrate hexa-hydrate ($Gd(NO_3)3.6H_2O$, Aldrich) as starting materials of 8YSZ and 5GDC. Each stoichiometric amounts of starting materials were dissolved in distilled water. An anhydrous citric acid ($C_6H_8O_7$, E. Merck) was dissolved in water and then was added to the solution. NH_4OH was added to the solution to control pH=5. The solution was heating at 150°C until gel formed. The gel was calcined at 800°C for 4.5 hours until fine 8YSZ white powder and 5GDC yellow powder were formed. Then, each powder with toluene solution as a medium was ultrasonicated for 30 hours with the frequency of 25kHz±50Hz and 39% amplitude.

Combination of 8YSZ and 5GDC was done in the form of composite and dissimilar bulk using 50:50 mass compositions. The composite of 8YSZ/5GDC was made by mixing the powder of 8YSZ and 5GDC. Then, the mixing powder was compacted and sintered at 1400^0C of temperature for 3 hours. The dissimilar bulk of 8YSZ/5GDC was made by compacting the piled up of the powder of 8YSZ and 5GDC. Then, the resulting pellet was also sintered using the same parameters.

In order to ensure the phase of each combination, an X-ray Diffraction characterization was carried out before and after sintering process. XRD data were collected with Cu Kα radiation in the range of 2θ from 10°-120°.

The conductivity of the samples was conducted using EIS (Electrochemical Impedance Spectroscopy). Pt electrodes were coated on both sides of the dense samples. LCR - inductance (L), capacitance (C), and resistance (R) meter measurement was carried out in dry air in a temperature range 400-600°C in frequency range of 42 Hz to 5MHz using HIOKI-3522. The ionic conductivity, the bulk resistance (Rb) and grain boundary resistance (Rgb) were measured using EIS Spectrum Analyser Software using LevMarq fitting and Amplitude function. The ionic conductivity was determined by the Arrhenius law subsection: $\sigma T = \sigma_o exp\left(\frac{-Ea}{kT}\right)$, where σ_o is the pre-exponential factor, E_a is the activation energy (in eV), k is the Boltzmann constant and T is in kelvin.

3. Results and Discussion

The XRD technique was used to determine the crystal structure of the composite 8YSZ/5GDC. Spectra of 5GDC powders were obtained with the range of 2θ from 10°-120° with 5°/s scanning rate. The black line in figure 1 indicated peaks belonging to GDC and YSZ before sintering. The analysis showed that the composite has been mixed well without any impurity. Diffraction patterns are matched with JCPDS 00-075-0161 for GDC which has a fluorite cubic crystal with space group Fm3m, and JCPDS 00-030-1468 for YSZ which also has a fluorite cubic crystal with space group Fm3m. After sintering, all the compositions exhibited a single-phase solid solution with a fluorite structure which is also similar before the sintering cases. However, there is a new found phase which indicated the reaction between 8YSZ and 5GDC after sintering process, as shown by the red line in figure. It means that there is a reaction and diffusion between 8YSZ and 5GDC during the sintering process at temperatures of 1400^{0}C. This reaction occurred in whole sides in 8YSZ/5GDC composite. Differently in dissimilar bulk, the reaction occurred in the interface of 8YSZ/5GDC.

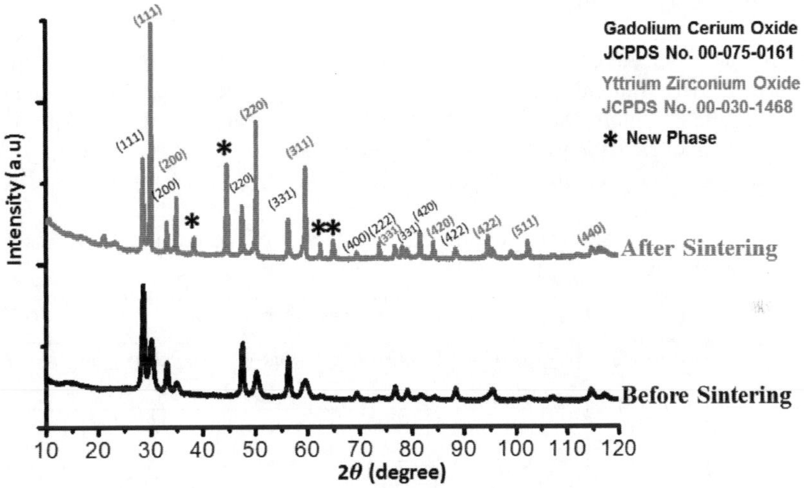

Figure 1. X-ray diffraction pattern of composite 8YSZ/5GDC (red line) before sintering and (black line) after sintering in range 10-120° of 2θ.

In order to investigate the reaction phenomena to its conductivity, the EIS was used to characterize the electrical behavior of them. The ionic conductivity measurement in combination 8YSZ/5GDC was conducted to determine the characteristics of its electrical properties in the form of composite and dissimilar bulk. Measurements were carried out at a temperature range of 400-600°C. The total impedance value was obtained by fitting the measured results with R series paralleled and the CPE (constant-phase elements) models.

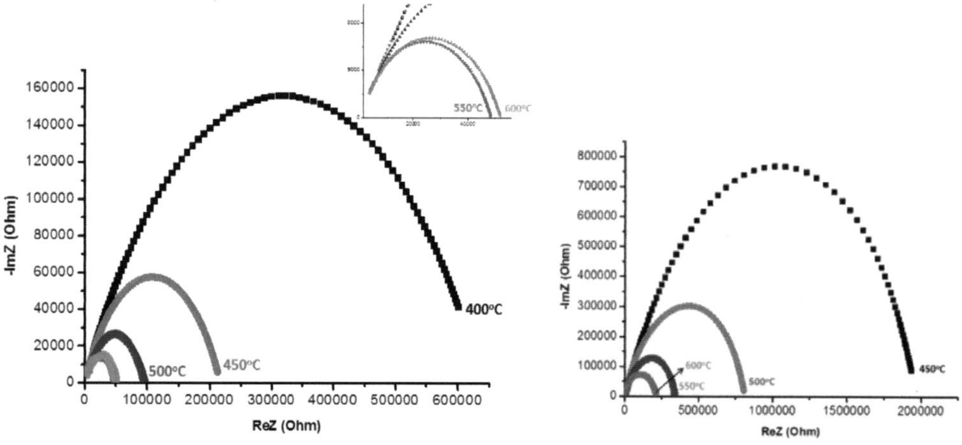

(a) (b)

Figure 2. Impedance plots of (a) composite 8YSZ/5GDC (b) dissimilar bulk 8YSZ/5GDC measured at temperature of 400-600°C

The Nyquist plot for both composite and dissimilar bulk of 8YSZ/5GDC showed the same trends that the resistance decreases with increasing of operating temperature. At a temperature of 400°C, the Nyquist plots of the dissimilar bulk of 8YSZ/5GDC generate no semicircle shape and tend to be random, so it become difficult to be fitted. This may be occurred because the ion oxide of the dissimilar bulk does not have enough energy to migrate in the operating temperature of 400°C.

Table 1. Resistance of bulk, grain boundaries and total of composite and dissimilar bulk of 8YSZ/5GDC at temperature range of 450-600°C

Temp. (°C)	Rb (Ohm)		Rgb (Ohm)		Rtot (Ohm)	
	Composite	*Dissimilar bulk*	Composite	*Dissimilar bulk*	Composite	*Dissimilar bulk*
400	6.73×10^3	-	6.38×10^5	-	6.44×10^5	-
450	6.10×10^3	9.72×10^5	2.17×10^5	1.00×10^6	2.23×10^5	1.97×10^6
500	5.24×10^3	2.21×10^5	9.72×10^4	5.92×10^5	1.02×10^5	8.13×10^6
550	4.79×10^3	1.00×10^5	4.85×10^4	2.41×10^5	5.33×10^4	3.41×10^6
600	8.65×10^2	8.01×10^4	5.23×10^4	1.32×10^5	5.32×10^4	2.12×10^6

Table 1 shows the value of Rb, Rgb, and Rtot for each composite and dissimilar bulk of 8YSZ/5GDC. It shows that resistance decreases with the increasing operating temperature of both the resistances of grains, grain boundaries and total. Then, the decreasing resistance causes higher ionic conductivity. The resistance of composite is smaller than dissimilar bulk. Thus, the value of ionic conductivity of composite is higher than dissimilar bulk.

Table 2. Ionic conductivity of composite and dissimilar bulk of 8YSZ/5GDC at a temperature range of 450-600°C

Temperature (°C)	Total Conductivity			
	Composite	*Dissimilar bulk*	8YSZ	5GDC
400	6.38 x10^{-7}	-	5.73 x10^{-6}	1.38x10^{-4}
450	1.84 x10^{-6}	2.86 x10^{-7}	4.97 x10^{-5}	1.69x10^{-4}
500	4.02 x10^{-6}	6.93 x10^{-7}	1.25 x10^{-4}	2.00x10^{-4}
550	7.72 x10^{-6}	1.65 x10^{-6}	4.39 x10^{-4}	2.29x10^{-4}
600	7.73 x10^{-6}	2.65 x10^{-6}	7.35 x10^{-4}	8.56x10^{-4}

Table 2 shows the ionic conductivity of composite and dissimilar bulk of 8YSZ/5GDC. The reason of selecting the combination of 8YSZ/5GDC in the form of dissimilar bulk is to protect decreasing the ionic conductivity by using an electron blocking layer. The reduction of Ce^{4+} to Ce^{3+} in GDC has caused lower the ionic conductivity [14]. With the 8YSZ as an electron blocking layer, it is expected that the ionic conductivity will become higher. The result showed the ionic conductivity of composite is higher than dissimilar bulk in composition 50:50. It means that dissimilar bulk is not effective enough to increase the ionic conductivity.

The highest ionic conductivity on this combination system is in the form of composite 8YSZ/5GDC. However, the ionic conductivity value is still lower than the ionic conductivity of 8YSZ and 5GDC themselves. The phenomenon has also been found in another report [15], that the reactions occurred at whole side (for composite) and at the interface (for dissimilar bulk) exhibit lower ion conductivity results by almost two orders of a magnitude than that of YSZ and GDC themselves. It caused by an extensive inter-diffusion between 8YSZ and 5GDC solid solution.

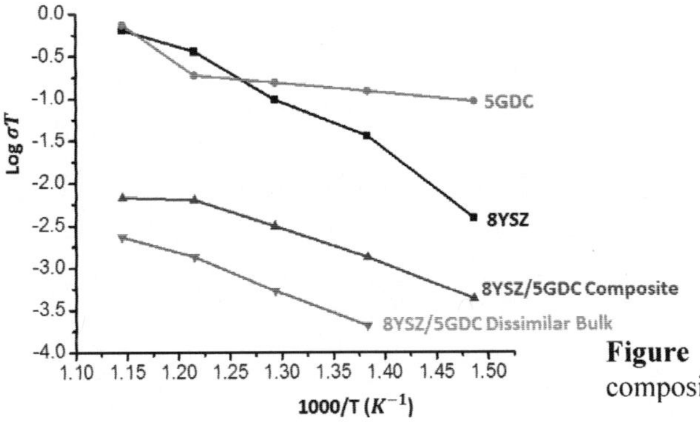

Figure 3. Arrhenius plot of 8YSZ, 5GDC, composite and dissimilar bulk of 8YSZ/5GDC

The activation energy of the composite and dissimilar bulk is obtained through the line slope values in the figure 3 chart through the arrhenius equation. The obtained activation energy values for composite and dissimilar bulk were 0.316 eV and 0.388 eV, respectively. The activation energy of composite is lower than dissimilar bulk which is also supported by its higher ionic conductivity. The activation energy value of the combination of 8YSZ/5GDC is located in the middle between the conductivity of 8YSZ and 5GDC themselves.

4. Conclusion

The combination electrolytes of 8YSZ and 5GDC in the form of composite and dissimilar bulk have been fabricated by sonochemistry-assisted sol-gel method. A new phase reaction formed between 8YSZ and 5GDC occurred after sintering process at temperature 1400°C. The ionic conductivity of composite is higher than dissimilar bulk. However, for both combinations, the ion conductivity is still lower than the both ion conductivity of YSZ and GDC themselves. The phenomenon occurred because of an extensive inter-diffusion between YSZ and GDC solid solution.

Acknowledgement

Authors would like to express the gratitude for Dra. Evi Yulianti, M.Si., Imam Wahyono, S.ST, and Drs. Bambang Sugeng for their helps in the characterizations, and the Head of Center for Science and Technology of Advanced Materials – National Energy Agency of Indonesia (Batan) for the support and encouragement.

References

[1] Campo L d, De Sousa Meneses D, Wittmann-Ténèze K, Bacciochini A, Denoirjean A and Echegut P 2014 *J. Phys. Chem. C* **118**(25) 13590-7
[2] Rivai A K, Kumagai T and Takahashi M 2008 *Prog. Nucl. Energy* **50**(2-6) 575-81
[3] Munggaran G P D, Fitriyani D and Rivai A K 2014 *Jurnal Fisika UNAND* **3**(2) 102-7
[4] Cho S, Kim Y, Kim J-H, Manthiram A and Wang H 2011 *Electrochim. Acta* **56**(16) 5472-7
[5] Dey T, Das Sharma A, Dutta A and Basu R N 2014 *J. Alloys Compd.* **604** 151-6
[6] Choi H, Cho G Y and Cha S-W 2014 *International Journal of Precision Engineering and Manufacturing-Green Technology* **1**(2) 95-9\
[7] Morales M, Roa J J, Capdevila X G, Segarra M and Piñol S 2010 *Acta Mater.* **58**(7) 2504-9
[8] Jee Y, Cho G Y, An J, Kim H-R, Son J-W, Lee J-H, Prinz F B, Lee M H and Cha S W 2014 *J. Power Sources* **253** 114-22
[9] Kwon T-H, Lee T and Yoo H-I 2011 *Solid State Ionics* **195**(1) 25-35
[10] Zuo C, Liu M and Liu M 2012 Solid Oxide Fuel Cells *Sol-Gel Processing for Conventional and Alternative Energy* (USA: Springer) pp 7-36
[11] Okkay H, Bayramoglu M and Faruk Öksüzömer M 2013 *Ultrason. Sonochem.* **20**(3) 978-83
[12] Bu Y-f, Zhong Q, Tan W-y, Zhou R-j, Song Y and Cai W 2013 *Mater. Sci. Semicond. Process.* **16**(6) 2058-62
[13] Okkay H, Bayramoğlu M and Öksüzömer M F 2013 *Ceram. Int,* **39**(5) 5219-25
[14] Jacobson A J 2010 *Chem. Mater.* **22**(3) 660-74
[15] Tsoga A, Naoumidis A and Stöver D 2000 *Solid State Ionics* **135**(1-4) 403-9

Enhancement of microwave absorption properties on modified kapton film by ion implantation process

S Purwanto[1], W A Adi[1] and Y Taryana[2]

[1]Center for Science and Technology for Advanced Material -BATAN, Puspiptek Area, Tangerang Indonesia 1314
[2]Research Center for Electronics and Telecomunication – LIPI, Bandung Indonesia

Email : setyo_p@batan.go.id

Abstract. Kapton-polyimide films have been modified by using ion implantation of Fe and Gd ions at dose 1×10^{17} ions/cm^2 and energy ions 30 keV. From X-ray diffractions (XRD) were verified the existence phases such as iron oxide, or Gd-oxide and graphitic carbon. Vibrating sample magnetometer (VSM) measurement the films showed a paramagnetic behavior. Microwave absorption (MWA) properties were measured by vector network analyzer (VNA) at X band range (8-12GHz). VNA measurement results showed the enhancement of MWA from -22dB for Kapton original film to -35dB for S11 vector direction and until -40 dB for S22 vector direction of modified Kapton film by Fe and Gd ions. It can be concluded that the nanostructure of embedded Fe-oxide or Gd-oxide nanoparticles on the near surface of Kapton film take a role and gives a rise the MWA properties of film.

1. Introduction

Research on Kapton-polyimide films $(C_{22}H_{10}N_2O_5)_n$ have many interest during recently decade. This kind of films was excellent thermal stability, chemical resistance and exceptional mechanical and electrical properties and many uses such in nuclear power plants, military aircraft and space shuttles [1, 2]. Ion implantation or irradiation is a powerful technique to modifying the films surfaces including electrical and magnetic properties [3]. Since modification due to ion implantation may be caused by two mechanisms: the doping effect and the bombardment effect. Therefore ion bombardment leads to significant change of the polymer structure as a substrate and can embed some impurity and fabricate a composite material. Implantation of metal ion to high fluence leads to the formation of nanoparticles [4]. There are some research about metal Fe, Co or Gd ions implanted into polymer films such as Polyimide (PI) or Kapton due to some physical properties such as Magnetoresistance [5], also modification electrical properties of polyimide (PI) by Xenon (Xe) ions [6]. Unfortunately the data related to microwave absorption properties have not been done yet. The aims of this research are to modify Kapton-polyimide film by using ion such as iron (Fe) or Gadolinium (Gd) and to investigate the relation between the existence of the phases, magnetic properties and microwave absorption.

2. Experimental

Pristine Kapton-polyimide $(C_{22}H_{10}N_2O_5)_n$ samples were purchased from Nilaco-Japan. The film thickness was 125 micrometer. Implanted ions of Gd and Fe were generated from Gd and Fe powder with purity 99.9%, respectively and purchased from Sigma-Aldrich. The 30 keV Gd and Fe ions were implanted into Kapton-Polyimide films within area 3.0 × 3.0 cm^2, with fluences or doses 1 × 10^{17} ions/cm^2 and ion current densities (20 μA/cm^2 during t = 200 minutes) . Implantation process has been done in the vacuum chamber by ion implanter at Center for Accelerator Technology-BATAN. The penetrating range of Gd and Fe ions into Kapton films was calculated to be 29.2 nm and 37.9 nm using the SRIM 2008 software. Verification of implanted samples were determined by means of X-ray diffraction (XRD) to verify the existence of carbon-rich phases, vibrating sample magnetometer (VSM) for M-H curve and vector network analysis (VNA) for microwave absorbance properties, respectively.

3. Results and discussions

3.1. Depth Distribution Simulation

Based on TRIM (the Transport of Ions in Matter) software calculation at the same energy ion 30 keV, only 39 nm and 29 nm at the near surface of film is supposed to be a layer strongly modified by the Fe ion and Gd ion bombardement, as shown in figure 1(a) and 1(b), respectively. V.N Popok [7] reported that only light narrow near surface to ~25 nm strongly modified by the ion bombardment and comprising composite metal/polymer material without any damaged on the top film surface. Furthermore Khabulin, et al. [8] determined the existence of formation nanogranular magnetic of Fe with the sizes ranges of 60 nm into Kapton film.

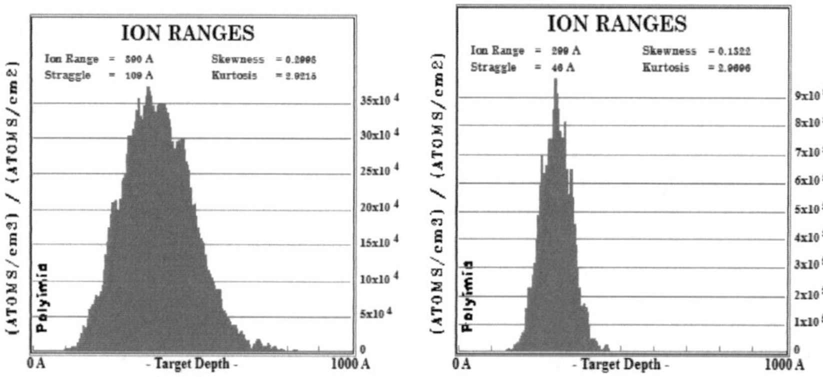

(a) (b)

Figure 1. TRIM simulation for Fe ion 30 keV implanted into Kapton film, with ion range 39 nm depth (a) and Gd ion 30 keV with 29 nm depth (b).

3.2. X-ray diffraction (XRD)

Since high doses of ion could promoted the formation of carbonaceous layer with the filling of metal factor up to 25% as reported by V.N. Popok [7], therefore post implanted Kapton-polyimide film exhibits a different XRD pattern as shows in figure 2 and figure 3. A new peak appears in the post implanted Kapton-polyimide XRD pattern at 2-Theta = 24.8° indicating the presence of a new lattice spacing with d=3.59 Å. This peak clearly corresponds to the carbonized film as reported by Chen, et al. [6]. Phase analysis of implanted samples has been performed with refinement of the XRD profile of the sample using GSAS software.

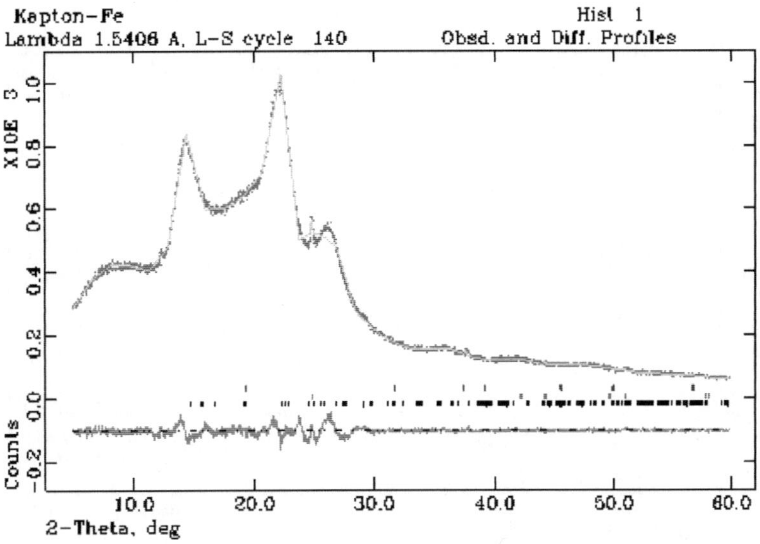

Figure 2. A typical refinement of the XRD pattern profile of Fe ion implanted Kapton film with phase identification, Kapton phase(black bar), graphitic Carbon phase (red bar) and Magnetite (blue bar).

The refinement results tabulated in table 1 and 2. In the case of Fe ion implanted Kapton film have three phases, namely a main phase of Kapton-PI 94.80 wt.% , a new phase of graphitic carbon 3.95 wt.% and Fe_3O_4 (magnetite) 1.25 wt.%, respectively.

Table 1. Phase analysis for Iron (Fe) ion implanted Kapton from XRD profile.

Name of compound	Phase	Mass fraction (wt.%)	Reference
Kapton	C-H-O-N	94.80	ICDD-96-720-0096
Carbon	C	3.95	ICDD-96-120-0018
Magnetite	Fe_3O_4	1.25	ICDD-96-900-2332

In the case of Gd ion implanted Kapton film also contain three phases of the Kapton-PI 95.83 wt.%, graphitic carbon 3.62 wt.% and Gd_2O_3 phase 0.55 wt.%, respectively.

Table 2. Phase analysis for Gd ion implanted Kapton from XRD profile.

Name of compound	Phase	Mass fraction (wt.%)	Reference
Kapton	C-H-O-N	95.83	ICDD-96-720-0096
Carbon	C	3.62	ICDD-96-120-0018
Gd-oxide	Gd_2O_3	0.55	ICDD-96-101-0339

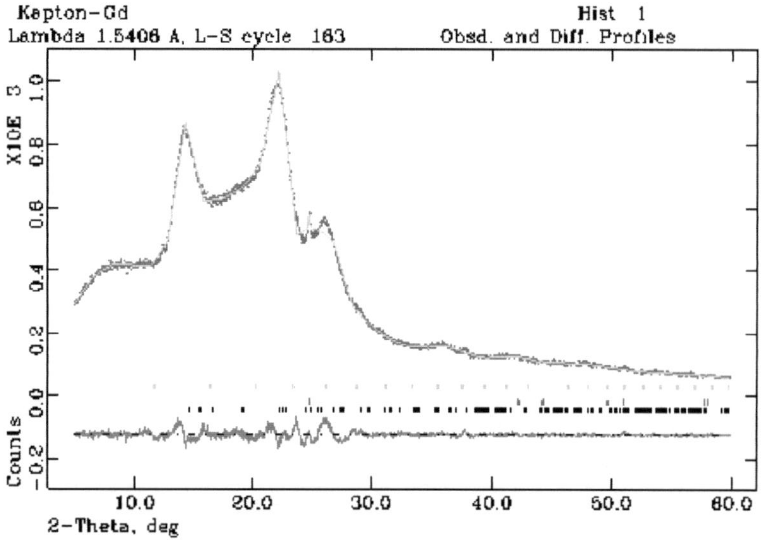

Figure 3. A typical refinement of the XRD pattern profile of Gd ion implanted Kapton film with phase identification, Kapton phase(black bar), graphitic Carbon phase (red bar) and Gd_2O_3 phase (blue bar).

3.3. Magnetic properties

Magnetic properties of both samples of Fe and Gd ion implanted ions were measured by means of VSM method up to external field H=1.0 Tesla, as shown in figure 4(a) and (b). Characteristic of both samples were exhibit paramagnetic behavior, with M_s=0.13 emu/g and 0.04 emu/g for Fe and Gd ions implanted samples respectively. These properties are consistent with reports from Murmu, et al. [9] for Gd_2O_3 and Leveneur, et al. [10] for iron-oxide nanoparticles. Since the samples were implanted above percolation doses 5×10^{16} ions/cm^2 with high ion current densities (20 μA/cm^2 during t = 200 minutes) which is a sort of equivalence to annealing. Therefore the granular metal layer on the top of surface film in the as-implanted sample consist of small oxides form of iron (Fe) or gadolinium (Gd) in paramagnetic state at room temperature [11]. The concentration for Gd-oxide compare to iron-oxide was lower due to the penetration depth of Gd ions shorter (only 29 nm) compare to Fe ions (39 nm), respectively. This data also is consistent with the results of the analysis phases of the XRD profile as tabulated in table 1 and 2.

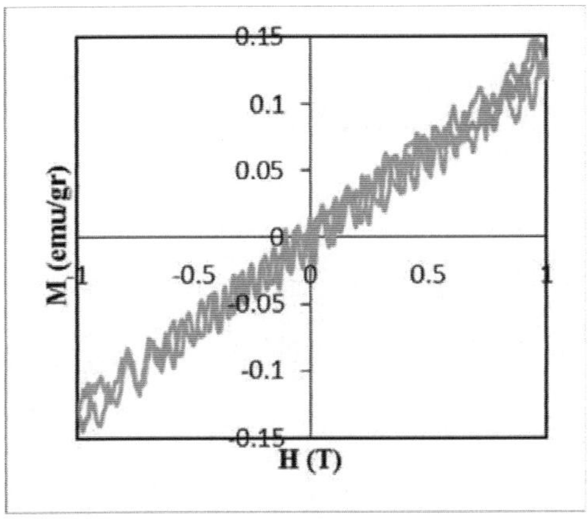

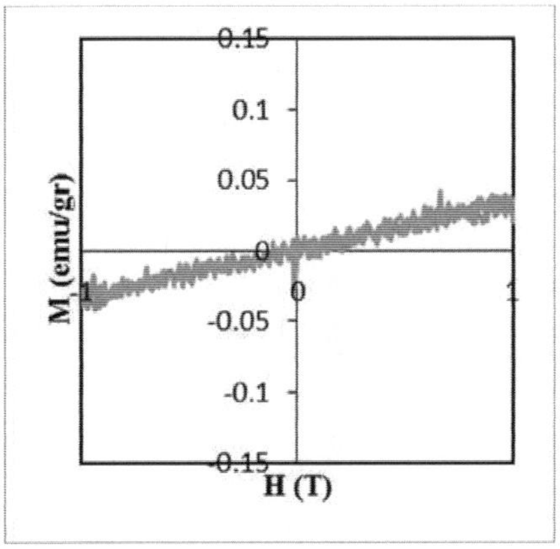

(a) **(b)**

Figure 4. Magnetization curve profile for iron ions implanted Kapton (a) Fe ions and (b) Gd ions implanted into Kapton at doses 10^{17} ions/cm^2.

3.4. Microwave absorption properties

Microwave absorbances of the samples were determined by VNA method and measure at range 8-12 GHz. In case of Kapton film, characteristic of reflection loss (RL) were shown maximum RL= -22 dB at 11.7 GHz, with three absorption frequency at 11.4 GHz, 10.0 GHz and 8.4 GHz. This properties are mainly come from dielectric loss of the pristine PI film. But, this characteristic were changed in post implanted Fe and Gd film as shown in figure 5. The optimum reflection loss in post implanted Kapton film has been increased to RL=-35 dB and resonance frequency shifted to 11.4 GHz. The frequencies of the RL peak move to the low frequency direction and the strength of absorption first frequency increases can be explained by the quarter wave theory [12]. This results were comparable to microwave absorbance of composites based on PANI or PET film [13]. We suggest paramagnetic nanogranular from implanted ion Fe and Gd at the near surface take a role in this phenomenon with form metal/polymer nanocomposite region, as well as the existence of a new phase carbonized film.

4. Conclusion

Low energy 30 keV Fe and Gd ions were implanted in Kapton-polyimide film at doses 1×10^{17} ions/cm^2 to create the nanogranular magnetic composite material beside a new carbonized phase. Both modified Kapton film have paramagnetic properties at room temperature. The implanted Fe and Gd ions were distributed homogenous and forming iron-oxide and Gd-oxide compound. Enhancement of microwave absorbance in modified Kapton film were occurred with reflection loss (RL) = -35 dB compare to -22 dB for pristine Kapton-polyimide.

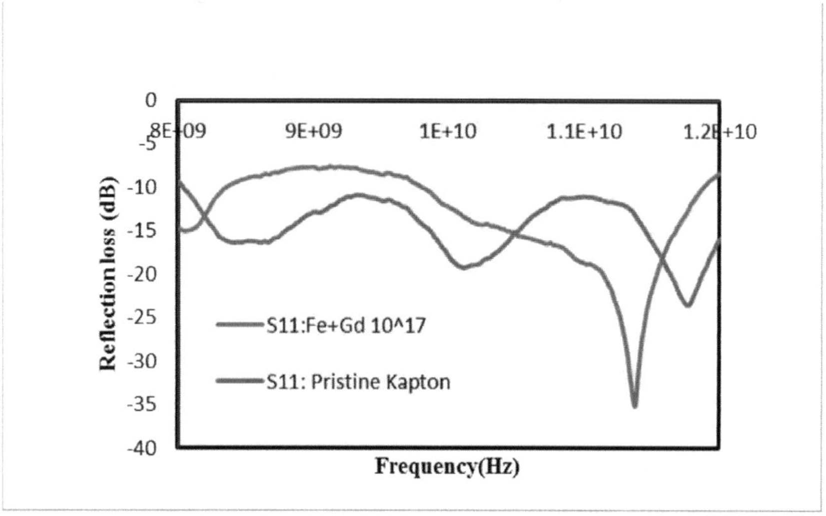

Figure 5. Vector Network Analysis profile for pristine Kapton-polyimide (red line) and post implanted Fe+Gd (blue-line)

References

[1] Simpler 2015 *Electron. Lett.* **51**(2) 127
[2] Teber A, Unver I, Kavas H, Aktas B and Bansal R 2016 *J. Magn. Magn. Mater.* **406** 228-32
[3] Popok V N. 2012 *Rev. Adv. Mater. Sci.* **30**(1) 1
[4] Popok V N, Nuzhdin V I, Valeev V F and Stepanov A L, 2014 *J. Mater. Res.* **29** 1-7
[5] Kharchenko A, Lukashevich M, Popok V, Khaibullin R, Valeev V and Bazarov V 2013 *Part. Part. Syst. Char.* **30**(2) 180–4
[6] Chen T, Yao S, Wang K, Wan H and Zhou S 2009 *Surf. Coat. Technol.* **203** 3718-21
[7] Popok V N 2014 *Rev. Adv. Mater. Sci.* **36**(1) 1–12
[8] Khaibullin R I, Popok V N, Bazarov V V Zheglov E P, Rameev B Z and Okay C 2002 *Nucl. Instrum. Methods Phys. Res., Sect. B.* **191**(1-4) 810–4
[9] Murmu P P, Kennedy J, Ruck B J, Markwitz A, Williams G V M and Rubanov S 2012 *Nucl. Instrum. Methods Phys. Res., Sect. B* **272** 100–3
[10] Leveneur J, Waterhouse I N, Kennedy J, Metson J B and Mitchell D R G 2011 *J. Phys. Chem. C* **115** 20978–85
[11] Rameev B, Okay C, Yildizm F, Khaibulin R I, Popok V N and Aktas B, 2004 *J. Magn. Magn. Mater.* **278** 164
[12] Li L Z, Wei J Z, Xia Y H, Wu R, Yun C and Yang Y B 2014 *Appl. Phys. Lett.* **105**(2) 022902
[13] Ni Q-Q, Zhu Y-F, Yu L-J and Fu Y-Q 2015 *Nanoscale Res. Lett.* **10**(1) 174

IOP Conf. Series: Journal of Physics: Conf. Series **817** (2017) 012003 doi:10.1088/1742-6596/817/1/012003

CNT coated thread micro-electro-mechanical system for finger proprioception sensing

A A Shafi[1] and D H B Wicaksono[1,2]

[1]Faculty of Biosciences and Medical Engineering, Universiti Teknologi Malaysia, 81310, Johor Bahru, Malaysia
[2]Dept. of Biomedical Engineering, Faculty of Life Sciences and Technology, Swiss German University, EduTown BSDCity, Tangerang 15339, Indonesia

E-mail: aisyahshafi@biomedical.utm.my

Abstract. In this paper, we aim to fabricate cotton thread based sensor for proprioceptive application. Cotton threads are utilized as the structural component of flexible sensors. The thread is coated with multi-walled carbon nanotube (MWCNT) dispersion by using facile conventional dipping-drying method. The electrical characterization of the coated thread found that the resistance per meter of the coated thread decreased with increasing the number of dipping. The CNT coated thread sensor works based on piezoresistive theory in which the resistance of the coated thread changes when force is applied. This thread sensor is sewed on glove at the index finger between middle and proximal phalanx parts and the resistance change is measured upon grasping mechanism. The thread based microelectromechanical system (MEMS) enables the flexible sensor to easily fit perfectly on the finger joint and gives reliable response as proprioceptive sensing.

1. Introduction

Spinal cord injury, arthritis and stroke may cause patients to lose their proprioceptive sensing. Their nerve and muscle function will degrade hence reducing the sensing of motion control and position stimuli. Hand is an important part of human body that controls our daily activities. Stroke causes muscle weaknesses, loss of range of motion and impaired motor control [1]. Stroke patients usually have hand disorder; their muscles are weak, decrease range of motion and loss motor control [1]. In order to move as a normal person, they need to undergo rehabilitation therapy or be assisted by a device or machine. There are several equipments and methods that have been developed to help patients improve their capabilities [2-4].

Some devices have been developed to assist and monitor the motion of impaired body parts [5]. Most studies utilized robots to assist patients with hand disorder [4-6]. However, the utilization of robot may create inconvenience for the patient. A glove based therapy that integrates virtual reality program was developed for rehabilitation training [2]. The system is quite complex, using cyberglove that is embedded with 18 strain gauges and custom preumatic actuator to provide force feedback to patient's fingertips[2].

Carbon nanotube that was discovered by Iijima [7] has created tremendous effect in electronic, mechanical and sensor fields as carbon nanotube possesses remarkably high stiffness and strength [8]. It also has superior thermal and electrical properties [9]. Carbon nanotube has been implemented in

Micro-Electro-Mechanical system (MEMS) field as it can sense, control and actuate in micro level and can generate effect at the macro level [10]. This micro scale system which merges electronics and mechanical characteristics have usually been fabricated in silicon [11].

Fabric can show reliable results as a wearable sensor since it is flexible and can contact conformally onto human body. Therefore, in this paper, thread or yarn was proposed as an alternative sensing substrate material since it is inexpensive, flexible, lightweight, and can be easily sewn. This thread based sensor will be integrated into fabric by sewing technique.

2. Materials and Experimental Method

A yarn of 100% cotton thread was purchased from Asli Pari 100% Cotton Heavy Duty Thread (No.40, Combed Gassed Merchendized), Pakistan with 32 twist/inch (TPI). Multiwalled Carbon Nanotube (MWCNT) in the powder form with carbon content more than 70% and surface resistivity less than 900 Ohm.sq, as well as Sodium Dedocyl Sulphate, $C_{12}H_{25}NaO_4S$ (SDS) surfactant and Sodium carbonate or soda ash, Na_2CO_3 were bought from Sigma Aldrich (USA).

Prior to coating MWCNT on cotton thread, the MWCNT was characterized using Transmission Electron Microscopy (TEM) to measure the number of layers of the MWCNT. The MWCNT powder was mixed with ethanol at weight ratio of 1:3 then, the mixture was ultrasonicated for 30 minutes at 100 rpm. 1 µl of the mixture was dropped on carbon grid (purchased from Ted Pella,inc) and dried at room temperature. Transmission Electron Microscope (TEM) imaging was then conducted on the MWCNT sample in Hitachi HT7700 TEM.

2.1. Pre-treatment of Cotton Thread

The threads function as substrate of the sensor. First, the threads underwent a scouring treatment to remove all the natural wax at the outer surface. This treatment improves the absorbency of the thread and removes impurities[12]. The scouring processes started by boiling 500 ml of distilled water. Then 10 g of soda ash was added to the boiled distilled water to produce soda ash solution at a concentration of $20 mg.ml^{-1}$. Then, 10 strands of 30 cm cotton threads were put into the mixture and stirred for 10 minutes. The threads were then rinsed with milipore water for several times until the pH of the rinsed water close to neutral level (pH 6.5-7.5).

2.2. Sensing Material Preparation

Prior to coating the cotton thread with MWCNT, dispersion of carbon nanotube was prepared by mixing 5 g MWCNT and 5g SDS to 50 ml of distilled water. The mixture was ultrasonicated for 30 min and was magnetically stirred for another 30 min.

2.3. Sensor Fabrication

The fabrication of thread sensor is illustrated in figure 1. The scoured threads were cut into 5 cm length and this thread was divided into gauge part (3 cm at the middle) and electrode part (1mm at each end). Then, the threads were dipped into 0.1 g/ml of carbon nanotube dispersion for 10 minutes. After that, they were dried at 60^0 C for 10 minutes. The dipping and drying processes were repeated 1-6 times based on number of coatings. The coated threads were then rolled with parafilm (elastic plastic) at the gauge part before continuing for another dipping repetition for electrode part. The electrode part of the sensor was dipped and dried for 10 times.

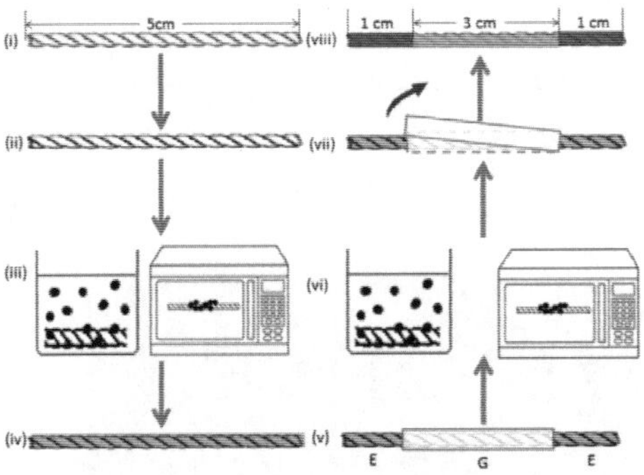

Figure 1. The fabrication process of the thread based MEMS. (i) Natural cotton thread initially coated with wax. (ii) After scouring, the wax at the surface is removed thus its absorbency was enhanced. (iii) The scoured thread was dipped in CNT dispersion and dried in oven. (iv) Cotton thread coated with carbon nanotube. (v) The middle part of the thread sensor was wrapped with parafilm to create electrode and gauge part. (vi) Dipping and drying process are repeated. (vii) Parafilm at the gauge part is removed. (viii) Thread sensor that consist of 3 cm gauge part and 1 cm electrode part at each end.

2.4. Sample characterization

Morphology of the carbon nanotube coated on the threads was characterized using Field Emission Scanning Electron Microscope (FE-SEM). The electrical resistivity of the sensor threads were measured using four point probe technique and were then converted into conductivity.

2.5. Sensor Calibration

The gold wire was attached to the CNT-coated thread using knot mechanism. Silver paste was placed onto the connection to improve the ohmic contact of the thread sensor and the wire. Then, the thread that was tied to the gold wire using silver paste was then placed into the oven at 60 ^{0} C for 1 hour to anneal the silver paste. To calibrate the thread sensor, the thread that was integrated with gold wire was positioned on the translational stage and clipped to the hold at both ends of the thread. The gold wire was then clipped by source meter probe to measure the change of resistance. Labview software was used to control the movement of the translational stage and reading of the sourcemeter (figure 2).

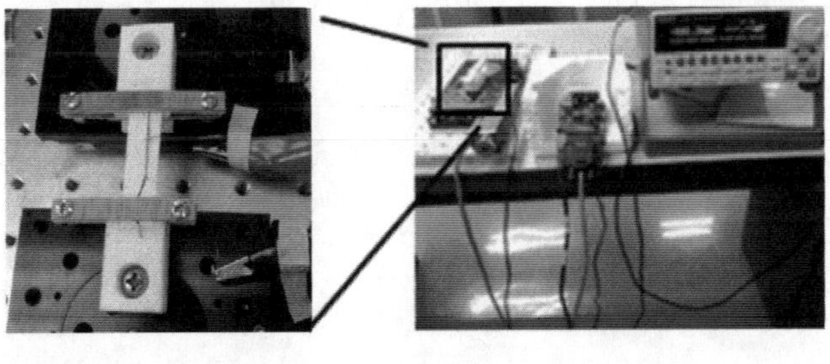

Figure 2. Setup of the calibration CNT-coated thread sensor

3. Results and Discussion

As seen in figure 3(a), the transmission electron microscope (TEM) image showed that the carbon nanotubes bundles are entangled to each other and individual tube of carbon nanotube has more than 10 layers of nanotube with an inner diameter of 5nm and an outer diameter of 15 nm. Carbon nanotubes in powder form are difficult to coat onto cotton thread, therefore, the powder carbon nanotube need to be mixed with aqueous solution to form carbon nanotube dispersion. As the van der Waal bonding between the carbon nanotubes are high, the carbon nanotubes tend to aggregate into bundles and it was difficult to produce good CNT dispersion [13]. Therefore, the surfactant (Sodium Dedocyl Sulfate) was used in this experiment to produce stable dispersion of solid in different medium [14]. The mixture underwent ultrasonication first to provide a high local shear to fray the carbon nanotube bundles and then create a gap between the nanotubes [13]. Afterward, the surfactant molecules will fill the gap and make the carbon nanotubes separated.

The scoured cotton thread was then dipped into stable carbon nanotube dispersion. Figure 3 (b) present the Scanning Electron Microscope (SEM) images of carbon nanotube coating on cotton thread. From the image, it can be seen that the carbon nanotube was coated on the cotton thread's surface. The high magnification of SEM image shows a high density coating of carbon nanotubes on the cotton thread's surface.

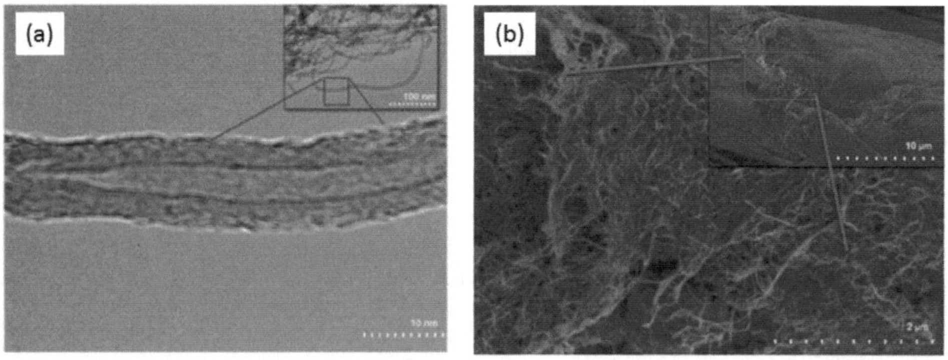

Figure 3. (a) Transmission Electron Miscoscope (TEM) Image of the Multiwalled Carbon Nanotube and (b) Scanning Electron Microscope (SEM) image of the carbon nanotube coated on cotton thread.

Surface resistivity of the thread sensor was measured using four point probe method, and was then converted into conductivity values. The conductivity of the thread sensors of the gauge part coated with 1 time dipping to 4 times dipping is depicted in figure 4. The graph show that the conductivity of the coated thread is increased as the number of dipping at the gauge part is increased. The electrical conductivity of the thread sensor started to reach steady state value after 4 times of dipping. This related to percolation theory that after the volume ratio reach certain value, the electrical property of carbon nanotube change from nonconductive to conductive state [15].

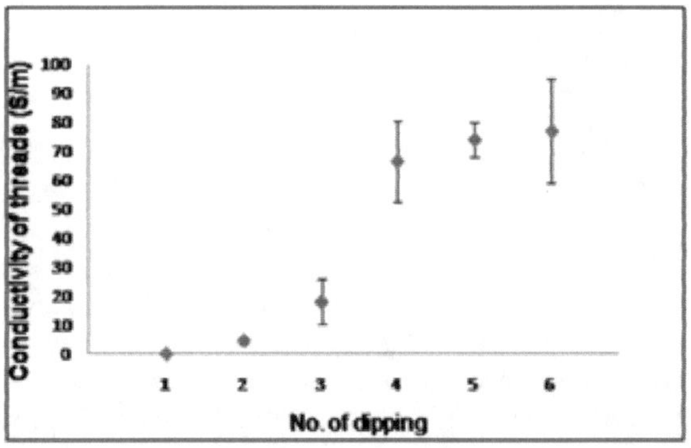

Figure 4. The conductivity of the thread sensor with the gauge part coated with 1-6 dippings.

A graph of relative resistance change (measured using sourcemeter) versus strain is showed in figure 5, as a result of the procedure described in 2.5. As seen in figure 5, the thread sensor shown positive piezoresistive properties as increasing strain increases the resistance of the sensor. Gauge factor is used as indicator for determining the strain sensitivity of the sensor. Gauge factor is defined as the ratio of relative change in the resistance of the sensor ($\Delta R/R_0$) to the applied mechanical strain, ε. A higher gauge factor means that the sensor is more sensitive to the changes in strain. The gauge factor of the sensor was obtained by measuring the slope of the graph. However, from the graph, the sensor that was dipped for 2 times has the highest gauge factor than other sensors. This phenomenon might be because of the structure of cotton thread is porous and the carbon nanotube was not well distributed on the cotton thread.

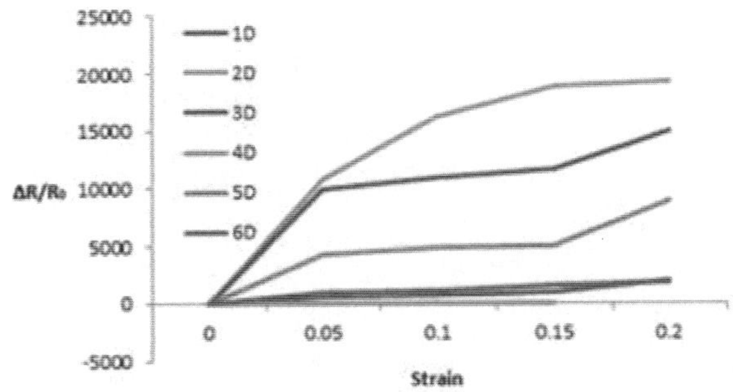

Figure 5. The graph of relative resistance change versus strain.

3.4. Sensor Testing

The sample of coated -thread was sewn on the glove to test the application of sensor. The sensor thread was sewn on the glove with 10 cm electrode part (10 dippings) and 25 cm gauge part (5 dippings) on the index finger between middle phalanx and proximal phalanx. The electrode part of the sensor was connected to the probe of sourcemeter that was integrated to the Labview. The finger was bent slowly and the resistance change was recorded, as shown in figure 6. The resistance is increased as the finger is bent. Therefore, this sensor can be utilized to monitor the bending ability of the stroke patient during rehabilitation.

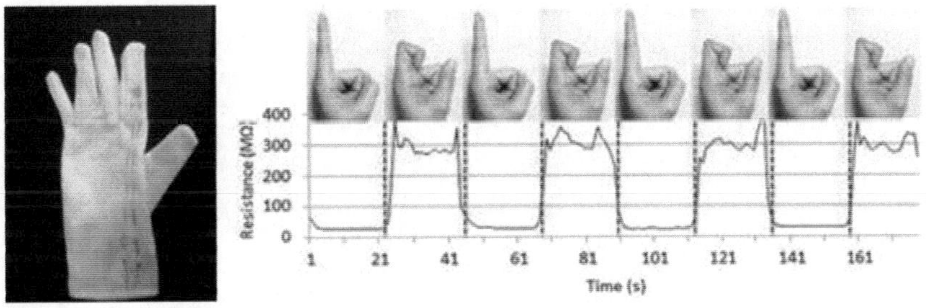

Figure 6 : The resistance change of the finger bending.

4. Conclusion

Hydrophobicity of carbon nanotube can be reduced by assist of surfactant. After immersed in carbon nanotube dispersion, absorbable cotton threads become conductive because the carbon nanotubes were attached on the cellulose fiber surface. This conductive CNT coated thread showed piezoresistive properties that changes its resistances when it bent. This condition can be applied for the proprioceptive sensor.

Acknowledgement

This works supported by Grant Universiti Teknologi Malaysia QJ130000.2545.05H32-EMBEDDED CNT-FIBER COMPOSITE FOR MONITORING GROWTH OF IMPLANTABLE SMART TISSUE EXPANDER. The authors thank Hi-Tech for the service of Field Emission Scanning Electron Microscope (FESEM), University Industry Research Laboratory (UIRL), UTM for the service of

Transmission Electron Microscope (TEM) and Science Officer of Faculty of Biosciences and Medical Engineering for the service of Scanning Electron Microscope (SEM).

References

[1] Gray V, Rice C L and Garland S J 2012 *Physiotherapy Canada* **64**(4) 415-26
[2] Jack D, Boian R, Merians A S, Tremaine M, Burdea G C, Adamovich S V, Recce M and Poizner H 2001 *IEEE Trans. Neural Syst. Rehabil. Eng.* **9**(3) 308-18
[3] Hernandez R A 2011 Thumb Wearable Page Turning and Key Stroke Assist Device U.S. Patent US D637763 S1
[4] Aisen M L, Krebs H I, Hogan N, Mcdowell F and Volpe B T 1997 *Arch. Neurol.* **54**(4) 443-46
[5] Marchal-Crespo L and Reinkensmeyer D J 2009 *J. NeuroEng. Rehabil.* **6**(20) DOI: 10.1186/1743-0003-6-20
[6] Huang V S and Krakauer J W 2009 *J. NeuroEng. Rehabil.* **6**(5) DOI: 10.1186/1743-0003-6-5
[7] Iijima S 1991 *Nature* **354**(6348) 56-5
[8] Li C, Thostenson E T and Chou T W 2008 *Compos. Sci. Technol.* **68**(6) 1227-49
[9] Thostenson E T, Ren Z and Chou T W 2001 *Compos. Sci. Technol.* **61**(13) 1899-912
[10] Partnership P F 2002 *An Introduction to MEMS (Micro-electromechanical Systems)* (UK: Prime Faraday Technology Watch)
[11] Mohammed A, Moussa W and Lou E 2008 *Sensors* **8**(4) 2642-661
[12] Ulum M F, Maylina L, Noviana D and Wicaksono D H B 2016 *Lab on a Chip* **16**(8) 1492-504
[13] Duan W H, Wang Q and Collins F 2011 *Chem. Sci.* **2**(7) 1407-13
[14] Vaisman L, Wagner H D and Marom G 2006 *Adv. Colloid Interface Sci.* **128** 37-46
[15] Mamunya Y 2011 Carbon Nanotubes as Conductive Filler in Segregated Polymer Composites-Electrical Properties *Carbon Nanotubes-Polymer Nanocomposites* ed S Yellampalli (Croatia: InTech Open) chapter 9

IOP Conf. Series: Journal of Physics: Conf. Series **817** (2017) 012004 doi:10.1088/1742-6596/817/1/012004

Electrophoretic deposition of zinc oxide on graphite drawn on paper

C S Sebastian, M U Herrera and A K G Tapia

Institute of Mathematical Sciences and Physics, College of Arts and Sciences, University of the Philippines Los Baños, Laguna, Philippines 4031

E-mail: agtapia@up.edu.ph

Abstract. Flexible substrates have many promising applications in sensing, electronics, and electromagnetic shielding and energy storage among many others. Paper can serve as substrate for these kinds of technologies offering a cheaper alternative. In this study, Zinc oxide (ZnO) was successfully deposited on graphite drawn on paper using electrophoretic deposition (EPD). Graphite from commercially-available pencil was drawn on paper. Graphite drawn on paper was used as electrodes for the EPD process. High-voltage power supply was used as source while ground ZnO in acetone was used as suspension in the deposition process. Scanning electron microscopy (SEM) and Energy dispersive x-ray spectroscopy (EDX) results reveal the deposition of ZnO on Graphite. In addition, the electrical contact of the ZnO-graphite interface showed Ohmic behaviour by two-point probe method.

1. Introduction

Zinc oxide has been used for many industrial applications such as catalysts, photocatalytic, photovoltaic, gas sensors, and diodes [1-5]. It is also low cost and nontoxic with a wide band gap of ~3.3 eV [3-5]. Traditionally, ZnO has been deposited on substrates with planar surfaces. However, several applications require deposition on non-flat surfaces. A possible approach for this is to deposit ZnO on a flexible substrate that can cover different types of surfaces. Paper has been used as substrate [6-14] for conducting strips using ink and organic materials. It is one of the most abundant manmade materials and very accessible. It is inexpensive, lightweight, and mechanically flexible. It can be bent and cut into different shapes and sizes to fit and integrate into many applications. Nowadays, commercial functional materials targeting low cost and large volume productions may require the use of inexpensive substrates such as paper.

There are methods to embed functional materials on paper by using binders or by chemical routes among others. The challenge in using such methods is when the application needs patterning and interfacing for electronic applications. Electrophoretic deposition offers advantages of being a versatile yet simple technique used to deposit materials on a variety of substrates. It only uses working electrodes, electrolyte, and power source [15-16]. Moreover, it has also tunable parameters (e.g. deposition time and applied voltage) to obtain enhanced characteristics of the deposited material. The electrodes need to be conducting. This is addressed by drawing graphite on paper. This pen-on-paper method is a straightforward process to draw patterns. The graphite drawn on paper was used as conducting strips for electrophoretic deposition. Graphite is composed of carbon atoms in a layered structure that exhibits high electrical and thermal conductivity. Graphite has been widely used as conducting filler for materials to improve conductivity [17].

In this paper, graphite was drawn on paper to serve as electrodes for electrophoretic deposition of Zinc oxide. The composition, morphology and electrical characteristics of the samples were studied using scanning electron microscopy, energy dispersive x-ray spectroscopy and electrical-probe method.

2. Experimental details

Electrophoretic deposition was carried out using graphite drawn on paper as conducting electrodes. ZnO powder was ground and mixed with acetone to obtain a 10 mg/mL concentration. This suspension was sonicated for 30 minutes to disperse the particles. ZnO was deposited using an applied voltage of 300 V and 500 V with deposition time of 3 min. After the deposition, the samples were heat treated at 60 °C for 1 hour.

The morphology and composition of the samples were observed using scanning electron microscope and energy dispersive x-ray spectroscopy, respectively. Two-point probe technique was used to obtain the current-voltage (I-V) characteristics of the samples.

3. Results and discussions

ZnO was deposited on graphite drawn on paper via electrophoretic deposition. Figure 1 shows the deposited ZnO on graphite drawn on paper.

Figure 1. Deposited ZnO on pencil graphite drawn on paper (0.5 in x 1.5 in).

Figure 2 shows SEM image of graphite drawn on paper. The film generally shows a uniform surface with flaked structures. The individual graphite sheets have typical planar structures. In commercial pencils, the graphite is mixed with binders to form and this gave variations in the SEM of graphite.

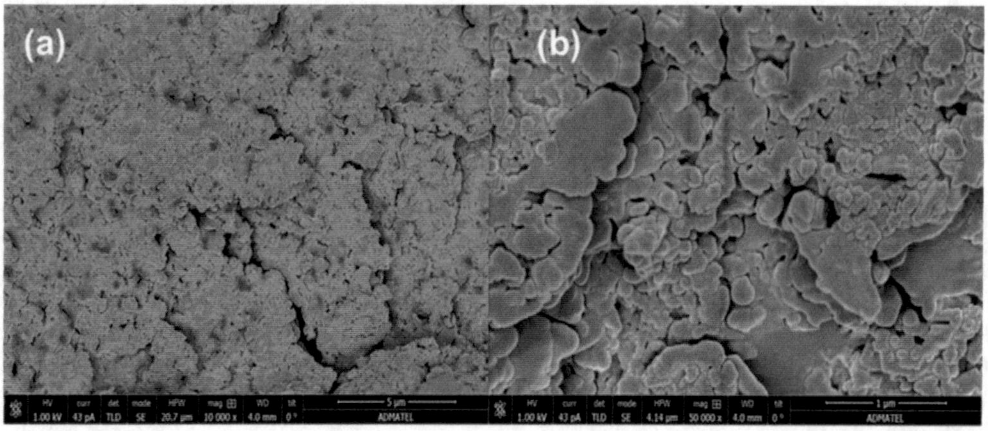

Figure 2. SEM images showing flat images of graphite drawn on paper at 10kx (a) and 50kx (b) magnification.

Figure 3 shows the SEM images of ZnO deposited on graphite drawn on paper synthesized by electrophoretic deposition at (a) 300 V and (b) 500 V both in 10kx magnification. SEM images showed deposited ZnO particles on graphite. It appears that film deposited with a larger voltage is more compact. On the other hand, figure 4 shows higher magnification SEM images of the samples fabricated at (a) 300 V and (b) 500 V applied voltages. More detailed and distinct structures of the deposited particles from the graphite substrate were observed.

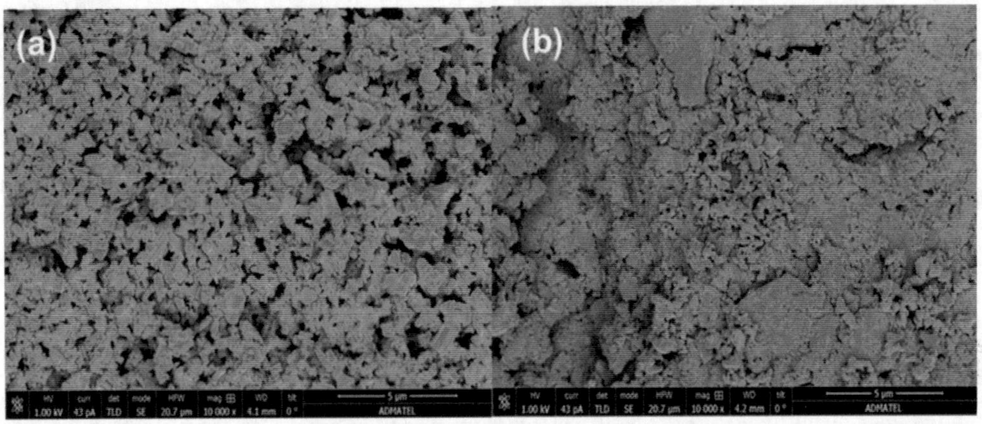

Figure 3. ZnO deposited on graphite drawn on paper using electrophoretic deposition with applied voltages of 300 V (a) and 500 V (b) both in 10kx magnification.

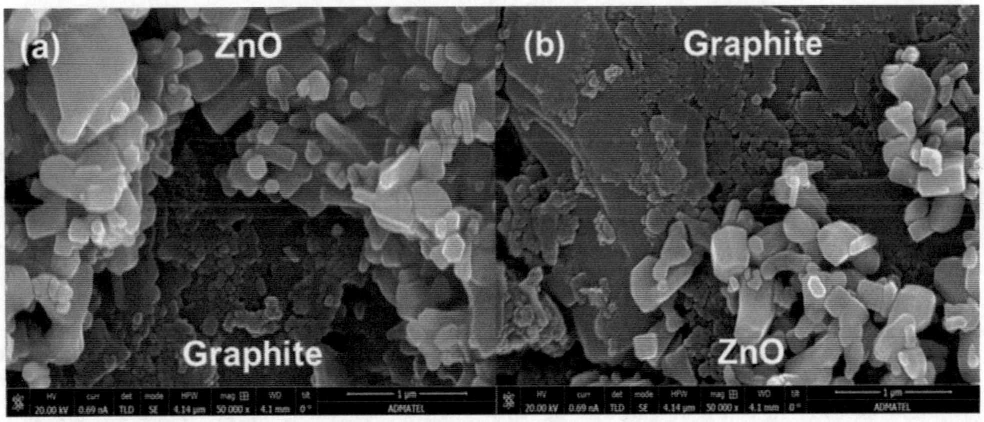

Figure 4. ZnO deposited on graphite drawn on paper using electrophoretic deposition with applied voltages of 300 V (a) and 500 V (b) both in 50kx magnification.

Furthermore, the structure and composition of the deposited ZnO on graphite drawn on paper were studied using SEM and EDX, respectively. Figure 5 shows the ZnO particles with varying shapes and sizes on the surface of the graphite substrate. It appears that there are no differences between the deposited particles. This suggests that the voltages used provided sufficient field to transport the particles from the suspension to the electrodes.

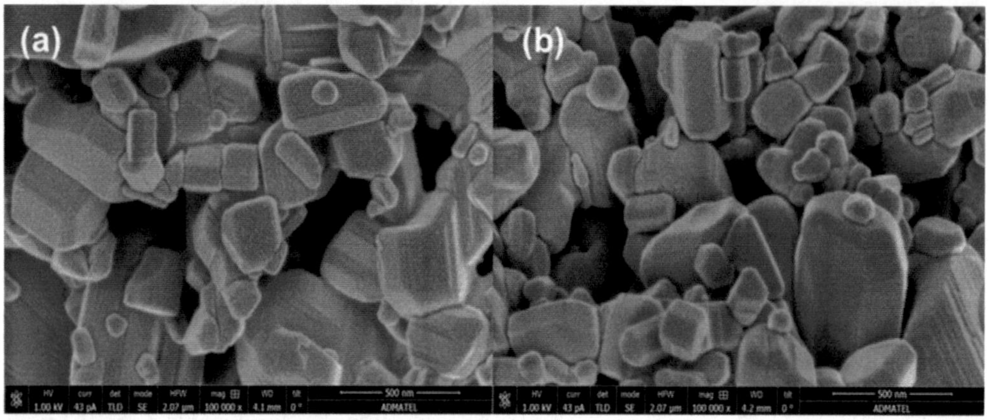

Figure 5. ZnO deposited on graphite using electrophoretic deposition at 300 V (a) and 500 V (b) both in 100kx magnification.

Table 1 shows the EDX results for graphite. The dominant composition is C and O. The elemental results for C give 58 wt.% proportion. This comes from the graphite drawn on paper because of the graphitic components. The elements Au and Pd comes from the conductive coating (gold palladium) of sample done for SEM characterization. The other compositions are due to sample contaminations at very low weight proportions.

Table 1. EDX results for Graphite used to be drawn on paper.

Sample	Element	Wt%	σ
Graphite (drawn on paper)	C	58.0	1.6
	O	19.0	0.8
	Au	10.5	0.4
	Ca	4.8	0.2
	Al	3.4	0.1
	Pd	2.1	0.2
	Si	2.0	0.1
	Fe	0.3	0.1

Table 2 shows the EDX results for deposited ZnO. Zn shows relatively large contributions which come from the deposited particles. The C from graphite still appears in the spectra but the relative composition is lower in contrast with that of table 2 which supports the deposition of ZnO on the surface of graphite. The presence of contaminants such as Si, Na, K and others at relatively low properties reflect impurities from the handling, heat treatment and characterization processes.

Figure 6 shows the current-voltage characteristics of the deposited ZnO on graphite drawn on paper using EPD with applied voltage of 300 V in the forward (figure 6(a)) and reverse (figure 6(b)) bias while figure 7 shows the I-V plots of the sample deposited at 500 V in the forward (figure 7(a)) and reverse (figure 7(b)) bias.The I-V measurements of the fabricated material were measured using two-point probe at room temperature. All plots show highly linear ($r^2 = 0.9996$) behavior. This is a characteristic of Ohmic behavior. The Ohmic contact implies the efficiency for charge transport. This suggests that patterning materials on graphite drawn on paper is possible using electrophoretic deposition.

Table 2. EDX results for the deposited ZnO at different voltages.

Sample	300 V			500 V		
	Element	Wt%	σ	Element	Wt%	σ
ZnO deposited on graphite drawn on paper using EPD	C	20.8	2.3	C	33.5	1.6
	O	15.6	0.6	O	23.9	0.7
	Zn	54.3	1.7	Zn	26.6	0.7
	Si	2.5	0.1	Si	6.1	0.2
	Al	1.8	0.1	Al	5.8	0.2
	Na	3.6	0.6	Na	3.2	0.3
	K	0.3	0.1	K	0.4	0.0
	Ca	0.6	0.1	Ca	0.3	0.0
	Fe	0.3	0.1	Fe	0.2	0.1
	Ti	0.3	0.1	Ti	0.2	0.0

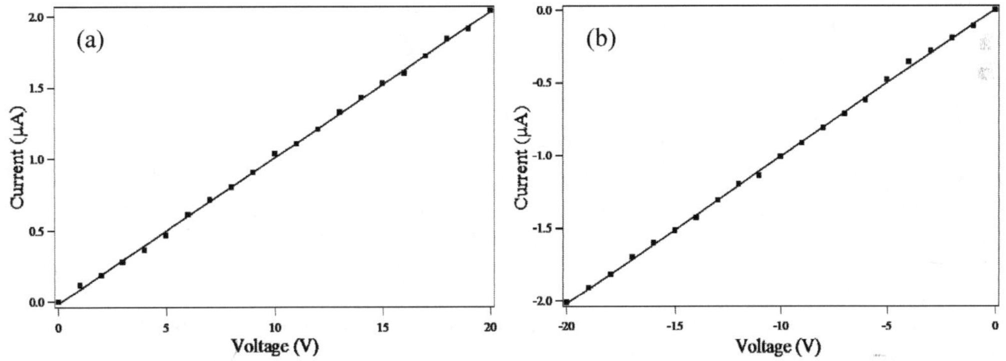

Figure 6. I-V plots of the deposited ZnO on graphite drawn on paper using EPD with applied voltage of 300 V in the (a) forward and (b) reverse bias.

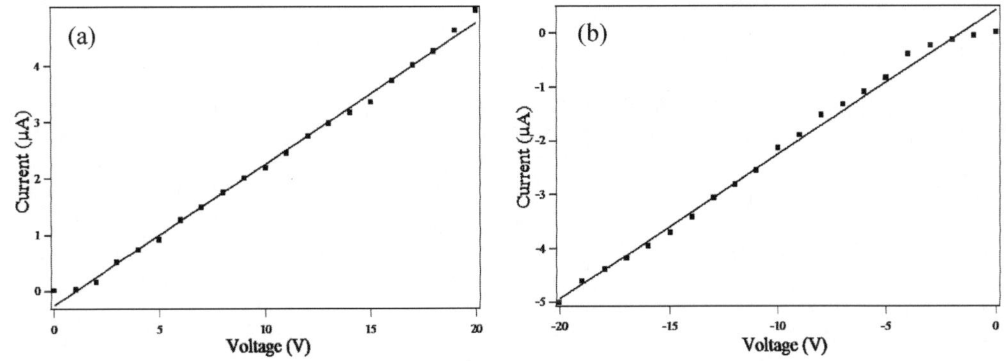

Figure 7. I-V plots of the deposited ZnO on graphite drawn on paper using EPD with applied voltage of 500 V in the (a) forward and (b) reverse bias.

4. Conclusion

In this study, we have successfully deposited functional material on paper using electrophoretic deposition. In order to have a low cost, mechanically flexible, and very accessible substrate, paper was used. Inexpensive and nontoxic ZnO was deposited on paper. The fabrication process, which is electrophoretic deposition, is very simple which includes controllable parameters (e.g. deposition time and applied voltage). Graphite was used as the intermediate material that is drawn on paper and was used as the electrodes in the EPD process. SEM-EDX observations confirmed the presence of ZnO accumulations on the surface of the graphite on paper. The current-voltage plots show a linear, Ohmic behaviour, confirming electrical contact between the deposited ZnO and graphite on paper.

Acknowledgement
This work was supported by the UPLB Basic Research Grant and the UP System Emerging Inter-disciplinary Research Program (OVPAA-EIDR C06-035).

References
[1] Khoza P B, Moloto M J and Sikhwivhilu L M 2012 *J. Nanotechnol.* **2012**
[2] Sulciute A and Valatka E 2011 *Mater. Sci.* **18** 318-24
[3] Schmidt-Mende L and MacManus-Driscoll J L 2007 *Mater. Today* **10** 40-8
[4] Nam G, Baek S and Park I 2014 *J. Alloy Compd.* **613** 37-41
[5] Shalyapina A Y, Soloveva A Y, Zaporozhets M A, Khokhlov E M, Plotnichenko V G, Savilov S V, Egorov A V, Nikolaichik V I, Buslaeva E Y, Rustamova E G, Avilov A S and Gubin S P 2013 *Russ. J. Inorg. Chem.* **58** 354-60
[6] Dogome K, Enomae T and Isogai A 2012 *Chem. Eng. Process.* **68** 21-5
[7] Yao B, Yuan L, Xiao X, Zhang J, Qi Y, Zhou J, Zhou J, Hu B and Chen W 2013 *Nano Energy* **2** 1071-8
[8] Ren T, Tian H, Xie D and Yang Y 2012 *Sensors* **12** 6685-94
[9] Eder F, Klauk H, Halik M, Zschieschang U, Schmid G and Dehm C 2004 *Appl. Phys. Lett.* **84** 2673-5
[10] Zschieschang U and Klauk H 2015 *Org. Electron.* **25** 340-4
[11] Yang L, Rida A, Vyas R and Tentzeris M M 2007 *IEEE Trans. Microwave Theory Tech.* **55** 2894-901
[12] Russo A, Ahn B Y, Adams J J, Duoss E B, Bernhard J T and Lewis J A 2011 *Adv. Mater.* **23** 3426-30
[13] Phan H, Dao D V, Dinh T, Brooke H, Qamara A, Nguyen N and Zhu Y 2015 *IEEE Int. Conf. on Micro Electro Mechanical Systems (MEMS)* **28** (Portugal: Estoril/IEEE) p 825-8
[14] Hu L, Choi J W, Yang Y, Jeong S, Mantia F L, Cui L and Cui Y 2009 *Proc. of the National Academy of Sciences of the United States of America* **106** 21490-4
[15] Hanaor D, Michelazzi M, Chenu J, Leonelli C and Sorrell C 2011 *J. Eur. Ceram. Soc.* **31** 2877-85
[16] Corni I, Ryan M P and Boccaccini A R 2008 *J. Eur. Ceram. Soc.* **28** 1353-67
[17] Wu X, Qi S, He J, Chen B and Duan G 2010 *J. Polym. Res.* **17** 751-7

Electrochemical performance of natural graphite coated by amorphous carbon using two step synthesis processes at various temperatures for anode material in Li-ion battery

F Rohman[1], S M Nikmah[2] and J Triwibowo[1]

[1]Research Center for Physics – LIPI, Kawasan Puspiptek Gd. 441 – 442 Setu Tangerang Selatan, Banten 15314, Indonesia
[2]Department of Physics, Faculty of Science and Technology – UIN Maulana Malik Ibrahim, Jalan Gajayana 1 Malang Jawa Timur, Indonesia

E-mail: fadlirohman@gmail.com

Abstract. Electrochemical performance of natural graphite as anode material in the Li-ion battery has been modified by coating this particle with amorphous carbon through two step synthesis process. Citric acid as the amorphous carbon source was mixed with natural graphite (NG) in the ethanol solvent at 80 °C using magnetic stirrer. In the first step, the mixture of NG and CA were dried at 350 °C for 5 hours under argon atmosphere to evaporate the solvent. This dried mixture was then sintered at different temperature i.e. 500 °C (labeled CNG500), 600 °C (CNG600) and 700 °C (CNG700) under argon atmosphere to form amorphous carbon layer on the surface of NG. The crystal structure and morphology of the particles were characterized by using XRD, SEM and TEM. Electrochemical performance and charge-discharge of amorphous carbon-coated graphite has been evaluated by cyclic voltammetry and WBCS 3000, respectively. Cyclic voltammogram showed the working potential and redox reaction peak of the sample. Charge-discharge data was obtained to determine the specific capacity of the sample at 0.1C.

1. Introduction

Li-ion battery has been used as energy storage for several applications such as mobile phone, notebook computer, gadget, cameras [1] and also for electric vehicles and hybrid electric vehicles [2] due to its high performance i.e. the energy density, safety, rate capability, low cost, sustainability [2], and cycle life [3]. Graphite was the promising and standard active material for anode in Li-ion battery over last two decades due to its excellent performance such as low operating potential, structural stability, moderate cost, and good life cycle [4].

The rate performance on the anode material of Li-ion battery depends on stability of the surface electrolyte interface (SEI) layer formed during charge-discharge cycle. One of the roles of SEI on graphite anodes is to ensure effective passivation of the active anode material against mechanical exfoliation and to impact kinetic stability against electrochemical reactions of the electrolyte in further charge-discharge cycles [5]. Surface modification on the graphite based anode materials have been reported by many researcher. This modification can achieve the high rate performance of the Li-ion battery. The surface chemistry on the interface of electrode–electrolyte has been controlled by modifying the graphite with introducing ion-functional groups, metal oxide, and carbon composite

[4].

Surface modification through by coating graphite with amorphous carbon was also reported by many researchers for enhancing the battery performance. Coal tar pitch (called CTP) was a kind of carbon source of coating layer on the graphite that enhance the rate performance of the battery [4]. CTP derived amorphous carbon coating could effectively decrease charge transfer resistance on the graphite electrode-electrolyte interface.

The characteristic of the coating material using amorphous carbon for the graphite anode in Li-ion battery should have several properties i.e.: isotropic properties, high carbonization yield, and uniform coating electrode. In this study, all the samples are prepared by a citric acid as a carbon source assisted using sol gel method dissolved in the ethanol. The temperature of heat treatment is varied and its effect on the electrochemical performance of the graphite is studied.

2. Experimental methods

2.1. Material preparations and characterizations

A natural graphite (uncoated), labelled UNG, was selected for raw material to evaluate surface modification as anode material for Li-ion batteries. Citric acid as a carbon layer source was dissolved in the ethanol solvent then mixed and stirred with UNG. Further, the samples were dried over night at 80°C in the oven. In the first step, these samples were heat treated at 350°C for four hours for evaporation process. In the second step, the samples were sintered at various temperature i.e. 500°C (CNG500), 600°C (CNG600), and 700°C (CNG700).

The crystal structure of the products was characterized by powder X-ray diffraction (XRD) (Rigaku, Japan) using Cu κα radiation λ = 1.541862 Å at 40 kV (30 mA) with continuous theta-2-theta in a range 20°-90°. The surface morphology of the products was observed using a scanning electron microscopy (SEM) (SU3500, Hitachi, Japan) operated at 20 kV accelerating voltages with different magnification. Transmission electron microscope (TEM) (Tecnai, FEI, the Netherlands) images and selected area electron diffraction (SAED) were performed at 200kV accelerating voltage to evaluate the coating product on the surface of the graphite.

2.2. Cell preparation and battery testing

The electrochemical performance of CNG was evaluated in the CR2032-type coin cell battery. A slurry of working electrode was prepared with 85% CNG, 5 wt.% acetylene black, and 10% Polyvinylidene Fluoride (PVDF) (Sigma Aldrich) dispersed in N,N dimethyl acetamide (DMAC). The homogenous anode slurry was coated onto copper foil using doctor blade and then dried at 80°C for four hours. The cell was assembled using $LiPF_6$ (1M) in ethylene carbonate and dimethyl carbonate (EC-DMC, 1:1 v/v) as electrolyte, Celgard micro porous polyethylene (PE) film as separator, and metallic lithium as the counter electrode prepared in an argon-filled glove box.

Electrochemical characteristics were measured in the cyclic voltammetry and charge discharge test using automatic battery cycler WBCS3000. Cyclic voltammograms were performed with the scan rate 0.1 mV/s in the voltage range of 0 to 2.5 volt. The assembled coin cell batteries were also charged and discharge at 0.1 C between 0–2.5 volt versus Li/Li^+ to determine the specific capacity at room temperature.

3. Results and discussions

3.1. Physical properties of material

The samples are heat treated at several temperatures 500, 600 and 700 °C for five hours in argon atmosphere to investigate the effect of heat treatment temperature on the electrochemical performance. Figure 1 shows the XRD patterns of graphite coated by amorphous carbon using citric acid at different temperature. According to data interpretation using Rietveld refinement method PDXL program, XRD patterns indicate that all the samples have hexagonal structures (lattice constant a = b = 2.4607 Å, c = 6.7096 Å, $\alpha = \beta = 90°$, and $\gamma = 120°$) with 194 : P63/mmc space group, graphite-2H phase name, and no impurity phases. This pattern also can be indexed as (002) crystal faces hexagonal (JCPDS, no. 75-1621). The results indicate graphite coated by amorphous carbon are successfully prepared. Figure 2 shows SEM images of the samples sintered at various temperatures in argon atmosphere. These images show a spherical shape and large number of micro-particles UNG and CNG with average size ranges from 2 to 20 μm. TEM images reveals the crystal area of graphite are covered by the amorphous carbon as shown in figure 3. SAED of the sample is also observed.

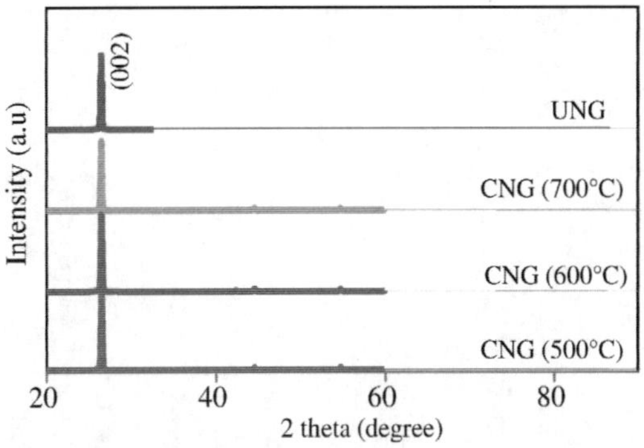

Figure 1. XRD patterns of uncoated natural graphite (UNG) and CNG for several heat-treatment temperatures.

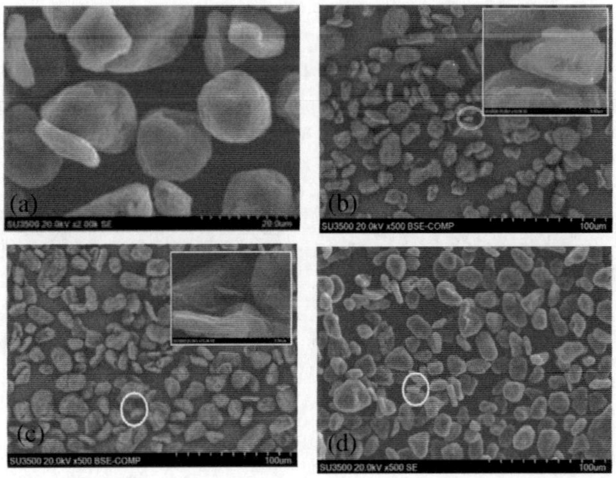

Figure 2. SEM micrographs of (a) uncoated graphite and CNG for several temperatures: (b) 500°C (CNG500), (c) 600°C (CNG600) and (d) 700°C (CNG700).

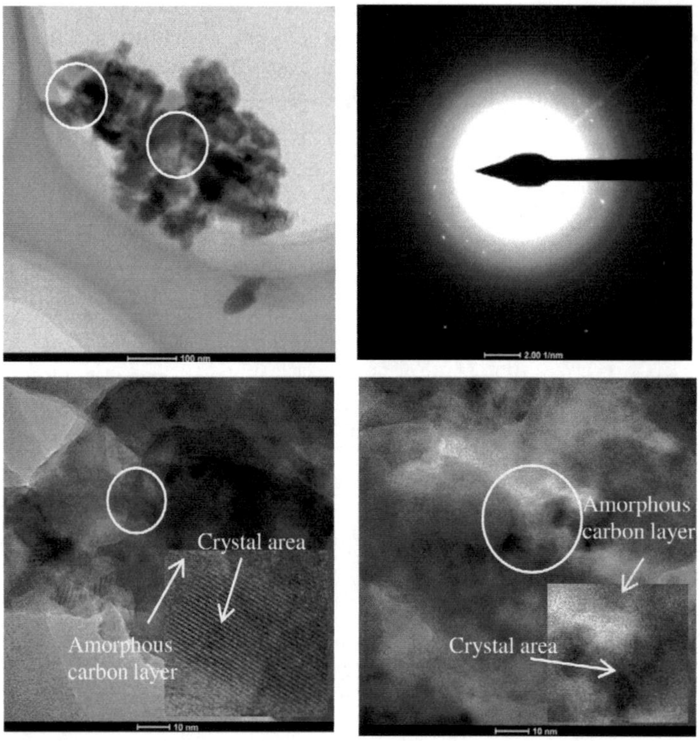

Figure 3. TEM images of graphite coated by amorphous carbon using citric acid.

3.2. Electrochemical characteristics of the sample

The cyclic voltammograms of UNG, CNG500, CNG600, and CNG700 measured with scan rate 0.1 mV/s at range 0 – 2.5 volts are shown in figure 4. In all samples, single cathodic and anodic peak is present and illustrates one electron exchange reaction through redox reaction during C_6 and LiC_6 transformation. CNG600 has the highest specific an oxidation peak current, nevertheless CNG700 has slightly higher an oxidation peak potential. It indicats that heat-treatment temperature has a considerable effect on the electrochemical properties of carbon coated graphite.

To further analyze the effect of coating on the charge-discharge capacity, UNG, CNG500, CNG600, and CNG700 are measured at 0.1 C as shown in figure 5. The highest capacity of carbon coated effect which is measured by half cells is 66 mAh/g (CNG500). The results show that coating amorphous carbon with heat treatment at 500 °C can improve the performance tested at 0.1C. The capacity of CNG600 and CNG700 almost smaller than UNG, indicating that too high heat-treatment temperature may decrease the reversible capacity of the composite coating graphite.

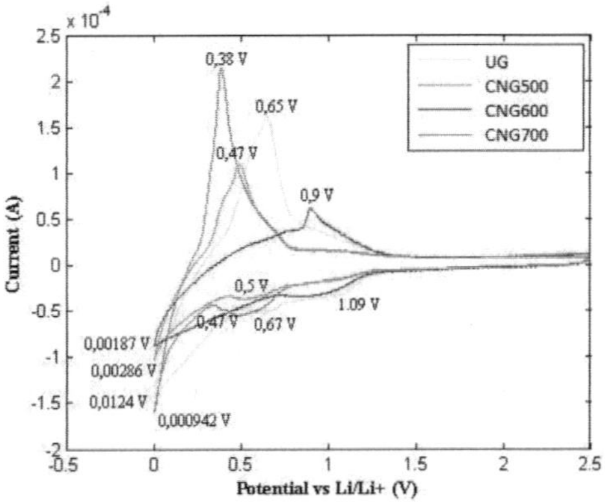

Figure 4. Cyclic voltammograms of uncoated graphite (UG) and CNG for several temperatures of 500°C (CNG500), 600°C (CNG600) and 700°C (CNG700) recorded with scan rate 0.1mV/s at range 0 – 2.5 volts.

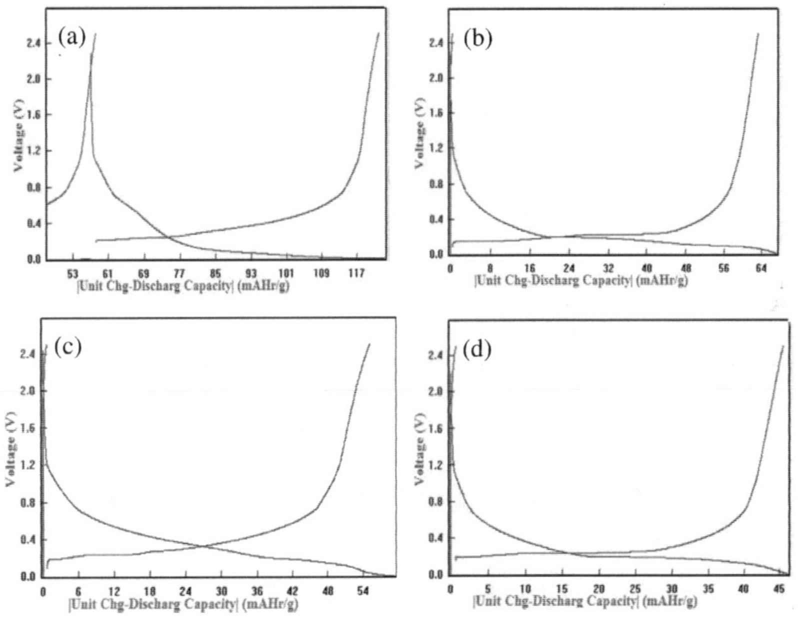

Figure 5. Charge discharge curves of (a) uncoated graphite and CNG for several temperatures: (b) 500°C (CNG500), (c) 600°C (CNG600) and (d) 700°C (CNG700) recorded for the current at 0.1C.

4. Conclusion

Effect of carbon coating layer on the surface of graphite were studied. Crystal structure and impurities effect were investigated as revealed from XRD characterization. Amorphous carbon and graphite area were observed through electron microscopy analysis. From the results can be concluded that amorphous carbon layer on the surface of graphite improved the electrochemical performance through charge-discharge capacity test which have the specific discharge capacity of 66 mAh/g.

Acknowledgements
The authors would like to acknowledge the Research Center for Physics – LIPI and Ministry of Research, Technology and Higher Education – Republic of Indonesia for financial support in this research under SINAS Research Fund Program year 2015.

References

[1] Ding C, Nohira T, Hagiwara R, Fukunaga A, Sakai S and Nitta K 2015 *Electrochim. Acta.* **176** 344
[2] Wu R, Xia G, Shen S, Zhu F, Jiang F and Zhang J 2015 *Electrochim. Acta.* **153** 334
[3] Miranda D, Costa C M and Lanceros-Mendez S 2015 *J. of Electroanal. Chem.* **739** 97
[4] Han Y-J, Kim J, Yeo J-S, An J C, Hong I-P, Nakabayashi K, Miyawaki J, Jung J-D and Yoon S-H 2015 *Carbon* **94** 432
[5] Gallego N C, Contescu C I, Meyer III H M, Howe J Y, Meisner R A, Payzant E A, Lance M J, Yoon S Y, Denlinger M and Wood III D L 2014 *Carbon* **72** 393

Production of activated carbon by using pyrolysis process in an ammonia atmosphere

N Indayaningsih, F Destyorini, R I Purawiardi, D R Insiyanda and H Widodo

Research Center for Physics, Indonesian Institute of Sciences, Kawasan Puspiptek, Tangerang Selatan, Banten, Indonesia

Email: nani005@lipi.go.id; naniksih@gmail.com

Abstract. Activated carbon is materials that have wide applications, including supercapacitor materials, absorbent in chemical industry, and absorbent material in the chemical industry. This study has carried out for the manufacturing of activated carbon from inexpensive materials through efficient processes. Carbon material was made from coconut fibers through pyrolysis process at temperature of 650, 700, 750 and 800°C. Aim of this study was to obtain carbon material that has a large surface area. Pyrolysis process is carried out in an inert atmosphere (N_2 gas) at a temperature of 450°C for 30 minutes, followed by pyrolysis process in an ammonia atmosphere at 800°C for 2 hours. The pyrolysis results showed that the etching process in ammonia is occurred; as it obtained some greater surface area when compared with the pyrolisis process in an atmosphere by inert gas only. The resulted activated carbon also showed to have good properties in surface area and total pore volume.

1. Introduction

Carbon is one of the materials that attracted most researchers' attention in recent decades, because it can be applied as smart materials. The development of technology enables the use of carbon for various applications, for instance electronic-textiles, super-capacitor, sensor, batteries [1, 2], biomedical [3], electrochemical devices, hydrogen storage, field emission devices, electronic devices [4], air, water and gas filtration [5, 6], conductive or reinforced plastics, structural composites, catalyst supports, and electrode fuel cell.

Carbon can be obtained from the decomposition of organic substances or biomass waste, such as coconut fibers, which is abundant and easily available. Decomposition process occurs usually by heating at a temperature corresponding to the thermodynamic conditions of the precursor to form charcoal, including by-products such as tar and gases. The process known as the carbonization or pyrolysis, conducted in the absence of air/oxygen [7] or in an inert gas condition [8, 9]. Nitrogen (N_2) or argon (Ar) gases are usually used in the pyrolysis process to obtain carbon [10]. The inert gas is a stable and non-reactive gas which causing the degradation of material without oxidation [11]. Form the references; it had conducted several ways to produce carbon with good characteristics in terms of high porosity and surface area [12-14]. Charcoal or carbon is then processed to form internal pores within the macro-, meso- and micro-pores in order to generate a larger surface area [15, 16].

Activated carbon has been extensively used as a material for energy storage due to large surface area, stable, easily polarized, amphoteric character, accessibility, relatively low cost and environmentally friendly [17]. It has been also used for filter of impurity elements [13] and as an absorbent [18]. Super-capacitor based carbon is used as a constituent of the electrode material in order to produce a large surface area. The larger surface area required for storage of cargo in the form of electrolyte ions, this occurred during the charging process, ions are stored at the interface of electrode/electrolyte [19].

The activation process involves the physical mechanism by providing gas (gasification) or insertion of a chemical compound, also involves heating at relatively high temperatures [14, 16]. Chemical activation has the advantage, namely low temperature process to generate higher yields and could control at the micro level [20], chemical agent used such as phosphoric acid, zinc chloride, potassium hydroxide [13, 14, 21], and sodium hydroxide [22]. However, the use of chemical substance to increase the carbon characteristic is time consuming, costly and hazardous for the environment. The use of ammonia in pyrolysis process is a breakthrough to decrease the pollutions. Ammonia (NH_3) contains nitrogen and hydrogen atoms which are abundant in the earth [23]. The active carbon from cellulose filter has been made using the pyrolysis process in NH_3 atmosphere. The mechanism is leading to a process of gasification of carbon to form methane (CH_4), the etching process resulting in a larger surface area compared to pyrolysis in nitrogen [24].

In this study, activated carbon from coconut fibers had been made through the pyrolysis process in an atmosphere of ammonia at various temperatures, to observe the effect on surface area.

2. Material and methode

2.1. Preparation of active carbon materials

Coconut fibers as raw marerial were subjected to degumming process by immersion in a 10% NaOH (Merck) solution 90-95 °C temperature for 2-3 hours then washed, drained and dried. In this study, the weight ratio of fiber and NaOH is 1:5. The above procedure is aimed to obtain the cellulose and hemicellulose of coconut fiber. Through this step, it also will reduce the lignin and rubber that contained in coconut fiber [25]. The lignin and rubber will cause a sticky brown liquid during the pyrolysis process and affect the furnace.

Active carbon was made through a pyrolysis process with two step processes, including a carbonization followed by an activation steps. Carbonization process is carried out in an inert gas atmosphere (nitrogen gas) at a temperature of 450 °C for 30 minutes, followed by activation process in an ammonia (NH_3) atmosphere at various temperatures of 650, 700, 750 and 800°C for 2 hours with a gas flow rate of 5 ml/min. The etching process is expected to occur during the pyrolysis process in an atmosphere of ammonia. This process is aimed to obtain a larger surface area and improve the properties of materials to produce activated carbon. Then, the carbon was grinded into powder for further sample characterizations.

2.2. Analysis and characterizations

An X-ray powder diffraction with a Cu tube anode an X-Ray Diffraction (XRD) Rigaku type SmartLab 3 kW was used to record the X-ray patterns of samples. Scanning electron microscopy (SEM from HITACHI SU-3500) and the samples Energy-dispersive spectroscopy (X-max, Horiba, Japan) probed at 4 points was used to visualize the surface morphology of the carbonized and activated products. The carbon content of these samples (charcoal that is produced) is also determined using scanning electron microscopy equipped with energy dispersive spectrometer. The EDS spectra showing elemental composition were obtained by scanning through the surfaces of the samples.

The surface areas measurements (m^2/g) of the activated carbon samples were made by low temperature nitrogen adsorption at range room temperature to 450°C, 1°C intervals, methods vacuum and flow on Brunauer Emmett Teller (BET) surface area analyser, by BET Nova 4200e with minimum pore volume 0.35-400 nm. The specific surface areas were estimated by the multipoint BET equation. The electrical conductivity was measured by using HIOKI 3522-50 LCR-meter.

IOP Conf. Series: Journal of Physics: Conf. Series **817** (2017) 012006 doi:10.1088/1742-6596/817/1/012006

3. Result and discussion

3.1. X-ray patternsof carbon

Microstructure of carbon in the charcoal of coconut fiber was analyzed using XRD. Figure 1 shows the diffraction spectra of the charcoal from coconut fiber prepared at temperatures of 650, 700, 750 and 800°C. In XRD pattern, there are two main peaks with some slope slightly; this indicates that the carbon has an amorphous structure. The first peak of the carbon, as the result of pyrolysis process at of 600, 750, 750 and 800°C, the 2-theta positions are 22.71, 23.65, 23.33, and 23.27°, respectively. Increasing temperature of pyrolisis affect the shift peak position close to graphite peak position about 26° [JCPDS PDF 75-2078]. EDS data also confirm that the dominant element is carbon, it also evidence some other impurities (Na, K, Si) however with amount less than 2 wt. %, as presented in tabel 1.

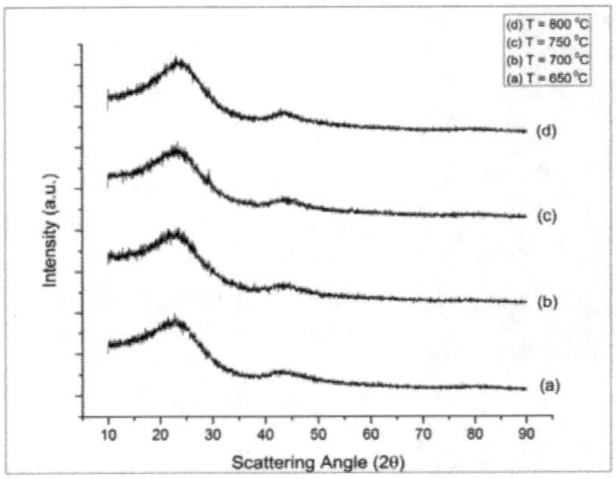

Figure 1. X-ray diffraction pattern of carbon prepared at variety of pyrolysis temperature.

Tabel 1. The EDS result of coconut fiber charcoal.

No	Element	%wt
1	C	85.68 ± 3.42
2	O	12.56 ± 2.97
3	Na	1.46 ± 0.49
4	K	0.25 ± 0.28
5	Si	0.14 ± 0.07

3.2. Surface morphology

The observation of carbon morphology using SEM is shown in figure 2 and 3. Hydrocarbon decomposition process on coconut fibers occurs during the pyrolysis in inert gas, and leaving the charcoal as by product. The morphology of charcoal is fibers shape with diameter of 100 –300 μm, consist of many parallel hollows like tube that are close to each other with diameters of 2 – 10 μm, and on the wall of the tube there are some porous with sizes around 1μm. The observations on carbon by pyrolysis results in an atmosphere of ammonia gas are presented in figure 3(b) and 4(b). The surface morphology showed finer than in nitrogen atmosphere. It seems some etching process on the surface of carbon is occurred, so the surface seems more tangled [24]. The elemental composition of carbon was also determined using scanning electron microscopy equipped with an energy dispersive spectrometer, the amount of carbon in the charcoal of coconut fiber was 85.68 ± 3.42 wt. %, the results showed that carbons are the dominant element in the charcoals.

3.3. Surface area analysis

The observations and measurements using BET surface area indicates that the total pore volume of carbon increased with increasing pyrolysis temperature, as shown in figure 4(a). This suggests that the increase in temperature, causing the more volatile compounds disappear, consequently the greater pore volume is obtained. Initially, the pyrolysis temperature rise causes an increase in the average of pore radius of carbon and the average of pore radius dropped again for pyrolysis temperature at 800°C, as presented in figure 4(b). Average pore size at a temperature of 800°C decreased, but the pore volume increases. This indicates that the degradation reaction continues during the pyrolysis process. Indeed, it does not add any large pore size; however it seems the number of new pores increase significantly. In observation of surface area, as presented in figure 5, the higher pyrolysis temperature, surface area rose in accordance with the total volume porosity.

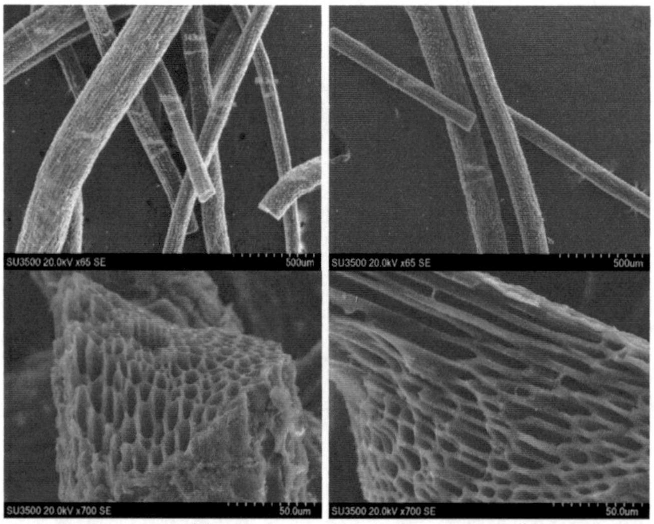

(a) (b)

Figure 2. Surface morphology of carbon from coconut fiber with non-degumming process and pyrolysis process at atmosphere: (a) Nitrogen, (b) Ammonia.

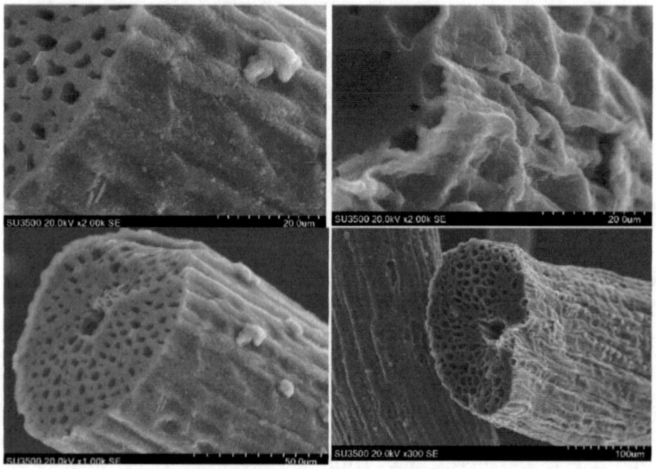

(a) (b)

Figure 3. Surface morphology of carbon from coconut fiber with degumming process and pyrolysis process at atmosphere: (a) Nitrogen, (b) Ammonia.

3.4. Electrical conductivity

The results of the measurement of the electrical properties of carbon powder using an LCR-meter show that carbon is quite conductive, as shown in figure 6. Pyrolysis temperature increases, the higher the electrical conductivity. At temperature of 800°C, electrical conductivity of carbon with ammonia gas treatment is slightly lower than the carbon with nitrogen gas atmosphere. This can be understood by observing on the SEM results (figures 2 and 3), which indicates that there is a change in morphology when linked with porosity. In the pyrolysis process using ammonia gas, giving the effect of etching on the surface of the sample, this led to an increase in porosity and surface area [26, 27]. From this result, the increase in porosity can give some effect on a decrease in the electrical properties.

The electrical conductivity of carbon is lower than graphite (output Fisher). It suggests that the electrical properties can be improved by raising the temperature pyrolysis.

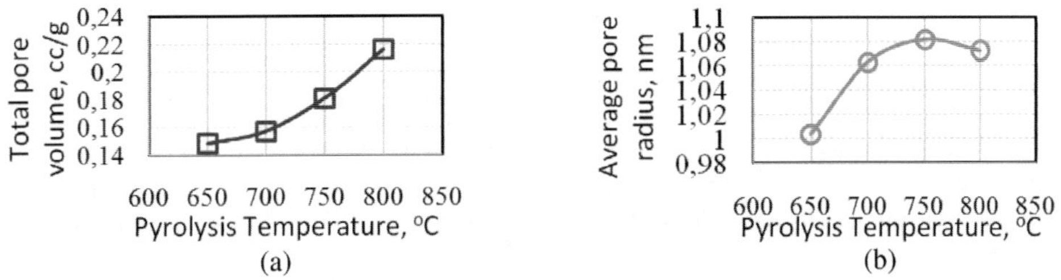

Figure 4. The total pore volume and average of radius pore of the carbon of coconut fiber with degumming process and pyrolysis process in ammonia gas.

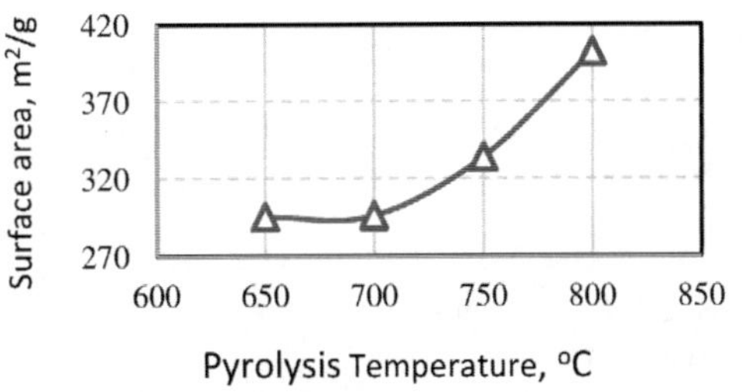

Figure 5. The surface area of coconut fiber carbon with degumming process and pyrolysis process in ammonia gas.

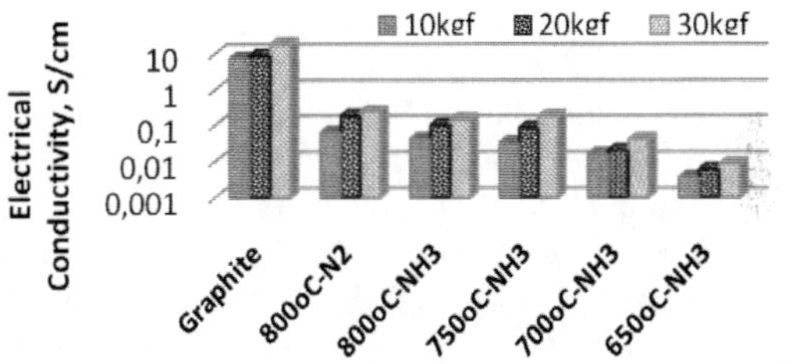

Figure 6. The electrical conductivity of thecarbon of coconut fiber with degumming process and pyrolysis process in ammonia gas.

4. Conclusions

In this study, activated carbon from coconut fibers have been made through the carbonization process in an inert gas and followed by pyrolysis process in an atmosphere of ammonia. Experimental data showed that the increase in temperature resulted in better properties of activated carbon. The activated carbon from coconut fibers is proven to have good properties in terms of surface area and total pore volume.

Acknowledgments

The authors gratefully acknowledge the financial support from Indonesian Institute of Sciences and the facilities provided by the Research Center for Physics—Indonesian Institute of Sciences.

References

[1] Dalton A B *et al* 2003 *Nature* **423** 703
[2] Baughman R H, Zakhidov A A and de Heer W A 2002 *Science* **297** 787-92
[3] Bianco A and Liu Z 2016 *Carbon* **97** 124-5
[4] Gehrke I, Geiser A and Somborn-Schulz A 2015 *Nanotechnol Sci. Appl.* **8** 1–17.
[5] Khajavi R, Bahadoran M M S, Bahador A and Khosravi A 2012 *J. Ind. Text.* **42** 219-30
[6] Fisher T, Hajaligol M, Waymack B and Kellogg D 2002 *J. Anal. Appl. Pyrolysis* **62** 331-49
[7] Vladimir K F, Vladimir A, Sinelshchikov, Sytchev G A and Zaichenko V M 2013 *Bionature 2013: The Fourth International Conference on Bioenvironment, Biodiversity and Renewable Energies (Lisbon)* (Wilmington: International Academy, Research and Industry Association) pp 29-32

[8] Ţiprigan O, Koos A A, Nemes - Incze P and Biro L P 2007 *Optoelectron. Adv. Mater. Rapid Commun.* **9** 617-20
[9] Health and Safety Executive 2011 *Using Electric Storage Batteries Safely* 1st edition (London: Health and Safety Executive Books) pp 1-10
[10] Saragih S A 2008 *Pembuatan dan karakterisasi karbon aktif dari batubara riau sebagai absorben* Master Thesis (Indonesia University) pp 11-2
[11] Ali E, Taherehand K S and Mansooreh 2011 *Res. J. Chem. Environ* **15** 433-7
[12] Rahman M M, Awang M, Mohosina B S, Kamaruzzaman B Y, Wan Nik W B and Adnan C M C 2012 *APCBEE Procedia* **1** 293-8
[13] Rashidi N A, Yusup S, Ahmad M M, Mohamed N M and Hameed B H 2012 *APCBEE Procedia* **3** 84-92
[14] Zhang M, He C, Liu E, Zhu S, Shi C, Li J, Li Q and Zhao N 2015 *J. Phys. Chem. C* **119** 21810-7
[15] Herawan S G, Hadi M S, Ayob Md R and Putra A 2013 *The Scientific World Journal* **2013** 1-6
[16] Frackowiak E and Be´guin F 2001 *Carbon* **39** 937-50
[17] Ilomuanya M O, Nashiru B, Ifudu N D and Igwilo C I 2016 *Journal of Microscopy and Ultrastructure* (In Press, available online 12 May 2016)
[18] Raymundo-Piñero E, Kierzek K, Machnikowski J and Béguin F 2006 *Carbon* **44** 2498-507
[19] Lillo-Ródenas M A, Cazorla-Amorós D and Linares-Solano A 2003 *Carbon* **41** 267-75
[20] Pandolfo A G and Hollenkamp A F 2006 *J. Power Sources* **157** 11-27
[21] Xu B, Chen Y, Wei G, Cao G, Zhang H and Yang Y 2010 *Mater. Chem. Phys.***124** 504-9
[22] Roney N, Llados F, Little S S and Knaebel D B 2004 *Toxicologcal Profile for Ammonia* (Atlanta: Agency for Toxic Substances and Disease Registry)
[23] Luo W, Wang B, Heron C G, Allen M J, Morre J, Maier C S, Stickle W F and Ji X 2014 *Nano Letters* **14** 2225-9
[24] Mathew S and Zakaria Z A 2015 *Appl. Microbiol. Biotechnol.* **99** 611-22
[25] Liu L, Yang Q and Shen J 2015 *Journal of Nanomaterials* **2015** 1-9
[26] Han C, Kim I T, Sun H J, Park G, Lee J J, Lee H K and Shim J 2012 *Int. J. Electrochem. Sci.***7** 8627-36

IOP Conf. Series: Journal of Physics: Conf. Series **817** (2017) 012007 doi:10.1088/1742-6596/817/1/012007

The role of Mg dopant on the morphological, structural and optical properties of Mg doped zinc oxide grown through hydrothermal method

P Susetyo[1], V Fauzia[1] and I Sugihartono[2]

[1] Department of Physics, Universitas Indonesia, Depok 16424, Indonesia
[2] Department of Physics, Universitas Negeri Jakarta, Jakarta 13220, Indonesia

E-mail: vivi@sci.ui.ac.id

Abstract. ZnO nanorods is a low cost II-VI semiconductor compound with huge potential to be applied in optoelectronic devices i.e. light emitting diodes, solar cells, gas sensor, spintronic devices and lasers. In order to improve the electrical and optical properties, group II, III and IV elements were widely investigated as dopand elements on ZnO. In this work, magnesium (Mg) was doped into ZnO nanorods. Samples were prepared firstly by deposition of undoped ZnO seed layer on indium thin oxide coated glass substrates by ultrasonic spray pyrolysis method and then followed by the growth of ZnO nanorods doped by three different Mg concentrations by hydrothermal method. Based on the morphological, microstructural and optical characterizations results, it is concluded that the increase of magnesium concentration tends to reduce the diameter of ZnO nanorods, increases the bandgap energy and decreases the UV absorption the luminescence in UV and visible range.

1. Introduction

In recent years, the properties of nanomaterial have been widely investigated by the scientists. Among semiconductor nanostructure, ZnO nanorods have attracted many scientists and engineers because of its wide application in optoelectronics, sensor and actuator, energy and biomedical [1]. ZnO is a low cost and eco-friendly II-VI semiconductors which is easily synthesized. ZnO has many advantages among those semiconductors, such as it can be grown at low temperature (<500°C) [2], wide direct band-gap energy (3.37 eV) [3] and large exciton binding energy (60 meV) [4].

Various methods have been introduced for the preparation of ZnO nanorods, such as vapour liquid solution deposition, pulsed laser deposition, spray pyrolisis and metal organic chemical vapour deposition (MOCVD) [2]. Each of these methods has their own advantages and disadvantages. Hydrotermal method is a preparation method which is simple, low cost and can be performed at low temperature without high vacuum condition [2].

Many researchers reported the properties of ZnO nanorods can be modified by the addition of doping elements. In 2009, Te Hua Fang et al showed that the magnesium addition into ZnO nanorods was able to increase the electrical conductivity [5], improved the optoelectrical properties in the ultraviolet detector [6] and also could increase the selectivity when it was applied as ethanol sensor [7].

This work reports the effect of three different concentrations of magnesium (0, 1 and 7 at. %) on the morphological, microstructural and optical properties of ZnO nanorods. Samples were grown on indium thin oxide coated glass substrates through deposition of a seed layer by ultrasonic spray pyrolysis method, and then followed by growth of ZnO nanorods by hydrothermal method.

2. Experimental

ZnO nanorods were synthesized through two steps. The first step was the deposition of seeding layers on the ITO subtrates using ultrasonic spray pyrolysis method. The seed solution consisted of 0.2 M zinc acetate dehydrate ($(CH_3COO)_2Zn.2H_2O$) dissolved in the deionized water. The solution was stirred at 400 rpm at room temperature, and then put into a container inside the commercial ultrasonic nebulizer, while a clean coated ITO glass substrate was put on a hotplate at the temperature of 550°C. The droplets of seed solution were then sprayed onto the substrates surface using the ultrasonic wave (1,7 MHz) for 10 minutes. The next step was the growth process of Mg-doped nanorod ZnO using hydrothermal method with three variation of magnesium concentrations (0, 1 and 7 at. %). The growth solution contained of 0.2 M zinc nitrate tetrahydrate ($Zn(NO_3)_2.4H_2O$), hexamethylentetramine ($C_6H_{12}N_4$) and magnesium nitrate hexahydrate ($Mg(NO_3)_2$ that were dissolved in the deionized water and stirred at 400 rpm until its temperature reached 60°C. The growth process was started by put the substrates into the growth solution and heated at 95°C in oven for 2 hours.

Morphological characterization was carried out using a field emission scanning electron microscopy (FE-SEM) Inspect F50, microstructural properties were investigated by Shimadzu XRD-7000 X-Ray Diffractometer and the optical properties were characterized by Thermo Fisher Scientific GENESYS 10S Ultraviolet Visible, Hitachi UV-Vis Spectrometer U-3900H and FLS920 Fluorescence Spectrometer.

3. Result and discussion

3.1. Morphological properties

Surface morphology of the ZnO nanorods is shown in figure 1. Generally, ZnO nanorods were growing overlap with each other. The growth direction of ZnO nanorods was not uniform and not too perpendicular to the ITO substrates compared to other previous results [8]. From this figure, it can be seen that the increase of magnesium concentration tends to reduce the diameter of ZnO nanorods. This may occur as a result of the substitution of Mg into ZnO with an ionic radius which was smaller than Zn ions. This result is different with the previous other result showing that the addition of Mg doping resulted in an increase of diameter of ZnO nanorods [5].

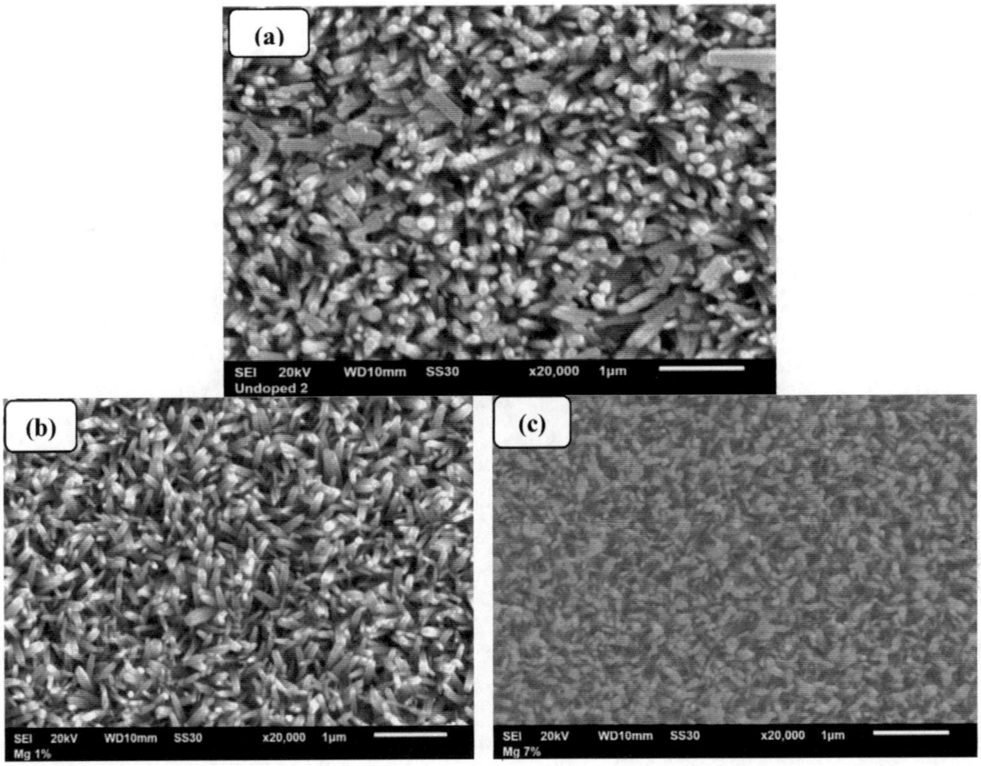

Figure 1. SEM images of Mg doped ZnO nanorods with a concentration of (a) 0%, (b) 1% and (c) 7 at. %

3.2. Microstructural properties

The XRD diffraction pattern of ZnO nanorods is shown in figure 2. The diffraction pattern shows the peaks of ZnO phase with the crystal planes of (100), (002), (102), (101), (110) and (103), while the other peaks come from the indium tin oxide substrate. In general, position of the peaks is same for three different samples. It indicates that the addition of Mg doping does not produce the other materials phase because Mg^{2+} ions were able to be completely substituted to the Zn^{2+}ions or as an interstitial site in the ZnO crystal [9].

The diffraction pattern which has many peaks also indicates that the direction of crystal growth of ZnO nanorod is not uniform. Furthermore, it can be observed that the addition of Mg dopant results in the decrease of peak intensities of (002), (102) and (103) crystal planes. This result has a close agreement with the SEM results that there is no any dominant growth directions of ZnO nanorods.

IOP Conf. Series: Journal of Physics: Conf. Series **817** (2017) 012007 doi:10.1088/1742-6596/817/1/012007

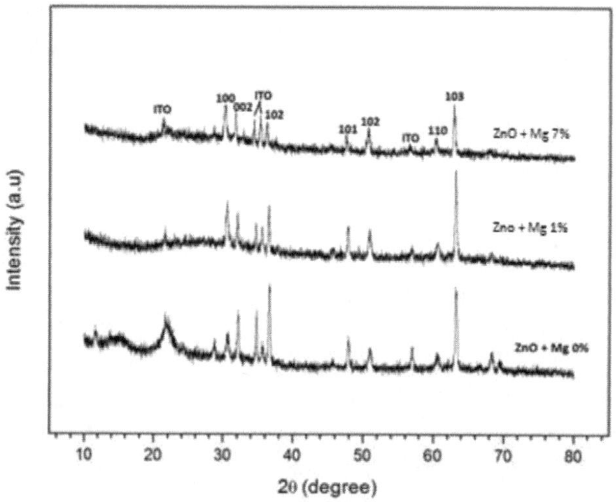

Figure 2. X-ray diffraction pattern of the Mg-doped ZnO nanorods

3.3. Optical properties

3.3.1. Absorbance and transmittance spectra. Figure 3 shows the absorbance and transmittance spectrum of ZnO nanorods. The absorbance spectra show that the intensity at a wavelength of 380 nm which is known as the absorption edge that correlated with band gap energy is sharply decreased. The addition of Mg dopant decreases significantly the absorption intensity in the ultraviolet region (300-380 nm). In the visible region, the addition of 1 at.% Mg leads to increase the absorbance intensity while the addition of 7 at.% Mg leads to decrease the absorbance intensity. The decrease of absorbance values in UV region could be related to the smaller diameter of nanorods that less effective for UV light absorption.

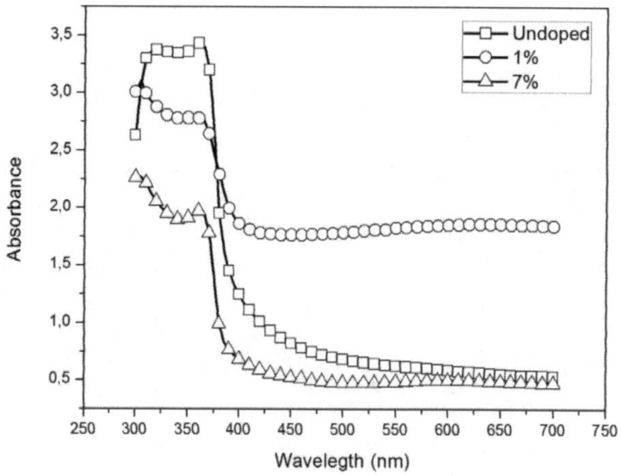

Figure 3. Absorbance spectra of Mg doped ZnO nanorods

IOP Conf. Series: Journal of Physics: Conf. Series **817** (2017) 012007 doi:10.1088/1742-6596/817/1/012007

3.3.2. Reflectance spectra and Tauc Plot graph. Figure 4(a) is the reflectance spectra of ZnO nanorods. This figure shows that for all samples, the reflectance intensity increases significantly at a wavelength of 380 nm. It is corresponding to the absorption edge as shown in Figure 3. The bandgap energy calculation of ZnO nanorods was performed based on the diffuse reflectance spectra by using Kubelka Munk equations[9]. The calculated data were then interpolated in Tauc plot as shown in figure 4(b). The result shows that the bandgap energy for undoped ZnO nanorods is 3.11 eV while the ZnO nanorods with Mg concentrations of 1 and 7 at.% are 3.18 eV and 3.15 eV, respectively. The increase of bandgap may occur due to the Moss-Burstein effects, namely the shifting of the Fermi energy to the conduction band due to the high concentration of dopant elements so that the energy gap becomes wider [10].

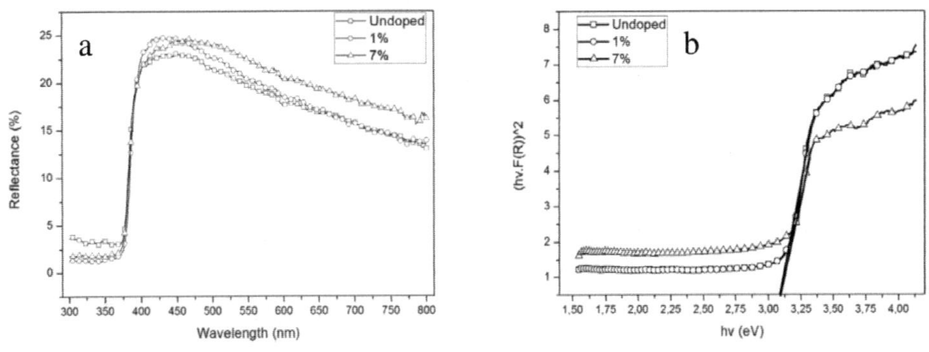

Figure 4. (a) Reflectance spectra and (b) Tauc Plot of Mg doped ZnO nanorods

3.4. Photoluminescene spectra

Figure 5 shows the photoluminescence spectra of ZnO nanorods. In this figure, it appears there are two peaks at the wavelengths of 400 nm and 630 nm. The luminescence intensity at lower wavelength represents the radiative transition from the recombination of the conduction band electrons with the valence band holes. The luminescence at a wavelength of 630 nm may occur due to many natural defects in ZnO nanorods. The increase of Mg concentration in ZnO nanorods appears to decrease the luminescence in the entire wavelength range.

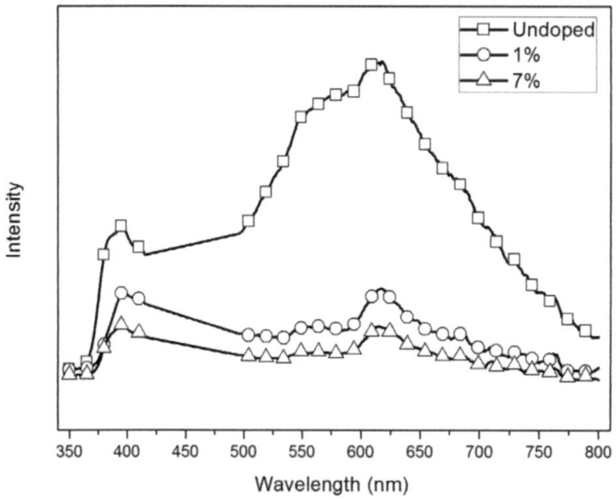

Figure 5. Photoluminescene spctra of Mg doped ZnO nanorods

4. Conclusion

In this study, ZnO nanorods have been successfully synthesized with the ultrasonic spray pyrolysis and hydrothermal method. ZnO nanorods are mostly growing overlap with no uniformity and not exactly perpendicular to the substrate in the growth direction. The increase of magnesium concentration tends to reduce the diameter of ZnO nanorods and decreases the intensity of almost all diffraction peaks of ZnO rods. The addition of Mg dopant of one and 7 at. % is found to decrease significantly the absorption intensity in the ultraviolet region (300-380 nm) and increases the bandgap energy from 3.11 eV into 3.18 eV and 3.15 eV, respectively. The photoluminescence spectra show there are two peaks at the wavelengths of 400 nm and 630 nm that represent the radiative transition from the recombination of electrons and holes and the presence of the natural defects in ZnO nanorods, respectively. The increase of Mg concentration in ZnO nanorods appears to decrease the luminescence in the UV and visible range.

References

[1] Wang Z L 2004 *J. Phys.: Condens. Matter.* **16**(25) R829
[2] Yi G C, Wang C and Park W I 2005 *Semicond. Sci. Technol.* **20** S22
[3] Zhong W W, Liu F M, Cai L G, Zhou C C, Zeng L G, Liu X Q and Li Y 2011 *J. Alloys Compd.* **509**(9) 3847-51
[4] Eskandari M, Ahmadi V and Ahmadi S H 2009 *Physica E* **42** 1683-86
[5] Fang T H and Kang S H 2010 *J. Alloys Compd.* **492**(1) 536-42
[6] Shewale P S and Yu Y S 2016 *J. Alloys Compd.* **654** 79-86
[7] Kwak C H, Woo H S, Abdel-Hady F, Wazzan A A and Lee J H 2016 *Sens. Actuators, B* **223** 527-34
[8] Ginting R T, Yap C C, Yahaya M, and Salleh M M, 2014 *J. Alloys Compd.* **585** 696-702
[9] Lestari A, Iwan S, Djuhana D, Imawan C, Harmoko A and Fauzia V 2016 *AIP Conf. Proc.* vol 1729 no 1 (New York: AIP Publishing) 020027
[10] Zeng W, Yang X, Shang M Xu X, Yang W and Hou H. 2016 *Ceram. Int.* **4**(8) 10021-29

Analysis of effective pulse current charging method for lithium ion battery

N Majid, S Hafiz, S Arianto, R Y Yuono, E T Astuti and B Prihandoko

Research Centre for Physics – Indonesian Institute of Sciences (LIPI)
Kawasan Puspiptek Gd 440-442, Serpong, Tangerang Selatan, INDONESIA 15314

E-mail : nurhalis.majid@lipi.go.id

Abstract. Pulse charging methods has been developed as one of the fast charging methods for Lithium ion battery. This technique applies the continuous constant current pulse with certain pulse width until the battery fully charged. In this research, four Lithium polymer batteries of same type and capacity were used and subjected by several current pulses as a variable. The phenomenon of capacity loss as an effect of charging method was analysed every ten charge-discharge cycles. Four batteries were charged using constant current (1C) for 30 minutes to fill half of the total capacity, which then continued by pulse current of different pulse width in order to reach full capacity of each battery. Constant current charging for one hour was also applied to each battery as a comparison with that of pulse current charging data. The similar degradation patterns on battery capacity were observed. Nevertheless, the percentage of capacity loss is different. In conclusion, this method can be considered as one of the effective charging method, owing to the smallest capacity loss and shorter charging time.

1. Introduction

Recently, research on Lithium ion battery is one of the promising research field which attracts considerable attention due to its contribution for solving the issue of dependency on fossil fuels. Lithium ion battery provides superior advantages over other kind of rechargeable battery owing to its extremely large charge capacity for a given volume and weight. One of the challenging problems in lithium ion battery research field is battery charging which include of charger and charging technique for maximizing the battery capacity. All of battery user need the charger with short charging time or fast charging device and small capacity loss cell. Unfortunately, the conventional charging technique induces degradation of battery capacity when large of charging rate or high current value is used. The most widely used conventional charging technique for Lithium ion batteries is constant current – constant voltage (CC-CV) technique, where the charging current is kept constant until the specified voltage threshold in the CC phase, which is then continued with the constant charging voltage until the charging rate below a certain threshold CV phase [1,2]. Conventional charging technique uses high value constant current (CC) for charging the battery to specified voltage value and continues with constant voltage until the significant degradation current value occurs. This method is usually used in mobile device charger. However, the battery state of charge (SoC) cannot fully reached 100%, which is the main disadvantage of this method. On the contrary, the state of fully charged can be achieved when we apply constant current technique with small value of current (depends on the battery type). Multi step constant current charging is also used for shortening charging time and prolonging the cycle life of the battery[3]. Unfortunately, this technique can only be used for small charging current.

Development on charging techniques for lithium ion battery has been carried out by researchers who works in the field of Battery Management System (BMS). There are a several techniques for charging the lithium ion battery. Each technique gives different result, which depends on the battery type[4]. In general, the life cycle is strongly influenced by the value of charging and discharging currents. High charging current can deteriorate life cycle at every SoC. Research for determining the effect of different charging method on aging process of the battery has been done by Monem et al[5]. They used three techniques for charging the high power 7 Ah $LiFePO_4$ battery, i.e. constant current (CC), constant current – constant voltage (CC-CV) and constant current – constant voltage with negative pulse (CC-CVNP). The demonstrated that the CC-CVNP technique with low amplitude and fewest number of negative pulses is more effective than that of other techniques in reducing the concentrating resistance and the diffusion time constant.

Pulse charging is one of widely-used fast charging method for lithium ion battery. This technique is the powerful method for charging lithium ion, which can maximize the battery capacity. This technique is clearly different with that of conventional method for charging the lithium ion cell. The strong point of this technique lies on the utilization of certain value of current or voltage to charge the battery at the specified time (usually in second and minute order). The condition of fully charged can be indicated by the extent of voltage degradation when the charging is at the rest phase or count of charging pulse[6].

There are many researches and patents which develop the pulse charging technique and charger. One of these is US patent by Hagino which develop the charging method for rechargeable battery[7]. In that patent, the open circuit battery voltage was measured during periods of suspended charging and charging was resumed when the measured voltage drops below a specified voltage. When the period of suspended charging is longer than a specified time, open circuit battery voltage is compared to a minimum voltage. When the open circuit battery voltage is higher than the minimum voltage and the period of suspended charging is longer than the specified time, the battery is considered fully charged. When the open circuit battery voltage is less than the minimum voltage, even though the period of suspended charging is longer than the specified time, the battery is assumed to be removed from the system.

In this study, we investigate the role of charging current pulse widths and analyse the effect of these treatments to the capacity loss on the lithium ion battery after the pulse charging were done at several cycles.

2. Experiment

Four Lithium polymer batteries (marked as cell A, B, C, and cell D) were charged and discharged using Battery Analyser BST8-10A30V of MTI Corporation.

Table 1. BTS8-10A30V Specification.

Item	Specification
Voltage measurement	Range: 0.03 – 30 V Accuracy: +/- (0.1% of reading + 0.05% range)
Current measurement	Range: 0.01 – 10 A Accuracy: +/- (0.2% of reading + 0.1% of range)
Time range	0 – 999 seconds
Cycle Measurement	1 – 9999 times
Input Impedance	≥ 10 mΩ
Channels	8 Channels

Battery Analyser BST8-10A30V specification in table 1 is the suitable equipment for analysing the capacity of various battery types, such as pouch cell, cylindrical cell, and coin cell. It can also be used for small and big capacity cell, for research or industrial scale. It has an accurate measurement system, for both voltage and current. It can be used for long large cycle value. In this research, we used lithium polymer battery model PL-063048-10C, which is assembled by AA Portable Power Corp. All of the batteries was charged to the final voltage of 4.2 V with current of 10 mA using constant current – constant voltage technique and discharged to 2.75 V using 10 mA constant current technique to determine the capacity of the battery. All of the batteries were then charged using different technique. All of the batteries were charged using constant current technique at 1C for one hour and discharged in constant current methods at 1C. Cell A, B, C, and D were then charged using constant current of 1C for 30 minutes to fill half of their full capacity and then stopped for 1 minute. After stop condition, cell A was continuously charged and stopped again for 2 and 1 minutes, respectively, for 15 cycles, cell B was continuously charged and stopped again for 4 and 1 minutes, respectively, for 8 cycles, cell C was continuously charged and stopped again for 6 and 1 minutes, respectively, for 5 cycles, while cell D was continuously charged and stopped again for 8 and 1 minutes, respectively, for 4 cycles. All of the cells were charged at 1C constant current with different pulse width for about 30 minutes to fill the battery completely. Cell A, B, C, and D were then discharged at 1C using the same method for 50 cycles. The capacity loss were evaluated every 10 cycles and compared with constant current charge methods. The cell specification is presented in table 2.

Table 2. Polymer Lithium ion battery specification.

Item		Specifications	Remark
Nominal Capacity		750mAh±5%	0.2C5A discharge，25°C
Nominal Voltage		3.7V	Average Voltage at 0.2C5A discharge
Standard Charge Current		$0.2C_5A$	Working temperature：0~40°C
Max Charge Current		$1C_5A$	Working temperature：0~40°C
Charge cut-off Voltage		4.2V	CC/CV
Discharge Current		Continuously $10C_5A$	Working temperature：0~60°C
Discharge cut-off Voltage		2.75V	
Cell Voltage		3.7-3.9V	When leave factory
Impedance		≤40mΩ	AC 1KHz after 50% charge,25°C
Weight		Approx 19g	
Storage temperature	≤1month	-10~45°C	Best 20±5°C for long-time Storage
	≤3month	0~30°C	
	≤6month	20±5°C	
Storage humidity		65±20% RH	

3. Result and Discussion

The effect of charging method on the capacity loss of four batteries were measured using two charging method, i.e. constant and pulse current charging. First, the capacity of batteries were measured using constant current – constant voltage (CC-CV) method at 10 mA. Since the average capacity was 700 mAh for all batteries, thus the 1C value was 700 mA. This value was then used as a charging and discharging current.

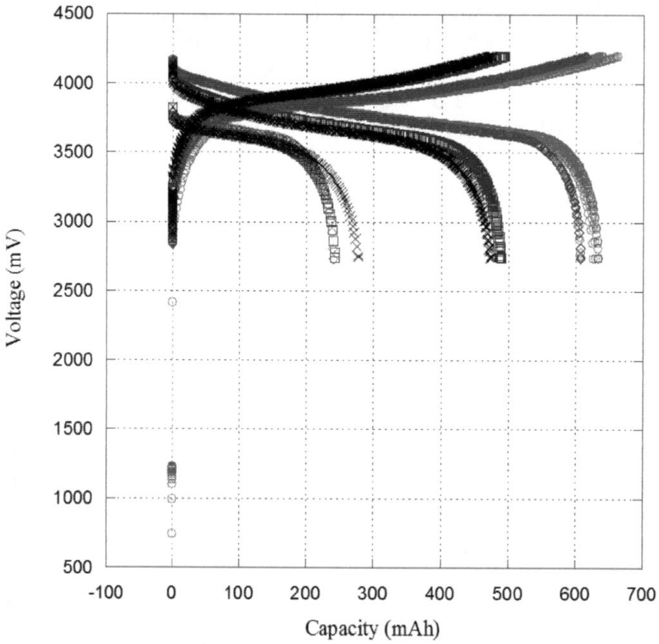

Figure 1. CC charging graph used 1C for cell (a) A, (b) B, (c) C, and cell (d) D.

In order to evaluate the charging method, the batteries were charged using constant current charge and discharged using 1C rate value. This value was then compared with that of initial capacity value. Figure 1 shows the charge-discharge plot of the batteries at 1C rate using constant current method for 1 hour. CC charging method affects different SoC values, which are smaller than the initial SoC value, i.e. 629.8, 487.5, 607.4, and 473.6 mAh for cell A, B, C, and D.

Different technique to charge lithium ion battery was effect the different SoC value to the battery. Figure 1 shows that CC technique only affects the increase of cell voltage to the cut off voltage, yet, the SoC of the cells can only reach 89.9%, 69.64%, 86.77%, and 67.66% for cell A, B, C, and D, respectively. When pulse charging method was applied with variation of pulse width, the SoC values reached 97.1% for cell A, B, C, and D. Although CC charging was applied for one hour, not all of the lithium ions released from anode could reach cathode. Furthermore, this charging methode increased the cell system entropy and temperature, which lead to the formation of gas bubbles within the battery.

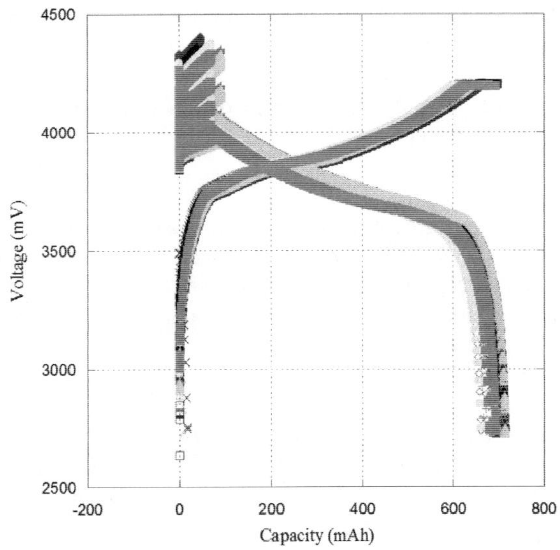

Figure 2. Capacity graph for different pulse width for cell (a) A, (b) B, (c) C, and cell (d) D.

Figure 2 shows the effect of different pulse width on the cell capacity. Cell A, B, C, and D were charged using constant current of 1C for 30 minutes to fill the half of their full capacity, which were continued by the stop state for 1 minute. After stop condition, cell A were continuously charged and stopped for 2 and 1 minutes, respectively until 15 cycles. Cell B were continuously charged and stopped for 4 and 1 minutes, respectively until 8 cycles. Cell C were continuously charged and stopped for 6 and 1 minutes, respectively until 5 cycles. While, cell D were continuously charged and stopped for 8 and 1 minutes, respectively until 4 cycles. The difference between the pulse width of current charging and stop state lead to the difference of total charging time for each battery. In this research CC charging by 1C for 30 minutes to fill half of the cell maximum capacity, followed by recharging with 30 minutes total charging time, means that the total charging time is 1 hour at 1C charging rate so that the battery SoC is 100%. It is found that after discharging, the filled capacity was 680 mAh or 97.1% of the maximum capacity. These results indicate that pulse charging is more effective for maximizing the cell capacity and charging time. The smallest charging time is obtained for pulse charging using 8 minutes pulse width and 1 minute stop time.

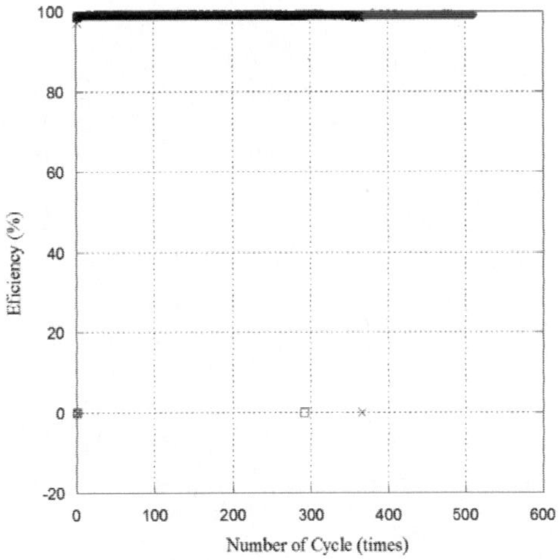

Figure 3. Efficiency as cycle function for (a) A, (b) cell B, (c) cell C, (d) cell D.

Although using the same charging time, pulse charging has higher efficiency for transporting more lithium ion anode to cathode. Charge-discharge efficiency of all pulse charging patterns were not changed for 500 cycles, as shown in figure 3. Figure 2 and 3 (pulse current charging method) show that no capacity loss occur for all cells. On the contrary, CC technique with high charging rate generates the capacity loss for the battery.

4. Conclusion

In this paper, the effective pulse charging method for Lithium ion batteries is analysed. We analyse the effect of pulse width current to the charging efficiency of lithium ion battery and consider the capacity loss effect. The result shows that charging using 8 minute pulse width current is the effective method for charging the lithium ion cell since it can reduce the charging time and decrease the capacity loss of the cell.

5. Acknowledgments

We acknowledge the support from Research Centre for Physics – Indonesian Institute of Sciences (LIPI) in facilitating this research and the Ministry of Science, Technology, and High Education (Kemenristekdikti) for funding support through SINas Konsorsium Baterai Lithium.

References

[1] Abdollahi A, Han X, Avvari G V, Raghunathan N, Balasingam B and Pattipati K R 2016 *J. Power Sources* **303** 388–98
[2] Hu X, Li S, Peng H and Sun F 2013 *J. Power Sources* **239** 449–57
[3] Ikeya T, Sawada N, Takagi S, Murakami J, Kobayashi K, Sakabe T, Kousaka E, Yoshioka H, Kato S, Yamashita M, Narisoko H, Mita Y, Nishiyama K, Adachi K and Ishihara K 1998 *J. Power Sources*. **75**(1) 101–07
[4] Keil P and Jossen A 2016 *J. Energy Storage* **6** 125–41
[5] Monem M A, Trad K, Omar N, Hegazy O, Mantels B, Mulder G, Bossche P V D and Mierlo J V 2015 *Appl. Energy* **152** 143–55
[6] Mitchell N 2000 Lithium Ion Fast Pulse Charger U.S. Patent 6 040 684
[7] HaginoT 1998 Pulse Charging Method for Rechargeable Batteries U.S. Patent 5 808 447

IOP Conf. Series: Journal of Physics: Conf. Series **817** (2017) 012009 doi:10.1088/1742-6596/817/1/012009

The effect of nitric acid and Hexadecyltrimethylammonium Bromide (CTAB) concentration on the morphology and plasmonic properties of gold nanorods

V Fauzia and H E Permana

Department of Physics, Fakultas MIPA, Universitas Indonesia, Depok 16424, Indonesia

E-mail: vivi@sci.ui.ac.id

Abstract. Metal nanoparticles have unique optical properties called Localized Surface Plasmon Resonance (LSPR). Gold nanorods have better plasmonic property for sensor application due to its longitudinal resonance mode is more sensitive to environment dielectric properties changes. Controlling the aspect ratio during the synthesis process becomes important because it would determine the plasmonic properties of the gold nanorods. In this study, gold nanorods have been successfully synthesized in the solution with three different concentrations of HNO_3 and CTAB. Increasing the concentration of HNO_3 and CTAB tends to increase the length, diameter, aspect ratio and yield of gold nanorods. All samples appeared to have a peak of transversal mode at the wavelength of 510 nm. However, the longitudinal mode peaks could not be seen in this observation.

1. Introduction

Nanoparticles have attracted much attention among researchers. It is well known that the size and shape of nanoparticles produced the unique physical and chemical properties which are very different from those in the bulk size, due to the high fraction of the atoms or molecules on the surface of nanoparticles. Unique characteristic of metal nanoparticles that has attracted many attentions is the effect of Localized Surface Plasmon Resonance (LSPR), which was then shortened as a plasmonic effect. This optical property arises as an effect of the resonance between the collective oscillations of conduction electrons of the metal nanoparticles with the electric field oscillation of the incoming light. In the case of gold and silver nanoparticles, this effect could be observed by the occurrence of absorption peaks in the visible light [1]. Metal nanorods show two plasmonic peaks, longitudinal and transversal peaks, where both have its own wavelength characteristic depends on the aspect ratio of nanorods [2]. Therefore, controlling the aspect ratio of nanorods during the synthesis process becomes important to study, since it would determine the plasmonic properties later on.

In general, the nanoparticles which were produced in solution would have relatively less homogeneous in shape and size. Seed-mediated growth method developed by Murphy was able to produce the gold nanorods with aspect ratio of 18 [3]. Some other studies has reported the effect of the NaOH, $AgNO_3$, HCl, H_2SO_4, H_3PO_4, and HNO_3 addition on the morphology of the nanoparticles [4]. It has been known that the surfactant plays an important role as a soft micellar template which could control the length and yield of nanorods [5]. Therefore, in this study we want to learn the effect of

HNO_3 and Hexadecyltrimethylammonium bromide (CTAB) addition as surfactant on the shape and yield of the produced gold nanorods.

2. Experiment

The synthesis method used in this research was conducted by modifying the seed-mediated growth method that was developed in our previous work [4]. In this method there are four kinds of solutions, i.e a seed solution A, consisting of 0.2 mL $C_6H_5Na_3O_7.2H_2O$ 0.025M and 19.8 mL $HAuCl_4.3H_2O$ 0.00025M. Then 0.6 mL of iced cold 0.1M $NaBH_4$ was added into seed solution and was left for 60 minutes at room temperature. Meanwhile, the growth solution consisted of three solutions, i.e. 4.5 ml solution B and C (4.5 mL of 0.01M $HAuCl_4.3H_2O$ 0.1125 mL, 25 mL of 0.1M and 25 µL $C_6H_8O_6$ 0.1M and 4.3875 ml CTAB 0.1M) and 45 ml solution D (1.1125 mL $HAuCl_4.3H_2O$ 0.01M, 250 µL $C_6H_8O_6$ 0.1M and 43.875 mL CTAB 0.1M). In order to study the effect of the concentration of HNO_3, 300 µL HNO_3 was added into solution D with three variations of concentration i.e. 0M, 0.02M and 0.05M. Meanwhile, to study the effect of the CTAB concentration, solution D containing 0.05M HNO_3 was made with three different CTAB concentrations, i.e. 0.05M, 0.1M and 0.2M.

The gold nanorods synthesis process was started by injecting 0.4 mL of solution A into solution B, and 0.4 mL of this mix solution was then injected into solution C. Afterwards, 4 mL this mixture was injected into solution D and then left it for 24 hours at room temperature to get the precipitation. In order to reduce the contamination of non-rod shaped particles and to get rid of the excess surfactant, the final solution was centrifuged twice for 20 minutes in 2000 rpm. The last precipitate was then dispersed in 1 mL deionized water.

The plasmonic effect characterization of the gold nanorods was conducted using Thermo Fisher Scientific GENESYS 10S UV Visible spectrometer. FESEM Inspect F50 was also employed to characterize the morphology of the gold nanorods.

3. Results and discussion

3.1. The effect of HNO_3

The SEM images of gold nanorods synthesized without and with three different HNO_3 concentrations were shown in figure 1. By observing these three figures, it was confirmed that the gold nanorods appeared as the white rods. Generally, HNO_3 influenced the number, length and the aspect ratio of the gold nanorods produced. The increase of HNO_3 concentration tends to increase the length and yield of gold nanorods.

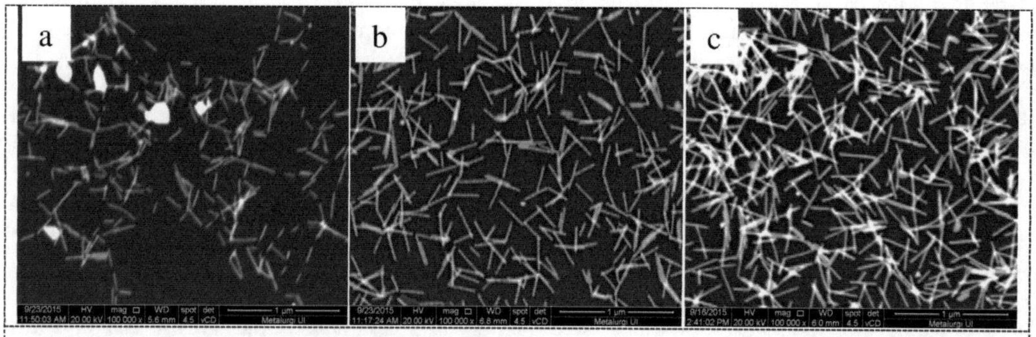

Figure 1. The FESEM images of gold nanorods synthesized (a) without and with HNO_3 concentration of (b) 0.02M, (c) 0.05M.

To obtain the quantitative data of the gold nanorods dimension, SEM images were taken for 70 pieces of gold nanorods and the length of each nanorod was measured manually. The summary of calculation is presented in table 1. The gold nanorods synthesized without HNO_3 has the average diameter of 20 nm, the average length of 118 nm, and the average aspect ratio of 5.9. It looks less homogeneous; it is reflected in the widening of length distribution (118 ± 37 nm). Gold nanorods synthesized with HNO_3 0.02M and 0.05M has the average diameter of 24 and 29 nm, the average length of 205 and 247 nm, respectively, and it has the same average aspect ratio of 8.5. They are relatively homogeneous in length, which is reflected from the narrow length distribution. From table 1, the increase of HNO_3 concentration tends to increase the length, the diameter and the aspect ratio of gold nanorods. It has similar trend with the previous results, however the formation of the high aspect ratio of gold nanorods, which was assisted by nitrate ions, still could not be explained clearly. [4]. From table 1, we also observe that the pH of the solution shows only minor change from 3.11 to 3.02. The pH of growth solution with addition of HNO3 may not be the major factor for the increase of nanorods length.

Table 1. The pH of growth solutions, the average of diameter, length, and aspect ratio of gold nanorods which were synthesized with three different HNO_3 concentrations.

HNO_3 (M)	pH	Diameter (nm)	Length (nm)	Aspect Ratio
0	3.11	20 ± 3	118 ± 37	5.9 ± 1.8
0.02	3.04	24 ± 3	201 ± 54	8.5 ± 2.0
0.05	3.02	29 ± 5	247 ± 60	8.5 ± 2.7

The plasmonic property of the gold nanorods which was obtained by using the UV-VIS absorption spectrum could be seen in figure 4. All samples appear to have a peak at a wavelength of 510 nm, which is confirmed as the transversal mode of the gold nanorods as reported by previously results [4, 6, 7]. All samples are seen to have only one peak at 510 nm. If we use the discrete dipole approximation method to predict the position of the longitudinal peak, the gold nanorods with aspect ratios of 5.9 and 8.5 should have longitudinal peak at the wavelengths of 984 nm and 1234 nm, respectively. In figure 4, the peak at 984 nm is not clearly seen, it could be caused by little amount of gold nanorods. Meanwhile, the peak at 1234 nm could not be observed because it is beyond the spectrum range of the instrument.

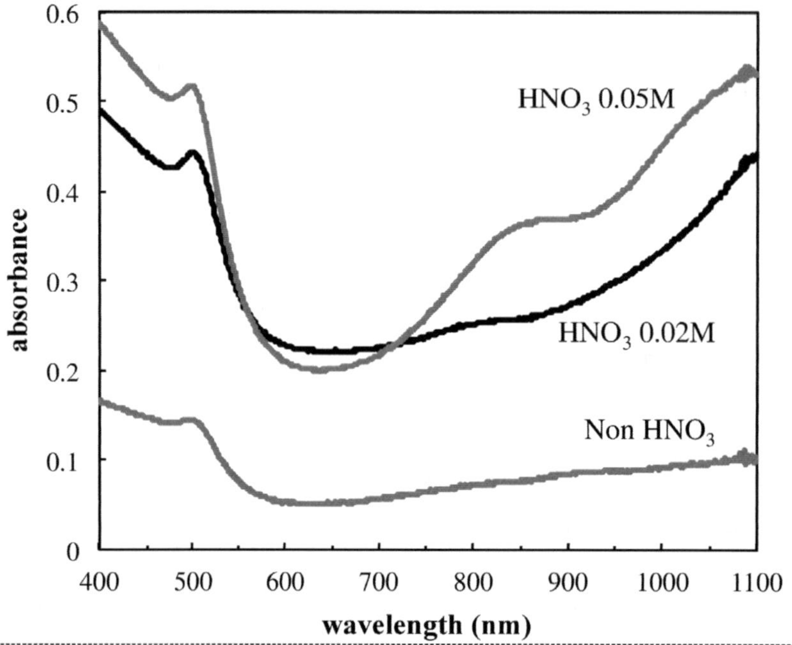

Figure 2. The absorbance spectrum of gold nanorods synthesized without and with HNO_3 concentration of 0.02M and 0.05M.

3.2. The effect of CTAB

Figure 3 shows the SEM images of gold nanorods synthesized with three different CTAB concentrations. It shows that the higher CTAB concentration, the longer gold nanorods were produced. However, the nanosphere and nanoplates also presented besides nanorods. Table 2 shows the quantitative data of the average of diameter, length, and aspect ratio of gold nanorods synthesized with three variations of CTAB concentrations. It is clearly seen that by increasing CTAB concentration the length of nanorods were also increased significantly from 273 nm to 526 nm, in the other side slightly the diameter were slightly reduced from 26 nm to 29 nm. This result corresponded to the aspect ratio, which increased from 10 to 18. These results are similar to those reported on previous researches, where the similar research was conducted by varying the CTAB concentration from 0.1M to 0.6M. Yoshiko et.al. reported that increasing the CTAB concentration leads to the enhancement of gold atoms number which was bonded with CTAB, therefore the longer gold nanorods could be produced [8]. Another analysis suggested that the surfactant concentration plays an important role to maximize the dissociation of bromide ions, so that much more formation of $AuBr_4^-$ ion complexes, which act as precursors for the gold nanorods, could be growth [5].

Table 2. The average of diameter, length, and aspect ratio of gold nanorods were synthesized with three different CTAB concentrations.

CTAB (M)	Diameter (nm)	Length (nm)	Aspect Ratio
0.05	27 ± 3	273 ± 43	10.3 ± 2.9
0.1	28 ± 3	323 ± 83	11.6 ± 2.9
0.2	29 ± 3	526 ± 164	18.0 ± 5.4

Figure 3. The FESEM images of gold nanorods synthesized with CTAB concentration of (a) 0.05M, (b) 0.1M, (c) 0.2M.

Figure 4 shows the UV-VIS absorption spectrum of the gold nanorods synthesized with three different CTAB concentrations. Similar with the result in figure 3, the transversal mode peak of the gold nanorods appears at the wavelength of 510 nm. For the sample with CTAB 0.05M there are other two peaks at the wavelength of 700 nm and 940 nm, the sample with CTAB 0.1M there are also two other peaks at the wavelength of 800 nm and 1000 nm and the sample with CTAB 0.2 M there is only one peak at 510 nm. The longitudinal mode peak position was also predicted using the discrete dipole approximation method. The gold nanorods with the aspect ratio of 10.3, 11.6 and 18.0 is predicted to have another peak at the wavelength of 1407 nm, 1532 nm, and 2146 nm, respectively. All these peaks position are beyond the measurement tool range, hence the peaks at wavelengths of 700, 800, 940 and 1000 nm could not be judged as the longitudinal mode peak, but those might appears as the resonance effect of nanoparticles besides nanorods.

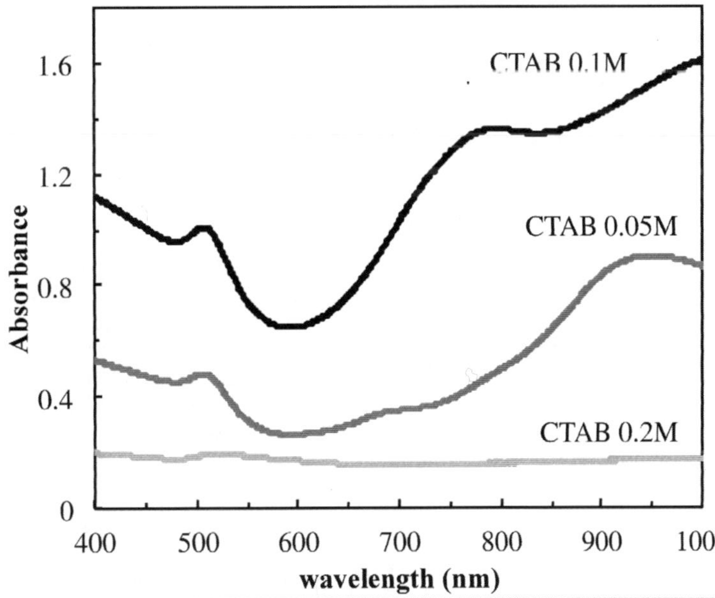

Figure 4. The absorbance spectrum of gold nanorods synthesized with CTAB concentration of 0.05M, 0.1M and 0.2M.

4. Conclusions

In this study, gold nanorods have been successfully synthesized in the solution with three different concentrations of HNO_3 and CTAB. Generally, the increase of HNO_3 concentration tends to increase the length and diameter; therefore the aspect ratio also increases from 5.9 to 8.5. All samples appear to have a transversal mode peak at the wavelength of 510 nm. However, the longitudinal peaks could not be seen in all samples. The mechanism formation of gold nanorods high aspect ratio, which was assisted by nitrate ions, also could not be explained clearly. Meanwhile, the increase of CTAB concentration significantly enhanced the length of gold nanorods from 273 nm to 526 nm which corresponded to the increase of aspect ratio from 10 to 18. The CTAB concentration plays an important role to enhance the number of gold atoms bonded with CTAB and to maximize the formation of $AuBr_4^-$ ion complexes which act as precursors for the gold nanorods growth.

Acknowledgment

This work was funded by Hibah PUPT 2016 from Ministry of Research, Technology and Higher Education Indonesia. The authors also would like to thank Prof. Dr. Muhamad Mat Saleh and Assoc. Prof. Dr Akrajas Ali Umar from Organic and Printed Electronic Laboratory (OPEL) IMEN Universiti Kebangsaan Malaysia for this beneficial research collaboration.

References

[1] Chen H, Kou X, Yang Z, Ni W and Wang J 2008 *Langmuir* **24**(10) 5233
[2] Lohse SE and Murphy CJ 2013 *Chem. Mater.* **25**(8) 1250
[3] Jana NR, Gearheart L and Murphy C 2001 *J. Phys. Chem. C* **105**(19) 4065
[4] Wu H, Chu H, Kuo TJ, Kuo C and Huang MH 2005 *Chem. Mater.* **17**(25) 6447
[5] Toussi SM, Zanella M, Abdelrasoul GN, Athanassiou A and Pignatelli F 2015 *J. Photochem. Photobiol. A* **311** 76
[6] Rioux D and Meunier M 2015 *J. Phys. Chem. C* **119**(23) 13160
[7] Cheng J, Ge L, Xiong B and He Y 2011 *J. Chin. Chem. Soc.* **58**(6) 822
[8] Takenaka Y, Kawabata Y, Kitahata H, Yoshida M, Matsuzawa Y and Ohzono T 2013 *J. Colloid Interface Sci.* **407** 265

IOP Conf. Series: Journal of Physics: Conf. Series **817** (2017) 012010 doi:10.1088/1742-6596/817/1/012010

Performance of gas diffusion layer from coconut waste for proton exchange membrane fuel cell

H Widodo, F Destyorini, D R Insiyanda and A Subhan

Research Center for Physics, Indonesian Institute of Sciences, Kawasan Puspiptek, Tangerang Selatan, Banten, Indonesia

Email: henr002@lipi.go.id; henrywidodo6@gmail.com

Abstract. The performance of Gas Diffusion Layer (GDL) synthesized from coconut waste. Gas Diffusion Layer (GDL), produced from coconut waste, as a part of Proton Exchange Membrane Fuel Cell (PEMFC) component, has been characterized. In order to know the performance, the commercial products were used as the remaining parts of PEMFC. The proposed GDL possesses 69% porosity for diffusion of Hydrogen fuel and Oxygen, as well as for transporting electron. With the electrical conductivity of 500 $mS.cm^{-1}$, it also has hydrophobic properties, which is important to avoid the reaction with water, with the contact angle of 139^0. The 5 x 5 cm^2 GDL paper was co-assembled with the catalyst, Nafion membrane, bipolar plate, current collector, end plate to obtain single Stack PEMFC. The performance was examined by flowing fuel and gas with the flow rate of 500 and 1000 $ml.min^{-1}$, respectively, and analyse the I-V polarization curve. The measurements were carried out at 30, 35, and 40°C for 5 cycles to ensure the repeatability. The results shows that the current density and the maximum power density reaches 203 $mA.cm^{-2}$ and 143 $mW.cm^{-2}$, respectively, with a given voltage 0.6 V, at 40°C.

1. Introduction

Proton Exchange Membrane Fuel Cell (PEMFC) is a device which converts hydrogen fuel into electrical energy via electrochemical process [1-4]. PEMFC consists of components, i.e. Nafion membrane, catalyst, Gas Diffusion Layer (GDL), bipolar plate, and end plate [1,3], which are assembled to become a stack. As a main part of PEMFC, GDL plays the essential role for transporting gas from the supply channel to the electrode. In PEM fuel cells, the GDL must also permit water to be transported from the electrode into the gas flow channels where it is moved out from fuel cell. The gas diffusion layer is a porous membrane that must permit transport of reactant gases to the electrode and provides a path for the product water to be removed from the catalyst layer [5,6]. The typical support material for GDL is carbon-based materials, such as carbon fiber [7], carbon nano-tube (CNT) [8], and carbon black [9]. These carbons can be synthesized from biomass waste, such as coconut waste, by manipulating the properties. Coconut waste is one type of natural resource, which contains complex organic compounds, i.e. hemicellulose, cellulose and lignin [10]. These complex organic compounds decompose at different temperature ranging from 200°C to 500°C [10], and produce charcoal with high carbon content [11]. In the present study, coconut waste was used as a precursor to make carbon paper for gas diffusion layer in PEMFC. If it can be utilized as a material for GDLs, the sustainability issues can also be resolved. The objective of this work is to investigate the performance of carbon paper made from coconut waste, which serves as gas diffusion layer in fuel cell stack.

2. Material and methode

1.1. Material

Coconut waste was used as a carbon precursors. Ethylene vinyl acetate (EVA) and Polyethylene glycol (PEG) obtained from Aldrich Chemical Co., Inc. (St. Lois, MO, USA) was used as a binder. Xylene as a solvent was obtained from Brataco Chemika. Teflon emulsion PTFE 30, obtained from Fuel Cell Earth LLC, was used as a hydrophobic agent of the carbon paper. Teflon emulsion PTFE 30 should be diluted to a concentration of 20% before it is coated on the surface of the carbon paper.

1.2. Preparation of carbon materials

The carbon materials were prepared from coconut waste and were used in two forms, i.e. carbon fiber and carbon powder. Both carbon materials were prepared by a two stage process: carbonization at 500°C and pyrolysis at 900°C, which is followed at 1300°C under N_2 atmosphere. The carbonization process produced charcoal with high carbon content. While, the pyrolysis process produced carbon materials with better physical properties, especially for the electrical conductivity. To produce the carbon fiber, coconut waste was cut into 2 mm size, before experience the carbonization and pyrolysis processes. While, the carbon powder was obtained by grinding the carbon material into a powder of about 74 μm size.

1.3. Preparation of carbon paper

Carbon paper was prepared by mixing the carbon material (carbon fiber:carbon powder = 7:1) with polymer EVA and PEG in xylene at 100°C for 2 hours to make slurry, which then casted on glass mold with dimension of 10 cm x 10 cm. The carbon paper sheet was dried at room temperature for 24 h. To enhance the hydrophobic property of carbon paper, it was immersed in 20 % PTFE suspension for 30 minutes and dried at room temperature overnight. Then, it was heated at 150°C for 30 minutes, followed by heating at 350°C for 30 minutes.

1.4. Characterization of carbon paper

Through-plane electrical conductivity of carbon paper was measured using HIOKI 3522-50 HITESTER LCR-meter. Porosity was determined by the kerosene density method using Archimedes principle in accordance with BS 1902: Part 1A standard. Hydrophobic properties were determined by the contact angle measurement using sessile drop test. For the contact angle measurement, a 50 μL water droplet was placed on the carbon paper surfaces. The tip of syringe position was directed close enough to the sample surface, while the images were captured. Furthermore, the droplet shape was analyzed using Bashforth and Adams tables [13] to determine the contact angle of the samples.

1.5. Single PEMFC operation measurenment

Each PEM fuel cell components have to be assembled into one stack. The first step is the attachment of GDL paste electrode with membrane, where the heat treatment at 100°C for 2 minutes with the applied force of about 2kgf is needed. The Membrane Electrode Assembly (MEA) was then moved to make sandwich-like sample with the rest of components, from the outside end-plate to gasket, bipolar plate, another gasket, then MEA, gasket, bipolar plate, gasket, and closed with end-plate. This stack was locked with rod at every edge by means torque of 2N.m.

3. Result and discussion

Polarization tests were carried out at 30, 35, and 40 °C, with the hydrogen fuel and gas oxygen flow rate of 500 and 1000 ml.min^{-1}, respectively. In the beginning, the measurement was carried out at 30 °C for five cycles in order to know the consistency of the polarization curves for this GDL. Figure 1 shows the polarization curve at 30 °C, where the obtained current density is 200 mA.cm^{-2} in ohmic region for 0.6 volt. While, the maximum power density is 145 mW.cm^{-2} at 30°C. When the temperature is raised to 35 °C and 40 °C, the current density in ohmic region become 189 and 196 mA.cm^{-2} as presented in figures 2 and 3, respectively. In fact, the fluctuation of current density at various temperatures is not significant since the cell was operated at low temperature. The maximum power densities are 141 and 143 mW.cm^{-2}, which was obtained at 35 and 40°C, respectively.

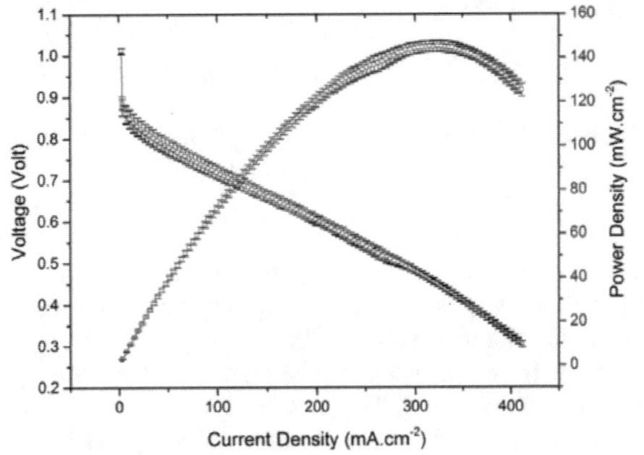

Figure 1. Polarization curve of PEM fuel cells with GDL from coconut waste at 30 °C.

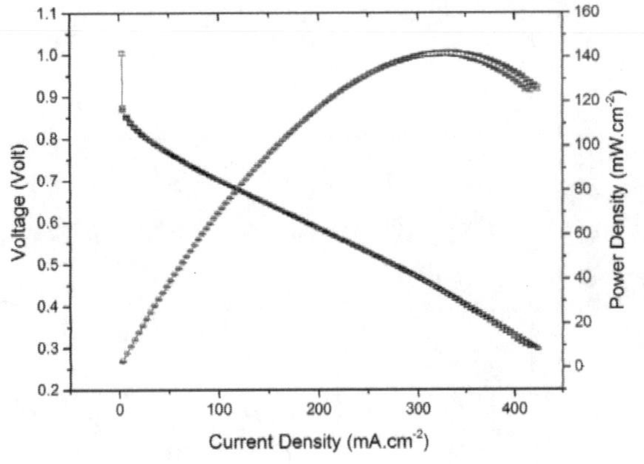

Figure 2. Polarization curve of PEM fuel cells with GDL from coconut waste for temperature 35°C.

IOP Conf. Series: Journal of Physics: Conf. Series **817** (2017) 012010 doi:10.1088/1742-6596/817/1/012010

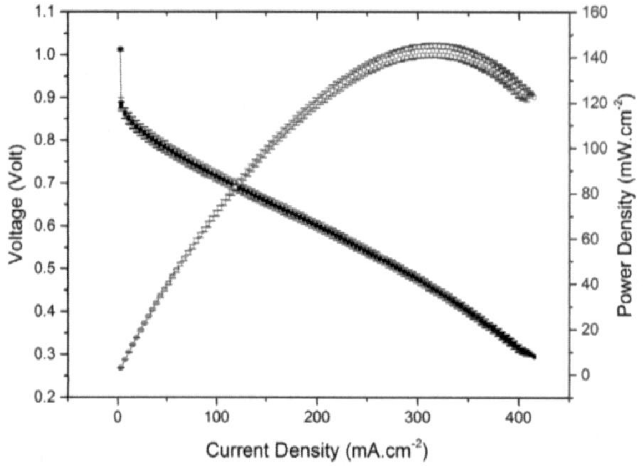

Figure 3. Polarization curve of PEM fuel cells with GDL from coconut waste for temperature 40°C.

Since we repeated the measurement for five times in five cycles, the average of the maximum density should be considered. The graph of power density versus temperature in figure 4 shows that the deviation of characteristic carbon as GDL at 30, 35, and 40 °C are 2, 1, and 2%, respectively. It can be seen that the carbon made from coconut waste is a promising material for Gas Diffusion Layer. The polarization curve looks relatively stable. The electrical conductivity could reach upto 500 $mS.cm^{-1}$, with porosity of about 69%. In addition, this material possesses good hydrophobic properties, which is shown by the contact angle of 139^0.

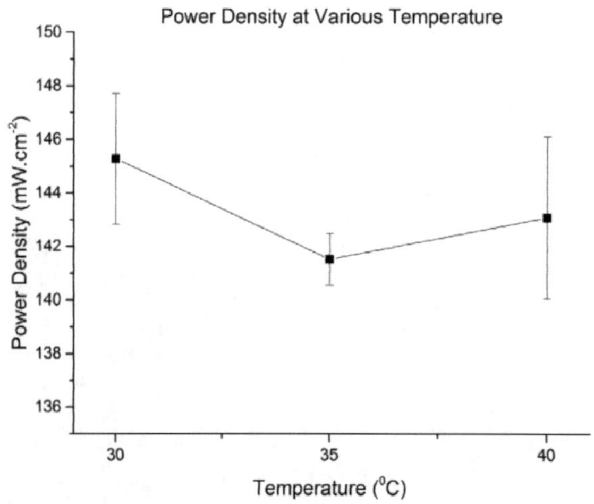

Figure 4. Power density of PEM fuel cells with GDL from coconut waste at different temperature

4. Conclusion

In this work, the carbon paper from coconut waste was synthesized, and its performance as a gas diffusion layer in fuel cell stack was examined. The characteristic is shown in polarization curve at various temperatures. The achieved power density is statistically stable with the deviation of 1-2%. In other word, the carbon from coconut waste is promising material for GDL of fuel cell.

Acknowledgments
The authors acknowledge the financial support from Indonesian Institute of Sciences and the facilities provided by the Research Center for Physics—Indonesian Institute of Sciences.

References

[1] Wan Z, Chang H, Shu S, Wang Y and Tang H 2014 *Energies* **7** 3179
[2] Feroldi D and Basualdo M 2012 Description of PEM fuel cells system *PEM Fuel Cells with Bio-ethanol Processor Systems* eds M S Basualdo, D Feroldi and R Outbib (London: Springer) p 49
[3] Litster S and McLean G 2004 *J. Power Sources* **130** 61
[4] Haile S M 2003 *Acta Mater.* **51** 5981
[5] Park S, Lee J and Popov B N 2012 *Int. J. Hydrogen Energ.* **37** 5850
[6] Cindrella L, Kannan A M, Lin J F, Saminathan K, Ho Y, Lin C W and Wertz J 2009 *J. Power Sources* **194** 146
[7] Mathur R B, Maheshwari P H, Dami T L, Sharma R K, Sharma C P 2006 *J. Power Sources* **161** 790
[8] Gao Y, Sun G Q, Wang S L and Zhu S 2010 *Energy* **34** 1455
[9] Shim J, Han C., Sun H., Park G, Lee J and Lee H 2012 *Trans. of the Korean Hydrogen and New Energy Society* **23**(1) 34
[10] Browne FL 1958 *Theories of the Combustion of Wood and its Control* Report No. 2136, (Wisconsin: USDA Forest Service)
[11] Indayaningsih N, Priadi D, Zulfia A and Suprapedi 2011 *Key Eng. Mater.* **462-463** 937

IOP Conf. Series: Journal of Physics: Conf. Series **817** (2017) 012011 doi:10.1088/1742-6596/817/1/012011

Analytical electron microscopy study of new class material Cr_2AlC for coating applications at high temperature environments

R Iskandar

Central Facility for Electron Microscopy, RWTH Aachen University, AhornStr 55, Aachen 52074, Germany

E-mail: iskandar@gfe.rwth-aachen.de

Abstract. The interactions between electron and specimens produce wide range of secondary signals from the specimen that can be used in analytical electron microscopy. These signals will give us chemical information and electronic structure information of specimen atoms. Two techniques which based on these secondary signals, X-rays energy-dispersive spectroscopy (XEDS) and electron energy-loss spectroscopy (EELS), are very powerful to be used for characterization materials. These two techniques are complementary each on the other. The former is very useful to quantify heavy elements and the other is very suitable to identify light elements as oxygen and carbon. In this report, we demonstrated the capability of these techniques to investigate microstructure evolution during the oxidation process of thin film Cr2AlC at 1320 °C. This thin film belongs to MAX phase, a new class material, and becomes a promising candidate for coating applications since it exhibits a good oxidation resistant.

1. Introduction

Since it has been invented almost 100 years ago, transmission electron microscope becomes a very important instrument for characterizing materials. For imaging, the modern TEM can achieve resolution up to 0.5 pm which is already smaller than atomic neighborhood distance. TEM offers other possibility to characterize materials. The electron specimen interaction produces secondary signals which can be used as addition to imaging techniques. These secondary signals contain chemical information and a lot of details of the specimen. The use of these signals on TEM is known by Analytical Electron Microscopy (AEM). AEM uses X-rays energy dispersive spectroscopy (X-EDS) and electron energy loss spectroscopy (EELS). With the help of better equipment, AEM opens very important possibility to explore the properties of specimen.

As one type of ionizing radiation, electrons capable to remove the tightly bound, inner-shell electrons from the attractive field of the nucleus by transferring some of its energy to individual atoms in the specimen. Secondary signals will be produced during electron-specimen interactions as seen in figure 1(a). One of these signals is produced from generated X-rays photons in beam-specimen interaction. This X-rays have energies characteristic for the different elements in the specimen and can be used in AEM analysis. These X-rays will be captured by energy-dispersive spectrometers (EDS). These spectrometers use lithium-drifted silicon Si(Li) solid state X-ray detectors, intrinsic or high-purity Ge (HPGe) EDS detectors or Si-drift detectors which was mainly be used in this work. This technique is known as X-rays energy-dispersive spectroscopy (X-EDS). Detail explanation about the principle of EDS can be found in literature [1, 2]. Unlike EDS in SEM system, the condition in TEM is more complicated. The geometry of specimen and also EDS detector requires more depth consideration. For quantitative analysis, the counts that collected by detector need to be statistically significant enough. Due to the geometry setup one needs to find the optimal detector position during the acquisition.

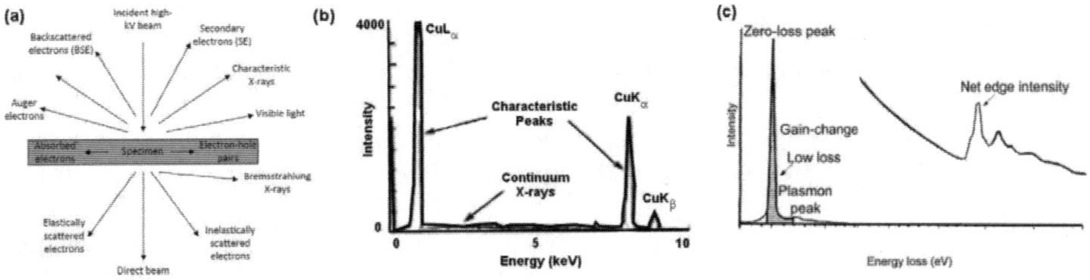

Figure 1. **(a)** Electron beam interaction with specimen, **(b)** EDS and **(c)** EELS spectrum [1, 3].

The second type of secondary signals comes from electrons which may have no lost or only small amount energy due to inelastic collisions. The energy distribution of electrons that have come through the specimen is analyzed by the technique called electron energy-loss spectrometry (EELS). These phenomena tell us many specific information of specimen atoms such as chemistry and electronic structure. These information reveal details of these atom's bonding, nearest-neighbor atomic structure and the free electron density. To get more detail explanation about EELS, reader is strongly suggested to read work from Edgerton [4].

EELS signals can also be used to produce an image from specific electrons. Unlike the conventional TEM, which uses all electrons that pass through the specimen, one technique that allows image formation with electrons that have undergone selected energy losses in the specimen has emerged recently. The technique, energy-filtered TEM (EFTEM), detects chemical contrast in specimens by adjusting an energy filter to accept only electrons that have lost energy in core ionizations of selected elements. Furthermore, the energy filter can also pass the zero-loss electrons, i.e. electrons that do not experience inelastic scattering. As a result, all bright-fields and diffraction patterns taken with energy-filter will only use purely elastically-scattered electrons. As a consequence, the chromatic aberration for thicker specimens and the inelastic background are eliminated. The chemical contrast can be used to map the element distribution in the specimen. There are two methods that are mainly used for elemental mapping, the jump ratio map or two windows technique [5] and the three windows technique [6].Detailed explanation for these two techniques can be found in the literatures [3, 5–9].

A modern TEM is a mandatory to utilize maximum results of AEM techniques. It must be equipped with analytical equipments such as energy dispersive X-rays and electron energy- loss spectroscopy. Furthermore it should also have scanning capability, known as scanning TEM (STEM), which allows the specimen to be scanned point by point with a small probe. In order to obtain a small probe with a diameter of ≈ 0.25 nm, a large aperture angle must be used. Two additional pairs of scan coils below the condenser lenses allow the raster process to be performed. Besides producing SE signal as in SEM instruments, the transmitted beam can also produce bright field, dark field and annular dark field (ADF) images. The latter mode is possible because of the presence of an annular detector which is positioned below the specimen that will collect electrons diffracted in large angles.

IOP Conf. Series: Journal of Physics: Conf. Series **817** (2017) 012011 doi:10.1088/1742-6596/817/1/012011

2. Experiments

A new class material, MAX phase Cr_2AlC is known as promising candidate materials for advanced technologies. It displays a unique combination of properties, including low density, high elastic modulus, easy machinability, excellent thermal shock resistance and damage tolerance. Unfortunately, only few descriptions about the oxidation behavior of Cr_2AlC are reported in the literature. In bulk systems, Lin et al. [10] have reported that Cr_2AlC has a parabolic rate constant kp = 6.8x10−10 kg2m−4s−1 at 1200 °C, which is 24 times lower than Ti_3SiC_2 at 1200 °C [11–14]. Later on, several extensive investigations on the oxidation resistance of bulk Cr_2AlC were reported by Lin et al. [10] and Tian et al. [15] which clearly show that Cr_2AlC exhibits an excellent oxidation resistance.

In the selective oxidation of a material containing Cr and Al, both Al_2O_3 and/or Cr_2O_3 scales can be formed. Cr_2O_3 forming alloys generally contain less than 2-3% Al and 15% or more Cr. On the other hand, Al_2O_3 forming alloys can be found in material with as low as 5% Al. Environments also play an important role. It has been known that at high-temperatures (above 1000 °C) Al_2O_3 will be formed more likely than Cr_2O_3 [16].

In bulk systems, Lee, et al. have showed that only a thin external Al_2O_3 oxide layer was formed after 336 h of oxidation at 1300 °C in air [17]. They have also showed the Al-depletion and Cr-enrichment have occurred underneath the Al_2O_3 layer. Microstructure investigations have revealed that the surface grains of α-Al_2O_3 were fine but the inner grains were coarse and more columnar. The α-Al_2O_3 has grown very slowly and the measured oxide thickness was only $\approx 27\ \mu$m. In this report, microstructure evolution of thin film Cr_2AlC oxidized at 1320 °C will be investigated to understand their oxidation behavior.

Detail for sample fabrication and also other supporting experimental for investigated material can be found in work of Hajas et.al [18].

Prior the AEM investigation were performed, microstructure analysis by means of SEM were done for each selected specimen. A field emission ZEISS DSM 982, equipped with EDS system from Oxford was used. To produce almost homogeneous TEM specimens thicknesses, all specimens are prepared as thin Lamelle by focused ion beam technique on FEI Strata 205. This technique produced lamellae of 12 - 15 μm length and 4-6 μm width with the thickness of less than 100 nm. The AEM investigation was performed on Zeiss Libra 200 FE. It has schotky field-emission source operated at 200 kV. It has also STEM capabilities which as described above are very important for AEM analysis. The microscope is also equipped with in-coloumn corrected omega filter, X-Flash EDS detector from Bruker and HAADF detectors from Fischione. All aquired data were processed using Esprit and DigitalMicrograph™ for EDS and EELS analysis respectively. Typical parameters used in this work are listed in table 1.

Table 1. Typical parameters used for AEM analysis

Parameter	Values		
	EDS	EELS	EFTEM
Mode of Operation	STEM	STEM	TEM
Type of measurements	spot, line scan, maps	spot	three windows
Specimen Tilt	5 °C	NC*)	NC**)
Acquisition time	180 - 3600 s	1 - 5 s	1 - 50 s /image
α / β	NC*)	17.43 / 7.99 mrads	17.43 / 18.50 mrads

3. Results and discussion

The Cr_2AlC thin films were oxidized at 1320 °C in TGA for three different total times: 4 min, 39 min and 282 min, with the resulting scale thicknesses of approximately 1 μm, 3 μm and 6 μm, respectively. In this report we concentrate our discussion for specimen with the thinnest oxide scale to get the benefit of AEM. Figure 2(a) shows the SEM cross sectional image of the specimen with 1 μm oxide scale which will be noted as Cr-1 specimen. A dense and continuous alumina scale with the thickness approximately 1 μm oxide scale is observed. An additional interlayer, presumably consists of chromium carbide, has formed below the scale interface. Small voids are also found in the Cr_2AlC film. In figure 2(b), detail morphology of Cr-1 cross-section can be seen from bright-field (BF) image. A dense and continuous alumima scale with a wavy interface can be seen clearly. SAED analysis identified Cr_2AlC as the main phase of the thin film along with Cr_3C_2 found below the scale interface.

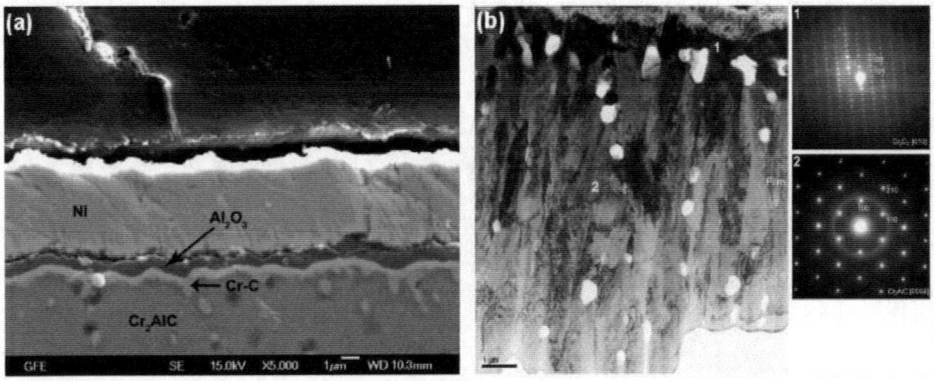

Figure 2. (a) The SEM cross-section of Cr-1 specimen. Electron beam interaction with the specimen. **(b)** BF image of the Cr-1 specimen. SAED patterns from region 1 and 2 show that the specimen consists mainly of Cr2C3 and Cr2AlC phases.

More detail investigations were performed by means EDS mapping as shown in figure 3. The STEM-ADF images taken from oxide scale of specimen Cr-1 with corresponding EDS elemental maps for Cr, Al, C, and O, respectively. EDS map revealed that chromium rich area, presumably chromia formed at outer part of the scale and scale-matrix interface respectively. Aluminum depletion can be seen from aluminum map at outer part alumina scale. Higher magnification of EDS maps show chromium oxide rich precipitates distribution on area close to the interface (figures 3(g) and (h)). Chrome-rich interlayer in size of 250 nm is clearly visible at upper part of the alumina scale. Additionally groups of chromium-rich precipitates were also formed at inner part of alumina scale. Unfortunately from EDS map, carbon distribution, especially in alumina scale could not be identified clearly.

Table 2 shows chemical analysis from areas marked in figure 3(a) and 4(a). It can be seen that EDS detected oxygen and include it into the calculation. It is because both C and O peaks locate close to each other at low energy. EFTEM map was used to overcome this problem (figure 4). From EFTEM elemental maps, it is verified that no oxygen was detected on matrix as that has been confirmed by EDS quantification. Moreover slightly carbon enrichment was also detected at the precipitates close to interface. Additionally, the chromium elemental map has also suggested that chromium was dissolved in the alumina scale.

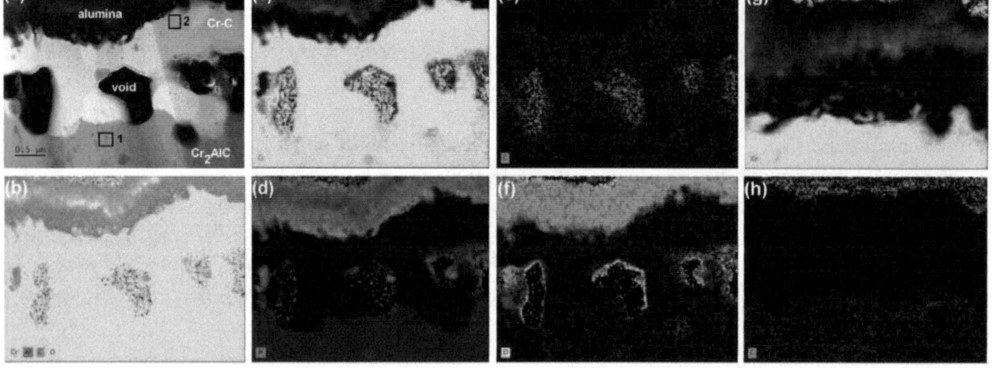

Figure 3. **(a)** STEM-ADF image of cross-section Cr-1 specimen. **(b)** STEM-EDS combinational and **(c-f)** EDS maps of Cr, Al, C and O. EDS map at higher magnification of Cr **(g)** and C **(h)** elemental maps.

Table 2. EDS and EELS quantification from selected areas

EDS					EELS				
Position	wt%				Position	at%			N
	Cr	Al	C	O		Cr	C	O	
Position 1	76.56	17.15	2.43	3.86	Position A	59.09	40.91	-	-
Position 2	85.20	0.26	3.00	11.54	Position B	27.09	23.06	19.46	30.40
					Position C	30.02	-	69.98	-

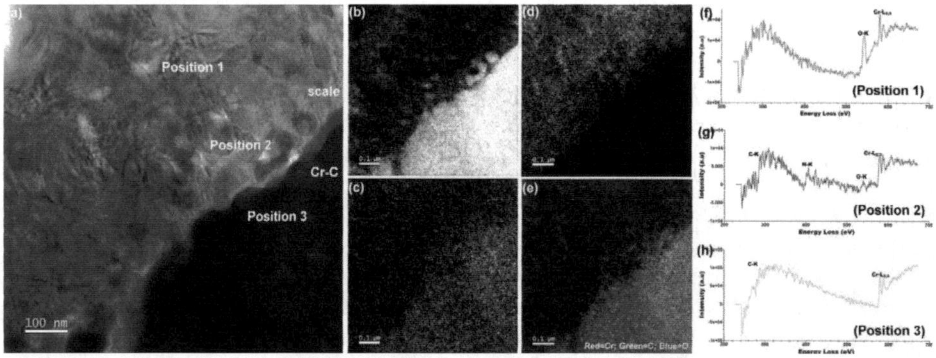

Figure 4. **(a)** EFTEM elemental maps of the scale region of the Cr-1 specimen and elemental maps of **(b)** Cr, **(c)** C and **(d)** O. **(f)** RGB map shows distribution of these elements more clearly (especially for carbon). **(f-h)** EELS spectra acquired from outer part of alumina, Cr-rich precipitates and film marked by position 1, 2 and 3 respectively.

To strengthen EFTEM elemental map results, STEM-EELS analysis was performed. Figure 4(f-h) show EELS spectra were taken from outer part of alumina, Cr-rich precipitates and film. Once again, on the matrix part, no presence of oxygen can be detected and quantification gives 59.09 wt% Cr and 40.91 wt% C. This results close to EDS-quantification if the concentration of oxygen is excluded. At Cr-rich area at outer scale, where aluminum depletion zone, by combining the aluminum EDS map with EELS quantification results, probably CrO_2 formed in this area. Interestingly, Cr-rich precipitates formed at inner scale, as addition to Cr, C and O, N was also detected by EELS spectra (figure 4.g) EELS quantification results suggested that probably a mixture of CrC and spinnel aluminum nitride.

4. Summary and conclusions

The use of Analytical Electron Microscopy (AEM) to investigate Material for high temperature applications was presented in this work. The STEM capability gives AEM possibility to reveal more information in higher spatial resolution than conventional TEM. The quantitative analysis using EDS is very straight forward and very suitable for elements with atomic number higher than Na (Z >11). Some extra precaution needs to be made such as a very thin specimen, when quantification will be done for light elements. On the other hand, EELS is very useful to identify as well as quantify light elements. But to get better EELS results, a very thin specimen is necessary. The EFTEM method can be used to overcome contamination problem which normally appears in STEM analysis. The acquisition time of EFTEM and the use of parallel beam are advantages for beam sensitive materials especially to reduce contamination problem.

Albeit EDS and EELS techniques have their pros and cons, they complement one another. At the end, it is clear that the AEM is an advanced technique to obtain more information on chemical analysis.

Acknowledgments

Authors thank for financial support from both German Research Foundation (DFG) and also German Academic Exchange Service (DAAD). Authors also thank to David Hajas from Institute of Material Chemistry RWTH Aachen University and his colleagues for providing the investigated specimens.

References

[1] Goldstein J et al 2003 *Scanning Electron Microscopy and X-rays Microanalysis* (New York: Springer US) chapter 4 pp 170-76
[2] Williams D B and Carter C B 1996 *Transmission Electron Microscopy A Textbook for Material Sciences* (New York and London: Plenum Press) chapter 1 p 7
[3] Brydson R D 2001 *Electron Energy Loss Spectroscopy (Microscopy Handbook)* (London: Taylor & Francis) chapter 5 p 69
[4] Egerton R F 1975 *Philos. Mag.* **31** 199-215
[5] Krivanek O L *et al* 1980 *Proc. 51st Annual Meeting MSA (San Fransisco)* pp 586-87
[6] Jeanguillaume C 1978 *Ultramicroscopy* **3** 237-42
[7] Hofer F et al 1995 *Ultramicroscopy* **59** 15-31
[8] Jäger W and Mayer J 1995 *Ultramicroscopy* **59** 33-45
[9] Schnenner M *et al* 1996 *Ultramicroscopy* **65** 95-9
[10] Lin Z J 2007 *Acta Mater.* **55** 6182
[11] Barsoum M W *et al* 1997 *J. Electrochem. Soc.* **144** 2508
[12] Li S *et al* 2003 *Mat. Sci. Eng. A-Struct.* **341** 112
[13] Sun Z *et al* 2002 *Oxid. Met.* **57** 379
[14] Yang S L *et al* 2003 *Oxid. Met.* **59** 155
[15] Tian W 2008 *J. Mater. Sci.* **43** 2785-91
[16] Quadakkers W J and Singheiser L 2001 *Mater. Sci. Forum* **369-372** 77-92
[17] Lee D B et al 2007 *Corros. Sci.* **49** 3926-34
[18] Hajas D E et al 2011 *Surf. Coat. Tech* **206** 591-98

IOP Conf. Series: Journal of Physics: Conf. Series **817** (2017) 012012 doi:10.1088/1742-6596/817/1/012012

Mechanical properties, microstructure, and biocompatibility of Ti-6Al-6Nb

G Senopati, C Sutowo and M I Amal

Research Center for Metallurgy and Materials
Indonesian Institute of Sciences, Building 470, Puspiptek Technological Park
Serpong, Tangerang Selatan, 15314, Banten, Indonesia

E-mail: gali002@lipi.go.id

Abstract. The alpha-beta Ti-6Al-4V alloy has been used as biocompatible materials due to its excellent mechanical properties, corrosion resistance, and lower specific strength than other metallic biomaterials such as stainless steel and Co-Cr alloy. However, the concern to the toxic element of vanadium led to the development of a new alpha-beta Ti6-Al-6Nb alloy. This paper presented microstructural, mechanical properties and biocompatibility of as cast Ti-6Al-6Nb alloy. The materials were characterized by scanning electron microscopy, optical microscopy, microhardness test, and X–ray diffraction. In addition corrosion test was performed by using Hank's solution at 37 °C and pH 7.4. The result showed that Ti-6Al-6Nb alloy could potentially used for biomedical application due to its good mechanical properties and corrosion rate.

1. Introduction

WHO Data in 2012 shows that increasing world population followed by increasing number of traffic accidents especially among young people in developing countries such as Indonesia [1]. This contributes to increasing number of implant materials demand. Besides, surgery process for bone exchange with implant materials and post operation also increase tremendously. Nowadays, researchers have been developing new implants for long term use (long term implantation/life time) of which the post-operation measure for implant change is not needed [2].

Titanium based implant material covers 60 % of the total implant in the world utility due to cheaper price and better mechanical properties than other implant materials such as Co-Cr-Mo. In the previous study, Ti-6Al-6Nb alloy was prepared by using arc melting furnace followed by thermomechanical treatment to study the influence of Nb content on β phase formation, whereas β phase formation process begin in casting process [3]. However, characteristics of as cast Ti-6Al-6Nb have not been investigated well. Cardozo reported in developing titanium for implant application each step has an important role, start from determining chemical composition, microstructural characteristics, heat treatment process, and mechanical properties of alloy [4]. Another study reported that microstructure and phase formed in casting process has important role for next process (thermomechanical processing) [5, 6]. In this paper as cast Ti-6Al-6Nb was characterized to study the microstructure of as cast Ti-6Al-6Nb and the correlation to its mechanical properties and biocompatibility.

IOP Conf. Series: Journal of Physics: Conf. Series **817** (2017) 012012 doi:10.1088/1742-6596/817/1/012012

2. Materials and Method

2.1. Material Preparation
Ti-6Al-6Nb was prepared from commercially available Ti (99.9 % purity), aluminum (99.9% purity) and niobium (99.5% purity). All materials were processed by arc melting with non consumable tungsten electrode on a water-cooled copper hearth. The alloy was prepared in ultra high purity argon atmosphere and the ingot was melted eight times to improve chemical homogeneity. The obtained ingot was aproximately 100 g. The casting was processed at Chiba laboratory Institute of Materials Research Tohoku University

2.2. Examination
As cast Ti-6Al-6Nb was sectioned by using cutting wheel machine with the dimension of 1x1x1 cm^3. The block specimen was mounted by using resin and mechanically polished via a metallographic procedure to a final level of 0.3 μm alumina powder and was then etched in solution of water, nitric acid, and hydrofluoric acid (80:15:5 in volume). The microstructure of the etched alloy was examined using Meiji Japan optical microscope. X–ray diffraction (XRD) measurement for phase analysis was conducted using Rigaku smartlab diffractometer, at room temperature using Cu Kα radiation and operating at 30kV/15mA. Hardness measurement was measured using Shimazu vickers hardness tester with a load 500 gf for 30 s. Corrosion test in simulated body fluid was tested to confirm the biocompatibility properties of Ti-6Al-6Nb alloy. The corrosion test was carried out in Hank's Solution of pH 7.4 at 37 °C to simulate the body fluid environment and body temperature. A standard three electrode cell were used in this test. Graphite was the counter electrode, saturated calomel electrode was the reference electrode and sample as work electrode. The total amount of Hank's Solution taken in the cell was about 500 ml. The potential scanning rate was 1 mV/s.

3. Result and Discussion
Figure 1 shows the microstructure of as cast Ti-6Al-6Nb alloy. The microstructures of as cast Ti-6Al-6Nb was lamellar with transformed beta and acicular alpha (light platelets). It was a result of nucleation and growth on crystallographic planes of the prior beta matrix [7]. This structure is similar as Ti-6Al-6V structure [8, 9].

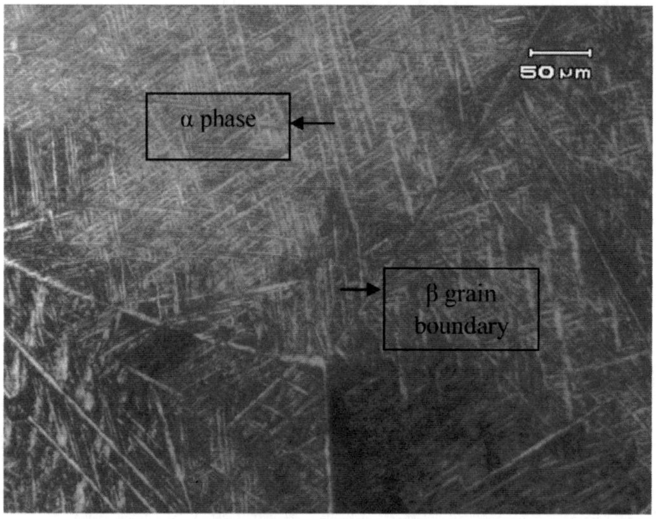

Figure 1. Microstructure of Ti–6Al-6Nb alloy.

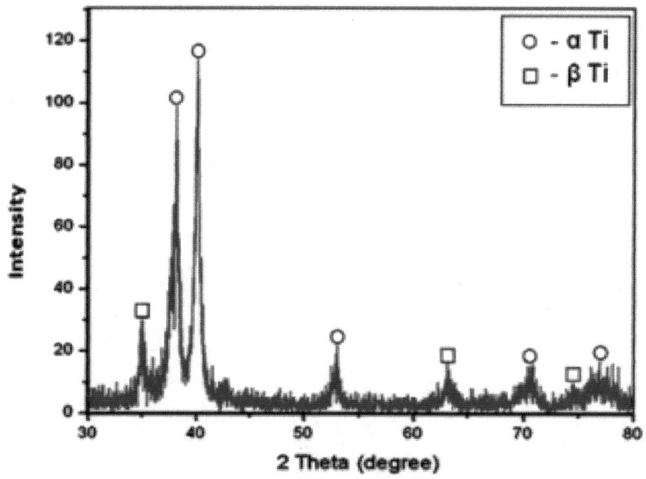

Figure 2. XRD Pattern of Ti-6Al-6Nb.

Figure 2 shows diffraction pattern of Ti-6Al-6Nb. The α and β phase were observed. It confirmed the metallographic result showed in figure 1. Figure 3 shows hardness value of as cast Ti-6Al-6Nb and another titanium alloy for biomedical purpose. As cast Ti-6Al-6Nb has hardness value 278 HV, higher than pure Ti (186 HV), Ti-15Mo-5Nb (246HV), and Ti-15Mo-10NB (257HV), but lower than Ti-5Sn-5Cr (503 HV). The mechanical strength of materials could influenced by solid solution hardening, high dislocation density, boundary hardening and precipitation hardening. In this case (Ti-6Al-6Nb alloy) has the higher hardness than pure titanium due to solid solution strengthening or crystal structure effect/phase (α, β, ώ) [6,11]. Another study by A. Kumar in Ti-6Al-4V alloy showed that the increasing of β volume fraction could increase hardness value in α + β system [12].

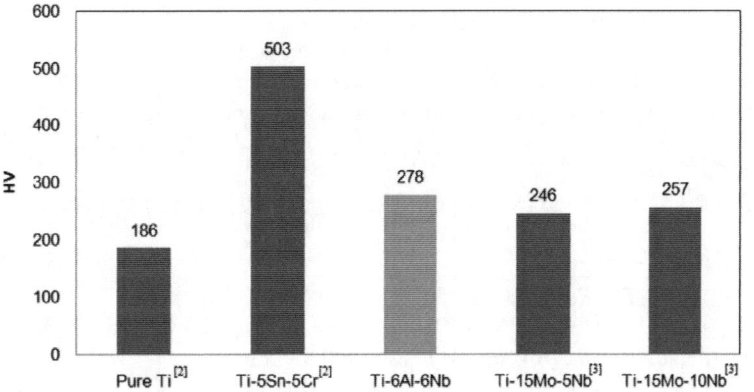

Figure 3. Hardness Properties of as-cast Ti-6Al-6Nb and some as-cast titanium alloy for biomedical application.

Determination the interaction of metal implant with body fluid environment is essential to understand implant stability in the human body. Figure 4 shows tafel diagram corrosion test of as cast Ti-6Al-6Nb in a simulated body fluid Hank's Solution. No hystersis loop was observed that indicated the as cast Ti-6Al-6Nb alloy had excellent pitting corrosion resistant. The corrosion rate of as cast Ti-6Al-6Nb was 0.0016 mmpy (mm/year), compared to another implant materials such as stainless steel 316L (0.004 mm/y), Co alloy (0.0014 mm/y), and Ti6Al4V (0.00018 mm/y) Ti-6Al-6Nb show better corrosion resistant than stainless steel 316L, similar to Co alloy, but lower than conventional Ti6Al4V [13].

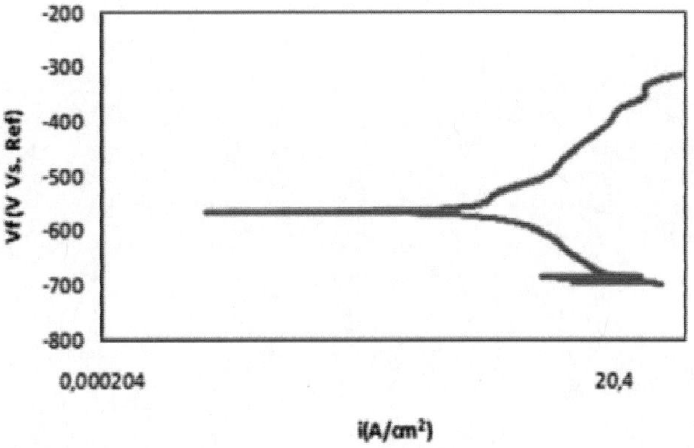

Figure 4. Tafel Diagrams Ti-6Al-6Nb alloy in Hank's Solution.

4. Conclusion

The microstructure, mechanical properties, and biocompatibility of as cast Ti-6Al-6Nb has been investigated in this study. The as cast Ti-6Al-6Nb has similar microstructure with conventional implant Ti-6Al-6V. The microstructure was lamellar, with α and β phase observed in the microstructure and in diffraction pattern result. The hardness of Ti-6Al-6Nb was higher than pure Ti and lower than β type alloy Ti-5Sn-5Cr. The corrosion rate of as cast Ti-6Al-6Nb was 0.0016 mmpy. The result showed that this alloy could pottentially used for biomedical application.

Acknowledgment

This work is financially supported by 2016 thematic program of Indonesian Institute of Sciences

References

[1] Sutanto S 2009 *IATSS Research* **33** 122-4
[2] Mohamed AH G *et al* 2013 *J Mech. Behav. Biomed. Mater.* **20** 407-15
[3] Yuswono Marsumi *et al* 2014 *Adv. Mat. Res.* **900** 53-63
[4] Cardoso F F et al 2014 *J. Mech. Behav. Biomed. Mater.* **32** 31–8
[5] Damon Kent *et al* 2013 *J. Mech. Behav. Biomed. Mater.* **28** 15-25
[6] Hsueh C H et al 2014 *Mater. Sci. Eng. A* **606** 157-64
[7] Gammon L M et al 2004 *ASM* **Vol 9** 899 -917
[8] Semlitsch *et al* 1985 *Biomed. Tech. (Berl).* **30** 334–39
[9] Das S *et al* 1998 *J. Med.* **50** 17-20
[10] Li J X *et al* 2015 *Trans. Nonferrous Met. Soc. China* **23** 692-98
[11] Galih S *et al* 2015 *AIP Conf. Proc. ISFAP 2015* **1711** 050005-1
[12] Anoop K *et al* 2012 *Adv. Mat. Res.* **585** 381-86
[13] Gurappa I 2012 *Mater. Charact.* **49** 73-9

IOP Conf. Series: Journal of Physics: Conf. Series **817** (2017) 012013 doi:10.1088/1742-6596/817/1/012013

Polyvinilidine fluoride (PVDF) nanofiber membrane for Li-ion rechargeable battery separator

H Widiyandari[1], A Purwanto[2] and S A Widyanto[3]

[1] Department of Physics, Diponegoro University, Jl. Prof. H. Soedarto SH, Semarang, 50275, Indonesia
[2] Department of Chemical Engineering, Sebelas Maret University, Jl. Ir. Sutami No. 36 A, Surakarta, Indonesia
[3] Department of Mechanical Engineering, Diponegoro University, Jl. Prof. H. Soedarto SH, Semarang, 50275, Indonesia

Email: h.widiyandari@undip.ac.id

Abstract. This study reported synthesis and characterization of PVDF (polyvinilidine fluoride) nanofiber membrane using electrospinning method for Li-ion rechargeable battery separator. Electrospinning equipment system consists of a DC high voltage (HV - DC), a controllable spinner and a plate collector. The effects of the applied voltage on morphological property, porosity and thermal property were systematically investigated. The application of DC voltage at a range of 13 to 17 kV resulted the one-dimension nanostructure of the PVdF nanofiber. The produced PVdF nanofiber membrane separators are evaluated to have a higher level of porosity (86 – 93%) and a good thermal shrinkage property in comparison to Polypropylene (PP) microporous membrane separator. The produced PVdF membrane separators were assembled into the $LiFePO_4$ cells and demonstrated high charge-discharge capacities at room temperature with the coulombic efficiency reaching 80 %.

1. Introductiom

A lithium-ion (Li-ion) rechargeable battery is an electrochemical-based energy converter device as an energy storage device as well. The main parts of the Li-ion battery include a negative electrode (anode) graphite, a positive electrode (cathode)-based on lithium, electrolyte and a membrane separator [1]. Separator is one of the most important parts in the development of Li-ion batteries. This separator must be able to get through the electrolyte ions but able to act as an electrical insulator so there is no contact between the anode and cathode. The type and material of separator are critical for enhancing the performance of Li-ion batteries such ascycling performance, power density, charge-discharge capacity, stability, and security [2]. A characteristic of a separator is determined by the chemical stability, mechanical strength, stable shape, wetability, porosity, permeability, thickness, thermal shrinkage, shut down, and cost. In general, separators in Li-ion battery are in the form of microporous membrane (pore size of less than 20 nm) made of a polymer membrane or nonwoven fabric mat. The polymer membranes made of polytetrafluoroethylene, polyethylene, polypropylene and polyvinyl chloride. Polivinylidene fluoride (PVdF) is a thermoplastic polymer crystalline that has physical properties such as high polarity, excellent thermal and mechanical properties, high affinity to

IOP Conf. Series: Journal of Physics: Conf. Series **817** (2017) 012013 doi:10.1088/1742-6596/817/1/012013

an electrolyte solution, good chemical stability, and excellent compatibility with electrodes that make PVDF potential to be used as a membrane separator [3].

Due to some superior properties owned by nanofiber structured material such as large specific surface area, small pore size and high porosity, making the nanofiber-based membranes as a promising candidate for a separator material for Li-ion battery applications with a high performance [4]. This study aims to produce a membrane separator based on a polymeric material of PVdF nanofiber using electrospinning method.

2. Method

2.1. Materials

Polyvinylidine fluoride (PVdF, Mw 534.000, Aldrich), N-N Dimethylformamide (DMF, Merck), Acetone (Merck) and n-butanol (Merck) were used as received without further purification. 1M lithium hexafluoroposphate (LiPF6), $LiFePO_4$, PVdF binder, acetylene black (AB), N-methyl-2-pyrollidone (NMP) were supplied by MTI corp., USA. The Celgard 2400 microporous polypropylene-commercial (PP-Comm., Celgard, USA) membrane separator was used for comparison.

2.2. Separator preparation

PVdF nanofiber membrane separator was prepared by electrospinning of PVdF solution. The PVdF solution was prepared by dissolving 10 wt% PVdF powder in acetone/DMF (3:7 by vol. ratio) as a solvent, which was stirred at 60 °C for 8 h.The polymer solution was fed through the syringe with the flow rate of 1.5 mL/h. The cone jet was formed at the tip of syringe by applying the DC high voltage to the body of metal syringe. The jets were then elongated toward the lower potential and the solvent was evaporated simultaneously to form the nanofiber. The tip-to collector distance was set at 17 cm to maintain the complete evaporation of the solvent. In this research, the DC high voltage was adjusted and varied from 13 to 17 kV. The electrospinning conditions are indicated in table 1. Figure 1 shows the schematic diagram of electrospinning system.

Tabel 1. The electrospinning condition of production of PVdF nanofiber membrane separator.

Sampel	Voltage (kV)	Spinning time (h)	TCD[a] (cm)	Flow rate (mL/h)
PVdF-A	13			
PVdF-B	15	1	15	1.5
PVdF-C	17			

[a]*Tip-to-collector distance*

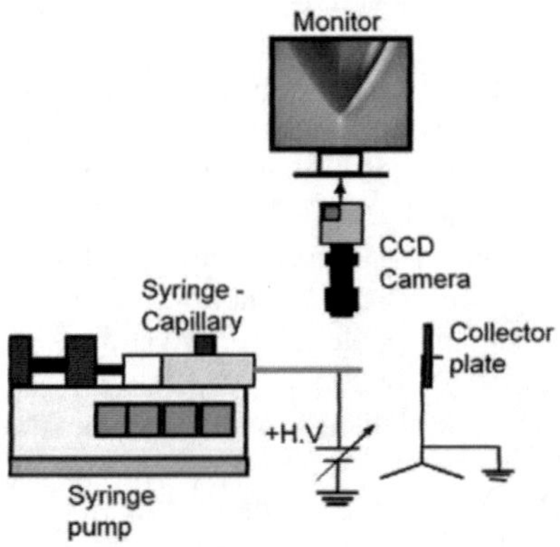

Figure 1. Schematic diagram of electrospinning system.

2.3. Characterization

2.3.1. Morphology of nanofiber separator. The morphology of the PVdF nanofiber separator was characterized using scanning electron microscope (SEM, JEOL JSM-6360 LA, Japan). The electron energy was adjusted at 15 keV.

2.3.2. Porosity of nanofiber separator. The porosity of nanofiber was measured by soaking the dried nanofiber with a measured weight of W_{dry} into the n-butanol (Merck) for 2 h then wiped using a filter paper to obtain their wet weight (W_{wet}). The porosity of PVdF nanofiber separator was obtained by Eg. (1)[5].

$$Porosity\ (\%) = \frac{W_{wet} - W_{dry}}{\rho_b V_{dry}} \times 100\% \quad (1)$$

where ρ_b is the density of n-butanol of 0.81 g/cm^3.

2.3.3. Thermal property of nanofiber separator. The thermal property of the PVdF nanofiber separator was obtained by measuring the shrinkage ratio of the nanofiber. The same dimension of nanofibers were heated at 105°C and 130°C separately in the oven for 8 h. The shrinkage ratio was calculated by Eq. 2 [5].

$$Shrinkage\ ratio\ (\%) = \frac{L_0 - L_1}{L_0} \times 10 \quad (2)$$

where L_0 is the length before heat treatment and L_1 is the length after heat treatment.

2.3.4. Performance evaluation. The performance of PVdF nanofiber as a separator in the Li-ion battery was evaluated by measuring the charge-discharge test. The charge-discharge of $LiFePO_4$ cells containing liquid electrolyte-soaked membrane (PVdF and PP-commercial) were carried out using a split test cell (MTI corp. USA). The main components of a cathode is $LiFePO_4$ as an active material, acetylene black (AB) and PVdF as binders. The anode and electrolyte are carbon and $LiPF_6$, respectively. Electrochemical performance of secondary lithium battery was analyzed using the eight channel battery analyzer (MTI corp, USA).

3. Result and Discussion

3.1. Morphology and Porosity of Separator

The photographs of produced PVdF nanofiber membrane separator are shown in figure 2. Figure 3 shows the morphology of PVdF nanofiber with different DC high voltage at the range of 13 to 17 kV. Adjusting the appropriated voltage is an important condition during nanofiber formation. The applied voltage produces electrostatic force around the needle conductor which acts to overcome the surface tension of the polymer solution [6].

The performance of the cells is strongly influenced by the ionic transport through the separator. It is closely related to the pore size and porosity of the membrane separator. Separators commonly have a thickness of 10 – 30 micron and a pore size of 0.1 – 1 micron. The porosity of the PVdF nanofibers and PP microporous membranes are indicated in table 2. The porosity of PVdF prepared at 13 kV applied high voltage shows highest porosity of 93% and overall the porosity of PVdF nanofiber membrane separator is comparable and even higher than PP microporous. This high porosity is due to the nanofiber that was produced in this condition has uniform and good fiber morphology without any beads or agglomerated parts as shown on the SEM images. Porosity is a measure of the void (i.e. "empty") spaces in a material, and is a fraction of the volume of voids over the total volume.

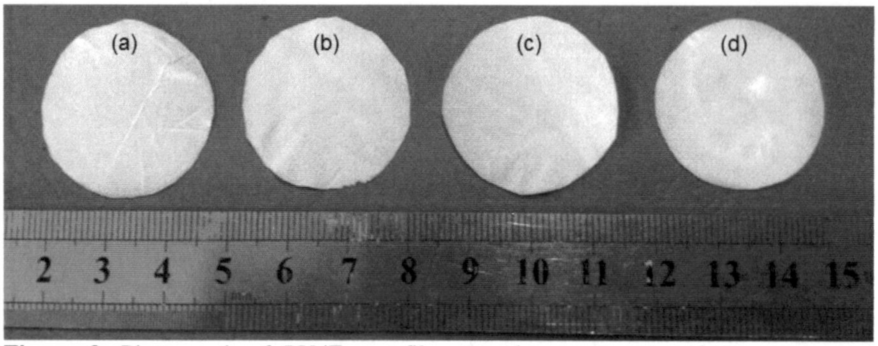

Figure 2. Photograph of PVdF nanofiber membrane separator that was spun at different high voltage, (a) 13, (b) 15, (c) 17 kV and (d) PP-comm.

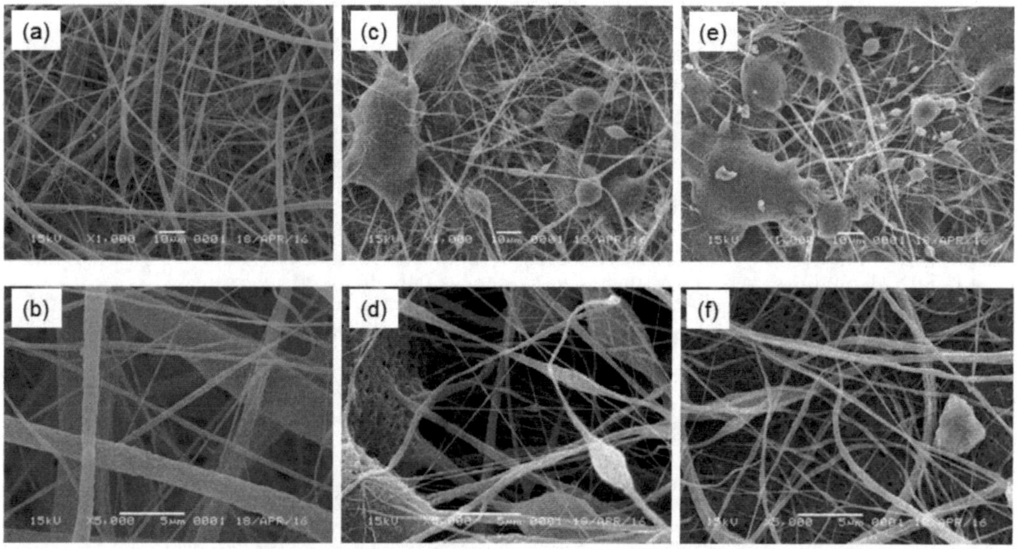

Figure 3. Low and high magnification SEM images of PVdF nanofiber membrane separator that were spun at different high voltage, (a, b) 13, (c, d) 15 and (e, f) 17 kV.

Tabel 2. Porosity properties of PVdF nanofiber.

Sample	Porosity (%)
PP- commercial	87
PVdF-A	93
PVdF-B	89
PVdF-C	86

3.2. Thermal property

In order to investigate the thermal stability of the nanofiber membrane separator, the PVdF nanofiber membrane and PP-Celgard microporous were heated at 105°C and 130°C for 8 h in an oven and its thermal shrinkage ratio (tabel 3) was measured. PP microporous membrane often shrinks at an elevated temperature due to the crystalline of PP decreases with curliness of molecular chain. This phenomenon could lead to a serious safety concern when used as a battery separator [5]. To prevent the internal short circuit failure, it is necessary that some separators with no or minimal thermal shrinkage which could be found at PVdF nanofiber membrane. Therefore the PVdF nanofiber membrane has a good potential as a separator of Li-ion battery.

Tabel 3. The shrinkage ratio of PVdF nanofiber and PP Celgard microporous.

Sample	at 105°C (%)	at 130°C (%)
PP-commercial	5	24
PVdF-A	3	11
PVdF-B	1	10
PVdF-C	0	4

3.3. Cycling performance

The cycling performance (i.e., discharge capacity of cells as a function of cycle number) of $LiFePO_4$ battery using PVdF nanofiber and PP microporous were investigated, in which the cells were cycled at constant charge/discharge current density (1C = 170 mA g^{-1}) rate. Figure 4 shows the charge (a) and discharge (b) of $LiFePO_4$ battery using PVdF nanofiber and PP microporous at first three cycles. The discharge capacity decay of $LiFePO_4$ battery using PVdF membrane separator is better than PP-microporous due to the ionic transport in PVdF membrane separator is better than PP-microporous. In addition, the coulombic efficiency of the Li-ion battery using PVdF separator reaches 80%, whereas PP-microporous of about 60% [figure 5]. The battery characteristic such as charge-discharge performance depends on the type of materials for the cathode and the anode. Separator membrane can perform in the ion transporting. The measurement to the battery performance shows that the PVDF nanofiber based- membrane has a good potential to be used as separator in the Li-ion battery.

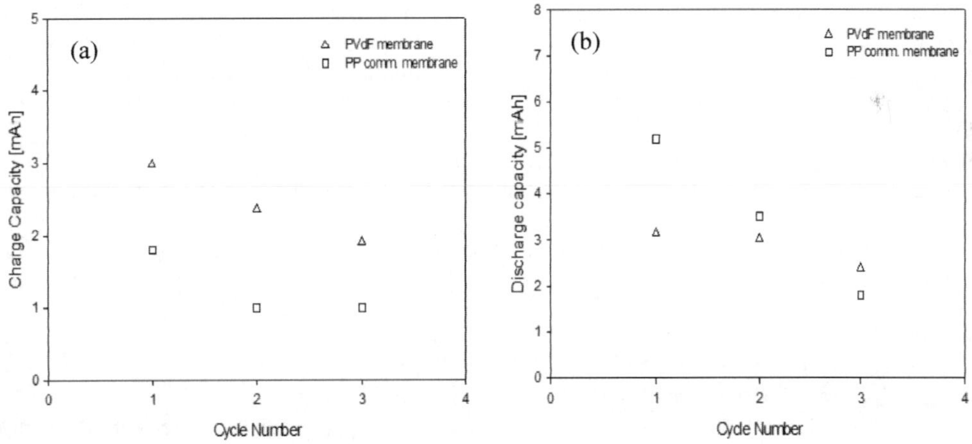

Figure 4. (a) Charge capacity and (b) discharge capacity of $LiFePO_4$ cells containing PVdF nanofiber and PP-microporous at 1C rate.

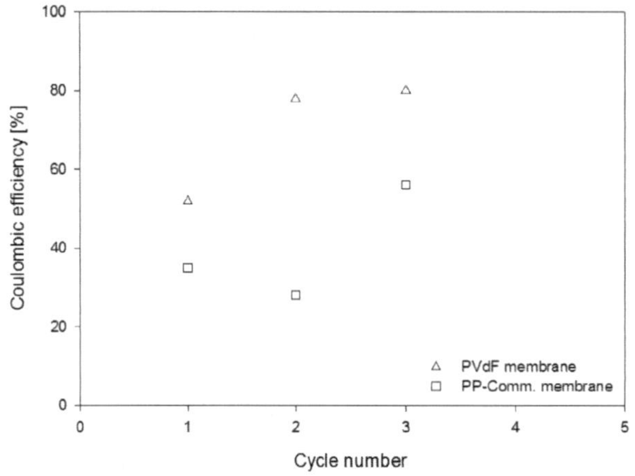

Figure 5. Coulombic efficiency of $LiFePO_4$ cells containing PVdF nanofiber and PP-microporous at 1C rate.

4. Conclusion`

The PVdF nanofiber was successfully prepared using electrospinning method for membrane separator of secondary Li-ion battery. The applied DC high voltage influenced the properties of produced nanofiber. The produced PVdF nanofiber showed high level of porosity (86-93%) in comparison to PP-commercial separator. The important point is that the PVdF nanofiber membrane separator provided substantial improvement in the thermal shrinkage. In the cell test at 1C rate, $LiFePO_4$ battery using PVdF nanofiber membrane separator gave a good performance.The discharge capacity decay in the first three cycling measurements showed that PVdF membrane separator relatively stable than PP-microporous membrane separator and it exhibited high coulombic efficiency at about 80%.

Acknowledgement

This research was supported by The Ministry of Research Technology and Higher Education Republic of Indonesia viaResearch Grant of PenelitianUnggulanPerguruan Tinggi (PUPT) with contract number: 176-07/UN7.5.1/PG/2016. The author also thanks to Iqbal Firdaus for conducting experiment in the some part.

References

[1] Goodenough J B and Kim Y 2010 *Chem. Mater.* **22** 587
[2] Zhang S S 2007 *J. Power Sources* **164** 351
[3] Costa C M, Gomez Ribelles J L, Lanceros-Méndez S, Appetecchi G B, Scrosati B 2014 *J. Power Sources* **245** 779
[4] Zhang X, Ji L, Toprakci O, Liang Y, Alcoutlabi M 2011 *Polym. Rev.* **51** 239
[5] Zhang F, Ma X, Cao C, Li J and Zhu Y 2014 *J. Power Sources* **251** 423
[6] Navarro N H, Gonzalez V G, Moreno-Cortez I E , Garza-Navarro M A 2016 *Mater. Lett.* **167** 34

IOP Conf. Series: Journal of Physics: Conf. Series **817** (2017) 012014 doi:10.1088/1742-6596/817/1/012014

The effect of Cu dopant on morphological, structural and optical properties of ZnO nanorods grown on indium tin oxide substrate

J Marselie[1], V Fauzia[1] and I Sugihartono[2]

[1]Department of Physics, Universitas Indonesia, Depok 16424, Indonesia
[2]Jurusan Fisika, FMIPA Universitas Negeri Jakarta, Jl. Pemuda No. 10 Rawamangun 13220, Indonesia

E-mail: vivi@sci.ui.ac.id

Abstract. Zinc oxide (ZnO) is a promising semiconductor due to its tunable optical properties. Among semiconductor nanostructures, one-dimensional structure such as ZnO nanorods becomes the main focus due to their wide potential application. Cu doped ZnO has attracted much attention recently because of its interesting properties, such as room temperature ferromagnetism, multiferroics properties and green emission. In this work, we deposited ZnO nanoseeds on indium tin oxide coated glass substrates and then ZnO nanorods were grown and doped with three different Cu concentrations (1, 4 and 7 at. %) by hydrothermal method. The scanning electron microscope showed that ZnO nanorod were grown with a hexagonal shape with diverse growth orientation. The X-ray diffraction analysis reveals that the addition of Cu doping tends to decrease the crystallite size and the lattice parameter. Generally, Cu doping reduces the absorbance at ultraviolet wavelength region except on the ZnO nanorods with 4% Cu. The addition of Cu also increases the bandgap energy and decreases the luminescence intensity in UV and visible regions, that is a sign that number of radiative transitions and the natural defects have been reduced.

1. Introduction

Zinc oxide (ZnO) is one of semiconductor materials that is intensively investigated by many researchers. Pure zinc oxide has a large band gap energy of 3.37 eV and large exciton binding energy of 60 meV at room temperature[1]. ZnO has significant potential for ultraviolet light detector, solar cells, and Light Emitting Diode (LED). Among the ZnO nanostructures, ZnO nanorods shows excellent properties due to bigger ratio area to volume ratios compare to other structures. Zinc oxide (ZnO) has grown by many methods, such as chemical bath deposition[2], spray pyrolysis[3] and hydrothermal deposition [4]. Recently, the properties of ZnO nanorods were successfully modified by the addition of metallic dopants. The addition of Cu element dopant is reported produce stronger luminescence in the blue region and ultraviolet [5], affects the orientation of crystal growth [6], enhances the photocatalytic activity[7], increases the dielectric constant [8] and even affects the ferromagnetic properties [9]. In this work, we synthesized Cu doped ZnO nanorods by ultrasonic spray pyrolysis and hydrothermal methods on the indium thin oxide (ITO) coated glass substrates. The main focus of this research is to study the effect of the Cu on the morphology, structural, and optical properties of ZnO nanorods.

IOP Conf. Series: Journal of Physics: Conf. Series **817** (2017) 012014 doi:10.1088/1742-6596/817/1/012014

2. Experimental

ZnO nanorods were grown on ITO coated glass substrates through two steps synthesis process. The first step is the deposition of ZnO seed layers using ultrasonic spray pyrolysis (USP) method and the second step is the growth of ZnO nanorods by using hydrothermal method. First, the seeding solution was made by dissolving 0.4 M zinc acetate dehydrate ($C_4H_6O_4Zn.2H_2O$) with deionized water. The solution was placed in the ultrasonic nebulizer container and then sprayed for 10 minutes onto the substrates heated at the temperature of 450°C. Second, The growth solution contains 0.1 M zinc nitrate tetrahydrate ($N_2O_6Zn.4H_2O$), hexamethylenetetramine ($C_6H_{12}N_4$), deionized water with three different copper (II) nitrate hemi(pentahydrate) ($Cu(NO_3)_2.2,5\ H_2O$) concentrations (1, 4 and 7 at. %). The growth solution was stirred and heated at 60°C. The ZnO seed coated ITO substrates were then carefully immersed in growth solution and then heated at temperature of 95°C in the oven for 2 hours.

Morphological characterization was carried out using a Field emission scanning electron microscopy (FE-SEM) Inspect F50. From the SEM images, the diameter of ZnO nanorods was measured manually. The characterization of ZnO microstructure was performed using SHIMADZU XRD-7000 X-ray diffractometer. The optical properties characterization was conducted using Thermo Fisher Scientific GENESYS 10S UV Visible and FLS920 Fluorescence Spectrometer.

3. Results and discussion

Figure 1 shows the SEM images of ZnO nanorods doped with three different Cu concentrations. It was seen that the ZnO nanorods were grown with hexagonal shape. The undoped ZnO nanorods have random growth orientation and when Cu was added, the orientation of ZnO nanorods slightly changed to the direction of perpendicular to substrates surface. This result is similar with a previous work performed by Polat et al. [2]. The increasing of Cu concentration tends to reduce the diameter of nanorods from the approximately 128 nm into 83 nm. The decreasing of diameter ZnO nanorods may be due to the ion radius of Cu^{2+} ions (0.73 Å) that were substituted Zn^{2+} ions (0.74 Å) which have smaller ion radii[2]. The Cu dopant also tends to increase the density of the nanorods.

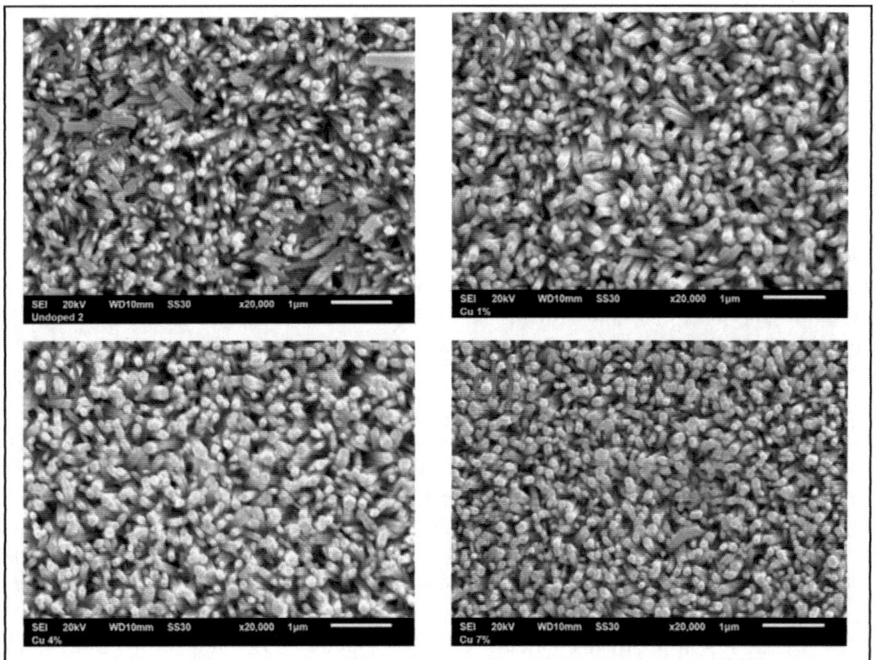

Figure 1. SEM images of ZnO nanorods with Cu doping of (a)0, (b)1, (c) 4 and (d) 7 at %.

Figure 2 shows the XRD pattern of ZnO nanorods with three different Cu concentrations. All ZnO nanorods samples show many diffraction peaks, indicating the ZnO nanorods were grown in random orientation, this agrees with the morphology as seen in figure 1. With the addition of Cu up to 7%, there are no new peaks found, it indicates that the Cu ions substitute the Zn ions and the role of Cu ions in ZnO structure as the deep acceptors bounding with oxygen ions [10].

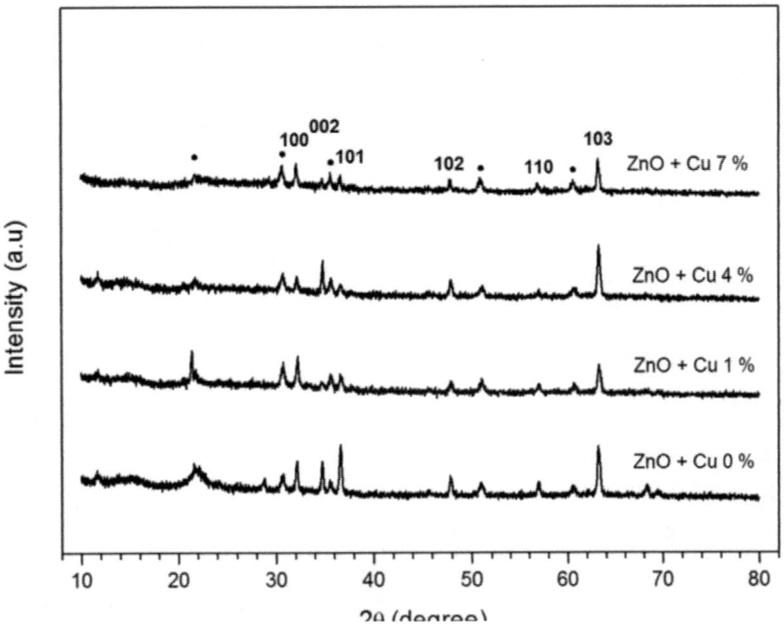

Figure 2. XRD pattern of ZnO nanorods doped Cu

Because of the intensity of (103) plane is dominant, hence (103) data are then processed to find the crystallite size, lattice parameters, density and volume crystal that all are summarized in table 1. The crystallite size was estimated from the Full Width Half Maximum (FWHM) data using the Debye-Scherrer formula [11]. Table 1 shows that generally the addition of Cu tends to decrease the crystallite size and we can confirm this result with the SEM results. The smallest crystallite size was achieved in ZnO nanorods doped with 4% Cu. With more Cu addition, we can see the increasing of crystallite size. This may be caused Cu ions did not only substitute the Zn ions but also filled the vacancy in ZnO lattice that made the structure got bigger. The density of ZnO nanorods also increased with the addition of Cu and this agrees with the SEM results. The lattice parameters also become smaller hence the volume of ZnO structure becomes smaller.

Figure 3 shows the absorption spectrum of Cu-doped ZnO nanorods. It is clearly seen that the absorption intensity at UV region is higher than the absorption at visible region, this transition is usually known as the absorption edge. This happened due to the wavelength at UV region has enough energy to excite the electrons to conduction band, while at the visible region, the energy is not enough for electron excitation hence the light will be transmitted or reflected. Figure 3 also shows that the addition of Cu doping concentration reduces the absorbance at ultraviolet wavelength region except the ZnO nanorods with 4% Cu that has the highest absorbance. Mittal et al [10] also reported the similar result where the addition of Cu tends to decrease the absorbance in UV region compared with undoped ZnO nanorods. The highest absorbance in ZnO nanorods with addition of 4 % Cu may be related to its morphology that has larger diameter and high density so it is more effective to absorb the ultraviolet light.

Tabel 1. Crystallite size, density, volume of Cu-doped ZnO nanorods

Cu concentration (at. %)	Crystallite size (Å)	Density (g/cm^3)	a(Å)	b(Å)	c(Å)	alpha	beta	gamma	Volume (Å^3)
0	192.6	5.66	3.250	3.250	5.200	90	90	120	47.717
1	144.1	5.68	3.247	3.247	5.209	90	90	120	47.575
4	105.6	5.67	3.251	3.251	5.210	90	90	120	47.695
7	158.4	5.68	3.248	3.248	5.200	90	90	120	47.584

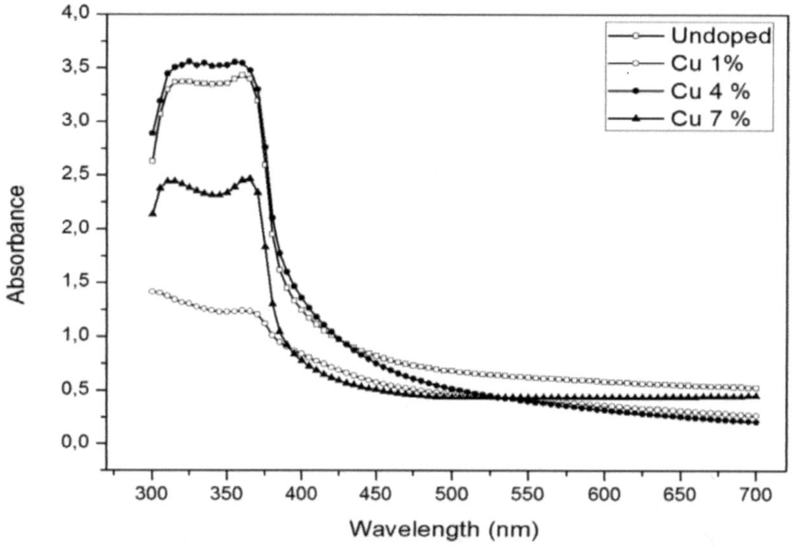

Figure 3. Absorption spectrum of Cu-doped ZnO nanorods

Figure 4 shows the reflectance spectrum of Cu-doped ZnO nanorods. There is a sharp transition at a wavelength 370 nm – 390 nm corresponding to the absorption edge in figure 3. The addition of Cu in ZnO nanorods tends to decrease the reflectance in all wavelengths region.

The bandgap energy of ZnO nanorods was estimated from the reflectance data by using Kubelka-munk formula and Tauc-plot [11], [12] as shown in figure 5. The bandgap energy for ZnO nanorods with additions of Cu 0%, 1%, 4%, 7% are 3.095 eV, 3.121 eV, 3.123 eV, and 3.147 eV respectively. The addition of Cu increases the bandgap energy of ZnO nanorods linearly. This may occur due to Moss-Burstein effect, where the addition of Cu shifts the Fermi energy level in the bandgap[13].

Figure 6 shows the photoluminescence spectrum of Cu-doped ZnO nanorods. We can see that the luminescence of all samples have two emission peaks at wavelengths of 400 nm and 620 nm. The first peak is well known due to radiative transition from recombination of electrons with holes in the valence band [14]. The other peak at 620 nm is usually caused by the presence of intrinsic defects such as Zn vacancy, Zn interstitial, oxygen vacancy, oxygen interstitial or impurities [2]. The intensity of both peak decreased with Cu addition. Undoped ZnO nanorods have the highest luminescence, indicating the sample contains a lot of defects and the addition of Cu tends to decrease the luminescence intensity, that is a sign that defects have also been reduced. While we compare with other works, Polat et al. [2], it has a different pattern where the addition of Cu tends to decrease the intensity in UV region and increase the intensity in the visible region.

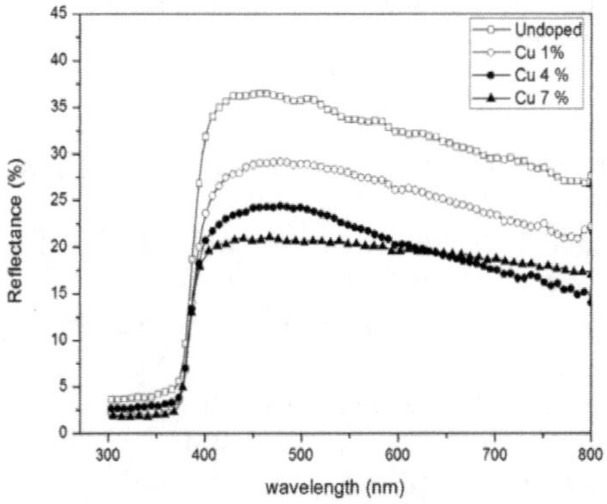

Figure 4. Reflectance spectrum of Cu-doped ZnO nanorods

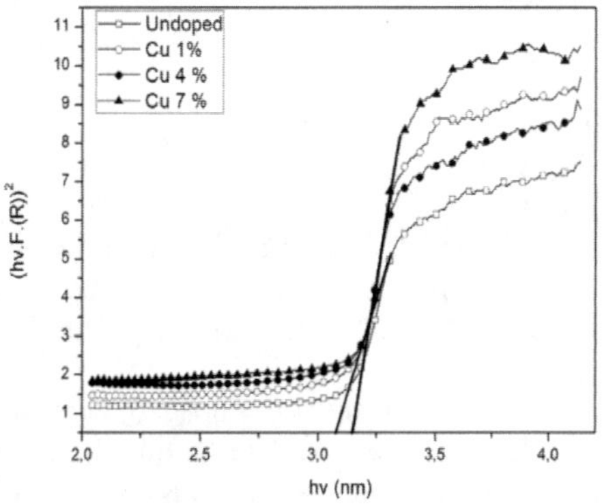

Figure 5. Tauc-plot of Cu-doped ZnO nanorods

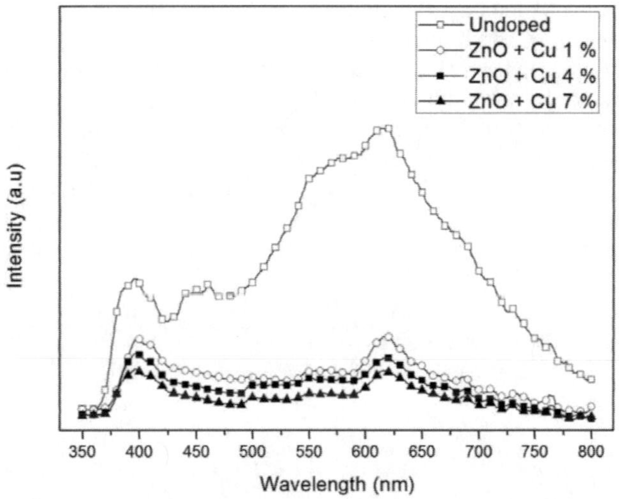

Figure 6. Transmittance spectrum of Cu-doped ZnO

4. Conclusion

ZnO nanorods with three different Cu concentrations (1, 4 and 7 at. %) have been successfully synthesized through ultrasonic spray pyrolysis and hydrothermal methods. Based on the results of structural analysis, morphological and optical properties characterization, it can be concluded that the addition of Cu slightly changed the growth orientation of ZnO nanorods to the direction of perpendicular to substrates surface and the diameter of nanorods was reduced. The addition of Cu also tends to decrease the crystallite size and the lattice parameter. Generally, Cu doping reduces the absorbance at ultraviolet wavelength region except on the ZnO nanorods with 4% Cu. The addition of Cu also increases the bandgap energy and decreases the luminescence intensity in UV and visible regions that is a sign that the number of radiative transitions and the natural defects have been reduced.

Acknowledgments
This works was supported by Hibah Publikasi International Terindeks untuk Tugas Akhir Mahasiswa (PITTA) 2016 Universitas Indonesia. The author also would like to thank Prof. Dr. Muhamad Mat Saleh and Assoc. Prof. Dr Akrajas Ali Umar from Organic and Printed Electronic Laboratory (OPEL) IMEN Universiti Kebangsaan Malaysia for this beneficial research collaboration.

References
[1] Musbah B, Wang D, Wang J, Li Q, Sun J, Yan Y, Yu Q and Shujie J 2015 *Thin Solid Films* **9**(1) 316–19
[2] Polat I, Yılmaz S, Altın I, Bacaksız E and Sökmen M 2014 *Mater. Chem. Phys.* **148**(3) 528–32
[3] Kenanakis G and Katsarakis N 2014 *Mater. Res. Bull.* **60**(1) 752–59
[4] Ramimoghadam D, Hussein M Z and Taufiq-Yap Y H 2013 *Chemistry Central Journal* **7**(1) 136
[5] Yu J L, Lai Y F, Cheng S Y, Zheng Q and Chen Y H 2015 *J. Lumin.* **161** 330–34
[6] Kim K H, Jin Z, Abe Y and Kawamura M 2014 *Superlattice Microst.* **77** 101–07
[7] Mohan R, Krishnamoorthy K and Kim S J 2012 *Solid State Commun.* **152**(5) 375–80
[8] Iqbal J, Safdar N, Jan T, Ismail M, Hussain S S, Mahmood A, Shahzad S and Mansoor Q 2015 *J. Mater. Sci. Technol.* **31**(3) 300–04
[9] Tang G, Shi X, Huo C and Wang Z 2013 *Ceram. Inter.* **39**(5) 4825–829
[10] Mittal M, Sharma M and Pandey O P 2014 *Sol. Energy* **110** 386–97
[11] Sharma S, Mehta S K and Kansal S K 2017 *J. Alloys Compd.* **699** 323–33
[12] Lestari A, Iwan S, Djuhana D, Imawan C, Harmoko A and Fauzia V 2016 *AIP Conf. Proc.* vol 1729 (New York: AIP Publishing) 020027
[13] Cheng-Xiao P, Hui-Min W, Yang Z, Xing-Ping M and Bang-Jiao Y 2008 *Chinese Phys. Lett.* **25**(12) 4442–445
[14] Vijayaprasath G, Murugan R, Mahalingam T, Hayakawa Y and Ravi G 2015 *J. Alloy Compd.* **649** 275–84

Electrical properties of Al-, Cu-, Zn- rice husk charcoal junctions

L A Dahonog and A K G Tapia

Institute of Mathematical Sciences and Physics, College of Arts and Sciences, University of the Philippines Los Baños, Laguna, Philippines 4031

E-mail: ladahonog@up.edu.ph

Abstract. Rice husk in the Philippines is considered as an agricultural waste. In order to utilize the material, one common technique is to carbonize these rice husks to produce charcoal briquettes. These materials are porous in nature exhibiting electrical properties from carbon structures. In this study, rice husk charcoals (RHC) were deposited on different metal substrates (Al, Cu, Zn) via a simple solution casting method. The deposited RHC on metal substrates was observed using Scanning Electron Microscopy (SEM). The films were characterized using two-point probe technique and the I-V curves were plotted. Al-RHC films appear to deviate from an ohmic behaviour while Zn-RHC and Cu-RHC showed diode-like behaviours.

1. Introduction

Nowadays, carbon materials are of great interest because of its physical and chemical properties which are useful in applications such as battery electrodes [1], capacitors [2], gas storage [3], biomedical engineering [4], thermoelectric materials [2] and catalytic supports [5]. Carbon exists in many forms such as diamond, graphite and charcoal [3]. Among all these forms, charcoal is the most readily available.

Charcoal is a material derived from the carbonization of wood and other plant biomasses [6]. Charcoal is commonly used as adsorbent for the removal of organic dyes since it is resistant to alteration and degradation [7, 8]. Raw materials from plants such as wood, bamboo and rice hull or rice husks are considered to be exceptional precursor to produce carbon structures from their natural porous matrices. These porous carbon materials provide electrical transport from the fundamental properties of carbon structures [9]. The conductive property of charcoal is attributed to its π-electron system that originates from the conjugated bonds [10, 11]. The formation of intermolecular bonding and bridge structure between its pyranose rings during carbonization of charcoal also contributes to its conductivity [12].

Rice husks are considered agricultural waste in rice-producing countries after being separated from the rice grains during the milling process. These husks, as wastes, are either burnt or dumped [13]. Rice husks are composed of cellulose, hemicelluloses, lignin, silica and other organic compounds [14]. Because of its high silica content, rice husks were commonly used as fuel in power plants and as raw material for the synthesis of new compounds [15]. Properties of these poor crystalline carbon materials are not yet studied in detail but the presence of sp^2 orbitals enhances the possibility of using these materials for applications involving electrical conduction [11]. Some studies performed techniques

IOP Conf. Series: Journal of Physics: Conf. Series **817** (2017) 012015 doi:10.1088/1742-6596/817/1/012015

such as acid etching, hydrogenation and irradiation to fabricate a schottky diode from charcoals as hydrogen sensor [16].

In this study, rice husk charcoals were used as the active material in the preparation of rice husks charcoal junctions using different metals as substrate. Lastly, the structural, morphological and electrical properties of the fabricated junctions were characterized.

2. Experimental details

Preparation of metal substrates: Copper, Zinc and Aluminum sheets were purchased from B.E. Scientific Glass Instruments. These metal sheets were cut into 2.5 cm x 2.0 cm dimensions. Metal sheets were then polished using silicon carbide and were ultrasonically cleaned using ethanol. After the process, the metal sheets were placed in a beaker.

Preparation of RHC films: Rice husk charcoals were grounded using mortar and pestle. Powders were mixed with acetone having a concentration of 12.5 mg/ml. The RHC solution was then poured in a beaker containing the metal sheets. The solution was air dried until the acetone evaporated.

Characterizations: The morphology and the junction of the RHC films were observed using scanning electron microscope (SEM, JEOL Ltd., JSM-5310). For the electrical characterization, two-point probe method was used. The sizes of the particles were measured using Image Analysis and Processing in Java (ImageJ) software. Fourier Transform Infrared Spectroscopy (FTIR) was done to elucidate the molecular structures in the RHC.

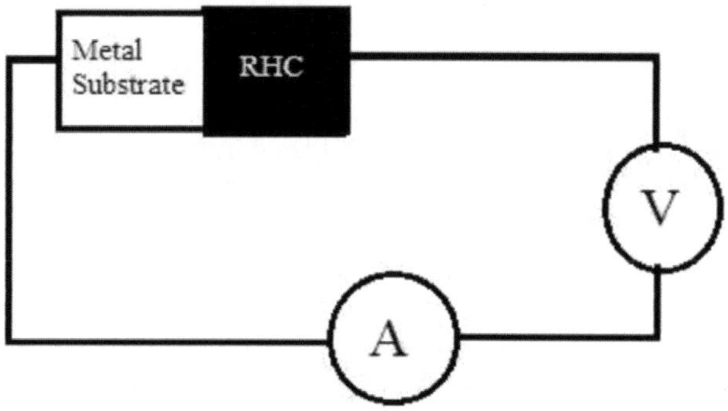

Figure 1. Schematic diagram for the two-point probe set-up.

3. Results and discussions

Figure 2 shows the FTIR spectra of the rice husk charcoals that were used in the fabrication of junctions. The characteristic peaks of the charcoal occur at the range of 1022-1044 cm^{-1}, 787-791 cm^{-1} and 1569-1593 cm^{-1}. The broad shoulder at the range of 1044 cm^{-1} corresponds to the presence of silica in the rice husk charcoal [11]. Sharp peaks that can be found at 787-791 cm^{-1} corresponds to the presence of C-H aromatic hydrogen while peaks at 1569-1593 cm^{-1} described the C=C stretching [17]. RHC is intrinsically insulating. However, the electrical transport in RHC is attributed to the carbon structures in the material. Although RHC is highly amorphous, the sp^2 orbital sites in graphite-like structures provide charge transport in the material.

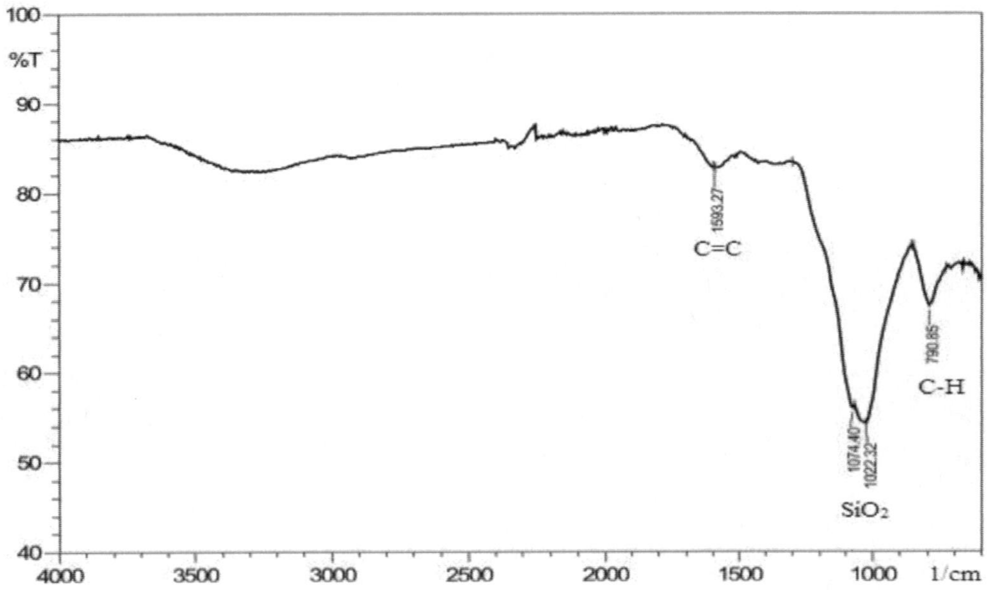

Figure 2. FTIR spectra of the rice husk charcoals.

Rice Husk Charcoal (RHC) was successfully casted on metal substrates to fabricate metal-insulator junction. Figure 3 shows the SEM image of the fabricated Zn-RHC film.

Figure 3. Cross-sectional area of the fabricated film.

IOP Conf. Series: Journal of Physics: Conf. Series **817** (2017) 012015 doi:10.1088/1742-6596/817/1/012015

Figure 4a shows highly heterogeneous structures of RHC. Grinding produced smaller particles with sizes following normal distribution as seen in the histogram plot in figure 4b. The particle size of the rice husk charcoals deposited was measured to be on the range of 1.874 to 7.774 μm having an average of 4.693 μm.

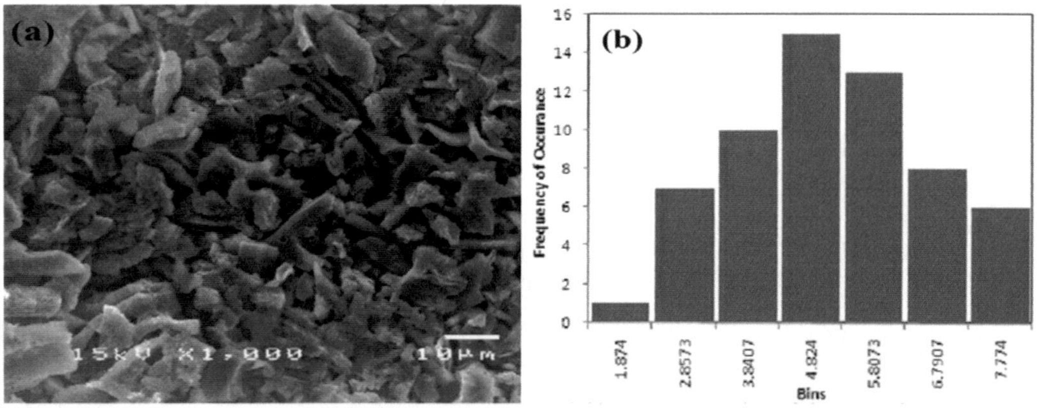

Figure 4. (a) SEM Image of the particles deposited. (b) Histogram plot of the particle sizes.

For the electrical characterization, forward bias configuration was done by connecting the positive terminal on the metal and the negative terminal on the RHC. Reverse bias configuration was done by interchanging the terminals. Figures 5 showed the current-voltage plots of Al-, Cu-, and Zn- RHC junctions obtained using two point probe technique. Figure 5a shows deviation from an ohmic behaviour for Al-RHC junction. The Cu- and Zn- RHC junctions showed diode-like behaviours as seen in figures 5b and 5c, respectively. The diode-like behaviour of the junctions can be attributed to the presence of potential barriers between organic material and the metal substrate. Based from the thermionic emission-diffusion theory of diode [18], the computed barrier height for Al-, Cu-, and Zn – RHC junctions was summarized in table 1. The estimated ideality factor computed for each junction was less than 1. Ideality factors estimated for Al-, Cu-, and Zn- RHC films are also indicated in table 1. The estimated ideality factors computed for each junction were less than 1. This suggests that there are other factors and mechanisms that could describe the behaviour of RHC-metal junction.

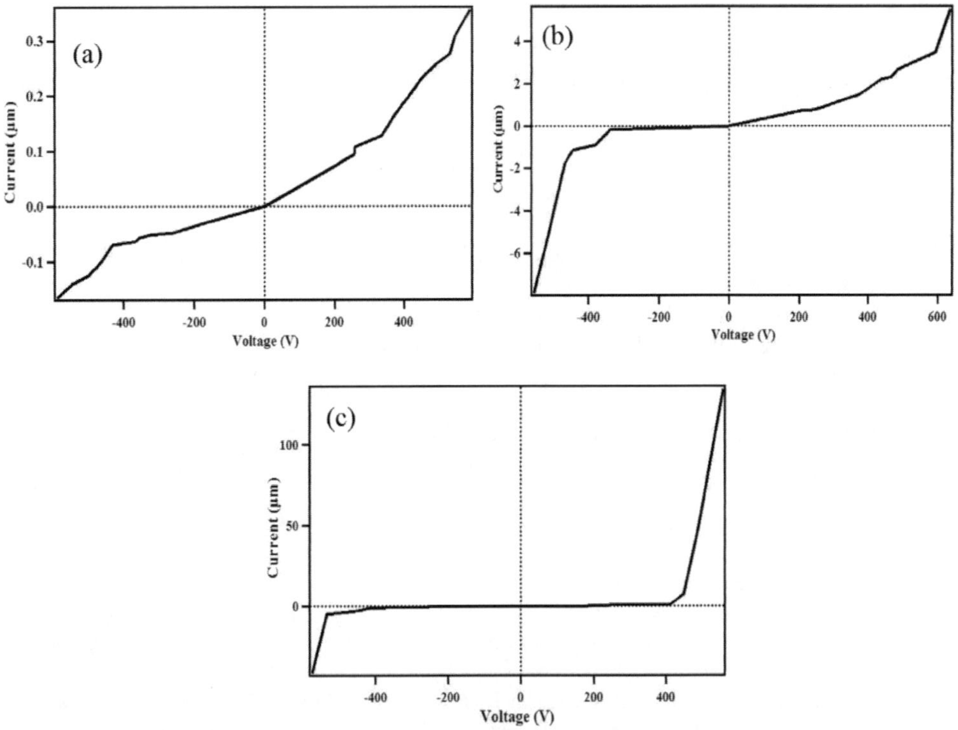

Figure 5. Current-Voltage plots for a) Al-RHC junction, b) Cu-RHC junction and c) Zn-RHC junction.

Table 1. Computed barrier height and ideality factor for RHC junctions.

Junction Type	Barrier Height (eV)	Ideality Factor
Al-RHC	0.623 eV	0.823
Cu-RHC	0.577 eV	0.889
Zn-RHC	0.645 eV	0.796

The amorphous nature of RHC involves hopping mechanism for charge transport. Based on the study of Novikov [19], the total energetic disorder in organic materials is due to randomly oriented dipoles present in the material. Also, the sizes, in micrometer range, of the charcoal particles suggest aggregation forming dense packing arrangement [3]. This provides larger interparticle interactions for conduction.

4. Conclusion

Al-, Cu- and Zn- RHC junctions were successfully fabricated via simple solution casting method using acetone as the dispersing solvent. Connections between the two materials were characterized using two-point probe technique. Current-voltage plots of metal-charcoal junctions showed interesting diode-like behaviours. Mechanism regarding can be further studied for future electronic applications.

Acknowledgments
This work was supported by the University of the Philippines Los Baños Basic Research Grant and the UP System Emerging Inter-Disciplinary Research Program (OVPAA-EIDR C06-035).

References

[1] Srivastava M, Kumar M, Singh R, Agrawal U and Garg M 2009 *J. Sci. Ind.* **68** 94-6
[2] Madroñero A, Robla J, Garcia-Hierro J, Gonzalez MS, Diaz P and Lopez A 2012 *J. Mater. Sci. Mater. Electron.* **23** 1470-76
[3] Rattanaweeranon S, Limsuwan P, Thongpool V, Piriyawong V and Asanithi P 2012 *Procedia Eng.* **32** 1100-06
[4] Paipetis A S and Kostopoulos V 2013 *Springer* **188** 1-20
[5] Nitayaphat W and Jintakosol T 2015 *J. Cleaner Prod.* **87** 850-55
[6] Pastor-Villegas J, Pastor-Valle JF, Meneses Rodriguez M and Garcia Garcia M 2006 *J. Anal. Appl. Pyrolysis* **76** 103-8
[7] Mohammed H, Kadhim B and Mohammed A 2011 ***E**- J. Chem.* **8** 739-47
[8] Luo L, Yang Y, Xiao M, Bian L, Yuan B, Liu Y, Jiang F and Pan X 2015 *Chem. Eng. J.* **262** 1275-83
[9] Shishido M, Ogawa T, Goto H, Takahashi T and Iizuka H 2009 *Trans. Mater. Res. Soc. Jpn.* **34** 667-70
[10] Nishimiya K and Hata T 1995 *Wood Res.* **82** 34-6
[11] Kennedy L, Vijaya J and Sekaran G 2005 *Mater. Chem. Phys.* **91** 471-6
[12] Manabe T, Ohata M, Yoshizawa S, Nakajima D, Goto S, Uchida K and Yajima H 2007 *Trans. Mater. Res. Soc. Jpn.* **32** 1035-8
[13] Kumar A, Mohanta K, Kumar D and Parkash O 2012 *Int. J. Emerging Technol. Adv. Eng.* **2**
[14] Midhun CD, Begum PM, Joseph R, Joseph D, Kumar P and Ayswara EP 2013 *Int. J. Sci. Environ. Technol.* **2** 1027-35
[15] Masotti A and Caporali A 2013 *Int. J. Mol. Sci.* **14** 24619-642
[16] Culebras M, Lopez A, Madronero A, Mota C, Gomez CM and Cantatero A 2013 *Int. J. Eng. Sci. Technol.* **2** 649-56
[17] Theeba M, Bachmann RT, Illani ZI, Zulkefli M, Husni MHA and Samsuri AW 2012 *Malaysian J. Soil Sci.* **16** 89-102
[18] Brillson L J 2012 *Surface and Interface of Electronic Materials* (Weinheim: 2010 WILEY-VCH Verlag GmbH & Co. KGaA) pp 25-8
[19] Novikov S 2012 *Russ. J. Electrochem.* **48** 388-400

IOP Conf. Series: Journal of Physics: Conf. Series **817** (2017) 012016 doi:10.1088/1742-6596/817/1/012016

Effect of coating thickness of Al-doped $Li_4Ti_5O_{12}$ (LTO) on the electrochemical performance of half cell battery

S Priyono[1], G F Marpaung[2], T R Simbolon[2] and A Subhan[1]

[1]Research Center for Physics – Indonesian Institute of Sciences Kawasan PUSPIPTEK Gedung 442 Serpong, Tangerang Selatan 15314, Banten, Indonesia Telp. 021-7560570, Fax 021- 7560554
[2]Department of Physics- Faculty of Mathematics and Natural Sciences-University of North Sumatera Jl. Bioteknologi I Kampus USU Medan, 20155, Indonesia

E-mail: slam013@lipi.go.id

Abstract. Effect of coating thickness of Al-doped $Li_4Ti_5O_{12}$ (LTO) sheet on the electrochemical performance of battery cells is reported. A 0.025 mol Al-doped LTO powder was synthesized by solid state reaction method, for the anode sheet was made by the doctor blade method. The sheet manufacturing was carried out by varying the coating thickness from 100, 200, 300 and 400 µm to determine the effect on the electrochemical performance of the battery cell. The analysis of phase was conducted by X-ray diffraction (XRD); for electrochemical performances were conducted by electrochemical impedance spectroscopy (EIS), cyclic voltammetry (CV) and charge discharge (CD) method. XRD analysis showed that there are two phases formed, i.e. LTO and rutile phases (TiO_2) of 75.6% and 24.4%, respectively. EIS results showed that the thicker the coating layer, the conductivity of the sample become slower. Furthermore, the results indicate that the sample with thickness of 100, 200 and 300µm indicated sharp redox peak, and 400 µm showed wide redox peaks. The specific capacity and specific power of the half cell battery decreases with the increasing of coating thickness. The best cell capacity is obtained by sample with 100 µm thickness with specific discharge capacity of 134.3 mAh/g.

1. Introduction

Lithium batteries were firstly proposed in 1976 [1] and have been widely adopted for commercial since the early 1990s. Lithium batteries can be divided into three different groups, i.e. lithium metal, lithium-ion and lithium-ion polymer. Lithium metal batteries are primary batteries with non-rechargeable, while lithium-ion and lithium-ion polymer batteries are the rechargeable one. The main difference between lithium-ion and lithium-ion polymer, the electrolyte of lithium-ion is based on a lithium salt in an organic solvent (such as $LiPF_6$), while lithium-ion polymer use solid composite polymer (such as oxymethylene-linked polyethylene glycol, PEMO) [2]. Lithium ion batteries are the most promising energy storage device for both portable electronic, electric vehicles and large-scale energy storage. Lithium ion battery consist of four main part, including cathode, anode, separator and electrolyte. The performance of lithium ion battery depends on the properties of those materials. Therefore, lithium ion batteries areas have spurred extensive research to make new electrode materials with high safety level and good electrochemical performance.

$Li_4Ti_5O_{12}$ (LTO), which was first proposed by J.R. Dahn et al [3], has been suggested as potential alternatives material to replace graphite as anode. LTO, one of spinel structure materials, is promising anode material for lithium ion batteries because it has many advantages compared to the conventional anode like graphite. LTO has small lithium diffusion coefficient and experience large volume variation about 9%. There is no structure change under charge-discharge condition, so it is called as "Zero strain materials". This material has a good Li-ion intercalation/de-intercalation, good reversibility, and structural stability and does not form solid-electrolyte interphase (SEI) layer because of high potential around 1.5 V. Additionally, LTO has more lithium ion mobility than graphite, hence promising for high rate battery applications [3, 4].

Although LTO has many advantages for anode lithium ion battery, LTO shows poor electronic conductivity due to empty Ti 3d state with a band energy of about 2eV, and its electron conductivity is 10^{-13}-10^{-19} S cm^{-1}. However, when $Li_4Ti_5O_{12}$ (LTO-spinel) transforms into $Li_7Ti_5O_{12}$ (LTO-rock-salt) in Li-ion insertion process, its electron conductivity become 10^{-5}-10^{-3} S cm^{-1}. In order to improve electrochemical properties, LTO is doped by some metal ion. Doping by metal ion is the best way to increase electron conductivity. There are many choices of metal ion for doping that can be applied, such as: Mg^{2+}, Ni^{3+}, Al^{3+}, Cr^{3+}, Co^{3+}, Fe^{3+}, Mn^{3+}, Ga^{3+}, Zr^{4+}, Mo^{4+}, V^{5+}, Ta^{5+}, F^- and Br^- in Li, Ti or O site [5].

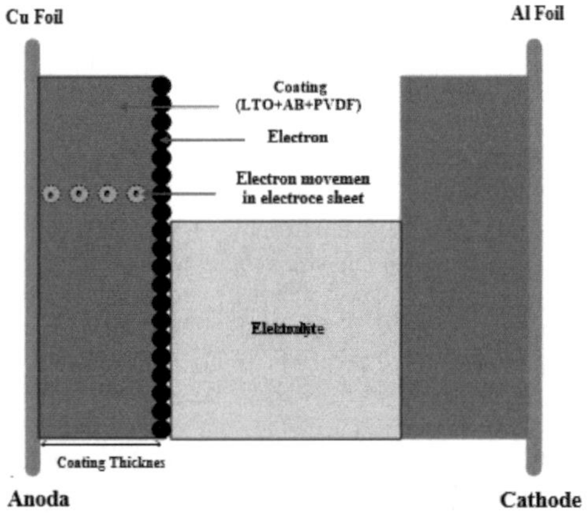

Figure 1. Illustration electron path from electrode/electrolyte interface to current collector.

In this research, Al^{3+} ion will be used as dopant to substitute a small portion of Ti^{4+} in LTO. There are so many research about Al^{3+} doped on LTO. Research done by Huang et.al [6] showed that Al^{3+} ion increased reversible capacity and cycle stability when charging and discharging. Lin, et al.[8] synthesized Al doped on LTO by using sol-gel process. The result showed that there is significantly increasing on discharge capacity and rate capability. Based on previous report, 0.025 mol Al doped LTO is the best result by showing cycling stability and lower polarization [7, 8]. In order to make battery cell, making electrode sheet is most important to be consider. Electrode sheet must have precise composition, good adhesive power, and high conductivity. The electrode sheet is made by mixed active materials with binder (i.e. PVDF), conductive carbon, and organic solvent (i.e. DMAC). There are some parameters that have to be considered in preparing electrode sheet, such as: material composition, mixing time, temperature process, mixing speed, viscosity, thickness, coating speed, drying temperature, and drying time.

Coating thickness will effect on cell performance, because it related with the length of electron path from electrode/electrolyte interface into current collector [9]. If the coating is thicker, electron path will become longer so that the efficiency of the battery cell decreases as the electron flow will be resisted in the active material. Additionally, the increased of thickness coating can also cause internal resistance increase and lead to self-discharge. The illustration of electron path from interface to the collector is shown in figure 1.

The power and energy density of the battery depend on coating thickness of the layer of active material on the current collector. Thicker coating of the electrode material layer provides higher energy because of the amount of the active ingredient which is great, but it also increases the length of ion diffusion in the material resulting in lower specific power. Instead, a thin layer electrode provides a short diffusion path so that a higher power density. But much less active material and consequently the specific energy decrease. In addition, optimization of the thickness of coating can minimize costs production [10, 11]. In this report, the effect of coating thickness on the electrochemical performance specifically on specific power and specific energy will be reported.

2. Experiment

Al doped LTO was synthesized by a solid-state reaction method. The 0.025 mol Al_2O_3 was used as Al ion to replace Ti ion in LTO. Stoichiometric $LiOH.H_2O$, TiO_2 and Al_2O_3 with technical grade used as raw materials. The raw materials were mixed and followed by screening #200. Finally, the product were sintered at 850°C for 4 h with heating rate of 10°C/min in air atmosphere to obtain active material of Al doped LTO. The crystal structure and phase of active material was identified and confirmed by X-ray diffraction (XRD) using Cu-Kα radiation.

The electrochemical properties of samples were measured using CR2032 coin cells. Active materials (Al doped LTO, 85 wt.%), poly-vinylidene fluoride (PVDF, 10 wt.%) and acetylene black (5 wt.%) were mixed in DMAC solvent until become slurry. The slurry was coated on Cu foil by varying thickness coating i.e., 100, 200, 300 and 400 μm become electrode sheets and the samples were called as G1, G2, G3 and G4, respectively. The cells were assembled using electrode sheets as working electrode, lithium metal as the counter electrode, and Celgard film as the separator. The electrolyte was $LiPF_6$ dissolved in organic solvent, i.e., ethylene carbonate (EC) and dietyl carbonate (DEC). The cells were assembled in a glove box filled with high purity argon gas and tested by LCR HIOKI to measure Electrochemical Impedance Spectroscopy (EIS). While, cyclic voltammograms (CV) and charge-discharge (CD) were recorded using WBCS battery tester. Cyclic voltammetry was carried out using scan rate of 0.16 mV/s from 0.7 - 2.8 V. Charge-discharge was measured at 0.1 C rate from 0.5 – 2.6 V to obtain capacity, specific capacity, and specific power.

3. Result and discussion

Figure 2 displayed the XRD patterns of the obtained Al doped LTO powder. There are two phases identified, i.e., LTO (75.6%) and rutile TiO_2 (24.4%) which denoted by filled diamond and circle shapes, respectively. The co-exist of two phases can be explained due to high heating rate when sintering process carried out; and in this work, there is no initial calcination process. The calcination process is usually needed to remove volatile from raw material while slow heat rate will make perfect reaction. However, from the XRD result, the percentage of LTO phase seems enough to be used as anode material. The Al doped LTO powder showed have high crystallinity as host of lithium ion, this benefit for charge-discharge of materials. The XRD peaks can be adjusted to pristine LTO spectra from the ICDD card No 00-049-020, at three main peaks at two theta of 18°, 43° and 36°.

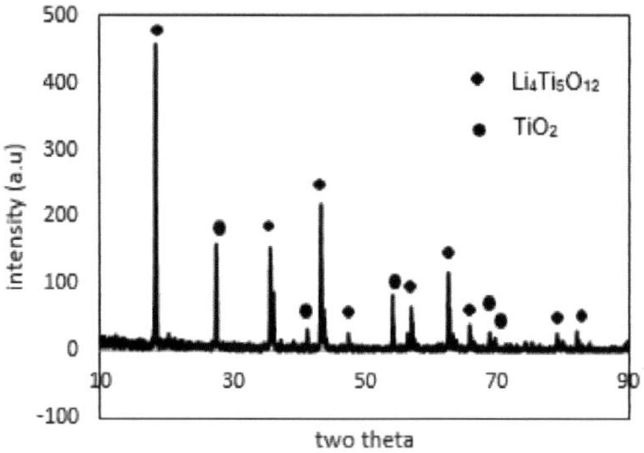

Figure 2. XRD of Al doped LTO.

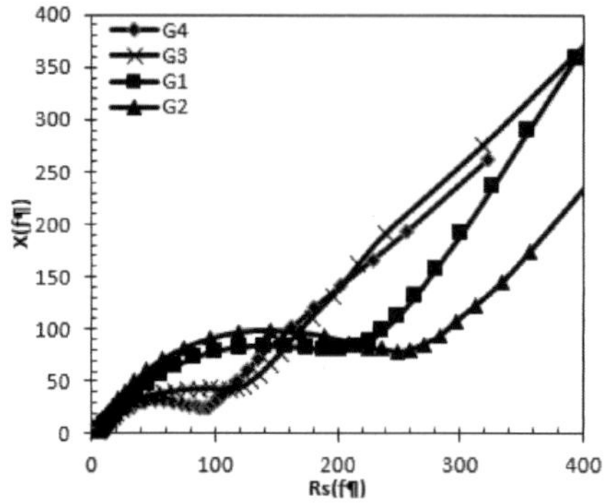

Figure 3. EIS spectra of Al doped LTO with thickness variation of electrode sheet.

Figure 3 showed EIS of all samples. The spectra showed semicircle at high to middle frequency and a slop line at low frequency. The intersection of the slope line with real axis refers to a bulk resistance (R_b), which reflects the electronic and ionic resistance of two electrodes and electrolyte/separator. The semicircle corresponds to the charge transfer resistance (R_{ct}) and its related double-layer capacitance (C_{dl}). The slopping line at low frequency represents the Warburg impedance, which is associated with lithium-ion diffusion into the LTO particles [12]. Table 1 showed resistance and conductivity for all samplesthat extracted from EIS spectra. The different of electronic resistance for all samples are quite small and it can be consideredsimilar. But, the different of ionic conductivity of all samples are quite big. This can be explained due to the increasing in layer active materialthickness causingthe semicircleplotbecome smaller; so that the resistance become small and the conductivity become high. The results also showed that the ionic conductivity is more dominant than electronic conductivity. Increasing in conductivity can be attributed from Al doped in LTO. The thicker Al doped LTO will make more contains Al, so increasing the final conductivity. But there is no significantly different in conductivity values because it has similar power value.

Table 1. Conductivity measurement from EIS spectra of Al doped LTO with thickness variation.

Samples	Real Thickness (μm)	R_e (Ω)	R_{ct} (Ω)	σ (S/cm)
G-1	80	9.84	226	3.185 x10^{-5}
G-2	120	4.86	259	4.324 x10^{-5}
G-3	170	5.91	124	11.938 x10^{-5}
G-4	280	6.71	94.8	23.950 x10^{-5}

Cyclic voltammetry test was carried out until 3 cycles by scan rate of 160 μV/s and range potential from 0.7 V to 2.8 V. Figure 4 showed the cyclic voltamogram of Al doped LTO with varying thickness of samples. All samples showed pair oxidation/reduction peaks with working voltage at around 1.5 V. The three first samples(G1, G2, and G3) showed some sharp and narrow oxidation/reduction peaks and this indicates rapid intercalation and deintercalation process. While, G4 showing wide oxidation/reduction peak indicates slow intercalation and deintercalation process. Increasing in coating thickness showed increasing in oxidation/reduction current because more active material involved in electrochemical process.

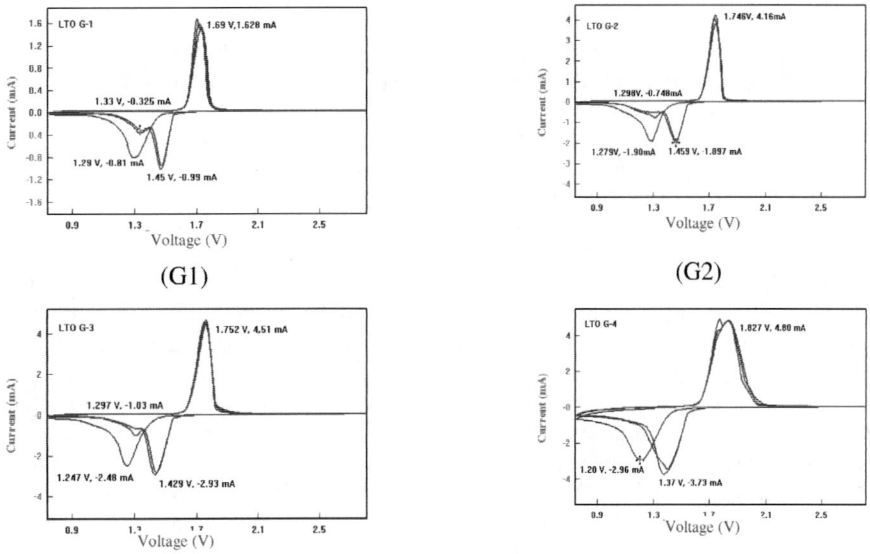

(G3) (G4)

Figure 4. Cyclic Voltamogram of Al doped LTO with varying thickness (G1) 100 um (G2) 200 um (G3) 300 um and (G4) 400 um.

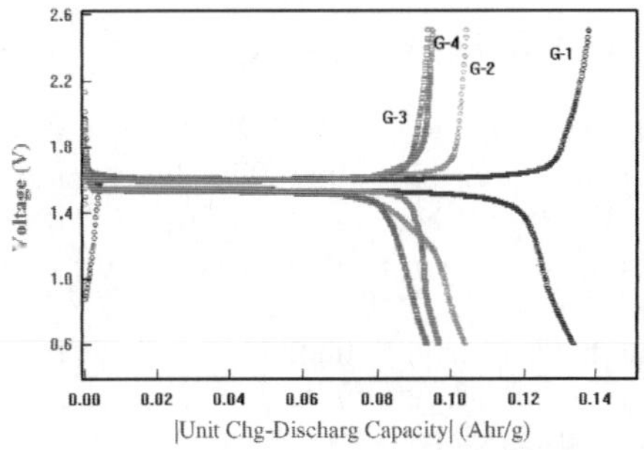

Figure 5. Charge-discharge curves of Al doped LTO with varying thickness (G1) 100 μm (G2) 200 μm (G3) 300 μm and (G4) 400 μm.

Table 2. Specific capacity and capacity of samples.

Sample	Specific Capacity (mAh/g)		Coloumbic Efficiency (%)	Capacity (mAh)		Power density (W/g)
	Charge	Discharge		Charge	Discharge	
G-1	137.08	134.30	98.0	0.3692	0.3617	1.28
G-2	103.64	103.39	99.8	0.6903	0.6886	1.11
G-3	93.65	93.07	99.4	0.8891	0.8836	0.80
G-4	94.73	96.09	101.4	2.0257	2.0548	0.36

Figure 5 showed charge-discharge test for all sample with rate 0.1C and it formed plateau potential at around 1.55 V. From the curves can be extracted specific capacity, coloumbic efficiency and capacity as shown in table 2. It explained that increasing thickness coating will decrease specific capacity because of Al contain and long ionic diffusion. Al contain in LTO will theoretically decrease specific capacity. From the results, all samples have excellent coloumbic efficiency above 98%. While, increasing thickness coating will increase capacity because of more active material involved in electrochemical process. Increasing thickness coating also causingthe difference between charge and discharge voltage become bigger. This indicates increasing in polarization. More active material will increase activation energy in initial electrochemical process, so the polarization increases.

Power density calculated by multipliedthe current density with working voltage. Power density is significantly affected by coating thickness of active material. Table 2 showed that increasing in coating thickness decrease power density,this due to the diffusion path of lithium become longer.

4. Conclusion

Al doped LTO (Al=0.025) was synthesized by simple solid-state method in an air atmosphere. XRD result shows LTO and TiO_2 rutile formed. The co-exist of these phases are caused by high rate temperature when sintering process carried out. Al contained in LTO not increase electronic and ionic conductivity significantly. The sample having thickness under 300 μm showed good and rapid intercalation and deintercalation process. They consistof sharp oxidation/reduction peak and lower polarization than other thickness. Increasing in coating thickness will increase capacity but decrease specific capacity and specific power. The best electrochemical performance was showed by G-1 with specific capacity about 137.08 mAh/g, 98% coloumbic efficiency and highest specific power.

Reference

[1] Whittingham M S 1976 *Science* **192**(4244) 1126-7
[2] Huggins R 2008 *Advanced batteries: materials science aspect* (New York: *Springer*)
[3] Zhenwei Z, Liyun C, Jianfeng H, Dunqiang W, Yan M and Yingjun C 2013 *Electrochim. Acta* **88** 443-6
[4] Kai W, Jun Y, Xiang YQ, Jin M X, Qian Q Z, Jing J and Quan C Z 2013 *Electrochim. Acta* **108** 841-51
[5] Xiaomei C, Xiangfeng G, Liping L and Guangshe L 2012 *J. Power Sources* **210** 297-302
[6] Shahua H, Zhaoyin W, Xiujian Z and Zuxiang L 2007 *J. Power Sources* **165** 408-12
[7] Jung S P, Seong H B, Yong I J, Bum Y N and Jae H K 2013 *J. Power Sources* **244** 527-31
[8] Priyono S, Triwibowo J and Prihandoko B 2015 *AIP Conf. Proc.* vol 1711 (New York: AIP Publishing) 060001
[9] Honghe Z, Jing L, Xiangyun S, Gao L and Battaglia V S 2012 *Electrochim. Acta* **71** 258-65
[10] Nordh Tim. 2013 *LTO as An Anode Material for Li Ion Batteries In Situ XRD and XPS Studies* Master Thesis (Sweden: Uppsala Universitet)
[11] Bruno S, Jurgen G 2010 *J Power Sources* **195**(9) 2419-30
[12] Feng N, Yan B H, Baohua L, Hongda D, Dengyun Z and Feiyu K 2012 *J. Alloys Comp.* **513** 524-9

IOP Conf. Series: Journal of Physics: Conf. Series **817** (2017) 012017 doi:10.1088/1742-6596/817/1/012017

Pi-pi* orbital transitions and photo-degeneracy of *C.acuminata* sensitized solar cells

T Abodunrin[1], A Boyo[2], M Usikalu[1], L Obafemi[1], O Oladapo[1], L Kotsedi[3], Z Yenus[3] and M Maaza[3]

[1]Covenant University, Canaanland, Nigeria
[2]Lagos State University, Ojo, Nigeria
[3]iThemba Labs, Western Cape, South Africa

E-mail: Temitope.Abodunrin@covenantuniversity.edu.ng

Abstract. Dye-sensitized solar cells (DSSCs) have acquired great prominence as favourable low-cost photovoltaics due to their ease of fabrication, all- year -availability, ease of obtaining raw materials and adjustable optical properties like transparency and colour. These advantages coupled with the ability to work under poor lighting makes them a suitable candidate for next generation of research. In this research, *C.acuminata*-sensitized photo anodes play an important role for achieving high performance since the porous metal oxide films provide a large specific surface area for dye loading and the possibility to extend the absorption threshold of past studies of sensitizers. The doctor blade method and high-temperature sintering were some of the methods used in the fabrication of the photo anode. A study of the performance of the *C.acuminata*-DSSCs with four different electrolyte sensitizers based on iodide redox mediator is determined. The result is DSSCs that exhibit a maximum power output of 39.37 W, fill factor of 0.7 and a power conversion efficiency of 0.6% under unfavourable sunlight intensity conditions and photo-degradation of about 37.5 % in absorbance after 425 suns.

1. Introduction

The Kola nut (Kola) a tropical African genus belonging to the *Sterculiaceae* family, has about one hundred and twenty-five species. The most prominent among the species are Cola acuminata (*C.acuminata*) and Cola nitida (*C.nitida*) due to their economic importance. Both are planted for their edible seed, *C.acuminata* is polycotyledonous while *C.nitida* is a dicotyledon fruit. *C.acuminata* and *C.nitida* contain caffeine, kolanin, and theobromine which make them natural stimulants. Aside from its chemical constituent, Kola is rich in phenol thus, it is an astringent.

These two species bear a striking resemblance, however, *C.nitida* is of higher commercial importance due to the higher demand both locally and export [1]. The molecular formula of *C.acuminata* is $C_8H_{10}N_4O_2$ this provides details of spontaneous redox chemical reactions. Since the first report of DSSC in 1991, remarkable progress has been recorded through the use of electrolytes, photo anode materials and choice of a sensitizer. The sensitizer has an influence on the power conversion efficiency of the DSSC and is crucial in the determination of the stability of the cell thus; this property is the objective of our inquiry. All through the developmental process, practical reforms using different dyes have been employed, but only ruthenium polypyridyl complexes gave reasonable outcomes. Ruthenium dyes, however, require elaborate purification during their preparation process and are very expensive. In order to avoid this setback, research focus has shifted to easily accessible and cheaper light harvesting molecules. All plants possess systems that harvest light using a conjugation of chlorophyll molecules for competent energy conversion. In natural chlorophylls are derivatives of modified bacteriochlorin units, this is the principle underlying several inorganic dyes which have been based on TiO_2 as photo anode in DSSC [2]. In this work, we report DSSC studies of organic *C.acuminata* dye based on TiO_2 photo anode. This framework is modified by amino groups bonding firmly with the TiO_2 matrix. In order to enhance light –harvesting chromophores, iodine base is added to four different DSSCs. Iodine redox reactions such as (I/I^-, I/Br^-, I/MnO^2, I/Cl^-), displace the amino functional group in a substitution to produce suitable and efficient DSSCs. The photoelectric parameters give the I-V plot, the incident photon to conversion efficiency (IPCE), maximum power (P_{max}), fill factor (FF) for the different electrolytes [4].

2. Materials and Method

C.acuminata leaves were harvested and air-dried until it assumed constant weight. The mass of the leaves was determined with Ohaus electronic balance before milling. This solid mass is aerated for 3-4 hours to discourage bacterial growth. 71 Kg of crushed leaves was soaked in 2,500 ml of methanol for 11 days. Liebig rotary evaporator was used to concentrate the *C.acuminata* dye for further use. Dye-sensitized solar cells were made by the doctor blade method of applying TiO_2 Anatase paste purchased from Solaronix onto indium doped tin oxide (TCO) glass substrate from Sweetmans Ltd of specification (75 mm X 25 mm) and resistivity 10 Ω/m^2. The wet TiO_2 paste was air dried for 1h, after which the active area of exposure (25 mm X 4.5 mm) was sintered at a temperature of 300^0C for 1h until TiO_2 cake turns brown and then white again. Initial preparation steps included; cleaning ITO in detergent solution and rinsing with distilled water. Concentrated HNO_3 was blended uniformly in TiO_2, in the molar ratio of 5:3, respectively, until the consistency of TiO_2 paste is like cake icing. Then the resulting colloidal suspension was applied to the conductive side of the ITO. The counter anode was coated with a thick layer of soot over a naked Bunsen flame. This procedure was repeated for KBr, KI, $KMnO_4$, and $HgCl_2$ electrolyte respectively [4].

The Genesys spectrophotometer (10S series) was used to obtain the UV/Vis spectrographs. The molar concentration of the sample was 1:100 g of *C.acuminata* dissolved in methanol of 99.8 % purity. Shimadzu FTIR spectrophotometer was used to determine the values for the Fourier transform in infrared within a range of 491-3821 cm^{-1}. The molar ratio used for the analysis was 1:100 [3,4].

3. Result and discussion

3.1. Phytochemical screening

C.acuminata reveals the presence of tannins, alkaloids, steroids, volatile oils, glycosides, saponin and cardiac glycosides. Alkaloids, tannins, steroids, glycoside and saponin glycosides high presence suggests a strong availability of materials for charge transport.

3.2. Photoelectric Result

The highest and lowest values of Isc were observed for KCl and KBr respectively. The same trend is repeated in the V_{oc} and η. However, the fill factor has the highest value of 0.70 for the KBr sensitized DSSC while KCl sensitized solar cell has FF value of 0.23. The V_{oc} in KCl is about 20 % more that obtained from KBr electrolyte as shown in Table 1. In this experiment, it is observed that the Cl^- is preferentially discharged for charge transport; this is due to favourable reaction with the iodine from I^-. Although the quality of The KCl cell is comparatively less than the KBr, the efficiency is higher by over 30 %. The factors responsible for the low FF could be the leaking electrolyte creating a high internal resistance and several trap and recombination sites within the titania framework consequently lowering the quality of the KCl/DSSC figure 1 in comparison with KBr/DSSC. The low values observed for KBr and KI is as a result of unfavourable chemical reaction and differential displacement of Br^- and I^- in the interface.

3.3. UV/Vis Spectroscopy
Figure 2 illustrates reveal wavelength peaks and highest absorbance in the near-ultraviolet region, the highest peak is within 650 to 660 nm wavelength corresponding to an absorbance index of 4.00 a.u; a π to π^* molecular orbital transition, this lies in the blue region of the electromagnetic spectrum. This is characteristic of porphyrin dye in the Q-band [3]. Table 1 shows the Cl^- causes the highest displacement. However, Br^- finds it difficult to overcome the chemical potential barrier but has a better quality (shown by the FF). The Vis region shows a loss in the initial activation energy (A.E) between 575 and 630 nm. A small peak at 625 nm shows a second π to π^* transition for an obvious minimum potential barrier, this indicates a strong C-H bond at this absorption peak (FTIR analysis). In the infrared region, *C.acuminata* reveals no observable spectral response in the far infrared region. Wavelength 525 to 575 nm shows rapid spectral responses which do not actually lead to an orbital jump for molecules of *C.acuminata* for 0-1 a.u, it may, therefore, be a gradual build-up of energy that was eventually responsible for the high peak. The absorbance of *C.acuminata* is about 0.4 a.u in the blue region; least within the infrared region; less than 0.1 a.u as illustrated in figure 2.

3.4. FTIR spectroscopy
C.acuminata contains the C-X bond in peaks 491.6 nm, 580.59 nm, 719.47 nm and 1166.97 nm, the functional group is the alkane, and the suffixes are iodine, bromine, chlorine and fluorine respectively; this affirms the CH_3 group in figure 1. The range of intensity for the four peaks is 22.01; the area indicates a medium appearance in which concentration of fluoroalkanes is highest of the four despite its relatively low intensity.

These represent surfaces for articulation either with other bonds or the Titania framework. Wavelength peaks 837.13 nm, 896.93 nm and 974.08 nm represent vinyl C-H bonds which vary from medium to strong appearance as shown by the intensity; it is higher than the C-X bond. The functional group, in this case, is alkene. Many essential molecules like amino acids and neurotransmitters are amine based with application in drug modelling and preparatory materials for dyes. Depending on how they are attached, they cause the desired lowest unoccupied molecular orbital (LUMO) to highest occupied Molecular Orbital (HOMO) transition. The aldehydes (C=O) that consist of ketones occur at 1722.49 nm are the most important functional group because of their influence of conjugation. Esters and Lactones comprising of derivatives of carboxylic acids appear at a wavelength of 1735.99 nm, occur in association with amides and serve as surfactants and phase transfer catalysts [4].

3.5. Surface Morphology
The framework in which *C.acuminata* dye lies is shown in figure C under an accelerating voltage of 15.0 KV and display magnitude of 500 reading through a diameter of 12.7 mm, the zinc oxide film has a thickness of 200 μm. The leaf-like structures show centers of nucleation and reaction around the interface. Figure 4 shows a contrast, there is a cluster of rods, cones from titanium reacting with the *C.acuminata* with the same sensitizer; KI. This facilitates the π to π^* transition of the methyl group in the visible region and an unidentified or a mixture of various bonds in the ultraviolet region; this could explain the several reaction spots (Figure 3), the summation of their individual Eigen value catalyses into the high-energy peak. Howbeit, charge transport could be impeded by the sites acting as traps or recombination sites resulting in the low output efficiency.

3.6. Photodegeneracy of C.acuminata dye
The rate at which *C.acuminata* decays or degenerate is determined as a function of the decrease in the absorbance of the dye over a given period of study. The rate of degeneracy may be faster or slower due to factors such as; increased ambient temperature, pH, and exposure to ultraviolet radiation. In *C.acuminata* dye, these conditions were kept constant, the only variable was time. The formula used for determining the rate of degeneracy expressed in percentage is given by equation 1.

$$D = \frac{A - A_o}{A} * 100 \quad (1)$$

Where:
D: rate of degeneracy
A: Initial Absorbance
A_0: Final Absorbance
From Figures 2 and 5, the rate of degeneracy is 37.5% in 24 months.

Table 1. Photoelectric parameters of *C.acuminata* with four different electrolytes

Electrolyte	I_{sc} (mA)	V_{oc} (V)	Pmax	η (%)	Fill Factor
KCl	1.666	101.4	39.368	0.566	0.23
KBr	0.112	5.0	0.392	0.018	0.70
KI	0.214	14.5	1.165	0.028	0.38
$HgCl_2$	0.653	11.0	2.242	0.043	0.31

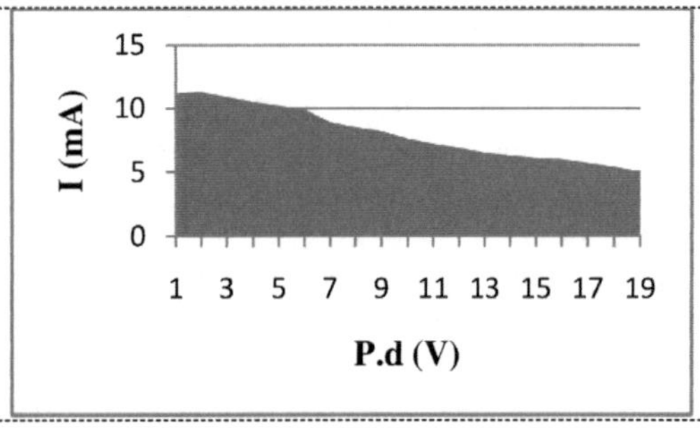

Figure 1. I-V of *C.acuminata*.

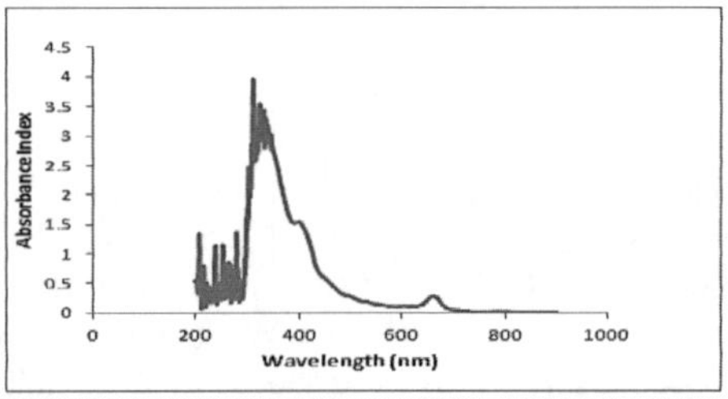

Figure 2. UV/Vis of *C.acuminata* @ Feb.2014.

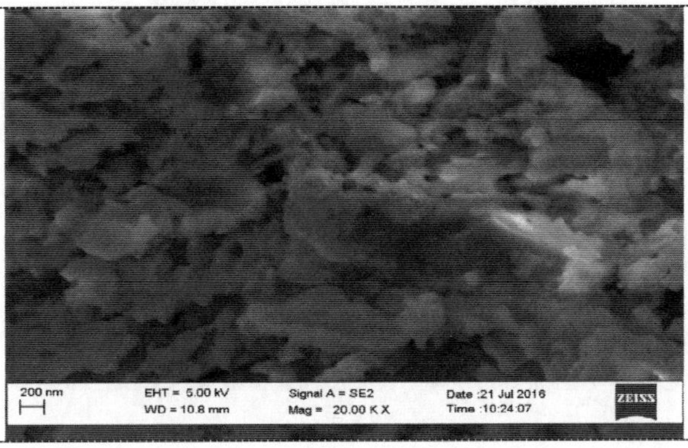

Figure 3. SEM Micrograph of *C.acuminata* on ZnO at depth of 1.0 nm.

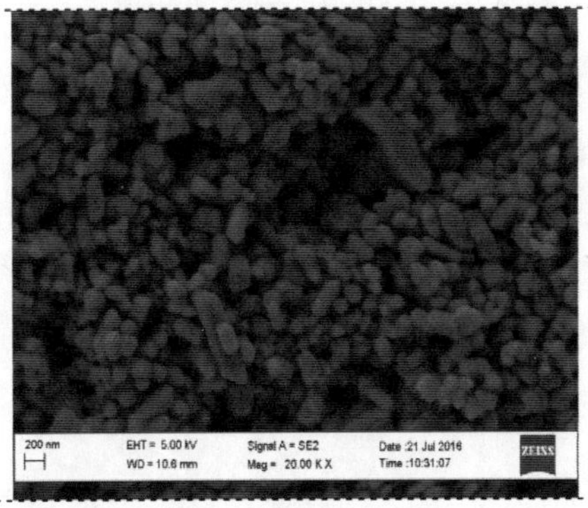

Figure 4. SEM Micrograph of *C.acuminata* on ZnO at depth of 1.0 nm.

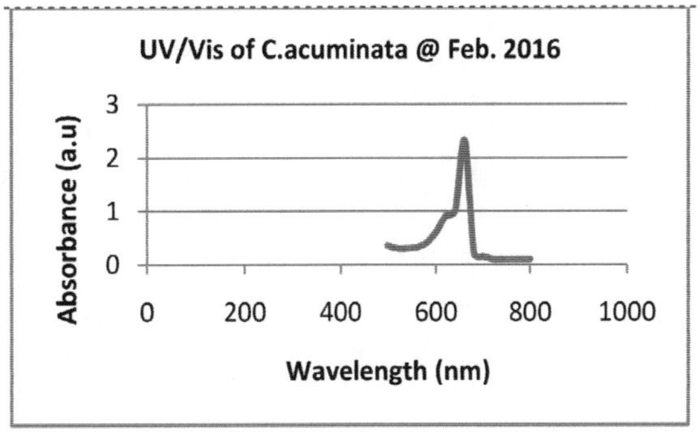

Figure 5. UV/Vis of C.acuminata @ Feb. 2016.

Table 2. The Fourier Transform in Infrared (FTIR) of *C.acuminata* in charge transport.

No.	Peak	Type of bond	Intensity	Corr. Inte	Base (H)	Base (L)	Area	Corr. Are
1	491.86	Iodoalkanes	82.077	0.731	493.79	482.22	0.955	0.033
2	580.59	Bromoalkanes	80.298	0.274	632.67	578.66	4.464	0.096
3	719.47	Chloroalkanes	76.334	4.079	727.19	690.54	3.055	-0.025
4	837.13	Para-disub. benzene	74.529	8.374	871.85	786.98	8.29	1.539
5	974.08	Monosub. alkenes	62.335	9.337	995.3	925.86	10.442	1.281
6	1082.1	Aliphatic amines	64.058	3.305	1103.32	1057.03	8.316	0.409
7	1166.97	Fluoroalkanes	60.067	7.64	1203.62	1132.25	13.844	1.738
8	1247.99	Ethers	58.769	8.093	1292.35	1217.12	14.709	1.64
9	1722.49	Aldehyde	45.794	7.877	1728.28	1683.91	10.039	0.799
10	1735.99	Esters and lactones	49.3	4.239	1830.51	1730.21	10.793	-6.087
11	2727.44	Ammonium ions	69.274	1.94	2750.58	2715.86	5.338	0.196
12	3421.83	Primary amines	47.622	0.82	3429.55	3417.98	3.678	0.031
13	3821.11	Secondary amines	84.135	5.153	3824.97	3813.4	0.665	0.1

Acknowledgments
This work was supported financially by the Covenant University Center for Research Innovation and Development Grant (CUCRID).The authors also appreciate the staff and technologists in Physics and Biochemistry laboratories of Covenant University for their assistance in this research.

References

[1] Nyamien Y, Adje F, Niamké F, Chatigre1 O, Adima A and Biego G H 2014 *Br. J. Appl. Sci. Technol.* **4** 4846-59
[2] Gratzel 2007 *Phil. Trans. R. Soc.* **365** 993-1005
[3] Jasim K E and Hassan A M 2009 *Int J Nanomanuf* **4** 242-47
[4] Boyo A, Kesinro O, Boyo H and Oluwole S 2014 *World Acad. Sci. Eng. Technol.* **8**(6) 924-6

IOP Conf. Series: Journal of Physics: Conf. Series **817** (2017) 012018 doi:10.1088/1742-6596/817/1/012018

Surface modification of ZSM-5 zeolite: effect of cation on selective conversion of biomass-derived oil

W B Widayatno

Research Center for Physics, Indonesian Institute of Sciences, Bld. 440/442, PUSPIPTEK Serpong, Tangerang Selatan 15314, Indonesia

Emails: wahy012@lipi.go.id; wahyubw@gmail.com

Abstract. This paper reports the surface modification of high silica ZSM-5 zeolite, particularly emphasizing the effect of cation type on selective conversion of biomass-derived oil. XRD spectra of the NaOH-treated HZSM-5 showed notable crystallinity decrease at specific crystal plane orientation. The N_2-physisorption tests confirmed mesoporosity evolution as NaOH concentration was increased. NH_3-desorption tests revealed a significant change on surface acidity which involved realumination and cation replacement processes. The utilization of untreated HZSM-5 as well as hierarchical NaZSM-5 for catalytic conversion of bio-oil showed the effect of cation type and mesoporosity on chemicals distribution. The untreated HZSM-5 showed high selectivity to aromatics, which degraded gradually due to deactivation and poisoning of the acid sites. Meanwhile, hierarchical NaZSM-5 showed high selectivity to phenolic compound, which became more stable for 0.4M NaOH-treated zeolite (Na04). The current findings provide an additional insight on the potentials of NaZSM-5 for bio-oil valorization.

1. Introduction

Bio-oil derived from biomass has been targeted as one of top issue in biomass utilization. Different approaches such as catalytic cracking over different zeolites [1], utilization of basic catalysts [2], hydrodeoxygenation [3], and hydrothermal processes [4] have been investigated to get more hydrocarbon from biomass conversion for fuel application. However, since bio-oil contains many chemicals compound, recovering a significant amount of certain chemicals from bio-oil should be more beneficial for many applications, rather than removing them to extract only the hydrocarbons.

Recently, some studies have already used new directions to get more valuable chemicals feedstock [5, 6] such as furfural and acetic acid [5], phenols [7, 8], succinic acid [9], alkanes and polyols [10] for various purposes, instead of only for fuel applications. As a material with high surface area and shape selectivity properties, zeolite has been used as a catalyst or catalyst support to produce certain kinds of chemicals [5, 11]. The microporosity of zeolite serves the separation of molecules based on their molecular size. However, it also induces transport limitation which might hinder complete decomposition of large molecule like sugar [12]. This can be overcame by controlling the crystal size [13], making zeolite composite with other mesoporous material [14], steaming [15], dealumination [15], and desilication [16].

Most zeolite studies for catalytic conversion of bio-oil, always used H-form (protonated) of low silica zeolite to get more hydrocarbons. Some post-treatment studies of zeolite e.g. desilication, also converted the treated zeolite into protonated or H-form [17-19] to obtain the active form for increasing the hydrocarbons yield. Meanwhile, the initial treated form e.g. alkali form-zeolite is rarely investigated. In addition, most of the existing studies only tested the performance of the fresh treated-zeolite. Other study considered the reusability of zeolite after thermal regeneration [17]. Yet, the catalytic activity of zeolite was degraded due to irreversible deactivation after repeated regeneration [20]. Thus, ensuring a good and stable catalytic activity without excessive regeneration cycle should be beneficial for practical purposes.

This study was aimed to elaborate the effect of surface cation on zeolite's surface, specifically the cation from post-treatment of zeolite, on the chemical distribution of converted bio-oil, as well as the reusability potential of it without thermal regeneration. In this study, the high silica ZSM-5 type zeolite was desilicated by alkali solution (NaOH) to increase its mesoporosity, and used in its alkali form as catalyst for selective conversion of bio-oil. For comparison, the performances of untreated HZSM and H-form of ZSM-5 after NaOH treatment were also evaluated.

2. Materials and Methods

2.1. Catalyst preparation and characterization

HZSM-5 zeolite (HSZ-890) from TOSOH Corporation with SiO_2/Al_2O_3 of 1500 was used without further treatment as a parent material. The desilication was conducted through chemical washing in 0.1, 0.2, and 0.4 M NaOH solution at stirring temperature of 50 °C for 30 mins to get Na-form zeolite. After that, the zeolite was washed using distilled water until neutral and dried overnight in drying oven at 110 °C. The parent and washed zeolites were then denoted as untreated, Na01, Na02, and Na04 for NaOH concentration of 0, 0.1, 0.2, and 0.4, respectively.

The structure of untreated and NaOH-treated zeolites were compared by X-ray Diffraction (Smartlab, Rigaku, Japan). The pore properties were examined using N_2 physisorption measurements (NOVA 4200e, Quantachrome Instruments, USA). The surface area of the samples were determined using Brunauer-Emmet-Teller method (BET), while the micropore surface area and volume were calculated using t-plot method from the isotherm profile. The acid site of the parent and post-treated zeolites were evaluated by comparing the NH_3-desorption profile of the samples with the calibration data in the machine (Belcat, Japan). The details of the measurement preparation can be found elsewhere [21].

2.2. Catalytic activity and selectivity

The biomass feedstock used for this experiment was the stem part of Japanese knotweed (Fallopia Japonica). Details of the feedstock properties and preparation step have been described elsewhere [22]. Pyrolysis and catalytic conversion were conducted simultaneously in a down-draft fixed bed reactor using the same arrangements as the previous study [21]. The experiment was carried out from room temperature to 600 °C with a heating rate of 80 °C/min, and held at that temperature for 30 min. In addition to the fresh one, the zeolite beds were also reused three times without regeneration to elaborate their reusability.

The chemical composition of liquid product was evaluated by Gas Chromatography and Mass Spectroscopy method (GC-2010 Plus and GCMS-QP2010 Ultra, Shimadzu Japan). The built-in NIST database was used as a reference to determine the chemicals in the bio-oil. The relative yield of each chemical class was calculated from the GC-MS peak area using the following formula:

$$\text{Rel. yield} = \frac{\text{specified peak area}}{\text{total peak area}} \times \frac{\textit{catalytic oil yield}}{\textit{non-catalytic oil yield}} \quad (1)$$

3. Results and Discussion

3.1 Structural properties of the zeolites

Some reports mentioned that for zeolite with Si/Al > 50, desilication treatment will lead to excessive dissolution of Si framework and the loss of intrinsic crystallinity [16, 23, 24]. Figure 1 presents the XRD profile of the untreated and treated zeolite which shows a similar result with the previous works. The loss of crystallinity was only observed for (101) crystal plane at a diffraction angle of 8° which became more significant as the NaOH concentration was increased. It was reported that for high silica zeolite, the extraction of Si was more effective from the crystal surface zone [23]. Thus, the decrease of peak at (101) crystal plane was correlated with the extraction of Si in the external surface of zeolite.

Table 1 presents the area-volume summary of the untreated and treated zeolite. It can be seen that the extraction of Si from the surface zone increased the mesopore surface area and gave more access to the deeper active site after NaOH treatment. As a result, its mesopore volume also increased to some extent.

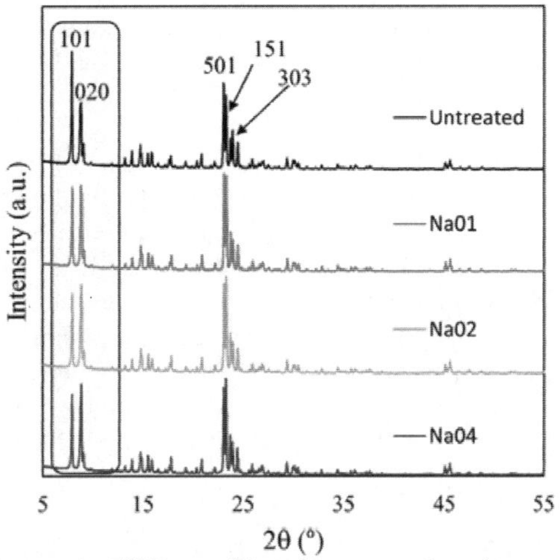

Figure 1. XRD profile of untreated and treated zeolite

Table 1. Area-Volume summary of the untreated and treated zeolite

Catalyst	S_{BET} (m^2/g)	S_{Micro} (m^2/g)	S_{Meso} (m^2/g)	V_{Total} (cm^3/g)	V_{Micro} (cm^3/g)	V_{Meso} (cm^3/g)
Untreated	308	279	29	0.19	0.14	0.05
Na01	335	300	35	0.22	0.15	0.07
Na02	325	279	46	0.23	0.14	0.09
Na04	331	271	60	0.26	0.14	0.12

It should also be noted that the BET surface decreased when NaOH concentration was increased to 0.2M, yet the mesopore surface area increased due to the extent of Si extraction from the framework which led to the enlargement of some old micropores. As a result, the neighbor micropores should collapsed and give extensive decrease of micropore surface area. When NaOH concentration was increased to 0.4M, the BET and mesopore surface area increased further, while the micropore surface area decreased slightly. It is highly possible that NaOH treatment only led to the enlargement of the existing micropores in the surface zone and gave less effect to the inner micropore channels, which contributes most of zeolite pore volume. The increase of mesopore volume is highly possible due to the increase of external surface roughness [25] or formation of new mesopore structure from the extracted Si and/or Al in the surface zone.

Further investigation by Groen et al. concluded that macropores were formed after NaOH treatment of high silica zeolite [26]. The opposite phenomenon was found in this experiment, as depicted in figure 2. The isotherm profiles from N_2 adsorption test of the NaOH-treated show clear combination of type I isotherm, which is denoted by the high uptake of N_2 at relatively low pressures, and type IV isotherm, which is denoted by the hysteresis of H4 type due to multi-layer adsorption in the range of 0.5 – 0.99 [27]. Type I isotherm shows the microporosity nature of ZSM-5, while the type IV isotherm confirms the existence of mesoporosity. There is no sign of reversible sorption properties as should be observed for non-porous or macroporous material (type II isotherm), which explain the further increase of mesopore volume while retaining its micropore volume. The different phenomenon compared with the works by Groen et al., can be attributed to the difference of the utilized treatment temperature which led to progressive extraction of Si and Al [23]. A similar explanation has been mentioned in other literatures [28, 29]. The NaOH treatment in this work was conducted at a lower temperature compared with the earlier work. Thus, the rate of Si extraction should be different and the extent of NaOH effect, which led to uncontrolled Si extraction, might be suppressed in the current experiment.

3.2 Acid properties of the zeolites

The NH_3-TPD profile of the zeolites in figure 3 presents the evolution of weak acid site (Low temperature peak, below 400 °C) and the significant decrease of stronger acid site (high temperature peak, above 400 °C). Along with the Si extraction, some of Al in the surface region will be also released from the framework and subsequently reinserted as Al extraframework on the external surface of zeolite [23, 26], which is shown by the significant decrease of high temperature peaks and the increase of low temperature peaks. This Al extraframework can serve as weak Lewis acid site in the surface region. Since the as-treated ZSM-5 was used directly without further activation, the dehydroxylation of Al extraframework into strong Lewis acid site, which favors coke formation, can be avoided.

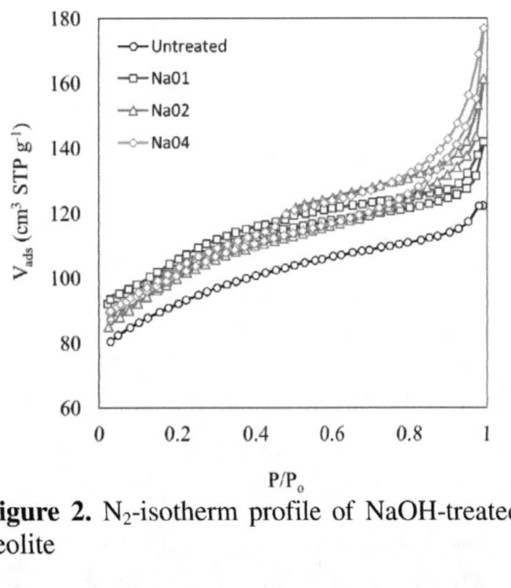

Figure 2. N_2-isotherm profile of NaOH-treated zeolite

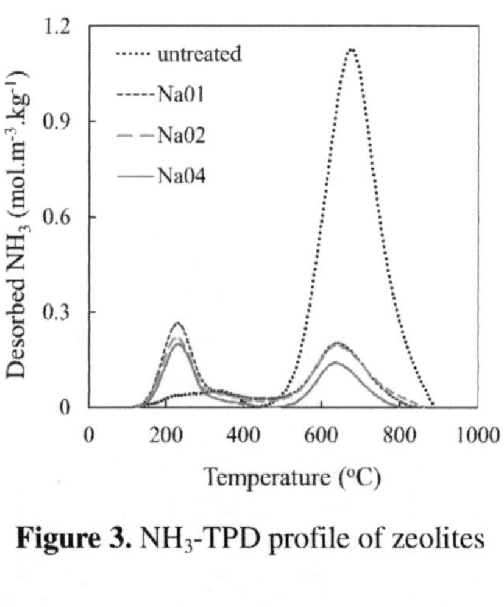

Figure 3. NH_3-TPD profile of zeolites

3.3 Catalytic activity and selectivity of the zeolites

Figure 4 presents the chemical distribution of bio oil after being converted. For the untreated HZSM-5, the highest product selectivity was aromatics compound (≈48%), as shown in figure 4(a), while the unwanted carboxyl compound can be reduced significantly. Since the dominant content of carboxylic compound in bio oil is acetic acid, the significant decrease of carboxyl compound is highly possible due to the decomposition of acetic acid which can poison the acid sites [30]. The poisoning of acid sites and the fact that HZSM-5 has high coking tendency of hydrocarbons due to its smaller channel resulted to deactivation of acid sites.

After 0.1 M NaOH treatment, some stronger acid sites in the inner structure of ZSM-5 were still remained which enabled strong adsorption of reactant and retain the conversion into aromatics. After several times being reused, those remaining stronger acid sites became inactive gradually due to hydrocarbon deposition and/or acetic acid poisoning. The deactivation of stronger acid sites led to the decrease of phenols dehydroxylation reaction and resulted to decrease of phenols conversion. Figure 4(b) shows the decreasing trend of aromatics when using Na01 catalyst, which implied the existence of remaining strong acid sites and their deactivation after several times reused.

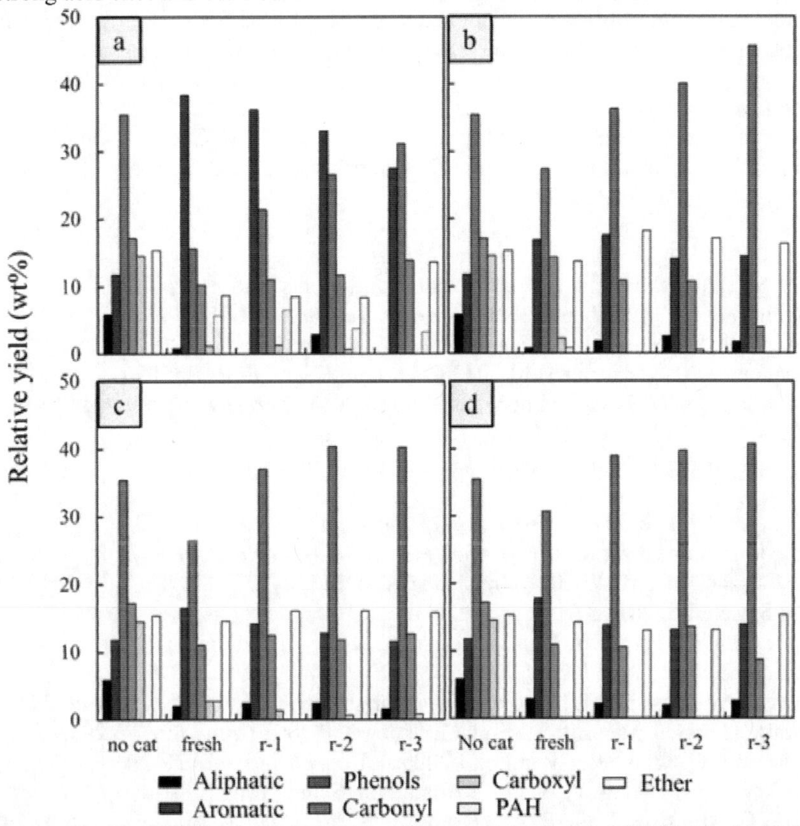

Figure 4. Chemical distribution of upgraded bio oil using (a) untreated HZSM-5, (b) Na01, (c) Na02 and (d) Na04.

Unlike Na01, Na02 catalyst showed different phenomenon, which is presented in figure 4(c). Phenols selectivity reached its maximum after the second reuse (r-2) and did not increase furthermore, while aromatics selectivity showed the opposite, due to the extent of NaOH treatment effect. As a result, more OH- concentration dissolved the extra framework in the surface region further, and the weak acid sites decreased.

When Na04 catalyst was used, no carboxyl peaks detected, neither for fresh catalyst nor also for the reused one, as shown in figure 4(d). Furthermore, the chemicals distribution showed the relatively stable selectivity for each chemical class after the first reuse which infers the increase of coking tolerance. It is then suggested that Na04 catalyst has the best performance, in terms of selectivity and stability over several times reuse. Such high selectivity of phenolic compound has also been reported by Bu et al. [7, 31]and Choi et al. [8]. However, the reusability of the system in this experiment showed some improvement.

4. Conclusions

Desilication of high silica ZSM-5 zeolite and its effect on the performance of selective phenols production derived from bio-oil have been elaborated in details. NaOH treatment on high silica ZSM-5 decreased the crystallinity at specific crystal plane orientation. The acid sites distribution changed due to realumination and cation replacement processes, resulted to the decrease of strong acid sites and evolution of weak acid sites. Untreated HZSM-5 showed high selectivity to aromatics, yet degraded gradually due to deactivation of acid sites. Meanwhile, NaZSM-5 showed high selectivity and reusability to phenolic compound, especially for Na04 catalyst. This shows that different surface cation on zeolite will produce different chemical distribution. These findings also provide information on the potentials of NaZSM-5 for bio oil valorization to obtain specific chemicals.

5. Acknowledgement

The author would like to thank Prof. Abuliti Abudula and Prof. Guoqing Guan from Hirosaki University, Japan and Dr. Jenny Rizkiana from Bandung Institute of Technology (ITB), Indonesia for helpful discussion during the experiment.

References

[1] Vitolo S, Seggiani M, Frediani P, Ambrosini G and Politi L 1999 *Fuel* **78** 1147
[2] Xu J, Jiang J, Sun Y and Chen J 2010 *Bioresour. Technol.* **101** 9803
[3] Wang Y, He T, Liu K, Wu J and Fang Y 2012 *Bioresour. Technol.* **108** 280
[4] Kruse A, Funke A and Titirici M M 2013 *Curr. Opi. Chem. Biol.* **17** 515
[5] Leng S, Wang X, Cai Q, Ma F, Liu Y and Wang J 2013 *Bioresour. Technol.* **149** 341
[6] Zhang X S, Yang G X, Jiang H, Liu W J and Ding H S 2013 *Sci. Rep.* **3** 1120
[7] Bu Q, Lei H, Wang L, Wei Y, Zhu L, Liu Y, Liang J and Tang J 2013 *Bioresour. Technol.* **142** 546
[8] Choi G G, Oh S J, Lee S J and Kim J S 2015 *Bioresour. Technol.* **178** 99
[9] Akhtar J, Idris A and Aziz R A 2014 *Appl. Microbiol. Biotechnol.* **98** 987
[10] Vispute T P and Huber G W 2009 *Green Chem.* **11** 1433
[11] Li Z and Savage P E 2013 *Algal Res.* **2** 154
[12] Lee H W, Choi S J, Park S H, Jeon J K, Jung S C, Joo S H and Park Y K 2014 *Energ.* **66** 2
[13] Tosheva L and Valtchev V P 2005 *Chem. Mat.* **17** 2494
[14] Egeblad K, Christensen C H, Kustova M and Christensen C H 2008 *Chem. Mat.* **20** 946
[15] Van Donk S, Janssen A H, Bitter J H and De Jong K P 2003 *Catal. Rev.* **45** 297
[16] Groen, Moulijn J A and Perez-Ramirez J 2006 *J. Mat. Chem.* **16** 2121
[17] Li J, Li X, Zhou G, Wang W, Wang C, Komarneni S and Wang Y 2014 *Appl. Catal. A* **470** 115
[18] Gayubo A G, Alonso A, Valle B, Aguayo A T and Bilbao J 2010 *Appl. Catal. B* **97** 299
[19] Graça I, Carmo A M, Lopes J M and Ribeiro M F 2015 *Fuel* **140** 484
[20] Vitolo S, Bresci B, Seggiani M and Gallo M G 2001 *Fuel* **80** 17
[21] Widayatno W B, Guan G, Rizkiana J, Du X, Hao X, Zhang Z and Abudula A 2015 *Bioresour. Technol.* **179** 518
[22] Widayatno W B, Guan G, Rizkiana J, Hao X, Wang Z, Samart C and Abudula A 2014 *Fuel* **132** 204
[23] Sadowska K, Góra-Marek K, Drozdek M, Kuśtrowski P, Datka J, Triguero J M and Rey F 2013 *Micropor. Mesopor. Mat.* **168** 195
[24] Verboekend D and Pérez-Ramírez J 2011 *Catal. Sci. Technol.* **1** 879
[25] Abelló S, Bonilla A and Pérez-Ramírez J 2009 *Appl. Catal. A* **364** 191
[26] Groen J C, Peffer L A, Moulijn J A and Perez-Ramirez J 2005 *Chemistry* **11** 4983
[27] Wang D, Zhang L, Chen L, Wu H and Wu P 2015 *J. Mat. Chem. A* **3** 3511

[28] Groen J C, Peffer L. A A, Moulijn J A and Pérez-Ramírez J 2004 *Colloid Surf. A* **241** 53
[29] Ogura M, Shinomiya S Y, Tateno J, Nara Y, Nomura M, Kikuchi E and Matsukata M 2001 *Appl. Catal. A* **219** 33
[30] Ramasamy K K, Gerber M A, Flake M, Zhang H and Wang Y 2014 *Green Chem.* **16** 748
[31] Bu Q, Lei H, Ren S, Wang L, Zhang Q, Tang J and Ruan R 2012 *Bioresour. Technol.* **108** 274

Improved properties of LiBOB-based solid polymer electrolyte by additive incorporation

C Ratri, Q Sabrina, T Lestariningsih and E Wigayati
Research Center for Physics, Indonesian Institute of Sciences
Kawasan Puspiptek Serpong Gd. 442 Tangerang Selatan, Banten 15314, Indonesia

E-mail: christin.rina.ratri@lipi.go.id

Abstract. Solid polymer electrolytes comprising of poly(vinylidene fluoride) (PVdF) and lithium bis (oxalato) borate (LiBOB) have been prepared using solution casting technique. Having an important role in lithium-ion battery system, electrolyte is required to have high ability to transfer lithium ions between electrodes. Safety aspect is the main reason for the development of solid polymer electrolyte as advancement from conventional liquid electrolyte. Nevertheless, solid polymer electrolyte generally has lower conductivities compared to liquid electrolyte. In this research, ceramic additives, as well as plasticiser materials, have been incorporated within the solid polymer electrolyte system to improve its conductivity. Addition of TiO2 filler has proven to increase ionic conductivity by two orders of magnitude. Further improvement was seen in the incorporation of PEG plasticiser, where ionic conductivity was enhanced by three orders of magnitude.

1. Introduction

A particular urgency has been raised from the rapid advancement of electrical/electronic technologies. Portable electronic devices are in requirement of energy storage system in order to sufficiently power them. Lithium-ion batteries have been widely developed since the early 1990's because of its high capacity and notably longer shelf-life compared to the predecessors such as NiMH and NiCd batteries.

A lithium-ion battery consists of positive and negative electrode where intercalation of lithium ions and exchange of electron takes place. A separator exists between those electrodes to prevent short-circuit. The movement of lithium ions is governed by electrolyte, which acts as a transfer media. Early implementation of lithium-ion battery used liquid electrolyte, i.e electrolyte salt dissolved in carbonate-based solvent.

Along the way, it has been discovered that liquid electrolyte in the lithium-ion battery system possessed disadvantages [1–3] such as the potential of solid electrolyte interface (SEI) layer formation caused by reaction which decreases intercalation potential of Li-ion between electrodes. In addition, reliability in high temperature is highly required because lithium-ion battery is widely used in heated environments such as electric vehicle or oil and gas application.

Solid polymer electrolyte system employs polymer host as a solid solvent for metal cations. Electrolyte salts and additive materials are dissolved in the polymer matrix. In this experiment, poly(vinylidene fluoride) or PVdF was used as the main polymer matrix because of its relatively high dielectric constant [4] compared to other polymers. In the early years of the development of polymer electrolyte, Wright et.al [5] utilised poly(ethylene oxide) or PEO as the matrix due to its desired properties i.e proper blending and good mechanical characteristics. These properties were found to be in accordance with research conducted by M. Jacob [6], whom experiment yielded in conclusion that blending of PVdF and PEO as a polymer matrix in solid polymer electrolyte system resulting in 80:20 composition as optimum ratio. Proper blending, good plasticity, and free-standing film was obtained. Any other composition resulted in unstable and fragile thin film. In this experiment, incorporation of additive materials [7] was observed to understand how they affect crystallinity level and ionic conductivity of solid polymer electrolyte system.

2. Experimental methods

2.1. Materials

Electrolyte salt lithium bis(oxalato) borate (LiBOB, Sigma Aldrich), was used after heating treatment to remove water. Poly(vinylidene) fluoride (PVdF, Sigma Aldrich), was used as polymer matrix. Interchangeably, polyethylene glycol, (PEG, Merck), and polyethylene oxide (PEO, Merck), was used in addition to PVdF. Additive materials titanium oxide (TiO_2, Merck) and aluminium oxide (Al_2O_3, Merck), were used without further purification. As the solvent, N,N-dimethylacetamide (DMAC, Merck) was utilised as the other materials have excellent solubility in DMAC compared to other solvents such as N-methyl pirrolidone, NMP, or dimethyl furan, DMF.

2.2. Sample preparation

Solid polymer electrolyte sample was prepared using solution cast technique. Polymer materials were dissolved in DMAC using magnetic stirrer on a hot plate. PVdF:LiBOB concentration was fixed throughout this investigation while additive content was varied as shown in table 1. Electrolyte salt and additive materials were subsequently added into the polymer solution until homogeneously mixed and a clear slurry has been formed. The slurry was cast onto a glass substrate and air-dried in atmospheric condition until a polymer membrane was obtained. To remove any possible remaining solvent, the membrane was heated in a vacuum oven for 24 hrs in 50°C temperature. A small section of the membrane was cut out for characterisation, then the rest of the sample was stored in a dry box with less than 40% relative humidity (RH) to maintain its properties.

Table 1. Sample Preparation of Solid Polymer Electrolyte using Various Additives.

	Content
Sample A	PVdF + PEO + LiBOB + TiO_2
Sample B	PVdF + PEO + LiBOB + Al_2O_3
Sample C	PVdF + PEG + LiBOB
Sample D	PVdF + PEO + LiBOB
Sample E	PVdF + PEO
Sample F	PVdF + LiBOB

2.3. Sample characterisation

Surface morphology was observed using Scanning Electron Microscopy (SEM) Hitachi SU-3500. To preserve sharpness and focus, low acceleration voltage of 5 kV was used. However, 20 kV acceleration voltage was employed while performing Energy Dispersive X-ray Spectroscopy (EDS) to examine element distribution. Phase shifting and change, as well as crystallinity level, was observed using X-Ray Diffractometer (XRD) analysis, using CuKα radiation between 2θ of 10° to 50°. Room temperature conductivity of the polymer electrolyte membrane was determined from AC-impedance measurement using Hioki 3520 LCR with data acquisition obtained over the frequency range of 1 Hz and 1 MHz. The membrane was sandwiched between two stainless steel plates acting as blocking electrodes.

3. Results and discussion

3.1. SEM

Figure 1 shows the surface morphology of solid polymer electrolyte membrane observed using SEM Hitachi SU-3500. Without salt and additives, surface of the polymer matrix appears to be smooth and flat as shown in figure 1(e). Addition of LiBOB electrolyte salt results in porous membrane across the surface (figure 1(f)).

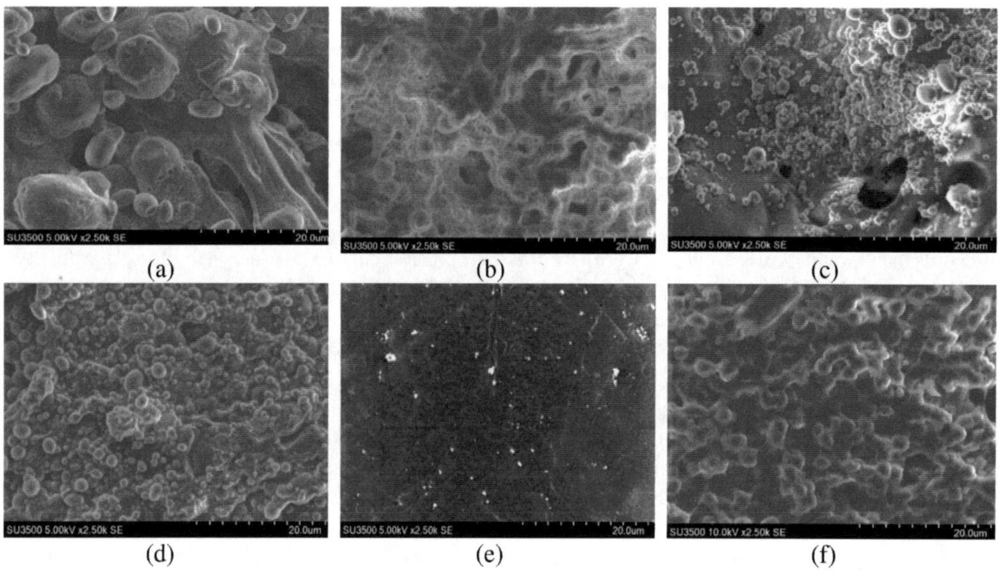

Figure 1. SEM pictures of (a) Sample A, (b) Sample B, (c) Sample C, (d) Sample D, (e) Sample E, and (f) Sample F

TiO_2 addition results in uneven blending of solid polymer electrolyte system, as shown in figure 1(a), in contrast, figure 1(b) shows homogeneously blended Al_2O_3 additive and LiBOB salt in the polymer matrix. The ionic conductivity of the solid polymer electrolyte system can be partly attributed to this phenomena; this would be discussed further in the next section. Surface morphology of sample C and D are presented in figure 1(c) and (d), respectively. The porous surface of the polymer membrane is observed on these samples, with grains of electrolyte salt dispersed throughout the layers.

3.2. XRD

Phase identification was performed using PDXL software; beta-PVdF phase was identified in all samples, as observed on 2θ of 20.47° and hkl of (220).

The prominent diffraction peak of porous film with TiO_2 and Al_2O_3 are shown in figure 2(a) and 2(b), respectively, indicating an unsuccessful blending of TiO_2 and Al_2O_3 filler. In addition, it also indicates that the crystallinity levels of both samples are still relatively high, which impact to the ionic conductivity is in good agreement with the EIS measurement. High crystallinity level could hinder the lithium-ion movement between electrodes, hence decreasing the ionic conductivity of the solid polymer electrolyte system [8].

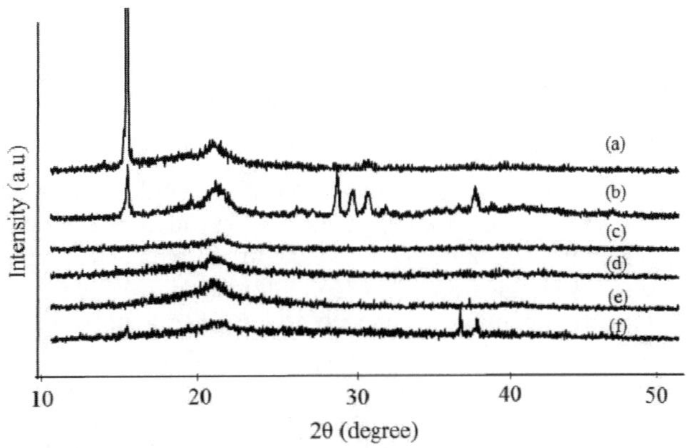

Figure 2. Diffraction pattern of (a) Sample A, (b) Sample B, (c) Sample C, (d) Sample D, (e) Sample E and (f) Sample F

The diffraction patterns of other samples are almost similar, demonstrating that the additive and electrolyte salt has been mixed homogeneously into the porous PvdF substrate (figure 2(c), 2(d), 2(e), and 2(f)). It is also observed that addition of PEG and PEO does not lead to higher crystallinity level of the solid polymer electrolyte system, which is desired properties from these additives.

3.3. EIS

Results of impedance measurement using LCR meter was presented in a Cole-Cole plot. Ionic conductivity of solid polymer electrolyte system was calculated using simple equation $\tau = t / (Rb * A)$, where τ is ionic conductivity, t is membrane thickness, Rb is bulk resistivity, and A is surface area of the membrane in contact with stainless steel plate of LCR meter, functioned as a blocking electrode. The value of Rb was taken from the intercept of the first semicircle on the X axis. Generally, Cole-Cole plot has two parts; first semicircle represents bulk/grain resistance while the linear line represents grain boundary resistance.

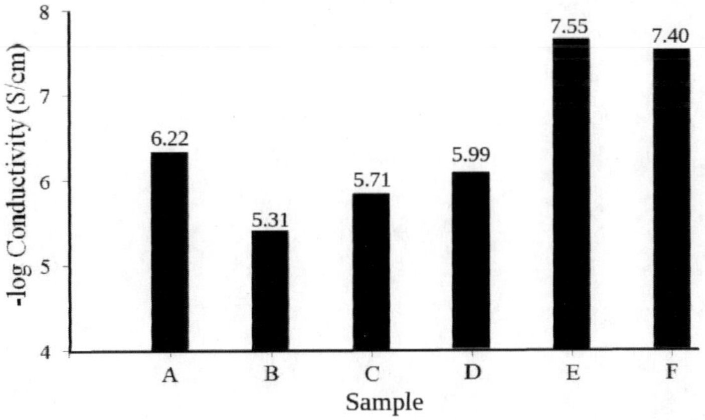

Figure 3. -log ionic conductivity of solid polymer electrolyte film sample

Figure 3 presents value of -log ionic conductivity of all sample. Highest ionic conductivity (lowest -log value) is indicated by sample B followed by sample C, D, A, and F. Naturally, sample E possess the lowest conductivity value as it only contains polymer substrate with neither electrolyte salt nor additive materials. Poorly blended sample A was causing high crystallinity of PEO phase, which in turn resulting in ionic conductivity that was less than expected. Solid polymer electrolyte system using

PEO and PEG without metal oxide additive materials does not appear to show a significant difference in terms of ionic conductivity, in accordance with XRD diffractogram which presents relatively similar pattern of these two samples.

4. Conclusions

Solid polymer electrolyte has been prepared via solution cast technique, producing a free-standing conductive membrane. From the characterisation performed on the sample, it can be concluded that incorporation of additive materials contributes highly to the ionic conductivity of the membrane due to the crystalline-amorphous phase transition. This condition would increase lithium-ion mobility between electrodes, hence improving the ionic conductivity of the solid polymer electrolyte.

Acknowledgements

The authors would like to acknowledge financial support for this project, from Ministry of Research and Technology and Higher Education of Indonesia through Insentif Riset Sinas grant budget year 2015.

References

[1] Li Y H, Wu X L, Kim J H, Xin S, Su J, Yan Y, Lee J S and Guo Y G *J. Power Sources* **244** 234–9
[2] Fasciani C, Panero S, Hassoun J and Scrosati B 2015 *J. Power Sources* **294** 180–6
[3] Johan M R, Shy O H, Ibrahim S, Mohd Yassin S M and Hui T Y 2011 *Solid State Ionics* **196**(1) 41–7
[4] Ma T, Cui Z, Wu Y, Qin S, Wang H and Yan F 2013 *J. Memb. Sci.* **444** 213–22
[5] Murata K, Izuchi S and Yoshihisa Y 2000 *Electrochim. Acta.* **45** 1501–8
[6] Jacob M 1997 *Solid State Ionics* **104** 267–76
[7] Nunes-Pereira J, Costa C M and Lanceros-Méndez S 2015 *J. Power Sources* **281** 378–98
[8] Klongkan S and Pumchusak J 2015 *Electrochim. Acta.* **161** 171–6

IOP Conf. Series: Journal of Physics: Conf. Series **817** (2017) 012020 doi:10.1088/1742-6596/817/1/012020

Activation energy determination in multi-frequency dynamic molecular interaction analysis of PEG 4000-Cristobalite composites using DMA

N A Fauziyah, T A Fadly, A Rosyidi, Mashuri and S Pratapa

Departement of Physics, Faculty of Mathematics and Sciences, Institute of Technology Sepuluh Nopember, Surabaya 60111 Indonesia

E-mail: suminar_pratapa@physics.its.ac.id

Abstract. Degradation activation energy of PEG-Cristobalite composites was investigated using Dynamic Mechanical Analysis (DMA) instrument at multi-frequency shear mode. The applied frequencies were 1, 10, 100, and 200 Hz. The cristobalite of as much as 20% and 40% by weight was used as the filler of the composites, while PEG 4000 was the matrix. Results showed that additional of cristobalite improved the storage modulus (G') of the composites almost five times of the pure PEG 4000. Moreover, such addition and higher applied frequency shifted glass transition temperature (T_g) to higher values. Furthermore, the multi-frequency measurement provided the degradation activation energy of the samples, where the highest value was reached by the 40% sample, i.e. 571.4 kJ. This result implied that there was a certain energetic barrier from the filler that the structural units of polymer need to surmount collectively before its viscous flow occured.

1. Introduction

Research focus in modern polymeric materials includes thermal stability as well as physical and mechanical properties at high temperature. The thermal stability of polymers is connected not only with the initial degradation temperature (T_g) [1], but also, with dislocations due to mechanic and thermal treatment [2]. In particular, the apparent activation energy (E_a) is of great importance aspect of degradation of such materials [3].

DMA technique is useful in characterizing polymer degradation as a function of frequency, temperature, time, stress or a combination of these parameters [4-6]. The dynamic mechanical responses of degradation are usually observed at temperatures near (or, in some cases, above) the melting point. As a result, the lifetime predictions based on these measurements can be determined. DMA response also depends on the physical or structural arrangement of phases such as interface, morphology and the nature of the constituents [4,5]. In particular, the effect of filler, fibre content, fibre orientation in fibre-reinforced composite materials has been extensively studied [7,8].

Cristobalite is a metastable silica polymorph. It exists at room temperature but easily transformed into quartz at high temperature. There is no study on the utilization of cristobalite as a filler in a composite system. This phase is very unique since it exhibits a negative Poisson's ratio [9-12]. Materials with a negative Poisson's ratio (called auxetic material) have a counterintuitive behaviour of becoming wider when stretched and thinner when compressed [12]. This behaviour has many beneficial effects on the materials' macroscopic properties (e.g., increased indentation resistance, a natural ability to form "dome-shaped" surfaces) that make them superior in many practical applications [10-11].

Research on polyethilene glycol (PEG)-based functional polymer-ceramic composites is relatively extensive. For example, SiO_2–PEG–3-isocyanato-propyltrietoxysilane nanocomposite was prepared as hydrophilic films [13]. Qian et. al. [14] also reported green shape-stabilized composite phase change material (PCM) of PEG/SiO_2 with enhanced thermal performance based on oil shale ash via temperature-assisted sol–gel method. They did not, however, determine the crystalline phase of SiO_2 in their studies. In our previous work, we reported the effect of quartz content on the PEG 4000-quartz composites using DMA in shear mode [15]. Quartz is a polymorph of silica which is chemically stable and exhibits positive Poisson's ratio. The dynamic mechanical responses using multi-frequency DMA of such composites are reported elsewhere [16]. However, the multi-frequency DMA measurements for PEG 4000-cristobalite composites have not been investigated yet.

In this report, we describe thermal degradation experiments of PEG 4000-cristobalite composites. The low melting point PEG 4000, i.e. T_m= 60-63 °C [15], [17] was used as the matrix, while cristobalite which was processed from natural silica sand was the filler. The activation energy determination was investigated through DMA measurements in shear mode at various frequency.

2. Experimental

2.1. Sample preparation

Cristobalite powder was prepared by purifying silica sand from Tanah Laut, Pelaihari, South Kalimantan. Polyethylene Glycol 4000 (Merck Schuchardt OHG 85662 Hohenbrunn, Germany) was used as matrix in this composite. Magnetic separation and immersion with HCl 2M was done to purify silica sand and get SiO_2 quartz powder. Next step, precursor quartz powder was copresipited by using NaOH. In order to get SiO_2 crystobalite (SC), copresipited powder was calcined at 950°C under the air for 8 hours. All of those phases were identified by using X-Ray Diffractometer (XRD). The composite samples were prepared by mixing SC powder (as much as 20% and 40% by weight) as the filler in PEG 4000 and then followed by casting in pellet forms. A pure PEG 4000 sample was also prepared for comparison.

2.2. DMA measurement

Composite samples were casted in small-dimension, 1 mm thickness and (5 mm × 5 mm) square which were compatible with the geometry and dimensions for DMA testing in shear mode. The experiment has been described elsewhere [15]. For each composite sample was heated at 50°C to increase samples necking after considering the theoretical melting point of PEG 4000. The mechanical properties of composite samples were measured using DMA Mettler Toledo instrument. From our initial experiments, the range temperature was about 25°C until 80°C with multi-frequency as oscillation responses (1, 10, 100, and 200 Hz). This range required to investigated dynamic responses from composites.

3. Results and discussion

Figures 1, 2, and 3 show comparative storage moduli (G') of the composite samples at various frequencies as a function of temperature, as well as those for pure PEG samples. In general, G' drops as temperature increases. The decline is very sharp around 50-65°C for all samples. A decline in G' is often associated with a reduction in viscosity and resistance of the materials toward deformation. In glassy region (see figure 1) is often related to the toughness with high viscosity. As incresing temperature continues increasing free motion of polymer chain. As heating continues, it will reach T_g (glass transition). In this region, T_g presents a major transition for drastically from glassy region to rubbery state. For thermoplastic, degradation continues after T_g because of light cross-links in polymer chain (it can be seen in rubbery plateau from figure 1) [18].

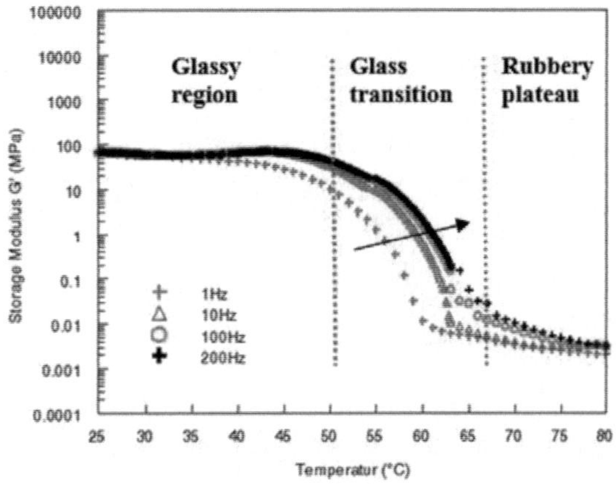

Figure 1. Storage modulus of pure PEG 4000 in multi-frequency (1, 10, 100, 200 Hz)

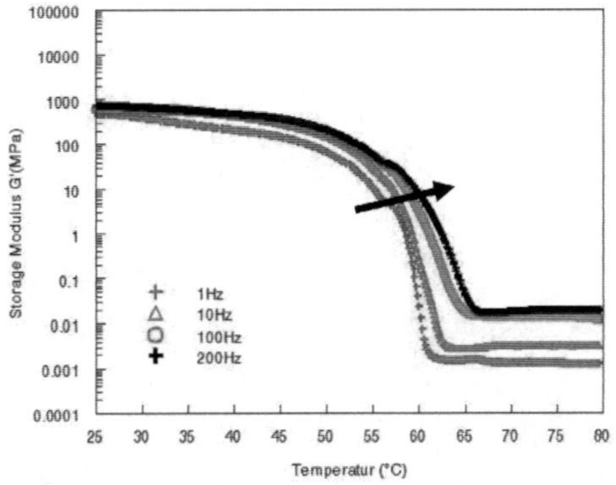

Figure 2. Storage modulus of composite 20wt% cristobalite in multi-frequency (1, 10, 100, 200 Hz)

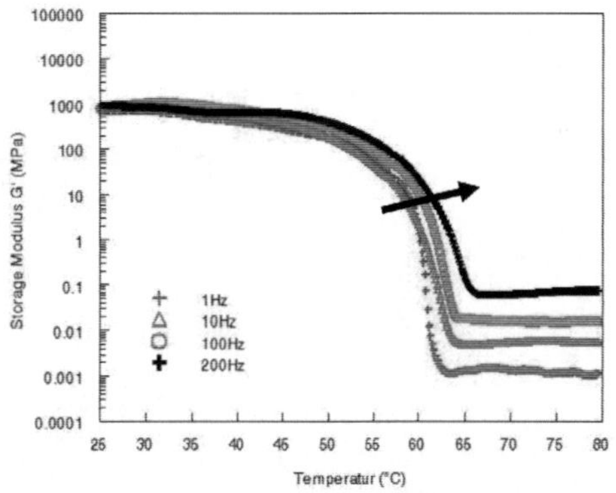

Figure 3. Storage modulus of composite 40wt% cristobalite in multi-frequency (1, 10, 100, 200 Hz)

As can be compared from figures 1, 2, and 3, the highest *G'* is achieved by 40 wt% cristobalite. Its *G'* value is five times of that of pure PEG. This phenomenon can be attributed to crosslinking between PEG and cristobalite. The effects of crosslinking on the various regions of the DMA curve are visible in the rubbery and glass transition regions. However, the effect is not visible in the glassy region. Thus, highly cross-linked thermoplastic polymer has much larger storage modulus indicating the tighter network structure and higher stiffness while that with light cross-links shows a considerably smaller storage modulus [7,19].

When various frequencies are applied, *G'* drop shifts to higher temperature for all samples – as shown by solid arrows in figures 1-3. The shift therefore indicates changes in Tg. The higher the frequency the higher T_g. T_g values for all samples are presented in table 1. The shift to higher temperature is a direct consequence of polymer to achieve (rubbery) equilibrium state. The polymer chain motions are constrained at high frequency but become more liberated at low frequency [18,20]. Consequently, higher T_g is required for a sample to achieve a mechanical equilibrium state at high frequency load.

Table 1. *Tg* value of PEG 4000-cristobalite in multi-frequency

	1Hz	10 Hz	100 Hz	200 Hz
PEG 4000	43.00	45.50	47.70	54.98
SC 20	47.00	49.34	53.75	57.56
SC 40	48.00	51.87	53.95	56.60

The resulting set of experimental *Tg* data (table 1) was used to determine the apparent activation energy (*Ea*) of material degradation. Using Arrhenius equation [21].

$$\ln f = \frac{Ea}{RT} \tag{1}$$

where f = applied frequency, R = general gas constant, and T = degradation temperature (here we used T_g), plots between ln(f) and 1000/T for all samples are presented in figure 4. *Ea* can be obtained from the slope of the plots. The Ea value for each sample is shown in figure 4, where in general increases with cristobalite content. Higher Ea value implies higher energy to initiate degradation [20]. The presence of cristobalite crystals inhibits the degradation in the composite. In particular, cristobalite prevents the free motions of the PEG polymer chains.

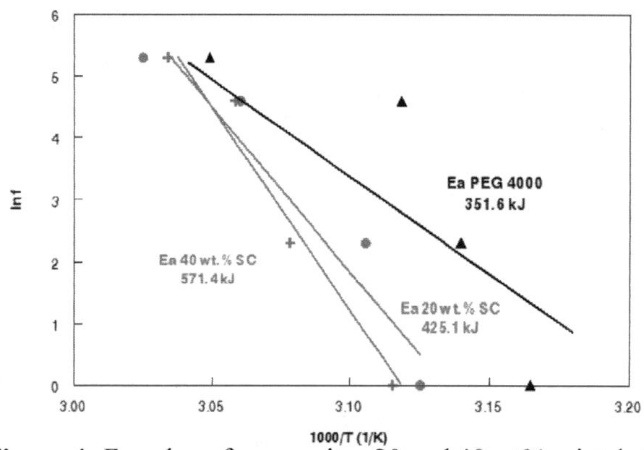

Figure 4. *Ea* value of composites 20 and 40 wt% cristobalite.

4. Conclusion

Since this study we can investigate dynamic mechanical behaviour by using multi-frequency DMA. Minimum degradation material energy (*Ea*) can be obtained by using arrhenius equation. The highest *Ea* value was reached by 40 wt% cristobalite. Adding cristoballite as filler is highly effective to increase degradation themperatur of composites.

Acknowledgements

The author acknowledges to Ministry of Research Technology and High Education of the Republic of Indonesia and LPPM ITS to support finance through program PUPT-2016 No. 01751/IT2.11/PN.08/2016. Also a great thanks to Departement of Physics, Institute of Technology Sepuluh Nopember (ITS) to support this work.

References

[1] Kissinger H E 1957 *Anal. Chem.* **29**(11) 1702–6
[2] Malyshev C 2014 *Annals of Physics* **351** 22–34
[3] Freeman E S and Carol B 1958 *J. Phys. Chem.* **62**(4) 394–7
[4] Jawaid M and Khalil H P S A 2011 *BioResources* **6** 2309–22
[5] Sreekala M S and Thomas S G G *P* 2005 *Polym. Compos.* **26** 388–400

IOP Conf. Series: Journal of Physics: Conf. Series **817** (2017) 012020 doi:10.1088/1742-6596/817/1/012020

[6] Khonakdar H A 2015 *Compos. Part B* **76** 343-53
[7] Saba N, Jawaid M, Alothman O Y and Paridah M T 2016 *Constr. Build. Mater.* **106** 149–59
[8] Sreenivasan V S, Rajini N, Alavudeen A and Arumugaprabu V 2015 *Compos. Part B* **69** 76–86
[9] Heimann R B 2010 *Classic and Advanced Ceramics* (Weinheim: Wiley-VCH Verlag GmbH & Co. KGaA) pp 71–81, 507–20
[10] Kimizuka H, Ogata S and Shibutani Y 2007 *Phys. Status Solidi B* **244** 900–9
[11] Yeganeh-Haeri A, Weidner D J and Parise J B 1992 *Science* **257** 650–2
[12] Grima J N, Gatt R, Alderson A and Evans K E 2006 *Mat. Sci. Eng. A* **423** 219–24
[13] Burunkaya E, Kesmez Ö, Kiraz N, Çamurlu H E, Asiltürk M and Arpaç E 2012 *Thin Solid Films* **522** 238–46
[14] Qian T, Li J, Ma H and Yang J 2015 *Sol. Energ. Mat. Sol. C.* **132** 29–39
[15] Fauziyah N A, Musyarofah M, Triwikantoro T, Mashuri M, Firdaus S and Pratapa S 2015 *Adv. Mater. Res.* **1112** 385–8
[16] Fadly T A, Fauziyah N A, Rosyidi A and Pratapa S 2016 *Experimental Approach to Activation Energy of Polymer Composites PEG 4000-Quartz by Using Dynamic Mechanical Analyser (DMA)*, AIP, submitted in ICESNANO conference
[17] Li J, Li S, Fan C and Yan Z 2012 *Colloids Surf., B* **89** 203–10
[18] Menard K P 2008 *Dynamic Mechanical Analysis (Secons edition)* (USA: CRC Press (Taylor Francis Group))
[19] Goertzen W K and Kessler M R 2007 *Compos. Part B* **38** 1-9
[20] Chartoff R P, Menczel J D and Dillman S H 2009 *Dynamic Mechanical Analysis (DMA)* eds J. D. Menczel and R. B. Prime (USA: John Wiley & Sons, Inc.)
[21] Baboo M, Dixit M, Sharma K and Saxena N S 2010 *Thermochim. Acta* **502**(1-2) 47–50

Controlled crystal phase of TiO_2 by spray pyrolysis method

B B Aji[1], S J Shih[2] and T Pradita[3]

[1]Research Centre for Minerals Processing, Indonesian Institute of Sciences, Jalan. Ir. Sutami Km 15, South Lampung 35361, Indonesia
[2]Department of Materials Science and Engineering, National Taiwan University of Science and Technology, 43 Sec 4 Keelung road, Taipei, 10607, Taiwan
[3]Department of Chemical Engineering, Serang Raya University, Jalan. Raya Serang-Cilegon Km 5, Banten,42116, Indonesia

Email: bramantyo.bayu.aji@lipi.go.id

Abstract: TiO2 in powder form was preferred due to its flexibility and easy handling. TiO2 powders with controlled crystal phase were successfully synthesised by ultrasonic spray pyrolysis from titanium tetraisopropoxide (TTIP) precursor. The effect of various acid solvents (nitric acid, acetic acid and citric acid) in relationship with crystal compositions have been well-studied. This experimental study suggested that the ratio of anatase phase achieved from TiO2 prepared by nitric, acid acetic acid and citric acid solvent are 85%, 39% and 27% respectively. Spurs-Myers equation used to determine the anatase-rutile phase. Crystal sizes were calculated using Scherer equation. Both of ratio and crystal size were estimated from corresponding X-ray diffraction pattern. Precipitation and chelating process were the keys to determine the last composition of TiO2.

1. Introduction

Titanium dioxide (TiO_2) was known as a semiconductor material with unique properties. In powder form, it offers major flexibility uses in catalyst [1], self-cleaning agent [2], fuel cell [3], and solar cell [4]. It has a proper band gap that allows electron-hole transfer which results in a strong oxidation features when illuminated with UV light [5]. Degussa P25 is widely known as a commercial TiO_2 powder which consist of 20% rutile and 80% mixed anatase and brookite phase [6]. Recent studies reported, anatase phase in TiO_2 enhances higher activity than pure phase of rutile structure [7]. Major constituents of anatase phase in a TiO_2 structure permit low energy in visible light absorption which can be a future alternative for solar cell applications [8]. Nevertheless, rutile phase is also preferred for a synthetized TiO_2 structure since it could be used to enhance the photocatalysist performance of TiO_2. However, there were insufficient results about controlling the composition ratio of anatase and rutile phases in a TiO_2 structure. Controlling the temperature process [9], and manipulation of hydrolysis rate [10] were the possible methods to control the crystal phase formation of TiO_2.

Spray pyrolysis is considered as a promising method since it can combine these methods from the preparation of metals, metal oxides, non-oxides and composite powders until produce particles with controlled composition and morphology, good crystallinity, and uniform size distribution, simply in one-step only [11]. Doeuff et. al reported carboxylic acids addition influence relate to the crystal phase, particle morphology and the particle size distribution [12]. The effect of various acids addition study could give better understanding in one-step TiO_2 synthesis. In this study, we conduct TiO_2 synthesis using ultrasonic spray pyrolysis with various acids addition as an additive solution.

2. Experimental

2.1. Powder preparation

TiO_2 powders from TTIP (98%, $Ti(OC_3H_4)_4$, Acros, US) precursor with various acid; nitric acid, acetic acid and citric acid ($C_6H_8O_7$) were prepared using a laboratory-scale SP electrostatic deposition system. The whole SP system has the same process with our previous work [13]. Initially, for TiO_2 with nitric acid solution, the TTIP was dissolved in the mixture of de-ionized water and nitric acid that instantly form a clear solution. For the second solution, TTIP was dissolved separately in de-ionized water with acetic acid and was stirred for 4 H to form clear solution. For the third solution, TTIP was dissolved in mix of citric acid and de-ionized water for 24 H to form clear solution. All solution were prepared for same concentration of 0.05 M. Afterwards, each solution will undergo ultrasonic SP process at 800°C.

2.2. Characterization

X-ray diffractometer (D2 Phaser, Bruker, Germany), with Ni-filtered Cu Kα radiation, was used to characterize the crystallographic structure of the TiO_2 and to determine the ratio of anatase-rutile phase. Field-emission scanning electron microscopy (FESEM, JSM-6500F, JEOL, Japan) was used to investigate the surface of TiO_2 particles and to determine the particles distribution. The particle distribution was conducted by measuring the average diameter of approximately more than 250 particles using the Image J software.

3. Results and Discussion

3.1. Crystallographic structure

Figure 1 shows XRD patterns for TiO_2 powders prepared using various acids solution. It can be seen that all samples show dual anatase-rutile phases of TiO_2 complex microstructure.

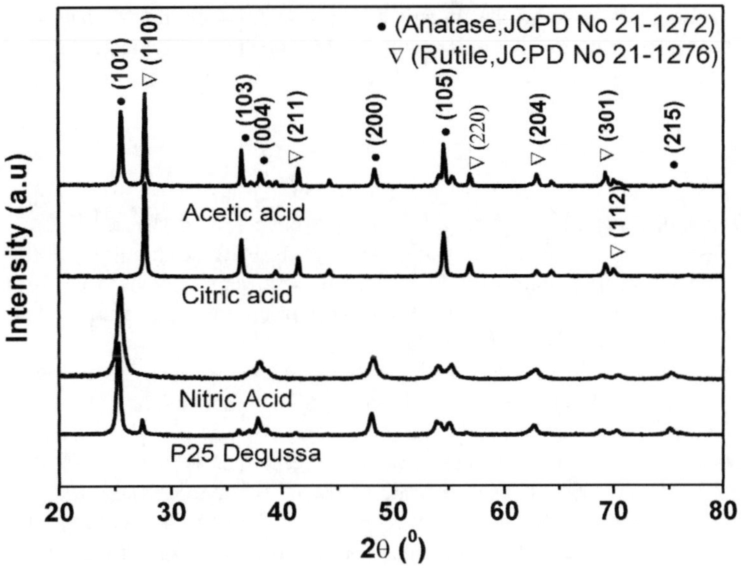

Figure 1. XRD patterns for TiO_2 powders prepared by using various acids.

Prepared powder with nitric acid has the sharpest XRD anatase pattern peaks that show composition dominantly consists of anatase phase as listed in table 1. Calculation of anatase-rutlie phase use The Spurr's-Myerss equation. By using Origin (version 8.1) software, the Spurr's-Myerss equation calculated from the selected areas under the highest normalized peak intensities of a diffraction pattern which represent anatase and rutile phase, are apply the Spurr's-Myerss equation calculation [14]. As shown in table 1 ratios of anatase phase formed with acid additives of nitric acid, acetic acid and citric acid were 85 %, 39%, and 27 %, respectively.

Table 1. Physical properties of prepared TiO_2 powders vs various acids additive.

Acid additive	Crystalline Size (nm)	% anatase	% rutile
Citric acid	28.3 ± 1.9	27	72
Acetic acid	34.6 ± 7.1	39	61
Nitric acid	10.0 ± 3.5	85	15

In the previous work, it found that TTIP with acetic acid fully decomposed with a temperature reaction at 700^0C [15]. Either the temperature reaction or the hydrolysis rate can be applied in order to control the crystal phase. As reported by Birnie et al, Acetic acid can prevent early precipitation of titanium particle due to reduce rate of hydrolysis relevant as below reaction [16].

$$Ti(OC_3H_7)_4{+}4CH_3COOH \rightarrow Ti(CH_3COOH)_4{+}4C_3H_7OH$$

Formation of TiO_2 in solutions is distinguished into three steps, hydrolysis, condensation, and solvent evaporation [17]. In the case of acid as additive, acids can act either as catalyst to accelerate the hydrolysis, or as chelating agent to prevent hydrolysis rate. Carboxylic acids such as acetic acid and citric acid, play role in chelating process. During hydrolysis, chelating agent will prevent precipitation but due to ligand effect, formation of particle nucleus will also get slower than condensation process. It makes atom sharing and dissolution only occurs after a colloidal formation in the stable phase. Atom sharing in TiO_6 octahedran via hydrolysis mainly occurs at atomic surface and will effect in anatase phase formation. Meanwhile, shared atom via chelation is processed at edge position and contributes to rutile phase formation.

3.2. Particle morphology

Generally, TiO_2 powders prepared various acid additives have similar shapes. Theory of one-droplet-per-particle fulfilled with the spherical shape formation. It was identified during the particles formation that the solution has high solubility therefore solid particles could be formed. The use of acid was proposed since its enhance precursor solubility in an aqueous solution. The solubility also means ability of dispersed particles in the solution.

For the case of acetic acid addition, it seems that prepared powder has a spherical shape with smooth surface and narrow size as shown in figure 2 (b). The narrow size of this powder is due to formation of acetates ligand which is stable and serve monodisperse particle [12]. The effects of citric acid on TiO_2 powder is also shown in figure 2 (c). It seems that TiO_2 particles also have spherical shape with several fractal particles. The presence of fractal particles explains that the particles were formed in early precipitation due to chelating process. Chelating process in TiO2 preparation promotes the formation of rutile phase than anatase since polycondensation did not occurs [18]. As a conclusion, these acids have different roles. In nitric acid case, the acid was acted as catalyst that accelerates the polycondensation-precipitation reaction. The acidity of nitric acid provide a controlled transformation of anatase phase. Rather than as a catalyst, citric acid and acetic acid were preferred to act as the chelating agent, where during hydrolysis reaction the ligand change the behavior of TiO_2 particle formation at molecular level.

Figure 2. SEM images of TiO2 particles prepared from (a) TTIP + nitric acid, (b) TTIP + acetic acid, and (c) TTIP citric acid.

3.3. Particle size distribution

SEM images were examined in order to determine the particles surface morphologies and to get information of the TiO_2 particles size distribution.

In this study, the particle sizes were polydispersed in submicron to micron size scale. Figure 3(a) and (c) show similar behavior that particle size distributions have a normal distribution. Its reveals clear information that no significant effects between nitric and citric acids addition, particle growth has the same rate for both acids. However, in the case used acetic acid as shown in figure 3(b), histogram shows a right skewed distribution that means particle was formed due to "natural force". As reported by Matijevic, acetates ligand was effectively control the precipitates of anion that promote monodisperse particles [19]. In our case, the particle sizes are mostly in micron size, which only refer to one-particle-per-droplet mechanism.

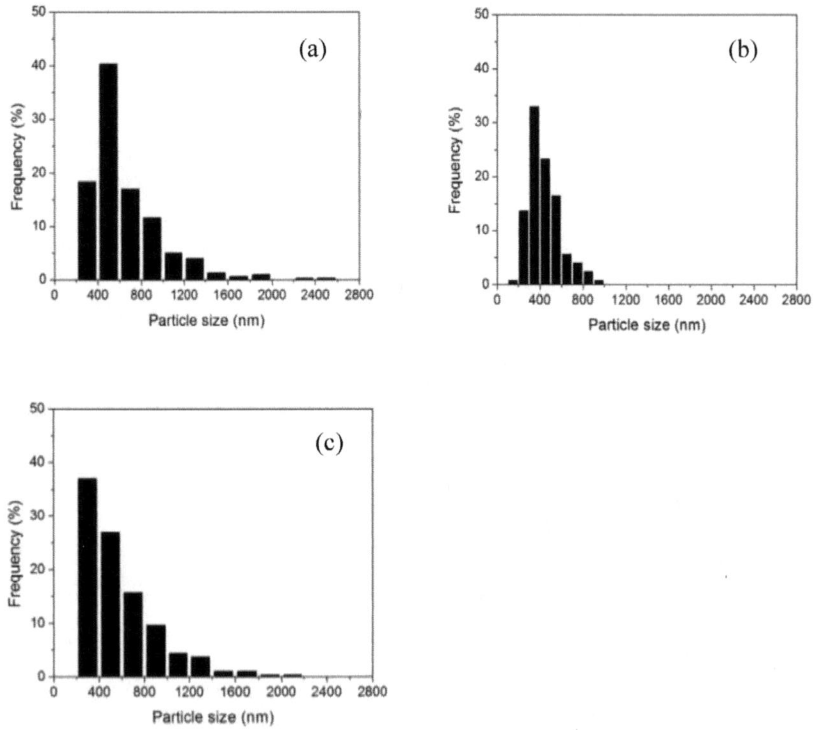

Figure 3. Particle size histograms of TiO2 particles prepared from (a) TTIP + nitric acid, (b) TTIP + acetic acid, and (c) TTIP citric acid.

Conclusion

Controlled crystal phase of TiO2 has been successfully prepared from titanium (IV) isopropoxide (TTIP) precursor within series of acids as additive solution. It suggested that, in one-step TiO2 synthesis both of temperature and hydrolysis contributes in transformation the anatase to rutile phase. The acidity property of nitric acid is the main reason that allows TiO2 particles precipitation due to accelerated the hydrolysis reaction. Meanwhile, for citric acid and acetic acid, chelating process mainly determine the final product.

Acknowledgment

The authors acknowledge the financial support from the National Science Council of Taiwan (Grant No. MOST 104-222-E-011-017).

References

[1] Faraji M, Amini M and Anbari A P 2016 *Catal. Commun.* **76** 72–5
[2] Cedillo-González E I, Riccò R, Montorsi M, Montorsi M, Falcaro P and Siligardi C 2014 *Build. Environ.* **71** 7–14
[3] Tang W, Chen X, Zhou T, Duan H, Chen Y and Wang J 2014 *Hydrometallurgy* **147–148** 210–6
[4] Zhao J, Yang Y, Cui C, Hu H, Zhang Y, Xu J, Lu B, Xu L, Pan J and Tang W 2016 *J.*

Alloys Compd. **663** 211–6
[5] Stashans A, Lunell S and Grimes R W 1996 *J. Phys. Chem. Solids* **57** 1293–301
[6] Ohno T, Sarukawa K, Tokieda K and Matsumura M 2001 *J. Catal.* **203** 82–6
[7] Wang S, Lian J S, Zheng W T and Jiang Q 2012 *Appl. Surf. Sci.* **263** 260–5
[8] Bhushan B, Jung Y C and Koch K 2009 *Philos. Trans. A. Math. Phys. Eng. Sci.* **367** 1631–72
[9] Zhang H and Banfield J F 2000 *J. Phys. Chem. B* **104** 3481–7
[10] Reyes-Coronado D, Rodríguez-Gattorno G, Espinosa-Pesqueira M E, Cab C, de Coss R and Oskam G 2008 *Nanotechnology* **19** 145605
[11] Messing G L, Zhang S-C and Jayanthi G V 1993 *J.Am.Ceram.Soc* **11** 2707–26
[12] Doeuff S, Henry M, Sanchez C and Livage J 1987 *J. Non. Cryst. Solids* **89** 206–16
[13] Shih S-J and Chien I-C 2013 *Powder Technol.* **237** 436–41
[14] Spurr R a. and Myers H 1957 *Anal. Chem.* **29** 760–2
[15] Shih S-J and Tzeng W-L 2014 *Powder Technol.* **264** 291–7
[16] Birnie D P 2000 *J. Mater. Sci.* **35** 367–74
[17] Gupta S M and Tripathi M 2012 *Cent. Eur. J. Chem.* **10** 279–94
[18] Yin H, Wada Y, Kitamura T, Kambe S, Murasawa S, Mori H, Sakata T and Yanagida S 2001 *J. Mater. Chem.* **11** 1694–703
[19] Matijevic E 1986 *Langmuir* **2** 12–20

Investigation on the path process of M doped ZnO (M = Gd, Mg) via the sol-gel method from dried gel to stable nanostructure powder

Suharno, B Soegijono, S Budiawanti and L Fadillah

Department of Physics, Faculty of Math and Sciences, University of Indonesia, Depok 16424, Indonesia

E-mail: sharno81@gmail.com

Abstract. Doping is one of the effective methods to modify the physical properties of ZnO material in order to extend its applications. An investigation on $Zn_{1-x}M_xO$ (M = Gd, Mg and x = 0.00, 0.03) nanopowders that have been synthesized by sol-gel method and sintered at 600 °C for 2 hours was reported. The decomposition process of the dried gel system was investigated by thermal gravimetric analysis (TGA) and the nanopowders with different heating temperature were studied using FT-IR spectroscopy. The crystal structure of the nanopowders after sintering at 600 °C was obtained using X-ray diffraction (XRD). The TGA curves of the samples showed the various weight loss regions corresponding to the removal of starting materials and no weight loss was observed in the temperature range of 300 to 800 °C which corresponded to the phase-crystallization step. The FTIR spectra showed that ZnO band was assigned to the stretching frequency at 669 cm^{-1} while Gd/Mg doped ZnO was at 668 cm^{-1} and 666 cm^{-1}. From the XRD studies, the crystal structure of the samples indicated single phase ZnO crystalline and confirmed hexagonal wurtzite structure (space group of $P6_3mc$).

1. Introduction

Multiferroic materials show simultaneously ferromagnetic and ferroelectric properties. They lead to a variety of promising future device applications in spintronics and sensors [1]. ZnO materials show ferroelectric properties but possess mainly antiferromagnetism at room temperature. Due to the ferroelectric properties in ZnO, further research for suitable multiferroic materials is interesting in relation with this materials.

ZnO is known as an n-type semiconductor material which has direct band gap of 3.37 eV and exciton binding energy of 60 meV [2]. It is a good candidate for applications in the field of lasers, gas sensors, solar cells, optoelectronic devices and luminescent screens. Certain element doped to ZnO also show ferromagnetic materials [3-6]. Undoped or doped ZnO is also very attractive due to its unique properties and versatile applications, such as ultraviolet light emitters, piezoelectric devices, chemical sensors, and spintronics [7-11].

Different physical or chemical synthetic methods have been used to synthesize the ZnO nanoparticles such as thermal decomposition, thermolysis, chemical vapour deposition, sol gel, spray pirolysis, precipitation, vapour phase oxidation, thermal vapour transport, condensation, co-precipitation, and hydrothermal [12-18]. The ferromagnetism in ZnO can be related to vacancies like Zn or O defects, Zn interstitial sites and lattice distortions [19]. However, ferroelectricity rises due to off-center distortion of hexagonal structure. In order to achieve room temperature ferromagnetism, the doping of various transition metals has been carried out in ZnO [20].

The different in saturation of magnetization values by non-magnetic ion dopants has been reported. Lin et al. [21] in 2007 studied the co-doping of Li and Co in ZnO films. It was observed that Li co-doping led to the increase in ferromagnetism due to the Li defects that mediated indirect exchange interactions. Yang et al. [22] in 2008 reported the multiferroic properties of Cr-doped ZnO films. Lim et al. [23] in 2013 studied the coexistence of both ferromagnetism and ferroelectricity in Mg-doped ZnO. Sharma et al. [24] in 2014 observed the multiferroicity and magnetoelectric coupling in (Li, Ni, and Mg) doped ZnO. The observation of multiferroic behavior in non-magnetic ion-doped ZnO will be an important step towards fundamental understanding and novel applications.

In this paper, we studied the synthesize path using thermal analysis, FTIR, X-ray diffraction to obtain chemical information and crystal structure for $Zn_{1-x}M_xO$ (M = Gd, Mg and x = 0.00, 0.03), which were synthesized by sol–gel method.

2. Materials and methods

$Zn_{1-x}M_xO$ (M = Gd, Mg and x = 0.00, 0.03) has been synthesized by sol–gel method. An appropriate amount of zinc acetate $Zn(CH_3COO)_2.2H_2O$, magnesium acetate tetrahydrate $Mg(CH_3COO)_4.2H_2O$, and gadolinium nitrate $Gd(NO_3)_3.3H_2O$ have been dissolved in ethanol. The resulting solution was stirred at 80 °C for 2 hours until gel was formed. The obtained gel was then dried in heating oven at 150 °C for 1 hour and then grounded with mortar and pestle to obtain fine powder. The powder was then sintered at 600 °C for 2 hours.

The gel has been investigated by Thermal Analysis (TGA/DTA) with nitrogen at 20 ml/min and the temperature ranged from 50 to 1000 °C. Dried samples were studied by FTIR spectrometer in the range of 300-4000 cm^{-1}. Meanwhile, the crystal structure of the samples were investigated by X-ray diffraction (XRD) in which the 2θ was from 20 to 80° with Cu-Kα radiation source having wavelength of 1.54060 Å.

3. Results and discussion

3.1. Thermal analysis

As shown in figure 1, the exothermic reaction with the largest weight loss in the temperature range between 200 and 300 °C could be assigned to the collapse of gel. No more weight loss observed in the temperature range of 300-800 °C corresponded to the phase crystallization step. For temperature higher than 800 °C, undoped ZnO showed exothermic reaction while Gd or Mg doped ZnO showed endothermic reaction.

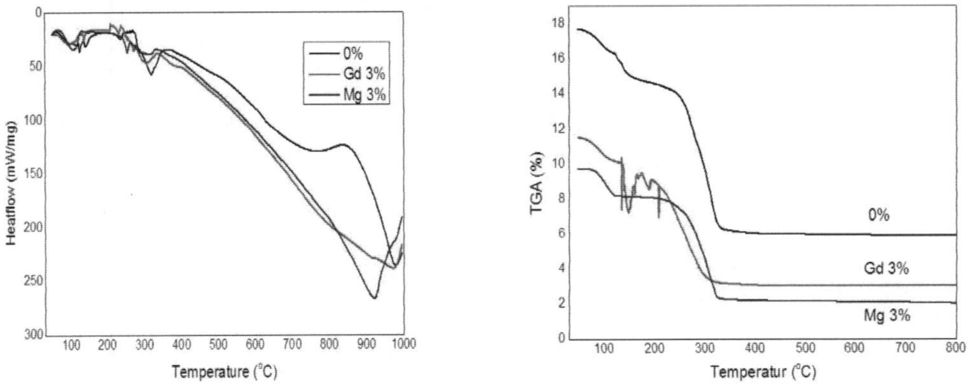

(a) (b)

Figure 1. The pattern of $Zn_{1-x}M_xO$ (M = Gd, Mg and x = 0.00, 0.03) analysed by (a) TGA and (b) DTA.

3.2. FT-IR studies

FTIR is a technique used to obtain information regarding chemical bonding in a material. The band positions and number of absorption peaks depend on crystalline structure, chemical composition and also morphology. FT-IR spectra of samples ZnO after drying at 150 °C are shown in figure 2(a) and IR peaks for the ZnO are shown in table 1. After heating at 600 °C, absorption band around 395 cm^{-1}, which correspond to Zn-O peak, occurred as shown in figure 2(b).

The absorption band at 3062 cm^{-1} corresponded to the O-H stretching vibration of water presence in ZnO. The broad transmission bands at 1556 cm^{-1} and 1425 cm^{-1} in both the samples were due to the carbonyl group of the carboxylate ions, which might remain adsorbed on the surface of ZnO. The ZnO band was assigned to the stretching frequency at 669 cm^{-1} for pure ZnO, 668 cm^{-1} for Gd-doped ZnO, and 666 cm^{-1} for Mg-doped ZnO. The vibration mode above 600 cm^{-1} was weak. The stretching of band appearing at 479 cm^{-1} confirmed the formation of rod shaped ZnO particles [25]. The strong vibrational mode observed at 525 cm^{-1} was assigned to the stretching vibration of ZnO [26]. The appearance of peaks in three different positions depending on the shapes of ZnO has been reported by Ashokkumar et al. [27] and Muthukumaran et al. [28].

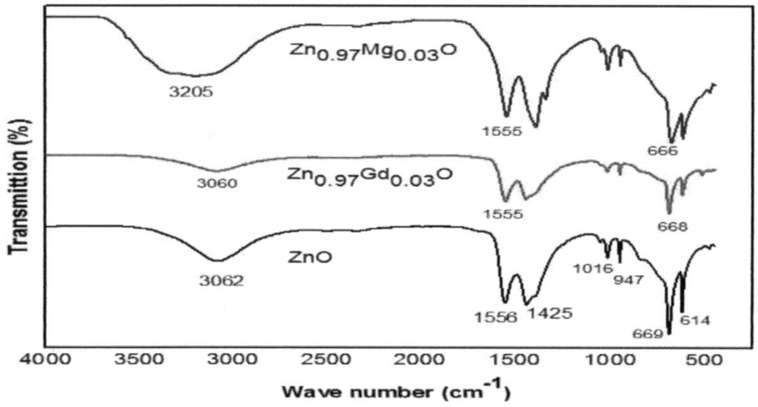

(a)

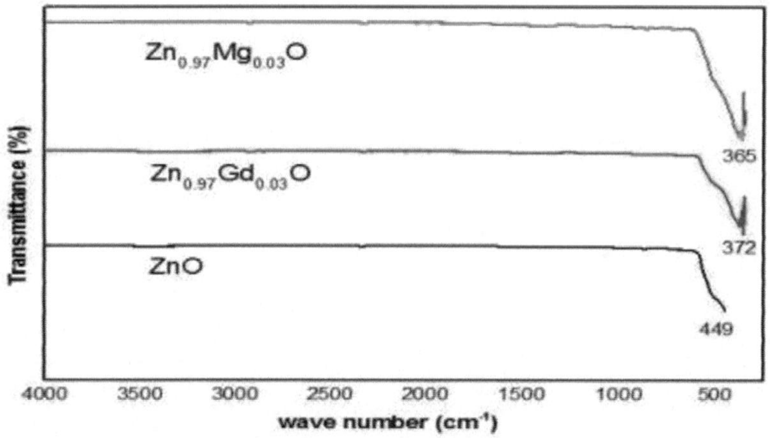

(b)

Figure 2. FT-IR spectra of $Zn_{1-x}M_xO$ (M = Gd, Mg and x = 0.00, 0.03) (a) before and (b) after heating at 600°C.

Table 1. IR peak of $Zn_{1-x}M_xO$ (M = Gd, Mg and x = 0.00, 0.03)

	Wave number (cm^{-1})		
Assignments	ZnO	$Zn_{0.97}Gd_{0.03}O$	$Zn_{0.97}Mg_{0.03}O$
O-H stretching	3062	3060	3205
Adsorbed surface of ZnO	1556	1555	1555
Zn-O stretching (before heating)	669	668	666
Zn-O stretching (after heating 600°C)	449	372	365

The characteristic IR peaks below 1000 cm^{-1} are very important to study the presence of Zn-O/Mg-O bonds. In addition, functional groups and absorption bands observed in the range of 420-600 cm^{-1} are attributed to the stretching modes of Zn-O [26].

3.3. XRD Analysis

The typical XRD diffraction peaks of $Zn_{1-x}M_xO$ (M = Gd, Mg and x = 0.00, 0.03) are shown in figure 3. The diffraction peaks in the XRD pattern clearly showed the crystalline with peaks corresponding to (100), (002), (101), (102), (110), (103), (200), and (112) planes. The diffraction peaks showed the crystal structure of $Zn_{1-x}M_xO$ (M = Gd, Mg and x = 0.00, 0.03) to be hexagonal wurtzite structure (space group $P6_3mc$) with preferred orientation along (101) plane for all the samples, which is the most stable phase of ZnO. ZnO_1 corresponded to the ZnO commercial and ZnO_2 corresponded to synthesized ZnO.

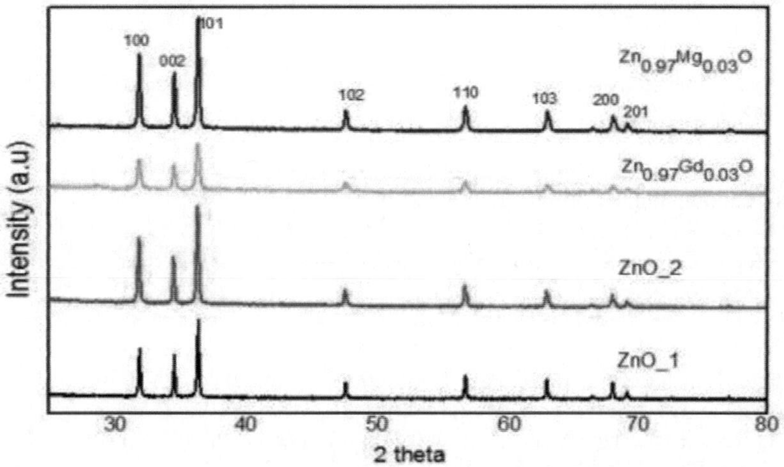

Figure 3. XRD pattern of ZnO_1, ZnO_2, and $Zn_{1-x}M_xO$ (M = Gd, Mg, x = 0.03).

The existence of broad peaks indicated that the average crystal size (D) was in the order of nanometer scale. It was evident from the XRD spectra that there were no extra peaks corresponding to Gd/Mg related secondary and impurity phases, which might be attributed to the incorporation of Gd^{3+} and Mg^{2+} instead of Zn^{2+} lattice site. The changes in cell parameters and peak shift towards higher 2θ value was observed in XRD analysis, thereby indicating successful substitution of Gd and Mg in host ZnO nanopowders. The determination of average crystal size for all samples were using Debye-Scherer's formula [29].

$$D = \frac{0.9\lambda}{\beta \cos\theta} \tag{1}$$

where λ is the wavelength of X-ray used (1.54060 A), β is the full width at half maximum (FWHM) of (101) plane peak of ZnO and θ is Bragg's diffraction angle. The crystal size decreased with doping of Gd and Mg on ZnO. This can be related to the larger ionic radii of Gd^{3+} (94 pm) and Mg^{2+} (72 pm) as compared to that of Zn^{2+} (74 pm). The average crystal sizes calculated from X-ray diffraction were 65.25 nm for ZnO, 51.97 nm for Gd-doped ZnO, and 58.06 nm for Mg-doped ZnO. These results suggested that the doping inhibited crystallite growth as observed in other systems [30]. The 2θ value, FWHM, cell parameters, *c/a* ratio, *d*-value, and average crystal size (D) are tabulated in table 2.

Table 2. The 2θ value, FWHM, cell parameters, and *c/a* ratio, *d*-value, and average crystal size (D) of $Zn_{1-x}M_xO$ (M = Gd, Mg)

Samples	2θ value (°)	FWHM (°)	cell parameters (Å)		*c/a* ratio	d-value (Å)	Average crystal size D (nm)
			a = *b*	*c*			
ZnO	36.25	0.2790	3.2493	5.2057	1.6021	2.4755	65.25
$Zn_{0.97}Gd_{0.03}O$	36.23	0.2024	3.2498	5.2060	1.6019	2.4757	51.97
$Zn_{0.97}Mg_{0.03}O$	36.00	0.2509	3.2503	5.2073	1.6020	2.4780	58.06

4. Conclusion

$Zn_{1-x}M_xO$ (M = Gd, Mg and x = 0.00, 0.03) nanopowders have been successfully synthesized by sol–gel method and sintered at 600 °C for 2 hours. Sintering at temperature between 400 to 800 °C gave a stable crystalite size. Doping process reduced particle size and ion OH in ZnO. The XRD investigation revealed the hexagonal wurtzite structure, which indicated single phase of ZnO nanopowders.

Acknowledgments

This research was financially supported by *Hibah Publikasi Internasional Terindeks untuk Tugas Akhir Mahasiswa Universitas Indonesia* through letter of decree no. 1997/UN2.R12/HKP.05.00/2016.

References

[1] Ramesh R and Spaldin N A 2007 *Nat. Mater.* **6** 21
[2] Kim Y S, Tai W P and Shul S J 2005 *Thin Solid Films* **491** 153
[3] Gao D, Zhang Z, Fu J, Xu Y, Qi J and Xue D 2009 *J. Appl. Phys.* **105** 113928
[4] Liu C, Yun F and Morkoc H 2005 *J. Mater. Sci. Mater. Electron.* **16** 555
[5] Cullity B D 1978 *Elements of X-ray Diffraction* (Massachusetts: Addison-Wesley Publishing Company, Inc)
[6] Motaung D E, Mhlongo G H, Nkosi S S, Malgas G F, Mwakikunga B W. Coetsee E, Swart H C, Abdallah H M, Moyo T and Ray S S 2014 *ACS Appl. Mater. Interfaces* **6** 8981
[7] Goldberger J, He R, Zhang Y, Lee S, Yan H, Choi H-J and Yang P 2003 *Nature* **422** 599
[8] Yan H, He R, Pham J and Yang P 2003 *Adv. Mater.* **15** 402
[9] Nomura K, Ohta H, Ueda K, Kamiya T, Hirano M and Hosono H 2003 *Science* **300** 1269
[10] Lee S Y, Shim E S, Kang H S, Pang S S and Kang J S 2005 *Thin Solid Film* **437** 31
[11] Konenkamp R and Word R C 2004 *Appl. Phys. Lett.* **85** 6004
[12] Harima H 2004 *J. Phys. Condens. Matter.* **16** 55653
[13] Niasari M S, Davar F and Khansari A 2011 *J. Alloy. Compd.* **509** 61
[14] Yang J, Feia L, Liua H, Liu Y and Yanga L 2011 *J. Alloy. Compd.* **509** 3672
[15] Yang Y, Chen H, Zhao B and Bao X 2004 *J. Cryst. Growth* **263** 447
[16] Hu J Q, Li Q, Wong N B, Lee C S and Lee S T 2002 *Chem. Mater.* **14** 1216
[17] Lao J Y, Huang J Y and Wang D Z 2003 *Nano Lett.* **3** 235
[18] Chauhan R, Kumar R and Chaodharya R P 2010 *J. Chem. Pharm. Res.* **2**(4) 178
[19] Savu R, Parra R, Jancar B and Elizario S A 2009 *J. Cryst. Growth* **311** 4102
[20] Kumar P, Kumar Y, Malik H K, Annapoorni S, Gautam S, Chae K H and Asokan K 2014 *J. Appl. Phys.* **114** 453

[21] Lin Y H, Ying M, Li M, Wang X and Nan C W 2007 *Appl. Phys. Lett.* **90** 222110
[22] Yang Y C, Zhong C F, Wang X H and He B 2008 *J. Appl. Phys.* **104** 064102
[23] Oh J Y, Lim S C, Ahn S D, Lee S S, Cho K K, Koo J B, Choi R and Hasan M 2013 *J. Phys. D Appl. Phys.* **46** 285101
[24] Sherma N, Gaur A, Kumar V and Kotnala R K 2014 *Superlatt. Microstruct.* **65** 308
[25] Raja K, Pamesh P S and Geetha D 2014 *Spectrochim. Acta A* **131** 183
[26] Singhal R, Voyles Y, Zhu Y, Jia W and Tomar M S *Nanotechnology* **18** 315606
[27] Ashokkumar M and Muthukuaran S 2014 *Opt. Mater.* **37** 671
[28] Muthukumaran S and R Gopalakhrisnan 2012 *Opt. Mater.* **34** 1946
[29] Kathirvel P, Manoharan D, Mohan S M and Kumar S 2009 *J. Optoelectron. Biomed. Mater.* **1**(1) 25
[30] Ueda K, Tabata H and Kawai T 2001 *Appl. Phys. Lett.* **79** 988

The effect of thermal and organic additive in morphology of ceramic based silicate

J Ginting, N Bangun, H Br Sembiring and N K Putri

Faculty of Sciences and Mathematic, University of Sumatera Utara, Medan, Indonesia

Email: nabilakarina97@gmail.com

Abstract. M-Silicate (M = Mg, Ca) has been prepared by exchange metal reaction from M-Chloride salts and sodium silicate. The resulting white solid of chloride salts then heated at 700, 800, 900 and 1000 ^{0}C. Due to increase the porosity of M-Silicate, 1,2-propanediol, oleic acid and glycerol were added, then formed M-silicates were heated at 800 ^{0}C. Then, obtained white solid M-Silicates were characterized by Scanning Electron Microscopy (SEM). SEM images show the variance of surface morphology when the temperature increases. The addition of organic compounds is involved in surface modification.

1. Introduction

Preparation of Silicate salts had been done in many sources and ways. The modification of its surface and properties could be achieved effectively with low-cost and the reusable silicate is mostly used as adsorbent [1-3]. Silicates are inorganic components which can be easily prepared into monolith or porous material [4]. The porosity of silicates can be enlarged for applications in industries. High temperature will increase its porosity and it classified as ceramics [5]. High temperature can change the surface texture, roughness and shapes of materials [6].

The combination of silicate with organic compound serves an opportunity to increase the stabilities of materials [7]. This combination of silicate and organic additive produce silicates with high medical stability [8]. Preparation of silicate material will improve its functional surface and ability to interact with different molecules that was obtained by combination of silicates and organic compound [9]. The combination of silica with organic compound addition, such as chitosan and methacylic acid, in sol-gel methods exhibit the formation of porous structures and improve its functional contact surface area [10].

In this paper, we report our finding about the effect of heat and organic additives which are glycerol, 1,2-propanediol and oleic acid in preparation of M-silicate (M= Mg, Ca). Oleic acid was used as to represent monounsaturated acid and long chain substance. While glycerol and 1,2-propanediol were used to compare the surface morphology, considering they were oleic acid derivative. The surface morphology was then observed by scanning electron microscopy (SEM).

2. Experiments

2.1. Materials and Instruments

IOP Conf. Series: Journal of Physics: Conf. Series **817** (2017) 012023 doi:10.1088/1742-6596/817/1/012023

2.1.1. Materials. $MgCl_2$, $CaCl_2$, sodium silicate, 1,2-propanediol, glycerol and oleic acid

2.1.2. Instruments. Surface morphology and elemental analysis were analyzed by SEM-EDX, (EVO MA-10 Bruker 129 ev Zeiss).

2.2. Preparation of Ceramic Based Silicate

2.2.1. Effect of Heating. Four kinds of $MgSiO_3$ were prepared by mixing an appropriate amount of $MgCl_2$ in aqueous solution with sodium silicate by stirring in room temperature for 3 h. Then, the precipitated white $MgSiO_3$ was heated in a furnace at 700 ^{0}C (1), 800 ^{0}C (2), 900 ^{0}C (3) and 1000 ^{0}C (4). The surface morphology and elemental analysis of (1), (2), (3) and (4) were characterized and analysed by Scanning Electron Microscopy (SEM) integrated with Energy Dispersive X-ray (EDX) as elemental analysis. The procedure was repeated to prepare four kinds of $CaSiO_3$ (5), (6), (7) and (8) by replacing $MgCl_2$ with $CaCl_2$.

2.2.2. Effect of organic additive. The other three $MgSiO_3$ were prepared with different methods with additions of three kinds of organic substances. An appropriate amount of $MgCl_2$ in aqueous solution was added to sodium silicate while stirred in room temperature. After 3 h, 10% 1,2-propanediol (9), glycerol (10) and oleic acid (11) was added then stirred for another 30 min. The precipitated white solid separated then heated in a furnace at 800 ^{0}C. The surface morphology and elemental analysis were measured by SEM-EDX.The procedure was repeated to prepare three other kinds of $CaSiO_3$ (12), (13) and (14) by replacing $MgCl_2$ with $CaCl_2$.

The summary of preparation ceramic based silicates is explained on table 1.

Table 1. Preparation of M-silicates (M = Mg, Ca)

Code	M-Silicates, M = Mg, Ca (M-$Cl_2 + Na_2SiO_3$)	Heated at	Organic additive (10%)
(1)	$MgSiO_3$	700 ^{0}C	-
(2)	$MgSiO_3$	800 ^{0}C	-
(3)	$MgSiO_3$	900 ^{0}C	-
(4)	$MgSiO_3$	1000 ^{0}C	-
(5)	$CaSiO_3$	700 ^{0}C	-
(6)	$CaSiO_3$	800 ^{0}C	-
(7)	$CaSiO_3$	900 ^{0}C	-
(8)	$CaSiO_3$	1000 ^{0}C	-
(9)	$MgSiO_3$	800 ^{0}C	1,2-propanediol
(10)	$MgSiO_3$	800 ^{0}C	Glycerol
(11)	$MgSiO_3$	800 ^{0}C	Oleic acid
(12)	$CaSiO_3$	800 ^{0}C	1,2-propanediol
(13)	$CaSiO_3$	800 ^{0}C	Glycerol
(14)	$CaSiO_3$	800 ^{0}C	Oleic acid

3. Results and Discussions

3.1. EDX spectra

The reaction between sodium silicate and M-Cl_2 (M = Mg, Ca) are shown below.

- $Na_2SiO_{3(aq)} + MgCl_{2(s)} \longrightarrow MgSiO_{3(s)} + 2\ NaCl_{(aq)}$
- $Na_2SiO_{3(aq)} + CaCl_{2(s)} \longrightarrow CaSiO_{3(s)} + 2\ NaCl_{(aq)}$

IOP Conf. Series: Journal of Physics: Conf. Series **817** (2017) 012023 doi:10.1088/1742-6596/817/1/012023

Energy Dispersive X-ray spectra represent the element that contains in M-silicates as the confirmation of exchange metal reaction had occur. The EDX spectra of M-silicates are shown in figure 1. The EDX spectra of Mg-silicate (a) that show the presence of O, Mg and Si elements confirm that Mg-silicate already performed while the EDX spectra of Ca-silicate (b) show the presence of Ca, O and Si.

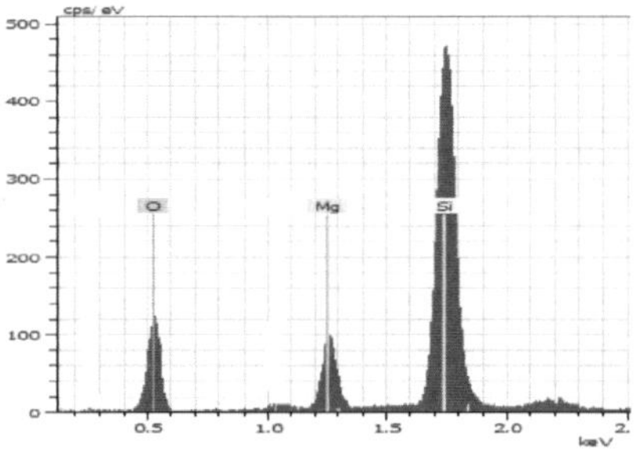

(a) Mg-silicate

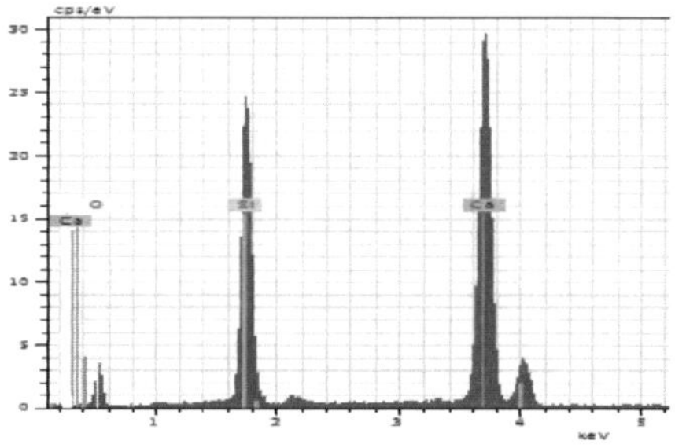

(b) Ca-silicate

Figure 1. EDX spectra of (a) Mg-silicate and (b) Ca-silicate

3.2. *SEM image*

3.2.1. Effect of heating. The SEM images of Mg-Silicate and Ca-silicate are shown in figures 2 and 3, respectively. The heat affected the change of surface morphology for both Mg-silicate and Ca-silicate.

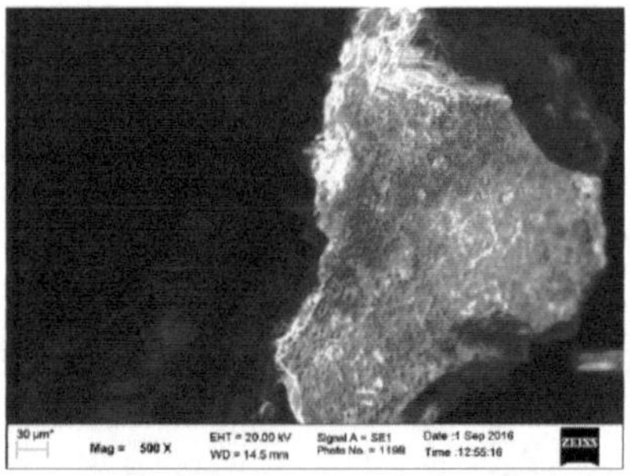

(a)

(b)

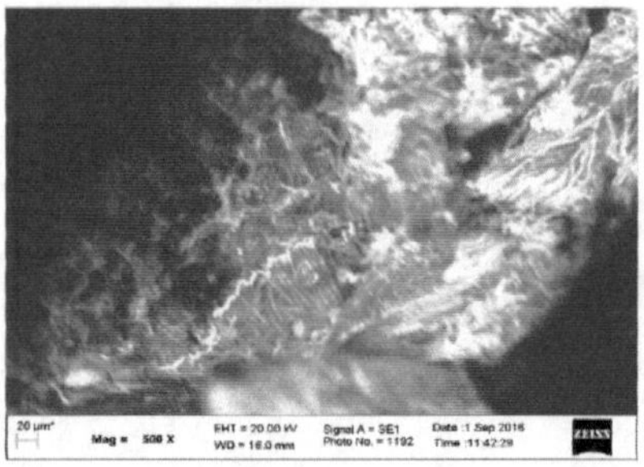

(c)

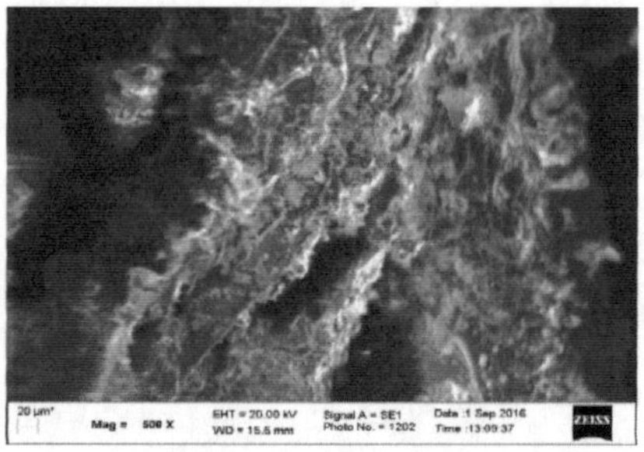

(d)

Figure 2. The SEM images of Mg-Silicate with temperature variation (a) 700 ^{0}C, (b) 800 ^{0}C, (c) 900 ^{0}C, (d) 1000 ^{0}C

IOP Conf. Series: Journal of Physics: Conf. Series **817** (2017) 012023 doi:10.1088/1742-6596/817/1/012023

From figure 2(a) displays the surface area with close pore. The surface is rough and has a hard texture. Figure 2(b) shows a rough surface with agglomeration and looks bumpy. Figure 2(c) shows an agglomerated surface. The fibre looks filling the surface of figure 2(c). While, figure 2(d) also show an agglomerated and rough surface.

The variation of temperature during preparation of Mg-silicate affected in surface morphology. Most of the Mg-silicates surface represent roughness and agglomeration, but the term of the effect is depending on each treatment temperature.

Ca-silicates morphology shows fibre and yarn surface. The surface of Ca-silicate displays a soft texture that is represented by figure 3. Figures 3(a) and 3(b) display a soft net surface with small porous. Figure 3(c) displays a wavy net surface with the form of rectangular porous in almost part while figure 3(d) represents the agglomeration and a very wavy surface. From both Mg-silicate and Ca-silicate images, the size of pore increases with the increasing temperature.

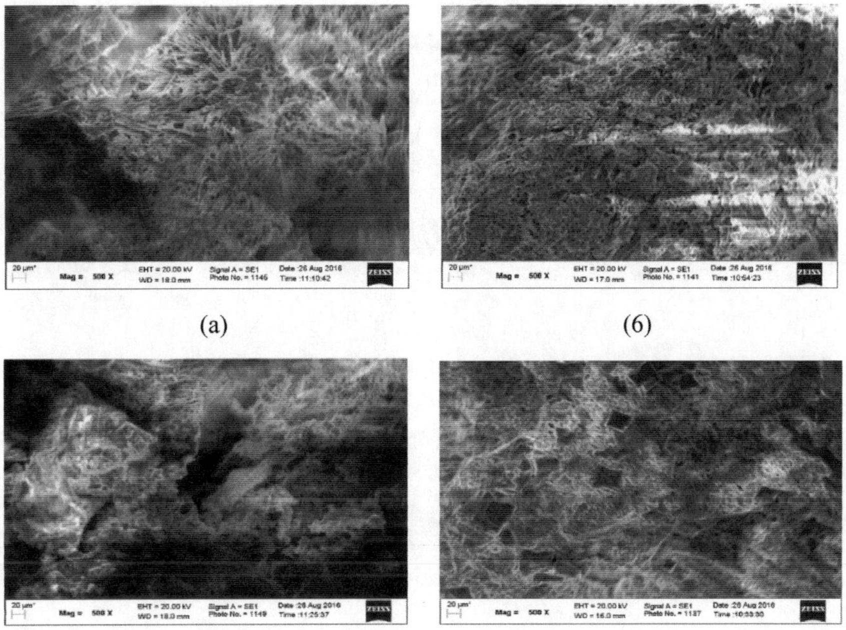

(a) (6)

(c) (d)

Figure 3. The SEM image of Ca-silicate with temperature variation (a) 700 ^{0}C, (b) 800 ^{0}C, (c) 900 ^{0}C, (d) 1000 ^{0}C

3.2.2. Effect of organic additives. To increase the porosity of M-silicate, organic substance addition such as 1,2-propanediol, glycerol and oleic acid were added to the mixture then M-silicate heated at 800 ^{0}C. M-silicate was characterized using SEM. The SEM images of Mg-silicate are shown in figure 4 while the SEM images of Ca-silicate are shown in figure 5.

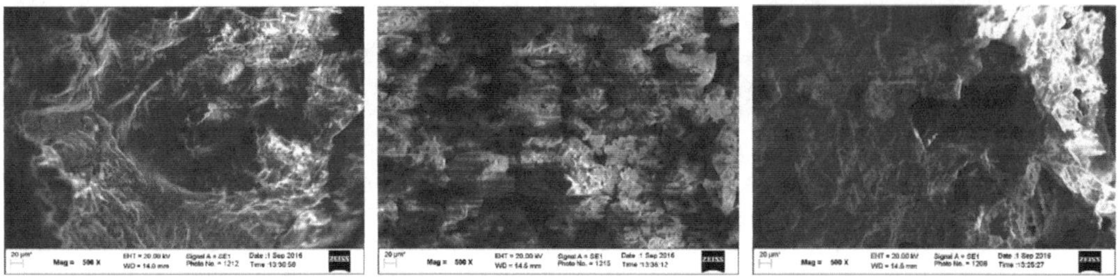

(a) (b) (c)

Figure 4. SEM image of Mg-silicate with organic additive variation (a) 1,2-propanediol, (b) glycerol, (c) oleic acid

(a) (b) (c)

Figure 5. The SEM image of Ca-silicate with organic additive variation (a) 1,2-propanediol, (b) glycerol, (c) oleic acid

Figures 4(a) and 5(a) have the same of large surface morphology and there is a hole like in the surface. Figures 4(b) and 5(b) show a shape of pores figures 4(c) and 5(c) have a detached part in the surface. The organic substance that had been added to the material leads to a difference characteristic in forming the surface area. 1,2-propanediol will produce wider surface, while glycerol will enhance the pore while oleic acid will split the surface.

4. Conclusions

Heat will affect the surface morphology of M-silicate (M= Mg, Ca). The porous size increase with the increasing of temperature. While organic substances addition changes the surface morphology of M-silicate (M= Mg, Ca), the results still depends of the properties of organic substance that used.

Acknowledgement

This research granted by University of Sumatera Utara Research Institution, Medan, Indonesia for 2016.

References

[1] Assawasaengrat P, Jintanavasan P, Kitchaiya P 2015 *Chemical Engineering Transactions* **43** 1135-40

[2] Zhang J P, Wang Q, Smith T R, Hurst W E and Sulpizio T 2016 *Biotechnology Progress* **21**(4) 1220-25

[3] Bangun N, Sembiring S B, Putri N K, Karo Karo J 2015 *Procedia Chemistry* **16** 349-56

[4] Hasanzadeh M, Shadjou N, Eskandani M. and Guardia M 2012 *Trends in Analytical Chemistry* **40** 106-18.

[5] Kingery W D, Bowen H K and Uhlmann D R 1976 *Introduction to Ceramics* (New York: Wiley-Interscience) p 690

[6] Li D 2015 *In situ High Temperature Surface Morphology Using 3D Profilometry* (California: Nanovea)
[7] Wang D, Zhang J, Tian X, Liu D and Sumathy K 2014 *Renewable and Sustainable Energy Reviews* **30** 85-104
[8] Yuan Y, Liu R, Wang C, Luo J and Liu X 2014 *Prog. Org. Coat.* **77**(4) 785-89
[9] Anirudhan T S and Rijith S 2012 *J. Environ. Radioact.* **106** 8-19
[10] Todorova E V, Chernev G E, Djambazov S P 2014 *Open Journal of Inorganic Non-Metallic Materials* **4** 35-43

IOP Conf. Series: Journal of Physics: Conf. Series **817** (2017) 012025 doi:10.1088/1742-6596/817/1/012025

Preparation and structural characterization of ZnO thin films by sol-gel method

D Aryanto[1], W N Jannah[1], Masturi[2], T Sudiro[1], A S Wismogroho[1], P Sebayang[1], Sugianto[2] and P Marwoto[2]

[1] Research Center for Physics, Indonesian Institute of Sciences, Serpong 15314, Tangerang Selatan, Banten, Indonesia
[2] Physics Department, Faculty of Mathematics and Natural Science Universitas Negeri Semarang, Sekaran-Gunungpati 50229, Semarang, Indonesia

E-mail: didik.aryanto@lipi.go.id

Abstract. ZnO thin films were prepared on glass substrate by sol-gel spin-coating method using zinc-acetate dihydrate as precursor. Effect of precursor concentration on the morphological and structural of the films was investigated. The diffraction patterns of X-ray diffraction (XRD) characterization indicated that all of ZnO thin films were polycrystalline with a hexagonal wurtzite crystal structure. The peaks were indexed to (100), (002) and (101) planes. Intensity of all diffraction peaks increased and became broader in full width at half maximum (FWHM) values with increasing precursor concentration. The calculation of texture coefficient (*TC*) indicated that ZnO thin films exhibited the preferential orientation growth along the *c*-axis. Increasing precursor concentration resulted in decreasing crystalline size and crystallization of the film. The lattice constants (*a* and *c*) and *d*-spacing also changed as function of precursor concentration. It was demonstrated by the bond length, volume per unit cell, lattice strain and dislocation density. The scanning electron microscopy (SEM) images of surface morphology of the films confirmed the results of XRD characterization. The grain size of ZnO thin films decreased as result of increasing precursor concentration. Cross-section of SEM images showed that the thickness of ZnO thin film increases from 149.4 nm to 447.7 nm with increasing precursor concentration. This works shown that morphological and structural of ZnO thin films prepared using sol-gel spin coating methods were strongly influenced by precursor concentration.

1. Introduction

In recent years, many researchers have been studied extensively wide-band gap materials such as zinc oxide (ZnO) because of their application in optoelectronic devices. ZnO has a wide and direct band gap (3.37 eV) at room temperature with hexagonal wurtzite structure [1-3]. Besides that, it has much larger exciton binding energy (60 meV) compared to gallium nitride (26 meV) at room temperature [1]. The other advantage of ZnO is simpler crystal-growth technology, resulting in a potentially lower cost for ZnO-based devices [4]. Therefore, ZnO is very interesting for various fields of industrial and high-technological applications.

ZnO thin films have elicited much interest due to their application, as transparent conducting oxides (TCO) [5] and the window layer in solar cell [6]. Applications of ZnO thin films as the film bulk acoustic resonator (FBAR) filter [7] and surface acoustic wave (SAW) device [8] have also been investigated. This is due to an excellent piezoelectric property of ZnO thin film and a high optical transmittance in the visible region. ZnO thin films can be prepared using various methods, such as magnetron sputtering technique [5, 9], molecular beam epitaxy (MBE) [10], metal-organic chemical vapour deposition (MOCVD) [11], pulsed laser deposition (PLD) [12], spray pyrolysis [13], ultrasonic spray [2] and sol gel process [3, 14]. However, the sol-gel method is more convenient among other methods, such as low cost, accurate compositional control, low crystallization temperature, homogeneity at the molecular level, and easy reproducibility [2].

In the sol-gel spin coating process, the characteristic of ZnO thin films was influenced by solvent, precursor concentrations, chemical for stabilizer, pre-annealing, post-annealing, annealing atmosphere and film thickness [1-3, 15]. Kim et al. prepared ZnO thin films using sol-gel spin coating method at various precursor concentrations (0.3-1.3 M) [16]. The pre-heating and post-heating temperatures of samples were 250°C and 650°C, respectively. The ZnO thin films with precursor concentration of 0.7 M have a higher peak intensity on the (002) orientation compared to the other samples. Another work by Kamaruddin et al. showed that ZnO thin film deposited on a glass by sol-gel spin coating method at different precursor concentrations (0.3-0.7M) and then followed by annealing at 450°C for 1 h [1]. The result showed that the grain size, root mean square (RMS) roughness, crystalline size, crystallinity and transparency of ZnO thin films increased with increasing precursor concentration. Both of previous studies used Isopropanol and 2-Methoxyethanol as solvent and monoethanolamine (MEA) for stabilizer. However, fabrication of ZnO thin films derived from sol-gel spin coating method with methanol and ethanolamine as solvent and stabilizer, respectively is still rare. The effect of precursor concentration on the thickness of films has also not been reported.

In this work, ZnO thin films from zinc acetate-dihydrate-methanol-ethanolamine solution were deposited on glass by sol-gel spin coating method. The concentration of precursor solutions was varied from 0.5 to 1 M. All films were pre-heated at 150°C for 1 hour and post-heated (annealed) in the furnace at 800°C for 1 hour. The influence of precursor concentrations on the morphology and structural of ZnO thin films was discussed.

2. Experimental Details

ZnO thin films were deposited onto glass substrate using sol-gel spin coating methods. The solution for ZnO thin films was prepared using zinc acetate dihydrate ($Zn(CH_3COO)_2{\bullet}2H_2O$), ethanol and ethanolamine ($NH_2CH_2CH_2OH$) as a precursor, solvent and stabilizer, respectively. The concentration of the precursor solutions containing zinc acetate dehydrate, ethanol and ethanolamine was varied from 0.5 M to 1 M and the molar ratio of zinc acetate dehydrate ethanolamine ($NH_2CH_2CH_2OH$) was maintained at 1:1. The resulting solution was stirred at room temperature for 1 h using a magnetic stirrer at 300 rpm. Finally, the solution was aged for 3 days before it was spin coated onto glass substrate. Prior to the deposition process, a glass substrate was cleaned sequentially with acetone, methanol and DI water in an ultrasonic bath. The deposition was carried out using a spin coater with a speed of 3000 rpm for 60 s. After coating, ZnO thin films were pre-annealed at 300°C for 1 h, followed by post-annealing at 800°C for 1 h in air atmosphere. The structural analysis of ZnO thin films was carried out using a Smartlab Rigaku X-ray diffractometer with a CuK_{α} (λ = 1.5406 Å) as an X-ray source. The lattice parameters, *d*-spacing, texture coefficient (*TC*), crystalline size (*D*), lattice strain (ε), dislocation density (ρ) were calculated from the XRD data. Morphology studies were carried out using a scanning electron microscope (SEM) (Hitachi High-Tech Co. Ltd, Japan).

3. Results and Discussion

The XRD patterns of ZnO thin films deposited on glass substrates using sol-gel spin coating method with various precursor concentrations are shown in the figure 1. The XRD patterns reveal that the three main diffraction peaks belong to ZnO with (100), (002) and (101) planes according to a JCPDS card No. ICDD-01-078-3315. No impurity peaks of Zn metallic are detected in the XRD patterns. The result indicates that the ZnO thin films are polycrystalline in nature with a hexagonal wurtzite crystal structure [1, 9]. Based on figure 1, all diffraction peak of ZnO thin films with the precursor concentration of 0.5 M have the lowest intensities. In general, the intensity of all diffraction peaks increases with increasing precursor concentration, as shown in figure 1. However, further analysis of XRD results indicates that the preferential crystal growth orientation is along (002) plane. The preferential crystal orientation can be obtained from texture coefficient (*TC*), which is calculated using the Equation as shown below, [9].

$$TC = \left(\frac{I_{(hkl)} / I_{r(hkl)}}{\left[1/n \sum I_{(hkl)} / I_{r(hkl)} \right]} \right) \tag{1}$$

where $I_{(\mathrm{hkl})}$ is the XRD intensity from the thin film, n is the number of reflections observed in the XRD pattern, and $I_{r(\mathrm{hkl})}$ is the intensity of the reference. A detail of the *TC* coefficient with variation of precursor concentrations is shown in Table 1. The value of *TC* indicates the maximum preferred orientation of the films along diffraction plane [9, 17]. As show in Table 1, all ZnO thin films have texture coefficients relatively (>1) higher value along the (002) plane. It was confirmed by the preferential orientation along the *c*-axis. The minimization of surface energy and internal stress influences the preferential orientation of the crystal [18]. The growth of the film along the *c*-axis as a result of the highest atomic density was found along the (002) plane. This could be related to the dominant crystal growth on the c-orientation. Previous work by Kamaruddin et al. shows a different result that the ZnO thin film has preferential growth along (101) planes with increasing precursor concentrations (0.3-0.7) [1]. This may be due to the different in the annealing temperature.

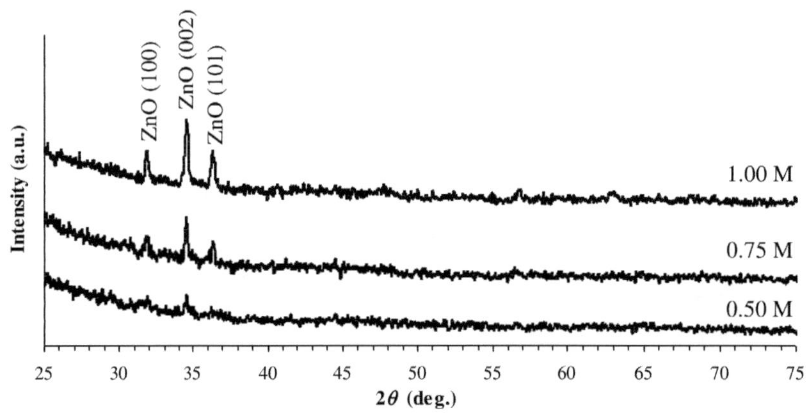

Figure 1. XRD patterns of ZnO thin films with various precursor concentrations and annealed at 800°C.

The lattice parameters, *a* and *c*, of hexagonal phase was estimated according to the peak position of phases and the following equation [15],

$$a = \frac{\lambda}{\sqrt{3}\sin\theta}\sqrt{h^2 + hk + k^2} \tag{2}$$

$$c = \frac{\lambda}{2\sin\theta} l \tag{3}$$

where λ is the X-ray wavelength of 1.5418 Å, θ is a Bragg diffraction angle of the (002) peak and $h\ k\ l$ is Miller indexes. The results are given in Table 1. The value of lattice constants (*a* and *c*) in this work changes as increasing precursor concentration with the variation different in the range of 0.004-0.027

Å, while the theoretical values (ICDD-01-078-3315) of a and c are 3.249 Å and 5.205 Å, respectively. The stress occurrence in the thin film was attributed to the different in the lattice constant [18].

The d-spacing of different crystal plane ($h\ k\ l$) for the hexagonal structure is determined using the equation [9, 15]

$$\frac{1}{d^2}=\frac{4\left(h^2+hk+l^2\right)}{\left(3a^2\right)}+\left(\frac{l^2}{c^2}\right) \tag{4}$$

The values of d-spacing in this work (as shown in Table 1) are larger than un-stress ZnO bulk (2.602 Å) [15], which indicate the presence of strain fields within the non-equilibrium grain boundaries inside of crystallite.

The Zn–O bond length (L) has been calculated using the following relationship [19],

$$L=\sqrt{\left(a^2/3\right)+\left((1/2)-u\right)^2c^2} \tag{5}$$

where u = $(a^2/3c^2) + 0.25$ is the potential parameter of the hexagonal structure. The volume (V) of unit cell of hexagonal system has been calculated from the equation [19].

$$V=0.866\times a^2\times c \tag{6}$$

As can be seen in Table 1, the Zn-O bond length increases with increasing the precursor concentration, which is the value in the range of 1.9651-1.9745Å. Compared to the results of previous studies, this Zn-O bond length is slightly smaller. Ilican et al., [20] reported that the Zn-O bond length is 1.9767 Å in the unit cell of ZnO and neighbouring atoms. The other work by Fang et al. [14], reported the Zn-O bond length is 1.9781 Å. In this work, the lattice volume of ZnO thin film increases with increasing precursor concentration. The volumes of ZnO thin films in this work are smaller than the result by Fang et al. [14]. Higher post-annealing temperatures affect the crystal volume of ZnO thin films. These also result in volume shrinkage of grain size, affecting the bond length and the lattice volume of ZnO thin films. The parameter process (pre- and post-annealing) during fabrication of ZnO thin film by sol-gel spin coating methods strongly influences the structure of ZnO thin films.

Table 1. Calculated parameters crystal of ZnO thin films with various precursor concentrations after annealing at 800°C.

Precursor concentration (M)	$h\ k\ l$	TC	d-spacing (Å)	a (Å)	c (Å)	L (Å)	V (Å^3)
0.50	(002)	1.00	2.5953	-	5.1907	-	-
0.75	(100)	0.88	2.7901	3.2217	-		
	(002)	1.50	2.5990	-	5.1980	1.9651	46.7237
	(101)	0.62	2.4663	-	-		
1.00	(100)	0.85	2.8088	3.2434	-		
	(002)	1.55	2.6004	-	5.2009	1.9745	47.3809
	(101)	0.60	2.4735	-	-		

Figure 2 shows the full width at half maximum (FWHM), crystalline size, lattice strain and dislocation density of (002) peak of ZnO thin film as function of precursor concentration. The FWHM increases from 0.14° to 0.26° with increasing the precursor concentration from 0.50 M to 1.00 M. This indicates that the crystal quality of ZnO thin film decreases with increasing precursor concentration. A smaller FWHM value indicates a better crystallization of the thin films [26]. The result of FWHM attributed to decrease in crystalline size. The crystallite size (D) of ZnO thin films was calculated using Scherrer's formula [5, 9]:

IOP Conf. Series: Journal of Physics: Conf. Series **817** (2017) 012025 doi:10.1088/1742-6596/817/1/012025

$$D = \frac{0.9\,\lambda}{\beta \cos\theta} \tag{7}$$

where β is the FWHM of the ZnO (002) diffraction peak. As shown in Figure 2, the crystalline size of ZnO thin films decreases with increasing precursor concentration. This might ~~be~~ due to the presence of internal stress and defects in the film. The stress relaxation at the grain boundaries results in the smaller crystalline size. This result is in close agreement with the results of SEM characterization which will be discussed later.

The lattice strain (ε) along the c-axis was calculated using the tangent formula [9]

$$\varepsilon = \frac{\beta}{(4\tan\theta)} \tag{8}$$

It affects the length of dislocation lines per unit volume of the crystal. The dislocation density (ρ) due to lattice strain can be expressed by the relation below [9].

$$\rho = \left[\frac{\sqrt{12}\,\varepsilon}{D\,d}\right] \tag{9}$$

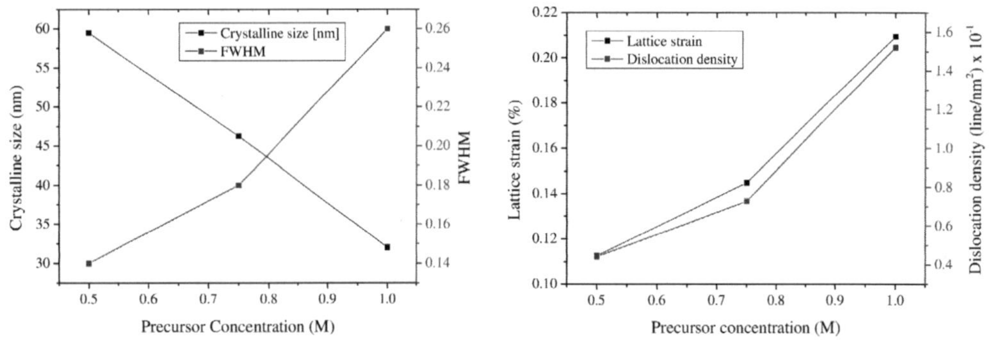

Figure 2. FWHM, Crystalline size (D), lattice strain (ε) and dislocation density (ρ) of ZnO (002) as function of precursor concentrations after annealing at 800°C.

According to XRD data, the lattice strain and dislocation density value increase as the precursor concentration increases from 0.50 M to 1.00 M. The increase in Zn content may lead to the stretch of lattice constant during annealing treatment. This can increase the strain, lattice energy and also diminish the driving force of the growth, resulting decrease in the particle size and crystallization. The results are enhanced by XRD results (Figure 1) and SEM image (Figure 3). Lattice mismatch between the substrate and film also contribute in the lattice strain and dislocation densities.

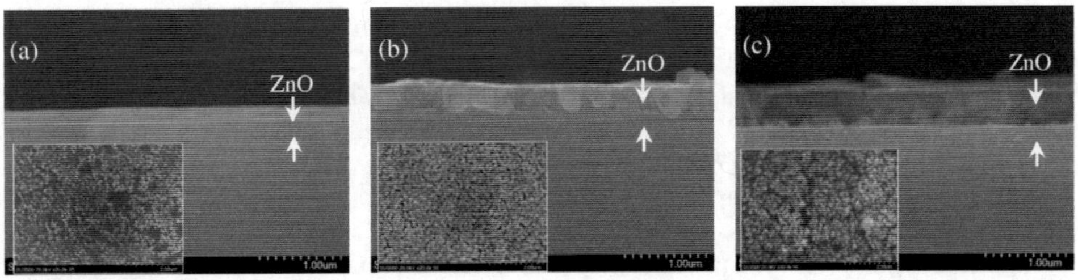

Figure 3. Cross-sectional and surface morphology images of ZnO thin films with precursor concentration of (a) 0.50M, (b) 0.75M and (c) 1.00 M.

The typical SEM surface and cross-sectional morphology of ZnO thin films deposited on a glass with different precursor concentrations are presented in Figure 3. The surface morphology and cross-sectional of SEM images shows that the grains of ZnO thin films becomes smaller and denser with increasing precursor concentrations, as shown on the work by Baneto et al. [21]. The different result is shown by Kamaruddin et al. [1], which the grain size of ZnO thin films increases slightly when the precursor concentration is increased from 0.3 to 0.6 M. In this work, the decrease in grain size seems to be related to a micro-densification effect. The number of nuclei of metal centers increases as an effect of increasing precursor concentration [13], which leads to the formation of a denser and compact structure of the ZnO thin film on the substrate surface. In addition, as show in Figure 3, an increasing precursor concentration causes an increase in the film thickness. The ZnO thin film with precursor concentration of 0.50, 0.75 and 1.00 M has an average thickness of around 149.4, 326.0, and 447.7 nm, respectively. The small crystallites were gathered to form a layer with increasing precursor concentration. Another reason, the increasing Zn^{2+} ions concentration as function of precursor concentration affects the formation of film thickness. The result of cross-sectional SEM images supports the XRD patterns that increasing the intensity of all diffraction peaks are influenced by the thickness of film.

4. Conslusion

ZnO thin films with various precursor concentrations were successfully deposited on glass substrates using sol-gel spin coating methods. All ZnO thin films are polycrystalline in nature with hexagonal wurtzite structure and have a preferential orientation along (002) plane. The lattice constant (*a* and *c*), d-spacing, bond length, volume per unit cell, crystalline size, FWHM, lattice strain and dislocation density of ZnO thin films changes are affected by the precursor concentration. An increase in the precursor concentration leads to increases in the thickness of ZnO thin film and the grain size of film becomes smaller. In summary, the structural and morphological of ZnO thin films strongly depend on the precursor concentration.

Acknowledgement

This work was supported by Indonesian Institute of Sciences under Competitive Research Grand Program. The authors are grateful to Laboratory of High Temperature Material and Coating-Research Center for Physics, Indonesian Institute of Sciences, Indonesia.

Reference

[1] Kamaruddin S A, Chan K-Y, Yow H-K, Sahdan M Z, Saim H & Knipp D 2011 *Appl. Phys. A* **104** 263
[2] Gahtar A, Rahal A, Benhaoua B and Benramache S 2014 *Optik* **125** 3674
[3] Jena I, Das S N, Mishra D K and Bose G 2014 *Int. J. Nano and Biomaterials* **5** 287
[4] Özgür Ü, Alivov Y I, Liu C, Teke A, Reshchikov M A, Doğan S, Avrutin V, Cho S-J and Morkoç H 2005 *J. Appl. Phys.* **98** 041301-1

[5] Yang W, Liu Z, Peng D-L, Zhang F, Huang H, Xie Y and Wu Z 2009 *Appl. Surf. Sci.* **255** 5669
[6] Tanaka K, Minemoto T and Takakura H 2009 *Solar Energy* **83** 477
[7] Dutta J, Singh A V, Singhal S and Upadhayay M D 2015 *IETE J. Res* **62**(1) 3-8
[8] Magnusson E B, Williams B H, Manenti R, Nam M-S, Nersisyan A, Peterer M J, Ardavan A and Leek P J 2015 *Appl. Phys. Lett.* **106** 063509-1
[9] Aryanto D, Marwoto P, Sudiro T, Birowosuto M D, Sugianto and Sulhadi 2016 *AIP Conf. Proc.* Vol 1729 No1 (New York: AIP Publishing) p 020039
[10] Ying M, Wang S, Duan T, Liao B, Zhang X, Mei Z, Du X, Gerriu F M, Fox A M and Gehring G A 2016 *Mater. Lett.* **171** 121
[11] Boukadhaba M A, Fouzri A, Sallet V, Hassani S S, Amiri G, Lusson A and Oumezzine M 2015 *Superlattices and Microstruct.* **85** 820
[1] Kodu M, Arroval T, Avarmaa T, Jaaniso R, Kink I, Leinberg S, Savi K and Timusk M 2014 *Appl. Surf. Sci.* **320** 756
[2] Shinde V R, Gujar T P and Lokhande C D 2007 *Sens.Actuators B* **120** 551
[3] Fang D, Lin K, Xue T, Cui C, Chen X, Yao P and H Li 2014 *J. Alloys and Compounds* **589** 346
[4] Lupan O, Pauporte T, Chow L, Viana B, Pelle F, Ono L K, Cuenya B R and Heinrich H 2010 *Appl. Surf. Sci.* **256** 1895
[5] Kim Y-S, Tai W-P and Shu S-J 2005 *Thin Solid Films* **491** 153
[6] Kim D-K and Kim H-B 2015 *Superlattices Microstruct.* **85** 50
[7] Raoufi D and Raoufi T 2009 *Appl. Surf. Sci.* **255** 5812
[8] Thirumoorthi M and Prakash J T J 2015 *Superlattices Microstruct.* **85** 237
[9] Ilican S, Caglar Y, Caglar M and Yakuphanoglu F 2008 *Appl. Surf. Sci.* **255** 2353
[10] Baneto M, Enesca A, Lare Y, Jondo K, Napo K and Duta A 2014 *Ceram. Int.* **40** 8397

IOP Conf. Series: Journal of Physics: Conf. Series **817** (2017) 012026 doi:10.1088/1742-6596/817/1/012026

Analysis of capacity fading effect on Lithium Cobalt cells caused by pulse current technique in fast charging methods

S Hafiz, S Arianto, R Y. Yunaningsih, N Majid and B Prihandoko

Lithium Battery Research Group, Research Center for Physics, Indonesian Institute of Sciences, Kawasan Puspiptek, Serpong, South Tangerang, Indonesia 15314

E-mail : sams012@lipi.go.id

Abstract. Charging a battery in a short time is important for portable devices. Many techniques have been developed to find out the proper method for fast charging. One of those techniques that has been patented in several fast charging methods is pulse current technique. This technique implements pulse current with adjusting pulse wide and voltage threshold in a certain values. In this paper, the capacity fading effects caused by the current pulse in lithium cobalt cells were investigated. The experiment was done by applying pulse current at high-level SOC to charge four cylindrical lithium cobalt cells. The Capacity of each cell was checked every 50 charge-discharge cycles. The experiment result shows that the changing capacity in each cell forms patterns alike. As if there was a slight increament on their capacities at first checking but rapidly decreasing at the next check. Then, their capacities continue to decrease slowly but the more often the charge-discharge cycling, the battery lifetime decreased. This research has provided analysis of pulse current effect on lithium cobalt capacity fading that should be noted as a reference in applying current pulse for fast charging methods.

1. Introduction

Fast charging is important in several lithium ion applications. The term of quick charge is commonly used in smartphone technology. A smartphone with built-in quick charge systems makes charging time shorter than conventional charging. In electric vehicle applications, users expect that the car can be charged in a short time.

There are many charging protocols which have been proposed that can be applied for fast charging, namely Single C-rate, Multi *C-rate* and Pulse Charging. Single *C-rate* applies a certain high constant current for charging until the target voltage of the cell is reached. Then the cell voltage is kept constant until the minimum charging current reached. Multi *C-rate* applies two or more high constant *C-rate* for charging [1]. And the last is Pulse Charging that applies a certain frequency and amplitude pulse currents or voltages for charging.

Pulse charging is one of the most widely developed charging methods for fast charging. Several researchers implemented fixed charge-discharge pulse frequencies and the others implemented varied charge-discharge pulse frequencies. Chen L R et al. claimed that varied charge-discharge pulse frequency technique gives better charging performance than conventional fixed frequencies [2]. The others only implemented charge pulse without discharge pulse, Keskin N and Liu H replaced discharge pulse phase by a short periods rest phase in their charging technique [3].

There are many analysis of pulse charging impacts to lithium-ion batteries. Several researchers revealed the relation of pulse charging to lithium chemistry behavior inside the cell and claimed that pulse charging is helpful in eliminating concentration polarization, increasing the power transfer rate, and lowering charge time [4]. Chiasserini et al. and Di Yin et al. claimed that the use of a pulse discharge profile with carefully selected amplitudes and rest periods can increase both specific energy and specific power of the cell. Furthermore, Despic A R et al. studies indicated that pulsing the potential of the electrode enables a control of the thickness of the equivalent diffusion layer and hence limits the concentration over-potentials [7]. There were also other researchers that had revealed pulse charging impacts to lithium-ion batteries by simulation and validated it to experimental test analysis [8]. Most of them agreed that pulse charging protocol has a positive impact to fast charging.

Temperature increment of cells becomes one of main concerns in lithium - ion cell application. Implementing fast charging usually causes a higher temperature beyond in standard charging. The temperature increment rates of cells vary based on lithium – ion chemistries. Cobalt base lithium – ion cell is one of poor thermal stability but offers higher electrochemical performance compared to other base-materials. Its temperature rises quicker compared to other chemistry base cells in a same charge-discharge procedure [9]. All of those literatures have not revealed undesirable battery states affected by pulse charging implementation. In this paper, we present several unwanted battery state possibilities affected by pulse charging implementation.

2. Experimental

Four LCO (cobalt-base lithium ion cells) commercial cells with a code "CGR17500" from Panasonic were used to test pulse charging implementation impact in low level SOC and high level SOC of lithium ion cells. Based on the datasheet, the cells have 830 mAh capacities in fresh condition and 2.5 to 4.2 Volt range operation of charge-discharge. In this test, all cells were in first use (not second use) but were more than 3 years old. Therefore, all cells need to be initiated because their properties were not like in the datasheet. First, the capacity of each cell was checked by a same charge-discharge procedure. The charge-discharge procedure was a CC-CV charging with 200 mA constant current, 4.2 Volt constant voltage and 50 mA cut-off current. The voltage characteristic of the cells and initial capacity of each cell are shown in figures 1(a) and 1(b), respectively.

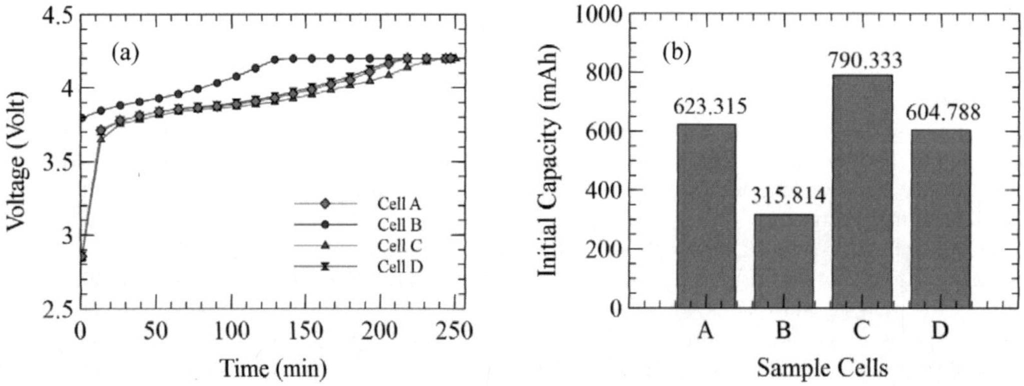

Figure 1. Verification of (a) the voltage characteristic and (b) initial capacity of each cell.

In accordance with cell datasheet, the safe operating area (SOA) of the cell is ranging between 2.5 and 4.2 Volt. Therefore, the maximum high level voltage (HVL) of the cell is 4.2 Volt. In this experiment, pulse charging test was carried out by taking high current charge-discharge with various frequencies. In the settings, cell voltage was kept below 4.2 Volt and pulse charging current was set 2000 mA at maximum.

In the present study, 16 Channel Battery Analyzer BTS8-16-10V2A-IR as programmable charge – discharge device and its software KTBS93en were utilized to record voltage (V) and current (I) data. Digital IC DS18B20 and Arduino as a dataloger were utilized in order to capture temperatures.

The first test was implementing pulse charging with low frequency (0.03 Hz) at high level SOC. This test was intended to investigate undesirable battery state possibilities at high level SOC in terms of voltage response and rising temperature. Before pulse charging current was applied, all cells were fully charged in normal CC-CV with a constant current 500 mA and a cut-off current 40 mA. After that, all cells were discharged 10 minutes by 1C-rate current, and then pulse charging is applied after rest for 30 minutes. The test procedure is illustrated by flowchart in figure 2(a).

The second test was intended to investigate undesirable possibilities at higher level SOC affected by pulse charging implementation. The test procedure was similar with the first test but only different in its pulse frequencies. Where the pulse frequency was set by 1 Hz and the discharge time period was shortened to 10 minutes. The test procedure is illustrated in figure 2(b).

The third test was intended to investigate fading capacity effects of pulse charging implementation at high level SOC. The procedure is similar with the first test, but only added looping procedure at pulse charging step. The total capacity of each cell was checked every 50 cycles. One cycle has two steps which are discharge 10 minutes and pulse charge steps. The procedure is illustrated in figure 3.

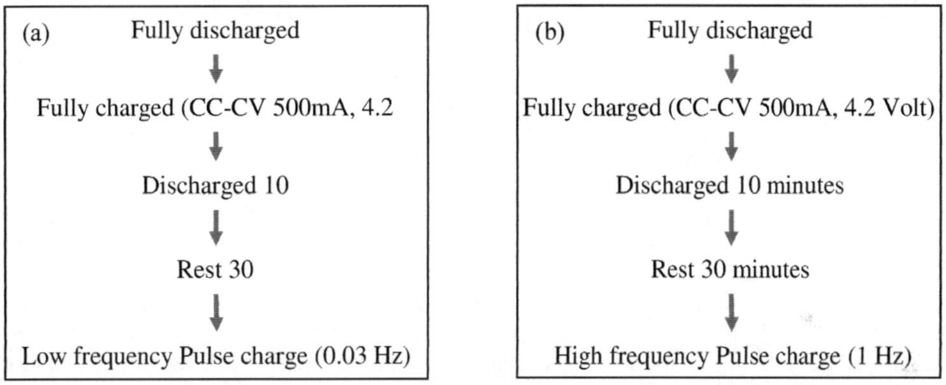

Figure 2. Flowcharts of the experimental procedure of (a) first test and (b) second test.

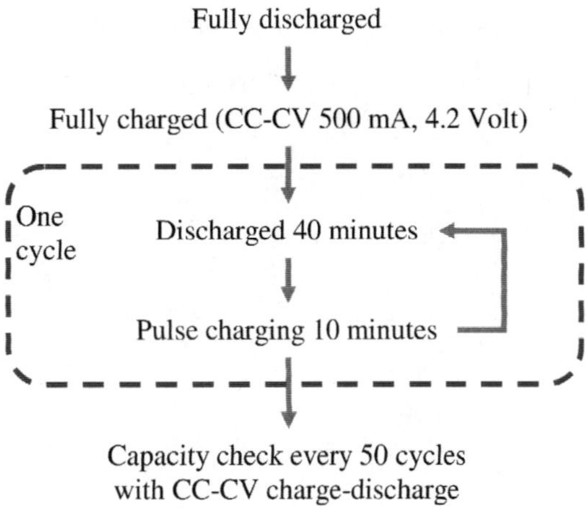

Figure 3. Flowcharts of the experimental procedure of third test.

3. Result and Discussion

3.1. Voltage responses

Control of pulse charging is not easy. As shown in figures 4(a) and 4(b), the voltage of both procedures were set not to exceed 4.2 Volt. But the voltage responses exceeded the maximum of preset voltage value (4.2 Volt). For the low-frequency pulses illustrated in figure 4(a), its response voltage is still in tolerable limits, whereas the high-frequency pulse described in figure 4(b) shows that its voltage response reached too far from safe limit values and even exceeded 5 Volt. This condition is quite risky and if allowed continuously, it could damage the batteries.

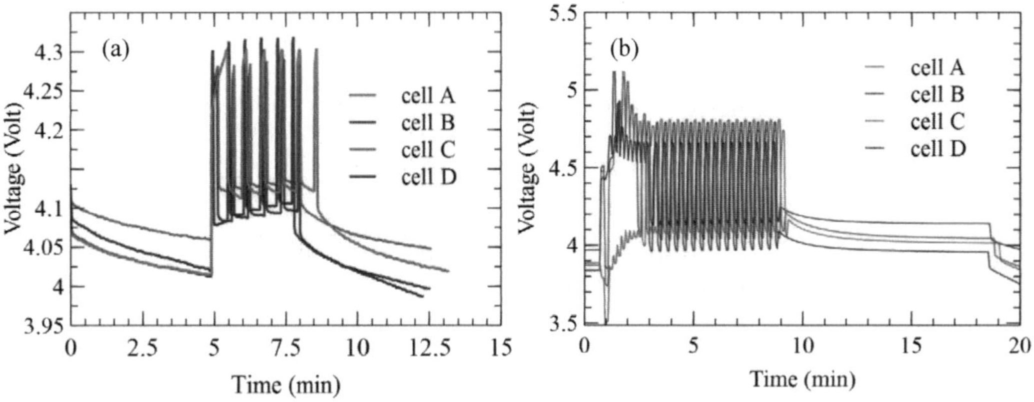

Figure 4. Voltage responses of all cells by (a) low frequency pulse and (b) high frequency pulse.

3.2. Rising temperatures

Implementation of fast charging technique usually causes an increasing temperature higher than in standard charging. Lithium cobalt is a battery type that has poorer thermal properties than other types like phosphate based lithium-ion [9]. Its temperature changes rapidly. 10 minutes of pulse charging periods was enough to make its temperature rose 10 °C higher relative to the room temperature. As shown in figures 5(a) and 5(b), the red marked step 3 below is 10 minutes pulse charging step that causes their temperatures change rapidly. The low frequency of pulse implementation caused the increment of cell temperatures to be higher than the implementation of frequency pulse.

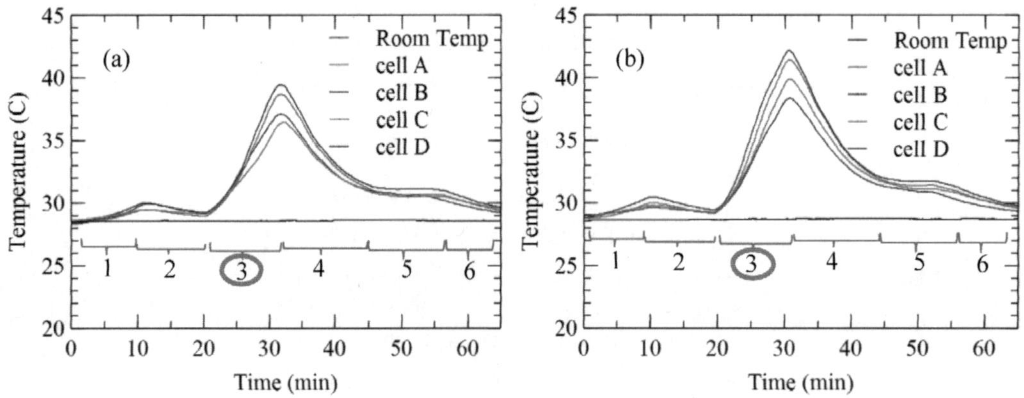

Figure 5. Temperature increment of pulse charge-discharge of step 3 (a) using 1 Hz and (b) using 0.5 Hz relative to the room temperature.

3.3. Capacity fade
Fading capacity caused by the implementation of pulse charging on the high level of SOC on each cell is rather unique, at the first 50 cycles checking, their capacities increased whereas after 100 cycles, their capacities declined continuously in significant values. Only in 450 cycles, the capacity of one cell (cell B) has reduced to less than half of its initial capacity. As shown in figure 6, the implementation of pulse charging is not always going to cause decreasing capacity dramatically; it depends heavily on how well the cell was fabricated.

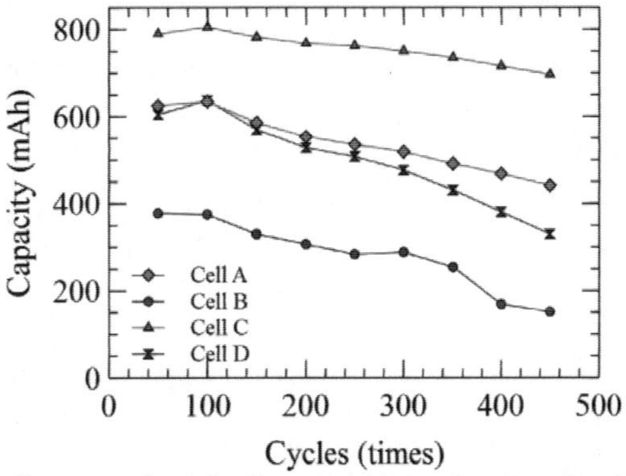

Figure 6. Fading capacity of cells caused by pulse charging implementation.

4. Conclusion
The implementation of pulse charging technique as an alternative of fast charging method is not always profitable. If the selection of pulse charging pattern is not appropriate to the type of batteries, it could damage the batteries. Excessive temperature increment and rapid capacity fade are the impacts. Implementation of pulse charging at high level SOC is improper technique, as it tends to cause uncontrollable over voltage of the lithium ion batteries.

References
[1] Awwad R, Ripan D, Arabi T and Hajj H 2012 *Int. Conf. on Energy Aware Computing* (USA: IEEE) p 1
[1]. Chen L R 2009 *IEEE T. Ind. Electron.* **56** 480
[2]. Keskin N and Liu H 2014 *IEEE Wireless Power Trans.* **24** 44
[3]. Li J, Murphy E, Winnick J and Kohl P A 2001 *J. Power Sources* **102** 302
[4]. Chiasserini C F and Rao R R 1999 *IEEE Wireless Communications and Networking Conf.* (Cat. No. 99TH8466) vol 2 (USA: IEEE) p 636
[5]. Yin M D, Youn J, Park D and Cho J 2015 *Ninth Int. Conf. on Frontier of Computer Science and Technology (FCST 2015)* (USA: IEEE) p 40
[6]. Despic A R and Popov K I 2002 *J. Appl. Electrochem.* **1** 275
[7]. De Jongh P E and Notten P H L 2002 *Sol. State Ion.* **148** 259
[8]. Nguyen J and Taylor C 2004 *Annual Int. Telecommunications Energy Conf.* (USA: IEEE) p 146

IOP Conf. Series: Journal of Physics: Conf. Series **817** (2017) 012027 doi:10.1088/1742-6596/817/1/012027

Comparative study of N719 dye on two different photo-anodes

T Abodunrin[1], A Boyo[2], M Usikalu[1], L Kotsedi[3], Z Yenus[3] and M Maaza[3]

[1]Covenant University, Canaanland, Nigeria
[2]Lagos State University, Ojo, Nigeria
[3]iThemba Labs, Western Cape, South Africa

E-mail: Temitope.Abodunrin@covenantuniversity.edu.ng

Abstract. Nano-composites of TiO_2 and ZnO were successfully prepared using the doctor blade application and high temperature sintering on indium-doped tin oxide (ITO) glass substrate. They were used as efficient photo anode in high performance dye-sensitized solar cells (DSSCs) assembled with N719 dye. The high-density frameworks of TiO_2 and ZnO were synthesized on separate ITO conducting glass using a facile and cost-effective two-step approach to compare the output efficiency. We report on the interfacial boundary relationships, charge – collection conversion efficiency and I-V characteristics of the DSSCs with different electrolytes. The TiO_2 photo anode demonstrated an enhanced solar-to-electrical energy conversion of approximately 5.41 % with KCl electrolyte which was far less than that of a ZnO photo anode with KCl electrolyte which had about 21 % increase under conditions of 1.5 AM. Because of the enhanced solar energy conversion of the ZnO photo anode, *l.arboreus* with KI sensitizer records 9.78 % is a promising candidate for large manufacture of high performance DSSCs modules.

1. Introduction

In the present day, photovoltaic technologies based on inorganic materials dominate the commercial sphere. They are however very expensive and require high temperature and specified environmental conditions for their preparation. Furthermore, materials like CdTe, are scarce, toxic and poisonous. Dye-sensitized solar cells (DSSCs) or organic photovoltaic conversely, are not susceptible to these problems. Although, the efficiencies of DSSCs is still presently less than pure inorganic based photovoltaic technologies, DSSC based on nano-structured electrodes has been subject of intensive research in the last decade. This is due to their special hybrid design which uses the phenomenon of photo electrics, a prospect for low cost electricity from the corresponding synthesis procedure and use of sunlight harvest [1]. Several elements in the intricate DSSC architecture can limit their performance and require an improvement simultaneously to obtain an optimal performance. Precisely, increased dye loading of internal surfaces of the photo-anode and improved conduction of the electrons through 2D metal oxide nano-structures are the specific objectives [2]. Nano-phase anatase TiO_2 is the most popular photo anode material in the DSSC context, but in most studies this material is used in spherical morphology or as faceted nanoparticles (NPs). It is to be noted that unlike ZnO, the crystal habit of TiO_2 does not produce normal progression of regular structures. Recent study has revealed that the photosensitizer is one of the most important materials in DSSCs, which is responsible in determining the performance of the solar cells. Basically, this framework consists of a TiO_2 nanoparticle film that offers a large surface area for adsorption of dye sensitizers, which absorbs light and transfers photo electrons into the conduction band of the TiO_2 after excitation. Law et al. compared the ZnO nanoparticle electron diffusion coefficient to that of TiO_2 nanoparticle film and obtained different efficiencies [3].

Presently, ruthenium dye complexes have power conversion efficiencies of over 11% under standard global air mass 1.5 and also show promising result in stability [4]. Nonetheless, their application is constrained by the high cost and scarcity of ruthenium. In recent times, interest in metal-free organic dyes as alternatives for noble metal complexes has been aroused due to their low cost, ease of manufacture, facile molecular design and high molar extinction coefficient. A large number of researches have been carried out on the molecular engineering of chromophores for higher capability of panchromatic light-harvesting. A variety of organic dyes, such as carbazole, cyanine, coumarin, thiophene, indoline, and fluorine based dyes have reportedly shown positive results [5].

Generally organic dyes used for efficient solar cells are required to possess broad and intense spectral absorption in the visible light region. Overall, DSSC is considered as a low cost and promising solution to solve the energy problem. The authors have shown that the output efficiency for DSSCs depends on the electrolytes used for sensitizing the solar cell and the choice of photo anode determines the photo electrochemical properties of composite films [6] .We therefore present, analyse and discuss these data in this paper.

2. The general components and function of a dye-sensitized solar cell

A dye-sensitized solar cell consists of five main parts: the conducting glass substrate usually fluorine or indium doped a photo anode, a dye which will be sensitized, an electrolyte and a platinum coated counter electrode. Ruthenium dyes have been used and are presently used as sensitizer, titanium oxide because of being readily available, non-toxic and cheap has become a preferred choice of photo anode, other materials such as zinc oxide, aluminium oxide are also suitable candidates for photo anode. Lastly the most common redox mediator used is the iodide/triiodide. In the initial process, the sensitizer D (equation (1)), absorbs quantized energy from the sun. This excites the sensitizer and causes it to inject an electron into the conduction band of the semiconductor thus, leaving the sensitizer in an oxidized state D^+ (equation (2)). The electron circuits the pathway provided by the semiconductor lattice to reach the back contact and the counter electrode through the external load to reduce the redox couple (equation (3)) which subsequently completes the circuit as it revitalizes the sensitizer (equation (4)).

$$D + \hbar\nu \rightarrow D* \quad (1)$$

$$D* \rightarrow D + e^- \quad (2)$$

$$I_{3^-} + 2e^- \rightarrow 3I^- \quad (3)$$

$$2D* + 3I^- \rightarrow 2S + I^{3-} \quad (4)$$

Some limitations to the favourable reactions above are illustrated by the following equations:

$$D* + e^- \rightarrow D \quad (5)$$

$$I^{3-} + 2e^- \rightarrow 3I^- \quad (6)$$

Equation (5) and (6) show undesirable reactions in which an injected electron either combines with the oxidized redox mediator or semiconductor. The output efficiency of a DSSC is determined by a harmonious blend of all the individual components, especially the spectral responses of the dye film on the semiconductor film [7].

The incident photon to current conversion efficiency (IPCE) is a vital factor for distinguishing a photovoltaic device. Specifically where the same architecture is employed with different devices, it is used to relate the performance of sensitizers. It is described as the integral sum of electrons produced per amount of incident photons expressed as a function of wavelength as shown in Equation (7).

$$IPCE(\lambda) = \frac{Photocurrent}{\lambda \times photonflux} \tag{7}$$

The total efficiency (ŋ) of the DSSC is determined by the photocurrent density (J_{sc}), open circuit voltage (V_{oc}), fill factor (ff) and intensity of light (I_s).

$$\eta_{global} = \frac{J_{sc} \times V_{oc}}{I_s} \tag{8}$$

The energy difference between the Nernst potential of the redox mediator in the electrolyte and the Fermi energy level of the photo anode is the open circuit voltage [8]. This is because of the differential discharge or kinetics arising from charge transfer and recombination path. An understanding of these devices and rates of reactions is crucial for upgrading and in the choice of efficient sensitizers for high efficiency DSSCs [2]. The fill factor (ff) is defined as the quality of the device expressed as the ratio of maximum power P_{max} output of the device to the theoretical maximum power. The value of ff lies between 0 and 1 and indicates electrochemical and electrical losses that may occur during the DSSC operation. Thus, the sensitizer delineates the performances of the device.

3. Materials and method

L.arboreus leaves were harvested and air dried until they assumed constant weight. The dried leaf was milled, weighed and allowed to aerate overnight. A 450 g of *L.arboreus* was soaked in 1000 ml of methanol to extract the dye according to established standard laboratory procedures. The conducting sides of glass substrate of FTO variety and resistivity 18 Ω/m^2 was coated with TiO_2 using the doctor blade method, following a hydrothermal method using a mixture of concentrated nitric acid, isopropyl alcohol and titanium oxide in a ratio of 5:3 respectively, a smooth paste was obtained. The coat was allowed to harden under conditions of 1.5 AM. Then the TiO_2 framework was sintered with Vecstar furnace at 300 °C for 90 min. The area of exposure was 2.5 mm X 2.5 mm, the characteristic colour change from brown to white indicated the process was complete [9].

4. Results and discussion

4.1. The absorbance, comparative efficiency of L.arboreus on different electrodes and microstructure of L.arboreus are discussed below.

4.1.1. Phytochemical Analysis. A qualitative phyto-screening reveals the presence of flavonoids, glycosides, saponins, steroids and terpenes in abundance. Proteins, tannins and resins are only moderately present while carbohydrates and alkaloids are in traces.

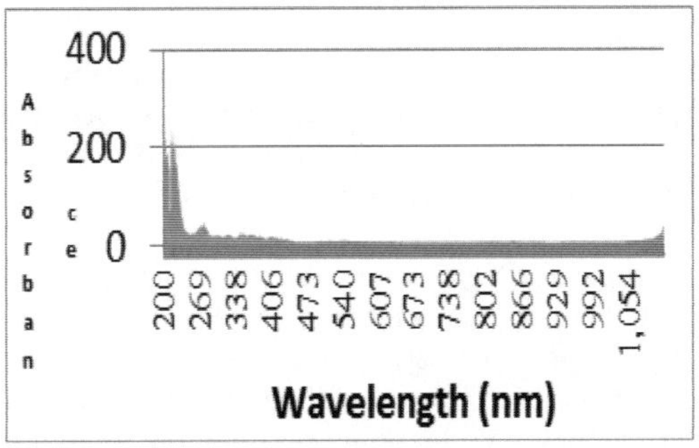

Figure 1. UV/Vis of *L.arboreus.*

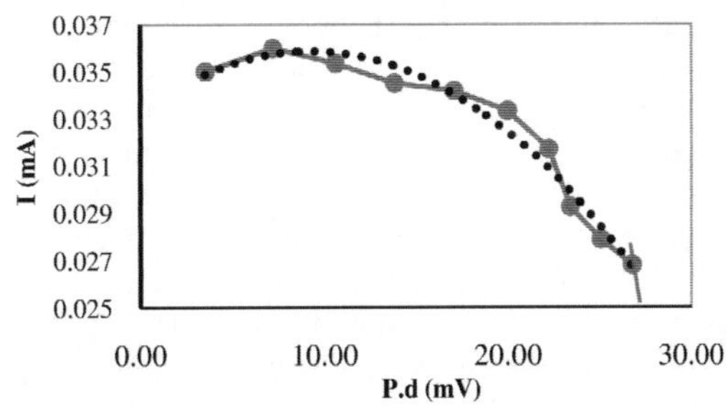

Figure 2. I-V of *L.arboreus* on ZnO with $HgCl_2$.

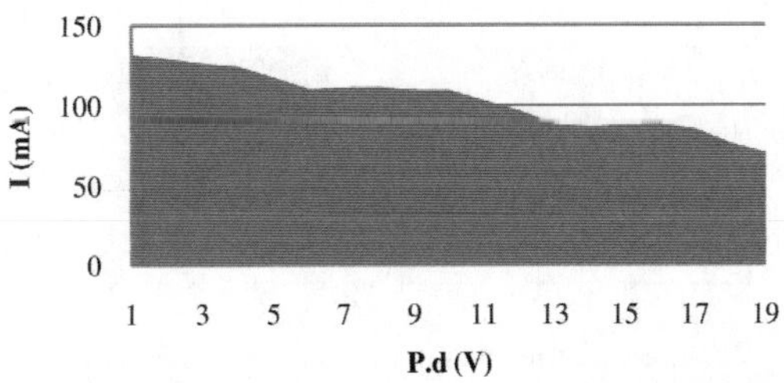

Figure 3. I-V of *L.arboreus* on ZnO with KI.

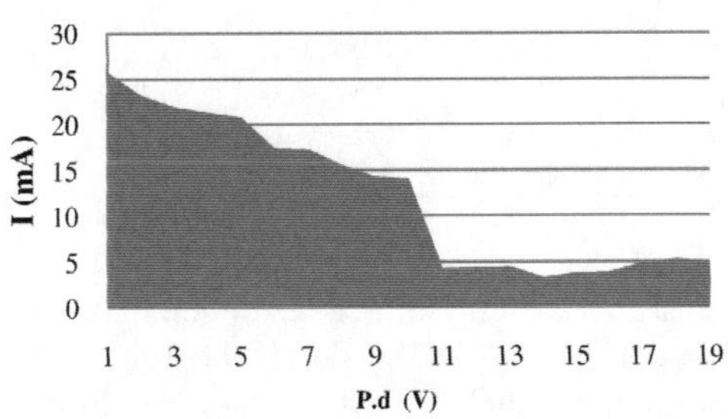

Figure 4. UV/Vis of *L.arboreus* on TiO_2 with KCl.

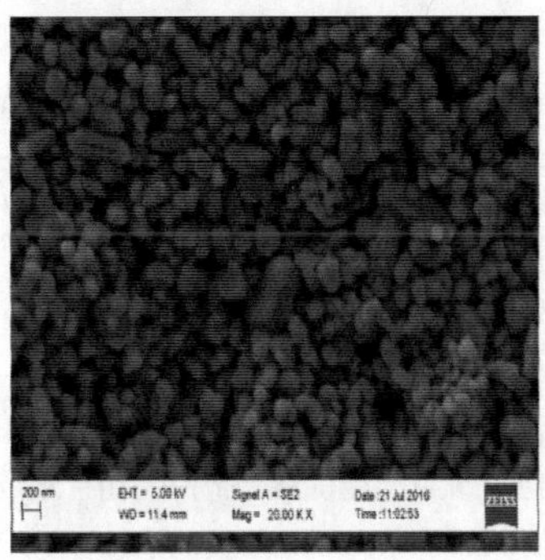

Figure 5. SEM of *L.arboreus* on TiO_2 with KBr.

4.1.2. UV/Vis spectroscopy. A function of absorbance between 200-1100 nm for *L.arboreus* reveals a peak absorbance at about 220 nm. This is a ruthenium dye characteristic. The absorbance has a value 2.0 a.u (on a scale of 100) in the blue region of the electromagnetic spectrum. The absorbance trend is least in the visible region and continues in this same pattern to the near infrared region. At the far infrared region, there is a marginal increase in absorbance as shown in figure 1.

4.1.3. Photoelectric Characteristics. The DSSC was connected in parallel to a variable load supply and a Voltcraft digital multimeter to obtain the photoelectric values. The result is shown on table 1 and the I-V curve illustrated in figure 2. *L.arboreus* based on TiO_2 and sensitized with $HgCl_2$ shows a close approximation to the theoretical value; a fill factor of about 0.8 was recorded. The chemical kinetics is not as favourable, as indicated by the I_{sc}, V_{oc} and power output due to poor charge transport and many trap sites. The effect of using KI and KCl as sensitizer is shown in figures 3 and 4 respectively on ZnO and TiO_2 photo anodes. The ZnO nano-composite is approximately ten times more efficient; output voltage about 7 times more than that of the TiO_2 nano-composite and maximum power output is about 9.78 % more in the ZnO nano-composite. The pattern is recurrent for all electrolyte sensitizers however, KI records the highest power output with the largest I_{sc} and V_{oc} values, this is a result of a more efficient I/I_3^- redox system and an excellent interfacial bonding at the TiO_2/*L.arboreus*/KI interface. The least efficient sensitizer was $HgCl_2$ due to poor excitation, inhibited charge transport, oxidation and weak spectral responses. The most efficient sensitizer on the ZnO photo anode; KI collaborates with KI/TiO_2 (table 1) this is attributable to the chemical kinetics, surface morphology and the strong interfacial bond between the ZnO/*L.arboreus*/KI as shown in figure 3. The least efficient sensitizer is KCl; its spectral response in ZnO is comparative to the result obtained for $HgCl_2$ in table 2. A 60 % increase was recorded from ZnO photo anode relative to the TiO_2 photo anode.

Table 1. Photoelectric values of *L.arboreus* based on TiO_2 Photo anode.

Sensitizer	Isc (mA)	Voc (V)	P max (W)	η (%)
$HgCl_2$	0.128	0.176	5.2×10^{-6}	0.0003
KBr	0.506	0.525	5.4×10^{-4}	0.0027
KI	7.800	12.5	1.85	0.98
KCl	2.600	5.30	0.0216	0.14

Table 2. Photoelectric values of L.arboreus based on ZnO photo anode.

Sensitizer	I_{sc} (mA)	V_{oc} (V)	P max (W)	η (%)
$HgCl_2$	0.937	70.5	23.3	0.66
KBr	3.200	11.5	0.426	0.37
KI	1.143	0.285	82.8	9.7
KCl	0.037	0.009	1.2×10^{-5}	3.3×10^{-6}

4.1.4. SEM Microscopy. The thickness of the TiO_2 film in figure 5 is 200 nm. The Zeiss electron microscope reveals *L.arboreus* on TiO_2 with KI sensitizer in figure 5; at a resolution of W.D of 11.4, 5.00 KV and magnification of 20.00KX. The nanoparticles are similar and form clusters. The gap in the clusters is significant for recombination, thus, obstructing the interfacial transfer of electrons. This explains the low output efficiency and poor power output for KBr and $HgCl_2$. The micrograph in figure 5 shows many trap sites amongst the rod-like and spherical nanoparticles. The sizes are more varied which limits effective charge transport, this accounts for its poor efficiency. The efficiency output would therefore be greatly enhanced if the nanoparticles are symmetrical in arrangement. This would reduce the rate of charge recombination and decrease the band gap energy thus exhibiting a synergistic effect.

5. Conclusion

The poor efficiency observed for TiO_2 dye-sensitized solar cell is probably as a result of several unfavourable inter-boundary reactions between the interface of the counter electrode and the TiO_2 photo anode. Nevertheless, the KI sensitizer still shows a promising yield. The reaction kinetics does not seem to favour the displacement of Cl^- as illustrated by the weak J_{sc} recorded for both KCl and $HgCl_2$. The output efficiency for ZnO/*L.arboreus*/KI was 9.7 %, the least efficient was ZnO/*L.arboreus*/KCl with a value of 3.6 X 10^{-6} %. TiO_2/*L.arboreus*/KI had a value of 0.98 %, the most efficient DSSC on TiO_2 photo anode while the least efficient TiO_2/*L.arboreus*/$HgCl_2$ has 0.00023 %. This reveals ZnO/*L.arboreus*/ KI as a promising candidate (IPCE of 21 % and 9.7%) which with positive synergistic sensitizer effect could even still be improved upon for large scale production of DSSC modules.

Acknowledgments

The research was funded by the Covenant University Centre for Research, Innovation and Development (CUCRID). The authors also appreciate the assistance of the staff and technologists in the Physics and Molecular Research laboratory of Covenant University and the technologists in Materials Research Department, iThemba Labs, Western Cape, South Africa.

References

[1] O'Regan B and Grätzel M A 1991 *Nature* **353** 737–40
[2] Han N, Wang F and Ho J C 2008 *Nanomaterials and Energy* **1** 4-17
[3] Dhas V, Muduli S, Agarkar S and Ogale S B 2011 *Solar Energy* **85** 1213-9
[4] Law M, Greene L E, Johnson J C, Saykally R and Yang P 2005 *Nat. Mater.* **4** 455 – 9
[5] Gao F, Wang Y, Zhang J, Shi D, Wang M, Humphry-Baker R, Wang P, Zakeeruddin S M, Grätzel M 2008 *Chem. Commun.* **44** 2635-7
[6] Yum J H, Jung I, Baik C, Ko J, Nazeeruddin M K and Grätzel M 2009 High efficient donoracceptor ruthenium complex for dye-sensitized solar cell applications *Energy Environ. Sci.* **2** 100-2
[7] Wang P, Zakeeruddin S M, Moser J E, Nazeeruddin M K, Sekiguchi T and Gratzel M 2003 *Nat. Mater.* **2** 402-7
[8] Zhang Y, Peng X, Zhang B, Liu L, Dong S, Lin L, Chen S, Meng S and Feng Y 2015 *Nano Res.* **8** 3830-41
[9] Abodunrin T, Boyo A, Obafemi O and Adebayo T 2015 *Mater. Sci. Appl.* **6** 646-55

Line profile analysis of ODS steels $Fe_{20}Cr_5AlTiY$ milled powders at different Y_2O_3 concentrations

A Afandi, R Nisa and K A Z Thosin

[1]Research Centre for physics, Indonesian Institute of Science (LIPI) Kawasan Puspiptek, Serpong, Tangerang Selatan, Indonesia 15315

Email: ahmad.afandi@lipi.go.id

Abstract. Mechanical properties of material are largely dictated by constituent microstructure parameters such as dislocation density, lattice microstrain, crystallite size and its distribution. To develop ultra-fine grain alloys such as Oxide Dispersion Strengthened (ODS) alloys, mechanical alloying is crucial step to introduce crystal defects, and refining the crystallite size. In this research the ODS sample powders were mechanically alloyed with different Y_20_3 concentration respectively of 0.5, 1, 3, and 5 wt%. MA process was conducted with High Energy Milling (HEM) with the ball to powder ratio of 15:1. The vial and the ball were made of alumina, and the milling condition is set 200 r.p.m constant. The ODS powders were investigated by X-Ray Diffractions (XRD), Bragg-Brentano setup of SmartLab Rigaku with 40 KV, and 30 mA, step size using 0.02°, with scanning speed of 4°min^{-1}. Line Profile Analysis (LPA) of classical Williamson-Hall was carried out, with the aim to investigate the different crystallite size, and microstrain due to the selection of the full wide at half maximum (FWHM) and integral breadth.

1. Introduction

ODS steels have been considered as the structural materials future nuclear reactor Gen. IV [1-3]. Primary features of the ODS steels are high oxidation resistance at high temperature [4], high creep resistance [5], and resistance to high dose of particle irradiation [6]. ODS steels are commonly produced by mechanical alloying (MA) of the constituent metal powder. MA is preferred due to its ability to develop alloy with fine grain. It is generally accepted that the finer the grains or particles, the stronger the materials.

Additionally, MA could introduce microscopic features such as hard precipitates, dislocation density enhancement, micro plasticity, which contribute to the superior mechanical properties. However, there is a trade of between high oxidation resistance, and the mechanical performance. Thus, to accommodate the oxidation resistance, and mechanical properties, additional rare earth materials are needed. ODS steels are developed with small amount of Y_2O_3, normally between 0.5-3 wt%, but sometimes can be added between 5-15 wt% [7].

The role of rare earth materials such as Y_2O_3 has not been very well documented, especially during the process of MA. With high milling energy, and long period of milling time, there were many investigations showed the formation of Y-Ti-O nano-precipitate [8]. In this study, the role of different addition of Y_2O_3 is investigated. XRD line profile analysis (LPA) is valuable methods to evaluate the crystallite size, and other crystal defects formation. Unlike Transmission Electron Microscopy (TEM) analysis, XRD could be carried out faster without sumptuous sample preparation. LPA based on XRD investigation takes into account the broadening of the diffraction peaks. The peak shape, the diffraction position, and the intensity of XRD profile give valuable information regarding the microstructure of materials and its feature.

The aim of this paper is investigating the crystallite size evolution, and lattice distortion using simple Williamson-Hall analysis. The LPA of XRD data was performed, and the determination of the peak shape was compared between Full-Width at half maximum (FWHM) and integral breadth.

2. Experimental Procedure

2.1. XRD measurement

The metal powder of Fe20Cr5AlTiY ODS steels was purchased from Sandvik Material Technology LTD, with the particle size of about 106 μm as received from manufacturer. The metal powder was manufactured by gas atomized process, with chemical composition as received shown at table 1 below.

Table 1. The chemical composition of Fe20Cr5AlTiY ODS Powders as received

Element	Cr	Al	Ti	Ni	Mn	Y	Cu	Fe
Mol%	17.2	6.6	0.58	0.47	0.36	0.097	0.097	balanced

MA process used planetary ball milled (PBM) of MTI Corp. There were 4 variances with different Y_2O_3 concentration, respectively of 0.5, 1, 3, and 5 wt%. The ball to powder ratio (BPR) was kept constant 15:1, using alumina ball of 10mm, and 20gr of mixed sample powder was placed inside tightly sealed alumina jars of 320 mL capacity. The MA was carried out for 24h, in which each cycle of MA took 4h, and 30 min paused to prevent excessive heating. After MA was finished, the ODS powder was examined by XRD using the machine of Smartlab, Rigaku operated at 40 KV, and 30 mA. Structural investigation used Bragg-Brentano setup, with Cu anode tube producing x-ray with λ=1.5419 Å. The scanning XRD covers 2θ from 20-120°, with scanning rate 4°min^{-1} and step size 0.02°.

2.2. Line profile analysis

The first approach of LPA was introduced by Scherer [9] which characterize the apparent domain size that is a volume-weighted quantity,

$$t = \frac{\gamma\lambda}{FWHM}\cos\theta \tag{1}$$

where γ is constant close to unity, FWHM is full width at half maximum of the peak, θ is the Bragg angle at [*h k l*] reflection. Though this formula was introduced in 1918, the Scherrer is still largely employed nowadays. Later, Wilson [10] defined the integral breadth of the apparent size:

$$< D >_v = \frac{\gamma\lambda}{\beta}\cos\theta \tag{2}$$

where the integral *breadth* $\beta = \frac{A}{I_o}$, A is the peak area, I_o is the height of peak intensity. If the broadening due to sole strain effect, Stokes and Wilson [14] defined the relation between integral breadth and strain as

$$\eta = \beta\cot\theta \tag{3}$$

$$e = \frac{\eta}{4} = \frac{\beta}{4}\tan\theta \tag{4}$$

where e is the maximum strain, and η is the apparent strain. In many cases, line broadening is caused by simultaneous strain and crystallite size, therefore, those two effects has been combined by Williamson and Hall [11]. The Williamson-Hall (WH) plot proposed the correlation that sample broadening is the sum of the broadening due to crystallite size, and lattice strain.

$$\Delta K = \Delta K^s + \Delta K^D \tag{5}$$

where ΔK^s and ΔK^D represent the broadenings affected by crystallite size, and lattice distortion respectively. If $\Delta K = \cos\theta(\Delta 2\theta)/\lambda$, and the diffraction vector $K = 2 \sin \theta/\lambda$, thus the linear plot between those value will have the intercept as the crystallite size ($\gamma / \langle Dv \rangle$), and the slope represent the lattice distortion (2η). The expression of $\Delta 2\theta$ can be replaced by FHWM or β, which will influence the determination of γ, 0.94 for FWHM, and 1 for breadth β. Despite WH plot is very simple to obtain the crystallite size and the microstrain, the application is still widely used [12-14].

Calculation of the broadenings require two basic inputs for LPA analysis, first is the position of the peak and the second is the precise value of the FWHM, and β. In this paper, the determination of the peak position, FWHM or β was done from the Rietveld analysis of the PDXL software. The phase found in the XRD profile from the ODS steels, are cubic Fe-Cr with space group number 229, *Im-3m*. From the Rietveld analysis, the background was simulated with Chebyshev I, and the shape function of *Pseudo-Voigt*, which is combination of the *Gauss* and *Lorentz*.

3. Results

Each of the XRD profile was analysed with Rietveld refinement, and limited parameters were varied such as U, V, W for the description of the shape function. The result of the Rietveld analysis is shown in figure 1. The selection of the shape function is crucial, since it determines the best fitting that can be obtained for the XRD data. The Pseudo-Voigt function is generally preferred due to the combination of Gauss and Lorentz that can simulate the wide broadening and peak intensity.

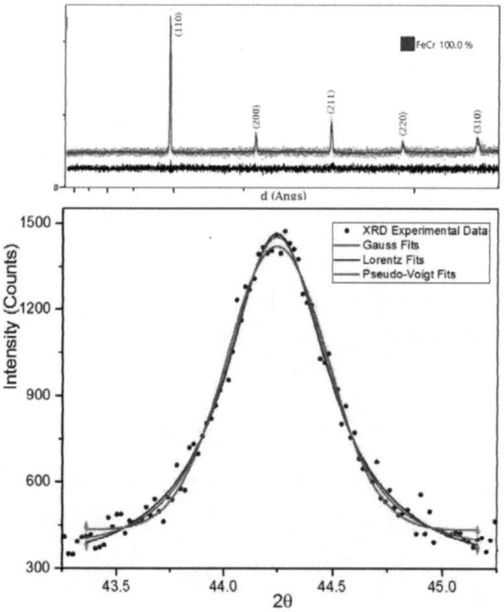

Figure 1. The Rietveld refinement of ODS Pure exhibiting 5 peaks, and the different shape functions.

The broadening of each XRD profiles is shown in the figure 2, using *Cagliotti function.* This function is described as $FWHM = \sqrt{U\ tan^2\theta + V \tan\theta + W}$, which the U, V, and W parameters taken from Rietveld refinement, and the θ is the position of the peaks.

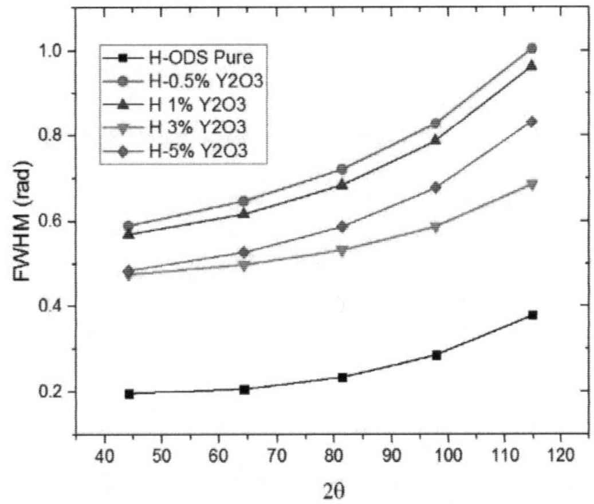

Figure 2. The broadening of XRD profile described by *Cagliotti Fuction*

The broadening can be firstly identified by the plot of *FWHM vs 2θ*, which explain the trend of the broadening. The higher FWHM corresponds to the smaller crystallite sizes, and the larger microstration. However, from figure 3 the most broadening is shown from the sample with 0.5 wt%, followed by 1 wt%, 3wt%, and 5 wt% of the Y_2O_3 concentration when subjected to the 24h mechanical alloying. It was generally accepted that applying the longer MA process, would result the smaller crystallite, and the larger microstrain [15].

Finally by applying the least square fitting of Eq. 5, it is obtained the WH plot as shown in figure 3 below.

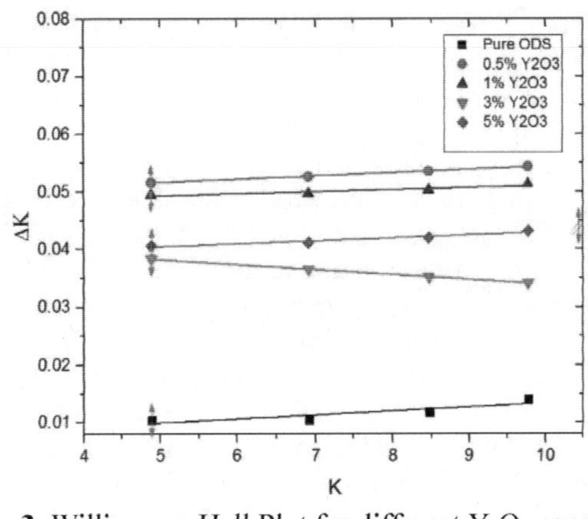

Figure 3. Williamson-Hall Plot for different Y_2O_3 concentration.

4. Discussion

The description of WH plot shows that the intercept corresponds to the crystallite size $\langle D \rangle_v$, while the slope corresponds to the microstrain (η). It was shown in the figure 1, the peak at d_{hkl}(310) has the lowest, and almost indistinguishable between the peak and the background. Therefore, for the calculation of the crystallite size, and microstrain the peak at d_{hkl}(310) was excluded. The calculation of the sample broadening has to include the contribution of the instrument broadening. In this study the instrument broadening was assessed with the sample standard, and compared to the broadening of the ODS pure. The sample without any significant broadening due to small crystal size and microstrain will have horizontal linear fit according to figure 3. Finally, after the subtraction of the instrument broadening, it is obtained the value the crystallite size, and microstrain as shown figure 4 below.

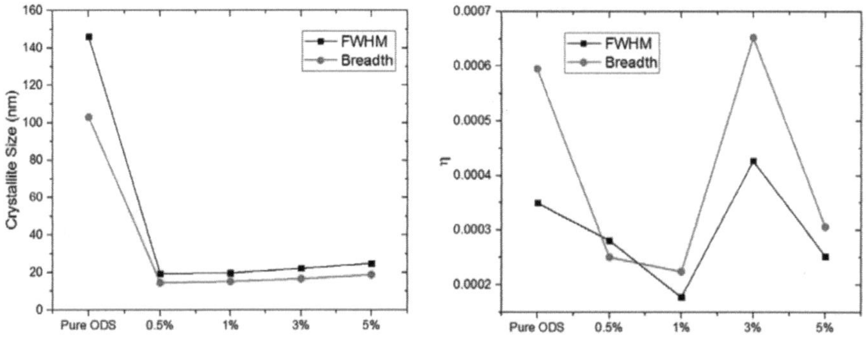

Figure 4. The trend of the $\langle D\rangle_v$ and η at different Y_2O_3 concentration

The lowest crystallite size was obtained at 0.5 wt% Y_2O_3, followed by 1%, 3%, 5% and ODS pure. Both FWHM and β exhibited similar trend for crystallite. However, the figure of microstrain requires further analysis since the source is mainly the dislocation enrichment, and other crystal defects that may present after mechanical alloying. The result of the W-H plots for all samples finally shown in table 2.

Table 2. The comparison of the crystallite size and microstrain due to FWHM and β (A_o/I)

Sample	FWHM		β (simple)	
(wt% Y_2O_3)	$\langle D\rangle_v$ (nm)	η (%)	$\langle D\rangle_v$ (nm)	η (%)
0	146.15	0.0349	102.96	0.0595
0.5%	19.23	0.0281	14.53	0.0250
1%	19.75	0.0177	15.17	0.0224
3%	22.12	0.0427	16.62	0.0653
5%	24.76	0.0252	18.76	0.0306

The correlation of the crystallite size and the broadening is shown in figure 5 using two primary peaks. In general the crystallite size from Breadth calculation is bigger while the microstrain is lower. The discrepancy shows that accurate determination of peak intensity and broadening is crucial, and in this case, FWHM term is more precise than simple Breadth.

The decreasing of the intensity is clearly visible at *dhkl* (110) especially the XRD profile with 1 *wt%* and 0.5 *wt%* Y_2O_3 is at the lowest. The loss of the intensity is mainly correlated to the decreasing crystallite size which is confirmed from table 2 above. The intensity and broadening from 3 wt% Y_2O_3 is significantly the largest, but the peak intensity of it is lower than 5 wt% Y_20_3 at *dhkl* (210). This particular behaviour explains the negative slope of WH plot in figure 3. Furthermore, the broadening of the sample after 24h, with various Y_2O_3 concentrations shows small discrepancy. Thus, it can be concluded that different concentration of Y_2O_3 influence the refinement of the crystallite, which hold the dominant source of the broadening.

The crystallite size is smaller with small amount of Y_2O_3 concentration. This phenomenon could be explained that the Y_2O_3 lattice does not stay intact during MA. During high energy milling, broken Y and O could possible interact with Ti, due to easy solubility to construct Y-Ti-O nanoclusters [16-17].

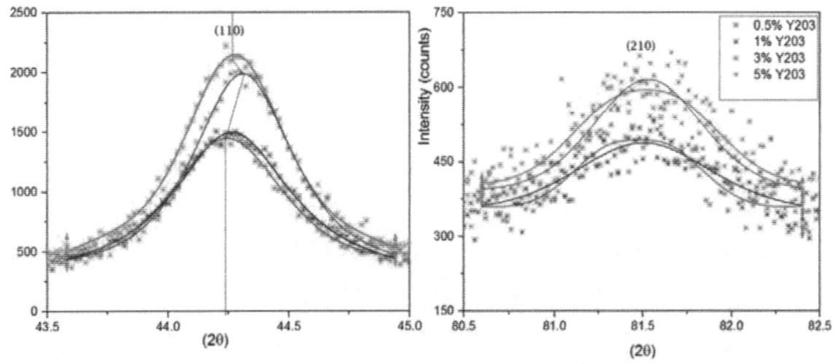

Figure 5. XRD profile stacking at (110) and (210) as the two primary peaks

The critical concentration of Y_2O_3 might be limited to 1 wt%, which is adequate to possibility form Y-Ti-O nanocluster. The size of nanocluster is around 10 nm, so then when the excessive Y2O3 concentration will not produce Y-Ti-O due to limited amount of Ti atoms. The detection of the Y-Ti-O with XRD is very difficult due to the limited resolving power, thus further analysis is needed using TEM to detect the diffraction of Y-Ti-O phases.

5. Conclusions

Line Profile Analysis based on XRD is useful method to get insight of the microstructure especially for the determination of the crystallite size and crystal defects. WH plot is adequate to characterize the crystallite size of the ODS powder after high energy milling. The calculation of the crystallite size using FWHM with Pseudo-Voight model is more accurate than simple Breadth. However, the microstrain is difficult to be investigated because the source of microstrain is not incorporated in the model.

References

[1] Ukai S, Harada M, Okada H, Inoue M, Noumura S, Shikakura S, Asabe K, Nishida T and Fujiwara M 1993 *J. Nucl Mater.* **204** 65-73
[2] Baluc N, Gelles D S, Jitsukawa S, Kimura A, Klueh R L, Odette G R, van der Schaaf B and Yu J 2007 *J. Nucl. Mater.* **367-370** 33
[3] Ukai S and Fujiwara M 2002 *J. Nucl. Mater.* **307-311** 749-57
[4] Kaito T, Narita T, Ukai S and Matsuda S 2004 *J. Nucl Mater.* **329-333** 1122-6
[5] Klueh R.L, Maziasz P.J, Kim I.S, Heatherly L, Hoelzer D.T, Hashimoto N, Kenik E.A and Miyahara K 2002 *J. Nucl Mater.* **307-311** 773-7
[6] Ukai S, Kaito T, Seki M, Mayorshin A A and Shishalo O V 2005 *J. Nucl. Sci. Technol.* **42** 109-22
[7] Kimura Y, Takaki S, Suejima S, Uemori R and Tamehiro H 1999 *ISIJ Int.* **39** 176–82
[8] Yamashita S, Ohtsuka S, Akasaka N, Ukai S and Ohnuki S. 2004. *Philos. Mag. Lett.* **84** 525-29
[9] Scherrer P. 1918 *Nachr. Ges. Wiss. Gottinge*. 98-100
[10] Wilson A C J 1963 *Mathematical Theory of X-ray Powder Diffractometry* (Eindhoven: Centrex Pub. Co.)
[11] Williamson G K and Hall W H 1953. *Acta. Metall.* **1** 22-31
[12] Khorsand Zak A, Abd. Majid W H, Abrishami M E and Yousefi R 2011 *Solid State Sci.* **13**(1) 251-56
[13] Prabhu Y T, Rao K V, Kumar V S S and Kumari B S 2014 *World J. Nano Sci. Eng.* **4** 21-8
[14] Mote V D, Purushotham Y and Dole B N 2012 *J. Theor. Appl. Phys.* 6:6
[15] Longzhou Ma, Bruce S -J K, Mary A A and Huang C C 2014 *KONA Powder Part. J.* **31** 146
[16] Hirata A, Fujita T, Wen Y R, Schneibel J H, Liu C T and Chen M W 2011 *Nat. Mater.* **10** 922
[17] Stoica G M, Stoica A D, Miller M K and Ma D 2014 *Nat. Commun.* **5** 5178

IOP Conf. Series: Journal of Physics: Conf. Series **817** (2017) 012029 doi:10.1088/1742-6596/817/1/012029

Green synthesis of silver nanoparticles using *Carica Papaya* fruit extract under sunlight irradiation and their colorimetric detection of mercury ions

M Firdaus, S Andriana, Elvinawati, W Alwi, E Swistoro, A Ruyani and A Sundaryono

Graduate School of Science Education, University of Bengkulu 38371, Indonesia

E-mail: lutfi@unib.ac.id

Abstract. We have successfully synthesized silver nanoparticles (AgNPs) by using aqueous extract of papaya (*Carica papaya*) fruit as bioreductant under sunlight irradiation without additional capping agent. Characterizations were done using UV-Visible spectrophotometry and Fourier Transform Infrared Spectroscopy (FTIR). The synthesized AgNPs have yellowish-brown color with surface plasmon resonance peak at 410 nm. Good selectivity of the AgNPs towards hazardous heavy metal of mercury ions in aqueous solution has been developed as a green environmental sensor. The presence of Hg(II) ions in the mixture changed the yellowish-brown color of AgNPs to colorless due to oxidation of Ag(0) in AgNPs to Ag(I) ions. Effect of samples matrix such as alkali metal, alkaline earth metal and transition metal ions were evaluated.

1. Introduction

The developments of silver (AgNPs) and gold nanoparticles (AuNPs) in recent years are significantly increased. They have potential applications in various fields such as chemistry, physics, biology, materials science, and medicine. AgNPs and AuNPs are highly efficient at absorbing and scattering light, different with many dyes and pigments, it has a color that depends upon the shape and size (10 – 100 nm) of the particles [1–3]. Due to its small particles size (~ 10^{-9} m) those results are in its large surface area. AgNPs and AuNPs have some unique optical, electrical, and thermal properties. The interaction of the AgNPs and AuNPs with light occurs due to the conduction electrons on the metal surface undergo an oscillation when excited by the light at specific wavelengths, known as surface plasmon resonance (SPR), this results in unusually strong scattering and absorption properties [4, 5].

In order to fabricate AgNPs from its salt $AgNO_3$, it usually referred to bottom up fabrication; the chemical reducing agents such as $NaBH_4$ are widely used. However, the chemical reductants are not environmentally friendly. Alternatively, biosynthesis of AgNPs using plants, marine organisms and micro-organisms that abundance in our environment is now established as an emerging area of nanoscience research. The green synthesized AgNPs have been widely used as an anti-bacteria and sensor for heavy metals detection in environment [6–8].

Mercury, which has density of 15.53 g cm^{-3}, is well-known as toxic heavy metals. Among its compounds, methylmercury $Hg(CH_3)_2$, is the most toxic one [9]. Due to its mobilities in living systems and abilities to cross cell membranes, it is often bio-accumulated in fish and other organisms [10]. So far, mercury analysis is performed using bulk instruments such as ICP and AAS, which are expensive and required laborious pretreatments. Here, we report a low-cost, portable and green protocol to analyze mercury in aqueous sample by means of AgNPs colorimetric method with sunlight irradiation assistance.

2. Experimental methods

2.1. Materials
Silver nitrate and other chemicals were supplied by Merck Ltd (Darmstadt, Germany). Papaya fruit was obtained from a local market in Indonesia. Double-distilled deionized water was used throughout the experiments. All glassware were washed with detergent 5%, HCl 4M, and water before use. A more detail washing and sampling procedures was published elsewhere [11].

2.2. Instruments
UV–Visible absorption spectra of the synthesized AgNPs were monitored using a BioSpectrometer (Eppendorf, Germany) in 270 – 700 nm range. Fourier Transform Infrared (FTIR) measurements were obtained on a Prestige 21 (Shimadzu, Japan). A Canon 30D digital camera (Tokyo, Japan) was employed to record the pictures.

2.3. Biosynthesis of silver nanoparticles
Fresh papaya fruits were washed with water to remove the dust. The 25 gram of fine-cut fruits was boiled with 50 ml of water at 80°C for 15 minutes. The extract was separated from the remaining small solids by filtration through Whatman no. 1 filter paper.

Silver nitrate with concentration 1 mM was prepared in Erlenmeyer flasks and the appropriate volume of fruit extract was added to the mixture under sunlight irradiation for 15 minutes. In order to optimize the AgNPs biosynthesis condition, a variation of different concentration of fruit extracts added into $AgNO_3$ solution were also conducted. The volume ratios of fruit extract to $AgNO_3$ were 1:5, 1:1, and 2:1. An aliquot of the mixture was taken periodically to monitor the progress of AgNPs biosynthesis by using UV–vis spectrophotometer. UV–vis spectra were recorded at a resolution of 1 nm. The pH of the reaction was kept at 4.5.

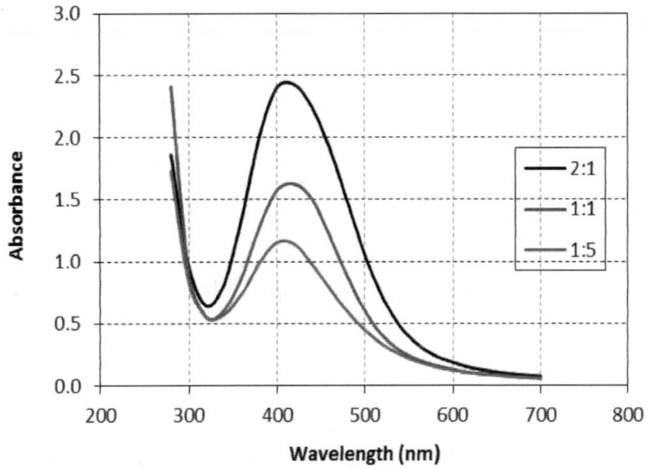

Figure 1. UV-Visible spectra of AgNPs fabricated by using papaya fruit extract at room temperature and with sunlight irradiation assistance. A peak absorbance of surface plasmon resonance was observed at 410 nm for all three different ratios of fruit extract to $AgNO_3$.

2.4. Colorimetric detection of mercury ions

An aliquot of freshly prepared AgNPs was put into transparent glass vials, then known concentration of Hg(II) and other metal ions (Al, Co, Cr, Cu, Fe, Mn, Ni, Pb and Zn) were added into the glass vials. Afterward, the absorption spectra were recorded using UV-visible spectrophotometer. The experiments of standards and samples were carried out under identical conditions to optimize the analytical performances. Simultaneously, the photographs were taken with a digital camera.

3. Results and discussion

Figure 1 shows UV-Visible spectra of AgNPs, which was fabricated using papaya fruit extract at room temperature with 15 minutes sunlight irradiation assistance. The best fruit extract to $AgNO_3$ solution ratio was 2:1. It is likely that lower concentration of fruit extract to $AgNO_3$ solution (1:5 and 1:1 ratios) was not sufficient to reduce Ag(I) quantitatively, thus more fruit extract exhibits better result as shown in figure 1. Colorless $AgNO_3$ was turned to yellowish-brown color of AgNPs colloid with an intense peak absorbance of surface plasmon resonance at 410 nm, which indicates a good dispersity of 35 – 50 nm size of AgNPs in water [12]. Without sunlight irradiation, it took seven days (vs 15 minutes) to synthetize silver nanoparticles under the same conditions mentioned above.

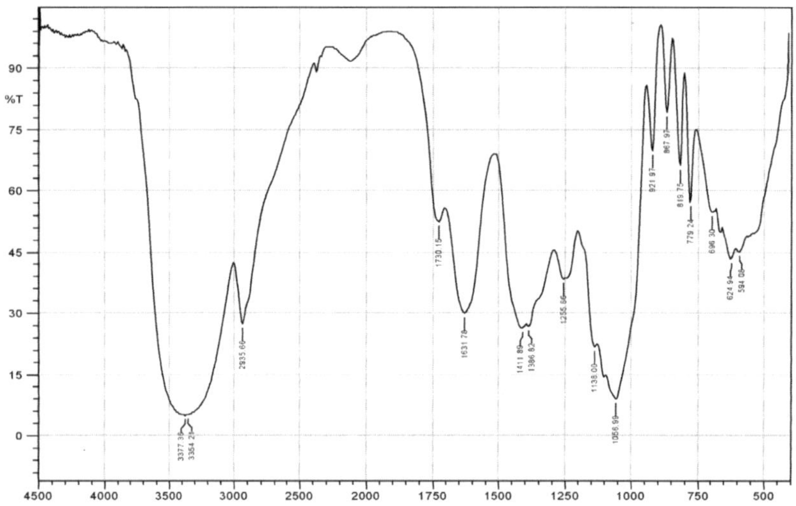

Figure 2. Fourier Transform Infrared (FTIR) spectra of dried powder of papaya fruit extract.

Table 1. Comparison of reagents for fabrication of NPs and its corresponding heavy metals detection.

Nanoparticles	Reagents	Detected heavy metals	References
	$NaBH_4$	Fe(III)	[16]
	Citrate	Cr(III), Cr(VI)	[17]
AuNPs	Citrate	Hg(II), Ag(I)	[18]
	Citrate	Al(III)	[19]
	L-tyrosine	Hg(II), Pb(II), Mn(II)	[20]
	$NaBH_4$	Mn(II)	[21]
	Citrus extract	Hg(II)	[22]
AgNPs	L-tyrosine	Hg(II), Pb(II), Mn(II)	[20]
	Soap-root plant	Hg(II)	[23]
	Papaya extract	Hg(II)	This study

High content of water-soluble organic compounds such as Vitamin C (ascorbic acid) and phenolic compounds in papaya fruit [13] may be attributed to its ability on reducing Ag(I) to Ag(0) in AgNPs green synthesis. Figure 2 shows FTIR spectra of papaya fruit extract that resemble those of ascorbic acid spectra [14]. Intense absorption bands at 3377 cm^{-1} can be assigned to O–H stretching of alcohols and/or phenols [15]. Absorption bands at 2935 cm^{-1} are characteristic of C–H stretching vibration. Medium bands at 1631 and 1056 cm^{-1} may result from the C=C ring stretching and C–OH bending vibration, respectively.

Table 1 shows the comparison of reagents that were previously used to synthesize nanoparticles and its corresponding heavy metals detection ability. Colorimetric method using AgNPs and AuNPs was used to analyze the concentration of various heavy metals such as Fe, Cr, Hg and Pb in aqueous environments at ppb level. The color changes of AgNPs and AuNPs were proportional to the heavy metals concentrations in sample, thus it is useful for qualitative and quantitative analysis.

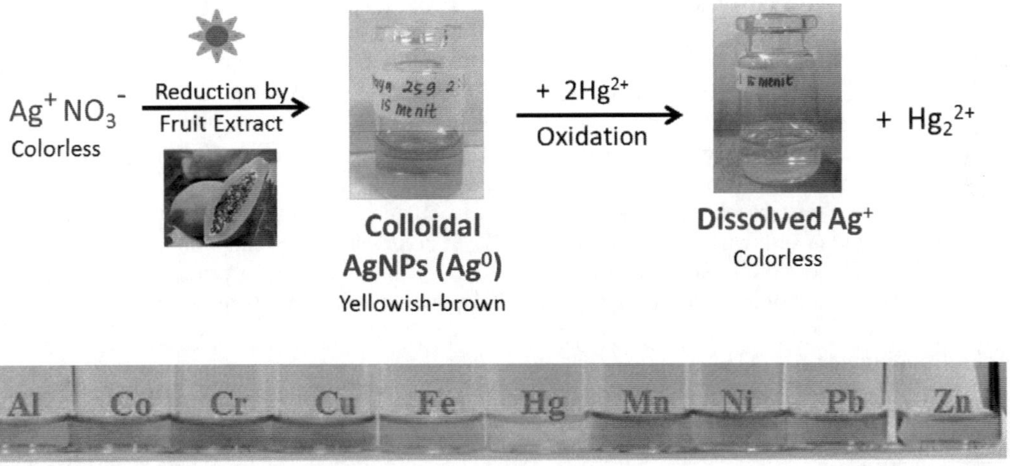

Figure 3. A schematic illustration of AgNPs formation using papaya fruit extract and Hg(II) colorimetric detection through redox reaction (upper scheme) and photograph of AgNPs selectivity test to various heavy metals (lower scheme).

A schematic mechanism of AgNPs formation using papaya fruit extract and Hg(II) colorimetric detection is shown in figure 3. Colorless $AgNO_3$ solution was turned to yellowish-brown color of AgNPs, and in the presence of Hg(II) ions, it turned back to colorless. The colorimetric detection was selective to only mercury ions. Other heavy metals, alkali and alkaline earth metals did not change the color of AgNPs. It is likely that reduction-oxidation reaction is the principle mechanism in AgNPs synthesis and mercury detection. As shown in table 2, the reduction potentials of Ag^+ and Hg_2^{2+} produced positive cell-potential (E cell) while other metals produce negative cell-potential. The redox reaction with positive cell-potential resultant will be favoured, as shown in equations (1) to (4).

$$E^{\theta}cell = E^{\theta}red - E^{\theta}ox \quad (1)$$

$$2Ag + 2Hg^{2+} \rightarrow 2Ag^+ + Hg_2^{2+} \qquad E^{\theta}cell = +0.12\ V \quad (2)$$

$$Ag + Cr^{3+} \rightarrow Ag^+ + Cr \qquad E^{\theta}cell = -1.54\ V \quad (3)$$

$$Ag + Na^+ \rightarrow Ag^+ + Na \qquad E^{\theta}cell = -3.51\ V \quad (4)$$

Table 2. Standard reduction potentials at 298°K. [24]

Reduction reaction	E^{θ} / V
$Au^+ + e^- \rightarrow Au$	+1.69
$Au^{3+} + 3e^- \rightarrow Au$	+1.40
$2Hg^{2+} + 2e^- \rightarrow Hg_2{}^{2+}$	+0,92
$Hg^{2+} + 2e^- \rightarrow Hg$	+0,86
$Ag^+ + e^- \rightarrow Ag$	**+0,80**
$Hg_2{}^{2+} + 2e^- \rightarrow 2Hg$	+0.79
$Fe^{3+} + e^- \rightarrow Fe^{2+}$	+0,77
$Cu^{2+} + e^- \rightarrow Cu+$	+0,16
$Cr^{3+} + 3e^- \rightarrow Cr$	-0.74
$Na^+ + e^- \rightarrow Na$	-2,71

4. Conclusion
In this paper, we report the first attempt to fabricate AgNPs using papaya fruit extract as bioreductant with sunlight irradiation assistance, which can be completed within 15 minutes. Water-soluble bioactive compounds from papaya fruit extract that act as bioreductants of Ag(I) to form AgNPs were most likely ascorbic acid and other phenolic compounds. The as-prepared AgNPs have yellowish-brown color with surface plasmon resonance peak at 410 nm. The presence of Hg(II) ions in the mixture changed the yellowish-brown color of AgNPs to colorless due to oxidation of Ag(0) in AgNPs to Ag(I) ions. The detection procedure of the developed method could be a promising tool as a low-cost and portable method for real-time qualitative and quantitative mercury analysis in environments.

Acknowledgments
We would like to thank anonymous reviewers for their valuable comments on the manuscript. Authors are grateful for the financial support from the Indonesia Ministry of Research, Technology and Higher Education (Kemenristek-Dikti).

References
[1] Vahabi K and Dorcheh S K 2014 *Biosynthesis of Silver Nano-Particles by Trichoderma and Its Medical Applications, in Biotechnology and Biology of Trichoderma* (Amsterdam: Nethterlands/ Elsevier) pp 393–404
[2] Radhakumary C and Sreenivasan K 2011 *Analyst* **136** 2959–62
[3] Chansuvarn W Tuntulani T and Imyim A 2015 *TrAC, Trends Anal. Chem.* **65** 83–96
[4] Alarcon E I 2015 *Silver Nanoparticle Applications*, ed Emilio Alarcon May Griffith Klas I. Udekwu (Switzerland: Springer International Publishing) p 146
[5] Gao Y X, Xin J W, Shen Z Y, Pan W, Li X and Wu A G 2013 *Sens. Actuators, B* **181** 288–93
[6] Jain D, Daima K H, Kachhwaha S and Kothari S L 2009 *Digest J. Nanomater. Biostructures* **4** 557–63
[7] Saxena A, Tripathi R and Singh R 2010 *Digest J. Nanomater. Biostructures* **5** 427–32
[8] Ramesh P S, Kokila T and Geetha D 2015 *Spectrochim. Acta. A. Mol. Biomol. Spectrosc.* **142** 339–43
[9] Nierenberg D W, Nordgren R E, Chang M B, Siegler R W, Blayney M B, Hochberg F, Toribara T Y, Cernichiari E and Clarkson T 1998 *The New Engl. J. Med.* **338** 1672–6
[10] Clarkson T W, Magos L and Myers G J 2003 *The New Engl. J. Med.* **349** 1731–7
[11] Firdaus M L, Darti P, Alwi W, Swistoro E, Sundaryono A and Ruyani A 2015 *AIP Conf. Proc.* vol 1677 (New York: Melvile/ AIP Publishing) p 110003
[12] Perez D P 2010 *Silver Nanoparticles*, ed Perez D P (Croatia: In-Tech)
[13] Gayosso-García Sancho L E, Yahia E M and González-Aguilar G A 2011 *Food Res. Int.* **44** 1284–91

[14] Panicker C Y, Varghese H T and Philip D 2006 *Spectrochim. Acta, Part A* **65** 802–4
[15] Schulz H and Baranska M 2007 *Vib. Spectrosc.* **43** 13–25
[16] Wu S P, Chen Y P and Sung Y M 2011 *Analyst* **136** 1887–91
[17] Chen W, Cao F, Zheng W, Tian Y, Xianyu Y, Xu P, Zhang W, Wang Z, Deng K and Jiang X 2015 *Nanoscale* **7** 2042–49
[18] Lou T, Chen Z, Wang Y and Chen L 2011 *ACS Appl. Mater. Interfaces* **3** 1568–73
[19] Chen S, Fang Y M, Xiao Q, Li J, Li S B, Chen H J, Sun J J and Yang H H 2012 *Analyst* **137** 2021–3
[20] Annadhasan M, Muthukumarasamyvel T, Sankar B V R and Rajendiran N 2014 *ACS Sustain. Chem. Eng.* **2** 887–96
[21] Zhou Y, Zhao H, Li C, He P, Peng W, Yuan L, Zeng L and He Y 2012 *Talanta* **97** 331–5
[22] Ravi S S, Christena L R, SaiSubramanian N and Anthony S P 2013 *Analyst* **138** 4370–7
[23] Farhadi K, Forough M, Molaei R, Hajizadeh S, and Rafipour A 2012 *Sens. Actuators, B* **161** 880–5
[24] Atkins P and Paula J de 2009 *Physical Chemistry* (UK: Oxford/ W H Freeman)

IOP Conf. Series: Journal of Physics: Conf. Series **817** (2017) 012030 doi:10.1088/1742-6596/817/1/012030

Study of LiBOB compound synthesis by vacuum process as lithium ion battery electrolytes

T Lestariningsih, E Wigayati, C Ratri and Q Sabrina

Research Centre for Physics, Indonesian Institute of Sciences
Kawasan Puspiptek Serpong Gd. 442 Tangerang Selatan, Banten 15314, Indonesia

E-mail: titik.lestariningsih@lipi.go.id

Abstract. Lithium bis (oxalato) borate or LiBOB is potential candidate to substitute $LiPF_6$ which has many problems in lithium ion batteries. Many studies have been synthesized of electrolyte salt LiBOB to improve performance as electrolyte lithium ion batteries. In this paper we have studied the synthesis of compounds LiBOB undergoing pre-heat treatment in a vacuum. LiBOB was synthesized by mixing technical grade raw materials $H_2C_2O_4.2H_2O$, LiOH and H_3BO_3. The mixture $H_2C_2O_4.2H_2O$ and LiOH was preheated at 60 °C for 2 h before adding H_3BO_3 in several time to be mortared in vacuum dryer, the mixture of the three starting materials was preheated in two steps at 70 °C for 6 h and the third step of preheating at a temperature of 100 °C. This powder was then characterized using XRD, FTIR and BET. The characterization results of LiBOB compared to commercial LiBOB powder. The XRD analysis results showed that the sample have formed LiBOB and LiBOB hydrate phase, while FTIR analysis results show the formation of functional groups of LiBOB. In addition, the BET results shows the surface area of synthesized LiBOB is 75.994 m^2/g, close the surface area of commercial LiBOB, i.e 108.776 m^2/g.

1. Introduction

Currently, lithium-ion batteries are the start-of-the-art power sources for its high voltage, high energy density, excellent cycle life, small self-discharge, and no memory effect, and good stability at high temperatures. The temperature stability of lithium-ion batteries depends mainly on the electrolyte component. The most widely used electrolyte salt, $LiPF_6$, hardly meet this need because of its poor thermal stability; $LiPF_6$ starts to decompose at 40 °C and will be totally decomposed at 80 °C. Lithium bis(oxalato)-borate (LiBOB) is regarded as the most promising electrolyte salt for the lithium-ion batteries. It exhibits the advantages of low cost, easy synthesis, and environmental friendliness [1]. LiBOB salt has many advantages compared to $LiPF_6$ including superior thermal stability, increased safety, and more stable solid electrolyte interphase (SEI) [2]. However, the poor salt solubility and the low conductivity of LiBOB in organic carbonates degrade the rate performances of cells using the LiBOB-based electrolyte [3]. LiBOB also has distinctive physical properties i.e hygroscopic and decomposes slowly in contact with air. At temperature above 300 °C LiBOB would be decomposed into Li_2CO_3, B_2O_3, CO_2, and CO [4].

The physical properties of LiBOB are interesting to observe as well as important for the research and development of lithium-ion battery. Many researches have been conducted about synthesis LiBOB such as ones that have been done by Bi-Tao Yu et al. [4] and Lestariningsih T et al. [5], who synthesized LiBOB from the analytical grade material by the method of solid state reaction (powder metallurgy).

Research concerning synthesis of LiBOB continued to be developed to get a quality product that is approaching the characteristics of the commercial product and also to improve the performance of lithium-ion battery, both from analytical grade as well as from technical grade materials. In this research, synthesis and characterization of LiBOB compound was performed by heat treatment under vacuum condition with reference to the research that has been done by Zhang Juan Li et al. [6]. As additional reference, heat treatment without employing vacuum condition refers to experiments conducted by Bi-Tao Yu [4] and Lestariningsih T. [5]. One of the important processes in the synthesis of LiBOB compound is the heat treatment. This process would remove water, carbon dioxide or other gases that have a chemical bond with the compound. Heating temperature that is too low will result in ineffective removal, while excess temperature will result in a decrease in surface area and pore size shrinking [7]. In the synthesis of LiBOB, overheating will result in decomposition reaction where Li_2CO_3, B_2O_3, CO_2 and CO would form. Bi-Tao reported stability of LiBOB in temperature range of 240-300 °C[4].

Purpose of this research is to study whether the synthesis of LIBOB from technical grade material under lower temperature and vacuum condition could generate salt LiBOB electrolyte with characteristics resembling commercial LiBOB. The use of technical grade materials is also employed in this research, to make use of abundant local materials therefore LiBOB salt can be synthesized with lower cost.

2. Experimental Methods

In the synthesis of LiBOB, lithium hydroxide (LiOH), oxalic acid dihydrate ($H_2C_2O_4.2H_2O$), and boric acid (H_3BO_3) were used as reactants. These materials were obtained in technical grade. The materials were homogenously mixed with grinding mill at 2:1:1 molar ratio. Solid state reaction method was employed in this research for the synthesis of LiBOB using established method. The reaction scheme for LiBOB synthesis is described as follows:

$$2H_2C_2O_4.2H_2O + LiOH.H_2O + H_3BO_3 \square LiB(C_2O_4)_2 + 9H_2O \quad (1)$$

In this study LiBOB salt was synthesized from technical grade raw materials under vacuum condition at temperature of 100 °C with reference to Zhang Juan Li et al.[6]. It was then compared with LiBOB salt synthesized without vacuum condition at temperature of 240 °C, referring to experiments by Bi Tao Yu et al. [4] and Lestariningsih T et al. [5]. Synthesis process without vacuum was performed in Thermoline muffle furnace while synthesis process in vacuum condition was done using oven Songling Chinese type SBVO-03. Powder crystallinity was scanned using X-Ray Diffractometer (XRD) Rigaku type Smart Lab, with CuKα target, wavelength λ = 1.5406 Å, and 2θ range of 10° to 80°. A diffractogram with diffraction peaks was obtained as a result and analyzed with PDXL software to collect formed phase information as well as qualitative analysis in form of crystal structure parameter. To confirm that LiBOB phase has been formed, LiBOB powder was also observed with FT-IR spectroscopy using Thermoscientific type Nicolet iS-10 to identify the functional group. FT-IR spectroscopy scanning was conducted within wave number range of 4000 to 500 cm^{-1} with ATR method. To study the pore structure of LiBOB compound, adsorption and desorption behaviour of synthesized LiBOB from technical and analytical grade was analyzed using Quantachrome Nova 4200e with BET method. Commercial LiBOB (Sigma Aldrich) was also analyzed as reference. LiBOB synthesis under vacuum condition as well as the process of with no vacuum follows these instructions as carefully as possible.

2.1. Synthesis of LiBOB under vacuum condition

In the synthesis of LiBOB, lithium hydroxide (LiOH), oxalic acid hydrate ($H_2C_2O_4.2H_2O$), and boric acid (H_3BO_3) were used as reactants. These materials were obtained in technical grade. The materials were homogenously mixed with grinding mill at 2:1:1 molar ratio. Heat treatment was performed using Songling Chinese type SBVO-03 oven in three steps; first step is heat treatment at temperature

60 °C held for 2 hours. After heating, the powder was mixed homogeneously with boric acid (H_3BO_3) using mortar. It was then heated under vacuum condition at temperature of 70 °C for 6 hours and then the temperature was raised to 100 °C and detained for 3 hours. The result was a white powder analyzed by FTIR, XRD and BET.

2.2. Synthesis of LiBOB without vacuum
In the synthesis of LiBOB, lithium hydroxide (LiOH), oxalic acid hydrate ($H_2C_2O_4.2H_2O$), and boric acid (H_3BO_3) were used as reactants. These materials were obtained in technical grade. The materials were homogenously mixed with grinding mill at 2:1:1 molar ratio. Heat treatment was performed using thermoline muffle furnace in two steps; first step at temperature 120 °C held for 4 hours, and subsequently at 240 °C held for 6 hours. The result was a white powder which was then analyzed by FTIR, XRD and BET.

3. Results and Discussion

3.1. XRD Analysis
X-ray diffraction patterns of synthesized LiBOB from technical grade raw materials with and without vacuum condition were presented on figure 1. Similar diffraction patterns were seen on both sample, showing similar phase formed in synthesized LiBOB with the process in vacuum (figure 1(a)), synthesized LiBOB without vacuum (figure 1(b)) and Commercial LiBOB (Sigma Aldrich) as reference (figure 1(c)). Based on the identification method using Hanawalt Table, the Miller index peaks of synthesized LiBOB with process vacuum showed the existence of LiBOB hydrate phase (based on ICDD 01-073-9447) and LiBOB phase (based on ICDD 00-056-0139), and the phase composition was found 94,6 % and 5,4 %, respectively. For synthesized LiBOB without process vacuum showed the existence of LiBOB hydrate phase (based on ICDD 01-073-9447), LiBOB phase (based on ICDD 00- 056-0139), and LiOH hydrate phase (based on ICDD04-010-3208) appeared as impurities. This is supposedly because the process of heating the sample has not been optimal. Where is the heating process to remove all unneeded substances from deep grain of salt LiBOB. The phase composition was found 13%, 59%, and 28% for LiBOB, LiBOB hydrate, and LiOH hydrate phase, respectively. While the synthesized LiBOB with the process in vacuum indicates results that already resembles salt LiBOB commercial. Research on synthesized LIBOB from analytical grade material without vacuum process generated 17.8% phase of LiBOB and 82.2% phase LiBOB hydrates [12]. Based on the refinement results of X-ray diffraction pattern, both sample possessed orthorhombic structure. The complete results of the X-ray diffraction refinement on the sample are shown in table 1.

Table 1. Crystallographic data of synthesized LiBOB sample from Rietveld refinement

Sample	Indicated phases	Lattice parameter and unit cell volume				Lattice system	Composition of the phase (%)	Density (g/cm^3)
		a/Å	b/Å	c/Å	V/Å^3			
LiBOB with vacuum process	LiBOB	6.501	7.512	12.079	143.9	Orthorhombic	5.4	2.053
	LiBOB hydrate	16.110	15.912	5.616	589.9	Orthorhombic	94.6	1.953
LiBOB without vacuum process	LiBOB	6.3698	15.918	5.616	1444	Orthorhombic	13.5	1.941
	LiBOB hydrate	16.132	15.918	5.616	637.7	Orthorhombic	59.0	2.020
	LiOH	7.314	8.543	3.197	185.5		28.0	1.469
LiBOB Commercial	LiBOB hydrate	16.110	15.912	5.616	1441	Orthorhombic	94.6	1.953

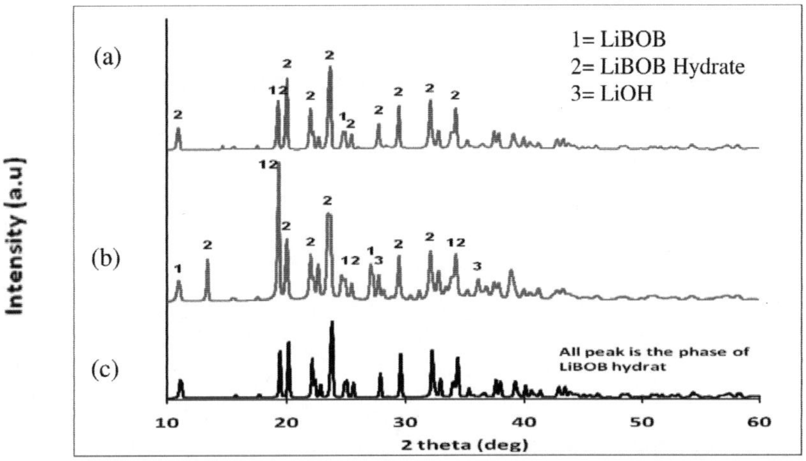

Figure 1. X-Ray diffraction patterns of (a) synthesized LiBOB with the process in vacuum (b) synthesized LiBOB without the process in vacuum and (c) commercial LiBOB.

3.2. FTIR Analysis

Characterization of LiBOB using FT-IR spectroscopy was aimed to identify the functional groups contained in the LiBOB compound. Functional groups of synthesized LiBOB with the process in vacuum and without the process in vacuum as well as the reference LiBOB commercial was presented in table 2. Based on the results of FTIR analysis shown in table 2, it indicates that the second sample of LiBOB synthesis results have demonstrated the existence of a functional group-forming compounds of LiBOB although there is still the presence of the big shift in wavelength which is not significant.

Table 2. Functional groups and absorption bands of synthesized LiBOB with and without vacuum process and commercial LiBOB as reference.

Functional groups	Wave number (cm^{-1}) Commercial LiBOB	Wave number (cm^{-1}) LiBOB without vacuum process	Wave number (cm^{-1}) LiBOB with vacuum process
OH	3441	3514.31	3511.37
C=O oscillate in phase and out of phase	1818. 1781	1812.9 1795.2	1819.6 , 1795.3
COO^- asymmetric stretch	1639	1633.36	1653.97
COO^- asymmetric stretch	1443	1447.91	1447.91
B-O stretch		1327.2	1347.8
C-O-B-O-C stretch	1307	1306.6	1306.6
C-O-C asymmetric stretch	1222	1227.15	1227.15
O-B-O symmetric stretch	1088	1085.85	1082.91
O-B-O symmetric and asymmetric stretch	998/983	994.6	991.66
COO^- deform	709	712.02	712.02
B-O deform	607	611.94	611.94
BO_4 bond	492/480	544.24	544.24

According to the literature that the absorption bands should appear on the LiBOB in wave numbers 1740 – 1820 cm^{-1} (C = O absorption bands) and the existence of C = O stretch, B-O at wave numbers (980 – 1370 cm^{-1}) [8-9] and the emergence of Lithium Oxalate in the sample in wave numbers around 1658 and 778 cm^{-1} [9].

3.3. BET Analysis

To determine pore structure of LiBOB compound, Brunauer-Emmett-Teller (BET) method was employed by calculating nitrogen gas adsorption to obtain surface area, average pore size and total pore volume of LiBOB compounds [10].

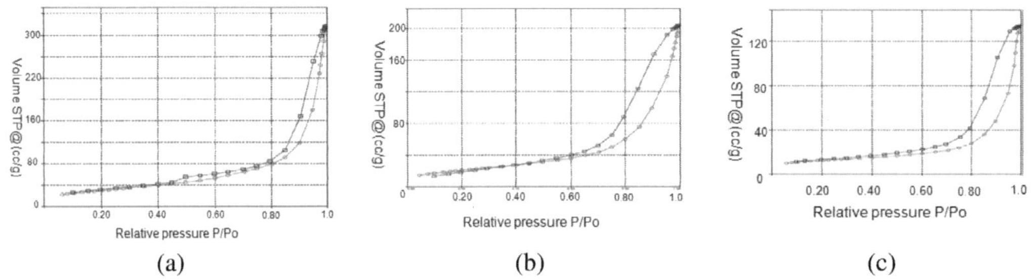

Figure 2. Linear isotherm graphic of (a) commercial LiBOB and synthesized LiBOB (b) with vacuum process (c) without vacuum process.

Linear isotherm of nitrogen adsorption-desorption system was shown in figure 2. It can be seen that the nitrogen adsorption isotherm of all LiBOB samples showed a similar pattern in which there was a rapid increase in relative pressure (P/Po). It started low then rose slowly in the middle and increased significantly at P/Po approaching one. This isotherm is a type IV isotherm, i.e. adsorption type of mesoporous solids with particle size of 2-50 nm [11]. It observed in figure 2(c) that adsorbed and desorbed gas at P/Po=0.4 were very low compared to figure 2(a) and 2(b). Similar behaviour was also observed at P/Po = 1. It was seen from table 3 that synthesized LiBOB without vacuum process had lower surface area albeit relatively similar pore size i.e. 19 nm due to lowest value of total pore volume. On the other hand, the surface area of synthesized LiBOB under vacuum condition was close to the surface area of commercial LiBOB. The small pore volume of synthesized LiBOB without vacuum process was caused by the impurities clogging the pores, as implied from the XRD data that there were phases other than LiBOB and LiBOB hydrate. These phases were the possible reason of the pore clogging, causing low surface area of LiBOB particle.

Table 3. Pore characteristics of LiBOB salt particles.

Sample	Surface Area (m^2/g)	Average pore size (nm)	Total pore volume (cc/g)
Commercial LiBOB	108.776	19.148	0.521
Synthesized LiBOB with vacuum	75.994	16.575	0.315
Synthesized LiBOB without vacuum	41.524	19.943	0.207

Table 4. Surface area against the phase composition of LiBOB hydrate.

Sample	Surface Area (m^2/g)	Composition of LiBOB hydrate phase (%)
Commercial LiBOB	108.776	100
Synthesized LiBOB with vacuum process	75.994	94.6
Synthesized LiBOB without vacuum process [12]	41.524	59

Table 4 showed that, when associated with the results of X-ray diffraction analysis, the larger the surface area of sample LiBOB then will form more LiBOB hydrates, because of larger surface area of hydrate compounds, therefore it will easily fit into the compound. LiBOB compound is hygroscopic.

When associated with the analysis result of XRD (table 1) and BET, low surface area and pore size of LiBOB synthesis without vacuum process was possibly due to the magnification of crystal grains (crystals inflate). For example, lithium hydroxide hydrate was formed during LiBOB synthesis without vacuum process resulted in larger size and growing volume. BET analysis showed that LiBOB without vacuum process have the smallest surface area, although the pore size is greater than LiBOB synthesized under vacuum condition. While prior research to synthesize LiBOB without vacuum process has produced LiBOB salt with similar characteristics to commercial LiBOB [12], this research used analytical grade raw materials which is more expensive than technical grade materials. In this research, LiBOB synthesis using technical grade raw materials under vacuum condition has obtained LiBOB compound with similar characteristics to commercial LiBOB products. It is expected for future research that technical grade material can be utilized to synthesize LiBOB with lower cost.

4. Conclusion

Synthesis of LiBOB from technical material with vacuum processes is better than synthesis LiBOB without vacuum process based on the XRD, FTIR or BET analysis results. LiBOB synthesized with vacuum process shows results that are closer to the characteristics of commercial LiBOB. The results of XRD analysis shows that synthesized LiBOB with and without vacuum process have formed LiBOB hydrate phase with composition of 94% and 59% respectively (commercial LiBOB has 100% LiBOB hydrate phase). The results of the BET analysis show that LiBOB synthesized in vacuum process have surface area of 75.994 m^2/g while LiBOB synthesized without vacuum process have surface area 41.524 m^2/g and the surface area of LiBOB commercial was 108.776 m^2/g.

Acknowledgements

The authors would like to acknowledge financial support for this project, from Ministry of Research and Technology and Higher Education of Indonesia through Insentif Riset SINas grant budget year 2015.

References

[1] Yu B T, Qiu W H, Li F S and Xu G X 2006 *Electrochem. Solid State Lett.* **91** A1-A4
[2] Yang L, Furczon M M, Xiao A, Lucht B L, Zhang Z and Abrahan D P 2010 *J. Power Sources* **195** 1698
[3] Abrahan D P, Furczon M M, Kang S H, Dees D W and Jansen A N 2008 *J. Power Sources* **180** 612
[4] Yu B T, Qiu W H, Li F S and Li L F 2007 *J. Power Sources* **174** 1012
[5] Lestariningsih T, Wigayati E M and Prihandoko B 2013 *Jurnal Telaah* **31** 39
[6] Juan Z L, Jing S, Giang L F, L WL 2010 *Journal of Central South University* **41**(1) 67-72
[7] Stanmore B R and Gilot P 2005 *Fuel Process. Technol* **86** 1707
[8] Aravindan V and Vickraman P 2009 *Polym. Eng. Sci.* **49** 2109
[9] Yang L, Furczon M M, Xiao A and Lucht B L 2010 *J. Power Sources* **195** 1698
[10] Haber J, Block J H and Delmon B 1995 *Pure Appl. Chem.* **67** 1257
[11] Gregg S J and Sing K S W 1982 *Adsorption, Surface Area and Porosity* 2nd Edition (USA: Academic Press)
[12] Lestariningsih T, Rina R C, Wigayati E M and Qolby S 2016 *AIP Conf. Proc.* vol 1711 (USA: AIP Publishing) 060005-6

IOP Conf. Series: Journal of Physics: Conf. Series **817** (2017) 012031 doi:10.1088/1742-6596/817/1/012031

The influence of Sb doping on the structural, optical and electrical properties of tin oxide thin film

M N Yusnidar[1], V Fauzia[1], D Handoko[1] and L Hanum[2]

[1]Department of Physics, Universitas Indonesia, Depok, Indonesia
[2]Research Center for Metallurgy and Material LIPI, Serpong, Indonesia

E-mail: vivi@sci.ui.ac.id

Abstract. Antimony-doped Tin Oxide ($Sb:SnO_2$), shortened as ATO, has become more popular due to their great technological importance. ATO was considered as important transparent conducting material for optoelectronic devices and sensors application because of its unique and special characteristic such as high transparency in visible region, and high electrons concentration and mobility. In this study, the $Sb:SnO_2$ thin film have been fabricated with low cost and simple ultrasonic spray pyrolysis method. SnO_2 thin film was deposited with three different Sb concentrations namely 1, 2, and 3 wt%. The structural, morphological, optical and electrical properties of this film have been analyzed by using X-ray diffraction, scanning electron microscopy, UV-VIS and four point probe instruments. Based on the optical and electrical properties characterization, the best concentration of Sb doping was 2wt% because its transmittance was above 80% at all wavelength range measured, and the electrical resistivity was quite low, at 0.663 x 10^{-3} Ω cm.

1. Introduction

Transparent Conducting Oxide (TCO) film is an important part of technologies which requires both, large area of high electrical conductivity and optical transmittance in the visible region [1]. Antimony-doped Tin Oxide ($Sb:SnO_2$), shortened as ATO, is the promising TCO material due to non toxic antimony characteristic and it is cheaper than some common used doping materials, such as indium and fluorine [2]. The transparent conducting $Sb:SnO_2$ film have huge potential application in solar cells, light emitting diodes, flat panel display and biochemical sensors devices [3, 4]. An important key to adjust both optical transparency and electrical conductivity in the oxides is by controlling band gap energy and improving the free charge carriers. Hence, the electrical conductivity of TCO should be improved by increasing the number of free charge carries through intrinsic and extrinsic defects, such as oxygen vacancies and extrinsic dopants, typically higher-valence metal cautions [5].

ATO thin film could be deposited using various methods such as pulse laser deposition, sputtering, chemical vapor deposition, sol-gel, hydrothermal, and spray pyrolysis [2]. Spray pyrolysis is one promising technique to produce a conductive and transparent thin film in a cost-effective way because it could be operated in low temperature without high vacuum instrument [6].

In this paper, the $Sb:SnO_2$ thin film were deposited using ultrasonic spray pyrolysis method with metal salts $SnCl_2.2H_2O$, $SbCl_3$ as the precursor and ethanol as the solvent. The $Sb:SnO_2$ film were kept at temperature of 450°C with three different Sb doping concentrations. Based on Bendjedidi [7], the film which was deposited below 450°C were amorphous, whereas for the higher temperatures the film crystalinity were increased hence the transmittance and electrical conductivity would be enhanced. In this paper, we would report the effect of antimony doping on the SnO_2 thin film properties such as morphology, crystal structure, optical and electrical properties.

2. Experimental

The precursor solutions containing 1M chloride dehydrate ($SnCl_2.2H_2O$) in ethanol and three different antimony chloride ($SbCl_3$) concentrations (1, 2, 3 wt %) were stirred for an hour at room temperature. The deposition of $Sb:SnO_2$ thin film was carried out using ultrasonic spray pyrolysis. The precursor solutions were placed into a container in the commercial ultrasonic nebulizer (1.7 MHz) and sprayed for 10 minutes onto the glass substrate which were heated on a hotplate at a temperature of 450°C.

The structural properties $Sb:SnO_2$ thin film were observed using a powder X-ray diffractometer XRD-7000 with Cu-Kα radiation in the angle range of 10°-80°. The morphology of the film was observed by scanning electron microscope (SEM) JEOL JED-2300. An optical spectrum of transmission, absorption, and reflectance at a wavelength range between 300 - 900 nm were monitored using GENESYS 10S UV-VIS Spectrometer and Hitachi UV-VIS Spectrometer U-3900H. The electrical properties were studied by using a Four Point Probe (FPP) 5000.

3. Result and discussion

3.1. Structural analysis

The X-ray diffraction patterns of $Sb:SnO_2$ thin film which were deposited at various Sb doping concentrations is shown in figure 1. All samples have five diffraction peaks at 26.61°, 33.92°, 37.99°, 51.84°, 61.95° and 66.05° corresponding to the (110), (101), (200), (211), (310) and (301) planes of tetragonal routile crystal structure of the SnO_2 phase. According to this observation, other phases, such as SnO, Sb_2O_3, and Sn_2O_3, did not exist in the deposited thin film. It indicated that the O atoms were replaced by Sb atoms in the $Sb:SnO_2$ thin film. For all samples, the intensity of (200) plane is more dominant than the intensity of other crystal planes, except for 1% $Sb:SnO_2$ sample. This low intensity is more likely appears as the result of a very thin layer was formed. Moreover, 1% and 2% $Sb:SnO_2$ samples show the amorphous phase at the 2θ range of 20-35°. This characteristic was also shown in the $Sb:SnO_2$ thin film deposited by RF magnetron sputtering with post annealing treatment [2] and chemical spray pyrolysis [8]. The reason for the coexistence of amorphous phase in $Sb:SnO_2$ thin film still remains unclear.

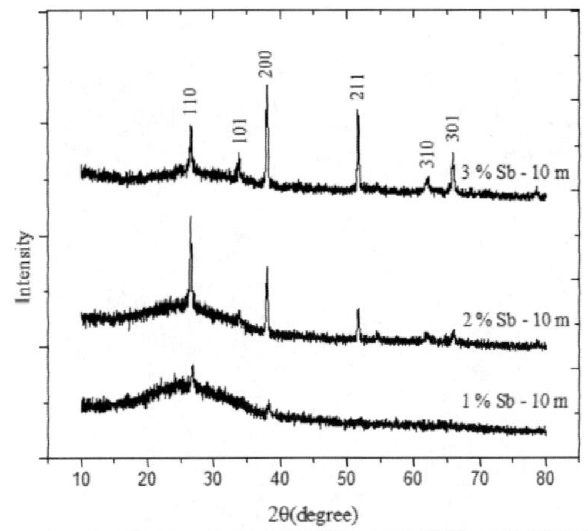

Figure 1. XRD patterns of $Sb:SnO_2$ thin film with three different Sb doping concentrations

IOP Conf. Series: Journal of Physics: Conf. Series **817** (2017) 012031 doi:10.1088/1742-6596/817/1/012031

All samples show the same trend for growing along the (200) direction. The full width at half maximum (FWHM) of (200) peak are 0.7454°, 0.3461°, and 0.4169° for the SnO_2 film with Sb doping concentrations of 1, 2, and 3 wt %, respectively. The increasing FWHM value means the decreasing of crystallite size and vice versa. The crystallite size of the SnO_2 film with Sb doping concentrations of 1, 2, and 3 wt%, is 9.8 nm, 15.0 nm and 10.1 nm, respectively. Based on this data, there is no linear relationship between the Sb concentrations with the crystallite size.

3.2. Morphological properties

Figure 2 shows the SEM images of Sb:SnO_2 thin film deposited on glass substrate at various Sb doping concentrations. It was seen that the surface morphology of the film is strongly dependent on the doping concentration. The film appeared to have uniform distribution without cracks. Except for 1%Sb:SnO_2 thin film, the layers were more likely made up of octahedron-shape grains. The average of grain size increased by enhanced the Sb concentration. This output was similar to the result reported by Yadav et al. [5] who also found that the average grain size of Sb-doped SnO_2 thin film was larger than the pure SnO_2 thin film. Figure 3 shows the cross-sectional view of the thin film, where all thickness of the ATO thin film was in the order of 400 nm.

Figure 2. SEM images of the SnO_2 thin film with Sb concentration of (a) 1, (b) 2, (c) 3 wt.%

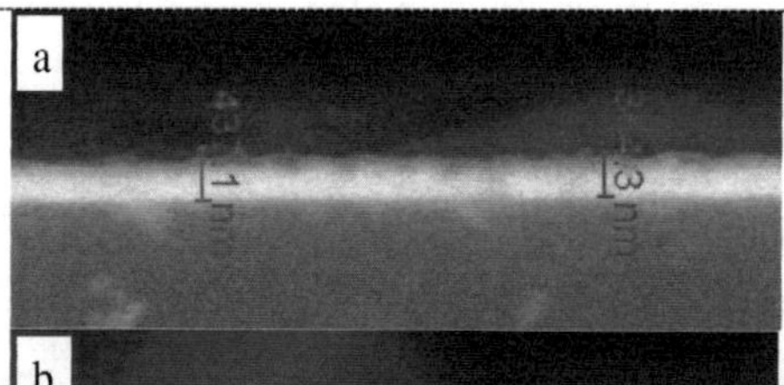

Figure 3. Cross sectional view of the SnO_2 thin film with Sb concentration of (a) 1, (b) 2, (c) 3 wt.%

3.3. Optical properties

Figure 4 shows the transmittance spectra of Sb:SnO_2 thin film with various Sb doping concentrations. According to the spectra, the addition of 1 and 2% Sb has increased the transmittance up to 80% at a wavelength range between 350 to 900 nm. This result meets the minimum transparency requirement for Transparent Conducting Oxide (TCO). The addition of 3% Sb has reduced the transmittance, down to 60%, because it could increase the rate of crystal defects. The huge defect might increase the light scattering, hence the transmittance would be reduced and the thin film became darker [9]. In contrast to the transmittance spectra result, the absorption spectra as seen in figure 4(b) shows that by increasing of Sb concentrations, the absorption was also enhanced at all wavelengths region.

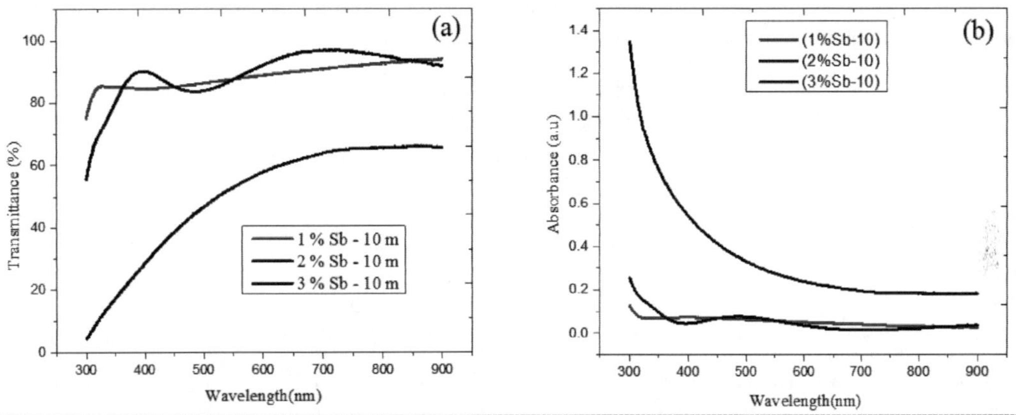

Figure 4. (a) Transmittance and (b) absorbance spectra of Sb:SnO_2 thin film with three different Sb doping concentrations

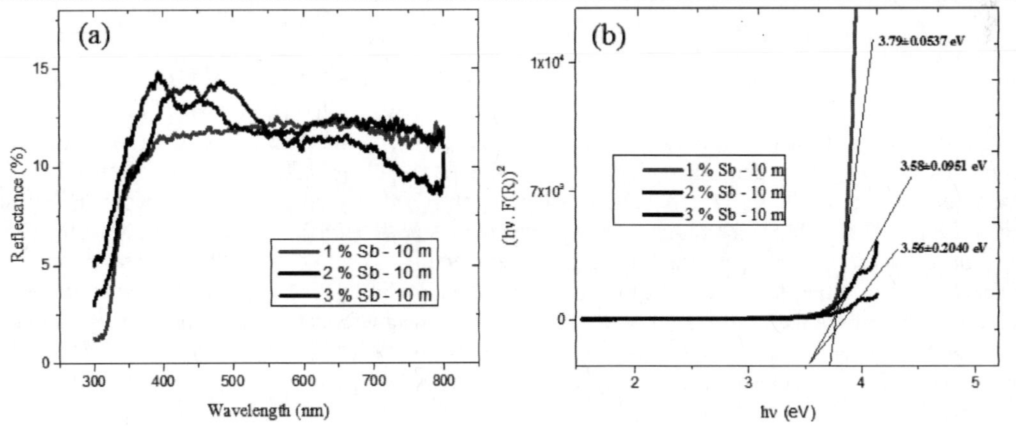

Figure 5. (a) Reflectance spectra and (b) tauc plot of Sb:SnO_2 thin film with three different Sb doping concentrations

The optical band gap of the Sb:SnO_2 thin film was determined using optical reflectance spectra and Tauc plot as shown in figure 5. The direct band gap for the Sb:SnO_2 thin film with Sb concentrations of 1, 2 and 3 wt.% are 3.79 ±0.0537, 3.58 ± 0.0951 and 3.56 ± 0.2040 eV respectively. The direct band gap has decreased by increasing the Sb concentrations. Yadav et al. reported that the band gap energy increased from 3.76 eV to 3.80 eV for the Sb:SnO_2 thin film 1 wt.% to 2 wt.% and then decreased to 3.72 eV for 3 wt.% [6]. The reduction of the band gap energy in this experiment is similar to those reported by Gurakar et al [10].

3.4. Electrical properties

The electrical resistivity values are tabulated in table 1. It shows that the increase of Sb concentrations could significantly reduce the resistivity, because Sb has been well known as an effective electrons donor, so that it might decrease the electrical resistivity of the thin film [5].

Table 1. The resistivity of SnO_2 thin film with three different Sb doping concentrations

Sb doping concentrations	Resistivity (10^{-3} Ω cm)
1%	165
2%	0.663
3%	0.1834

Based on the optical characterization, the SnO_2 thin film with Sb 1 and 2 wt.% appear to have higher transparency than those with Sb 3wt.%. However, the SnO2 thin film with Sb 1 wt.% comes out to have very high electrical resistivity. Therefore, it could be concluded that the optimum Sb doping concentration is 2 wt % because the transmittance is above 80% at wavelength range measured and the electrical resistivity as low as 0.663 x 10^{-3} Ω cm. The results is better than the result, which was recently reported by Yadav et al who have successfully obtained the ATO thin film with an electrical resistivity as low as 0,704461 x 10^{-3} Ω cm and the transmittance below 80% [5].

4. Conclusions

The antimony doped tin oxide thin film with three different Sb concentrations (1, 2 and 3 wt %) were synthesized by using ultrasonic spray pyrolysis method. Afterwards, the characterization and analysis were conducted to study the effect of Sb doping concentrations on the structural, optical and electrical properties. XRD pattern confirmed that the thin film was polycrystalline with tetragonal crystal structure. All samples show a preferential growth along the (200) crystal plane, while there is no linear relationship between Sb concentrations with the crystallite size. The SEM images showed that the film was made of continuous, uniform octahedron-shape grains without cracks. By increasing of Sb concentration, the grain size was increased; the transparency, direct band gap and the resistivity were decreased. Based on the optical and electrical properties characterization, the optimum Sb doping concentration should be used is 2 wt % because the transmittance is above 80% at all wavelength range measured and the electrical resistivity as low as 0.663 x 10^{-3} Ω cm.

Acknowledgements

This work was supported by Hibah Publikasi Internasional Terindeks untuk Tugas Akhir Mahasiswa (PITTA) 2016 Universitas Indonesia. The authors also would like to thank Prof. Dr. Muhamad Mat Saleh and Assoc. Prof. Dr. Akrajas Ali Umar From Organic and Printed Electronic Laboratory (OPEL) IMEN Universiti Kebangsaan Malaysia for this beneficial research collaboration.

References

[1] Ravichandran K and Philominathan P 2008 *Mater. Lett.* **62**(17) 2980
[2] Zheng H, Li L, Sun Z, Yu S and Luo W 2016 *Appl. Surf. Sci.* **362** 230
[3] Sharma S, Volosin A M, Schmitt D and Seo D K 2013 *Chem. Mater.* **1** 699
[4] Yang W, Yu S, Zhang Y and Zhang W 2013 *Thin Solid Films* **542** 285
[5] Yadav A A, Pawar SC, Patil D H and Ghogare M D 2015 *J. Alloys Compd.* **652** 145
[6] Babar A R, Shinde S S, Moholkar A V, Bhosale C H, Kim J H and Rajpure K Y 2011 *J. Semicond.* **32**(5) 1
[7] Bendjedidi H, Attaf A, Saidi H, Aida M S, Semmari S and Bouhdjar A 2015 *J. Semicond.* **36**(12) 2
[8] Orimi L R and Maghouli M 2016 *Optik* **127**(1) 263
[9] Babar A R and Rajpure K Y 2015 *J. Anal. Appl. Pyrolysis* **112** 214
[10] Gurakar S, Serin T and Serin N 2015 *Appl. Surf. Sci.* **352** 16

IOP Conf. Series: Journal of Physics: Conf. Series **817** (2017) 012032 doi:10.1088/1742-6596/817/1/012032

Effect of sonication process on natural zeolite at ferric chloride hexahydrate solution

T A B Prasetyo and B Soegijono

Department of Physics, Material Science Program, Universitas Indonesia, Depok, Indonesia 16424

E-mail: thomas.agus51@ui.ac.id; naufal@ui.ac.id

Abstract. Natural zeolite in $FeCl_3.6H_2O$ solution which was exposed with high intensity ultrasonic for 40 min, 80 min and 120 min has been studied. X-Ray Diffraction (XRD) pattern revealed changed pattern and new peaks due to sonication process. Sonication contributed into element composition changes and it had been found also during quantitative chemical microanalysis analysis (EDX) and indicates an increment of 320.81 % (ZAM2) Fe element. Noninvasive back scatter (NIBS) analysis of various ultrasonic times affected to the particle size distribution, surface area and pore analysis. By using density functional theory (DFT), we revealed some improvements such as 44.03% surface area and 67.25% pore radius. We believe that controllable ultrasonic processing in the ferric chloride hexahydrate solution will produce uniform natural zeolite physical and chemical properties as a candidate of adsorbent materials.

1. Introduction

In an attempt to improve Indonesian natural zeolite chemical properties, the authors undertaken characterization and modification of natural zeolite produced in Bayah, Indonesia, which has appeared to consist mainly of mordenite type crystalline matter [1], and its modifications by treating with an addition of $FeCl_3.6H_2O$ and sonification process for 40min, 80min and 120min and/or hydrothermal followed by microwave treatment. Bayah natural zeolites as various type adsorbates has been studied previously and this material has been standardized by Indonesian standardization body (SNI), as an adsorbent materials and is categorized as the material with the highest cation exchange capacity compared with other Indonesian natural zeolites from Sukabumi, Tasikmalaya and Cikalong [1,2].

High intensity ultrasonic processing is one of applied technology to produce Nano materials and chemical reaction process. High intensity ultrasonic liquid processor in operation produces micro bubles in between of the primary and secondary flow, this sonification effect can increase the physical and chemical properties of a material [3]. It has been studied and understood that high power ultrasonic irradiation is very effective in homogenizing the reactants in the suspension, leading to the improvement of reactivity of both solid and liquid by the simulation of their active surface [4]. Purification of Natural Zeolite by using ultrasonic technique is also applied by researchers to remove and separate impurities materials, this technique is believed to deliver a faster process and result [5].

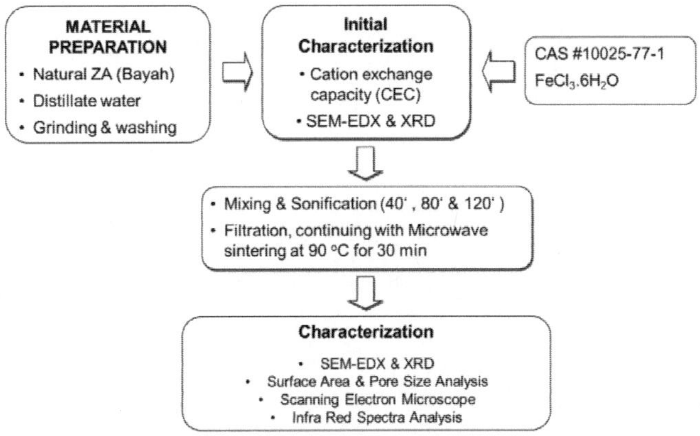

Figure 1. Research flow diagram

2. Experimental procedure

Natural zeolite produced in Bayah, Indonesia was grinded, homogenized and sieved below 150 μm (100 Mesh Tyler ™) before characterization and experimentation [6]. Ferric Chloride Hexahydrate ($FeCl_3.6H_2O$) produced by Merck Millipore, Chemical Abstracts Service (CAS) #10025-77-1 as specified composition was diluted in 100 ml distillate water.

To remove impurities, sieved natural zeolite were washed and stirred with distillate water, washing process was conducted four times with two hours long for each sequence [7]. Natural drying was performed prior to initial characterization to find out zeolite properties and involved phase, a natural zeolite sample coding is ZA (natural zeolite).

Reflux zeolite which is prepare was added with $FeCl_3.6H_2O$ as specified composition and dilute on 100 ml distillate water to become %wt composition of $ZA_{(66.7)}FeCl_3.6H_2O_{(33.3)}$. Synthesis of $ZA_{(66.7)}FeCl_3.6H_2O_{(33.3)}$ treated by high intensity ultrasonic liquid processor VCX750 at 20 kHz [8], 40 Amp for 40 min and sample code ZAM1, 80 min with sample code ZAM2 and 120 min with sample code ZAM3. Slurry product of sonification was precipitated and followed by Microwave heating of remain solid at temperature of 90°C for 30 min [7].

Particle size analysis was performed with dispersing 0.025 gr sample into aquades and evaluated with non invasive back scatter (NIBS) method.

X-ray powder diffraction analysis was performed using XRD analysis instruments (Philips PW3710). The samples of natural zeolites prior and after modification by adding $FeCl_3.6H_2O$ were mounted on holders then introduced for analysis. The source consisted of CoKα radiation (λ=1.78896Å), monochromator on secondary optics, 40 kV power and 30 mA current. Each sample was scanned within the 2θ range of 20–100.

JEOL-6510 SEM-EDS equipment was utilized to carry out photomicrographs observation of natural zeolites prior and after modification by adding $FeCl_3.6H_2O$. The quantitative chemical microanalysis analysis (EDX) also observed to the samples for element composition characterization of the analyzed volume.

The natural zeolites prior and after modification by adding $FeCl_3.6H_2O$ specific surface area was evaluated by the nitrogen gas adsorption method equipment (Quadrasorb-2 Quantachrome Instrument)

3. Result and discussion

Initial grinded and sieved zeolit (ZA) average particle size is 7154 d.nm. An addition of $FeCl_3.6H_2O$ solution and sonification process into ZAM1, ZAM2 and ZAM3 found that significant particle size reduction to averages of 5031 d.nm, 4298 d.nm and 3387 d.nm respectively, figure 2 shows that particle sizes are distributed into smaller size and proportional with sonification duration.

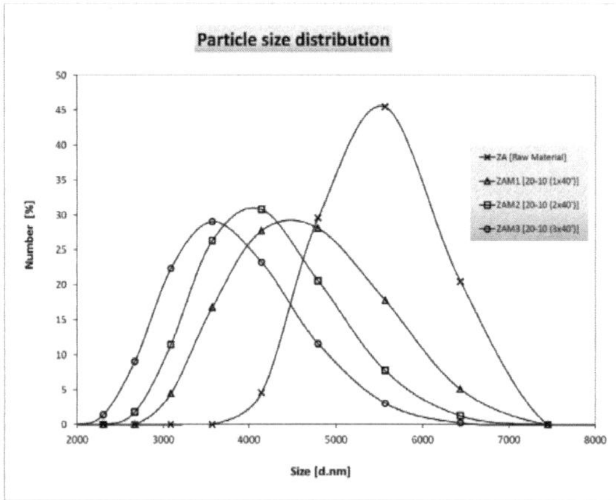

Figure 2. Particle size distribution of ZA and ZA + $FeCl_3.6H_2O$

The XRD analyses revealed a very complicated structure of the natural zeolite used, rietveld analysis by using GSAS software found that ZA sample majority consisting of mordenite and less clinoptilolite phase, according to the results of X-ray diffraction (XRD) studies and rietveld analysis of ZA sample found that mordenite phase unit cell formula weight 8101.269, density 4.583 gm/cm^3, weight fraction 0.99949 and clinoptilolite phase unit cell formula weight 4247.800, density 5.027 gm/cm^3, weight fraction 0.51158 x 10^{-3}.

Reference data from 99-100-6672 card to be used for mordenite phase unit cell parameters a = 18.3581 Å, b = 21.0641 Å, c = 7.5901 Å and $\alpha=\beta=\gamma=90°$ and space group Cmcm. Reference data from 96-900-1275 card to be used for clinoptilolite phase unit cell parameters are: a = 18.9119 Å, b = 7.5915 Å, c = 10.2398 Å and $\alpha=\beta=\gamma=90°$ and space group C1 2/m1.

New peaks have been found and observed on ZAM1 and ZAM2 samples, those peaks indicated that sonification process contributed into Fe element composition changes indicated more aerinite phase on ZAM1 and ZAM2 but less aerinite phase on ZAM3 due to less Fe element as seen on figure 3, this peaks also found on quantitative chemical microanalysis analysis (EDX).

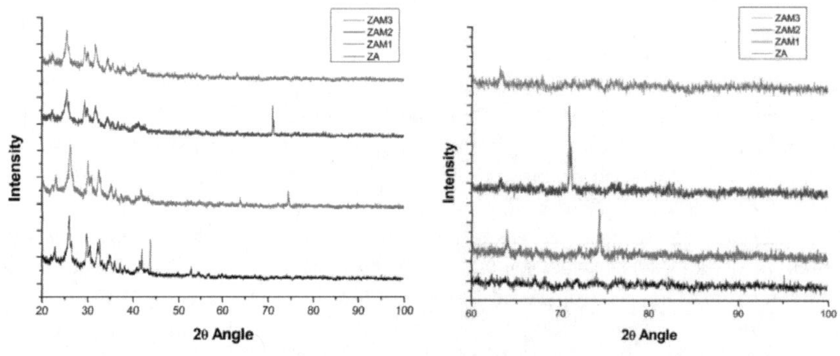

Figure 3. XRD Diffraction of ZA, ZAM1, ZAM2 and ZAM3

Photomicrographs observations of natural zeolites prior to and after modification by adding $FeCl_3.6H_2O$ were conducted, ZA sample showed that elements are bonded each other's and less porous [9]. Modified ZA samples with $FeCl_3.6H_2O$ and sonicated at 40 min, 80 min and 120 min showed that elements spread out and increasing porous as shown on ZAM1, longer sonicated process will create higher porous on ZAM2 and ZAM3 as shown in figure 4.

Figure 4. Photomicrographs observation of ZA(a) ZAM1(b), ZAM2(c) and ZAM3(d)

The quantitative chemical microanalysis analysis (EDX) observed to the sampel major element composition of the analyzed volume. Modified ZA samples with $FeCl_3.6H_2O$ with sonification process showed that significant increament of Fe element up to 160.07% (ZAM1) and 320.81% (ZAM2) as described on table 1, prolong sonicated duration did not influence %wt Fe element on ZAM3 due to saturated reaction of natural zeolite at ferric chloride hexahydrate solution, this process has been discussed on previous study on the relationship between ultrasonic sonochemistry elapse time with saturation process [10] .

Table 1. Quantitative element %wt of ZA and ZA + $FeCl_3.6H_2O$

Element (%wt)	ZA	ZAM1	ZAM2	ZAM3
Ca	2.29	1.41	1.10	0.92
Fe	2.98	7.75	12.54	9.41
Cl	0.00	1.50	2.16	1.33
Al	5.96	6.20	5.60	4.97

Additional element at sonification proceses is one of reaction conditions was determined by its inherent reaction parameters, compared to traditional energy sources, ultrasonic irradiation provides rather unusual reaction conditions (a short duration of extremely high temperatures and pressures in liquids) that cannot be realized by other methods [11].

IOP Conf. Series: Journal of Physics: Conf. Series **817** (2017) 012032 doi:10.1088/1742-6596/817/1/012032

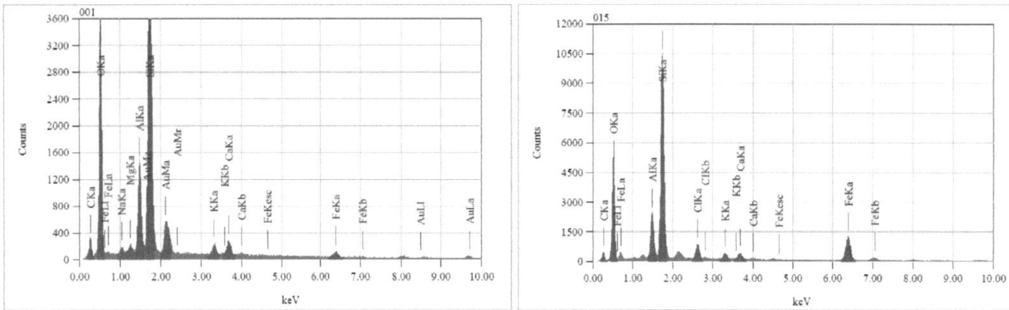

(a) (b)

Figure 5. Chemical microanalysis analysis (EDX) of ZA(a) ZAM2(b)

Surface area and pore analysis are conducted to ZA, ZAM1, ZAM2 and ZAM3 samples, Nitrogen gas utilized as analysis gas at bath temperature 77.3K and outgas temperature 300°C, thermal delay 30 second and outgas time 3 hours, reduction parameter are using Carbon adsorbent with Nitrogen adsorbate on temperature 77.35K and critical pressure 33.5 atm.

Table 2. Physical properties of ZA and ZA + $FeCl_3.6H_2O$

Description	ZA	ZAM1	ZAM2	ZAM3
Surface area (m^2/g)	78.7400	58.0400	50.3100	44.0700
Pore radius (Å)	24.9800	34.6300	36.1900	41.7800

By utilizing density functional theory (DFT) of Quantachrome's data analysis software which contains a comprehensive library of DFT and M.C. methods allows a micro/mesoporous analysis of carbon, zeolites and siliceous materials [12]. Table 2 represents the sonication lead time proportional with surface area reduction up to 26.29% (ZAM1), 36.11% (ZAM2) and 44.03% (ZAM3).

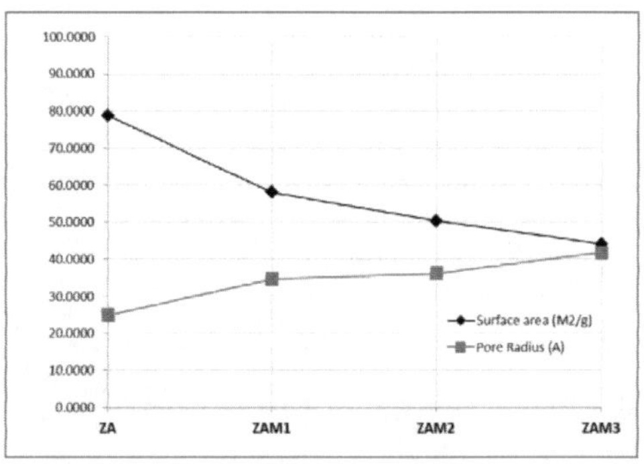

Figure 6. Surface area and pore radius of ZA and ZA + $FeCl_3.6H_2O$

As seen in figure 6 the increment of sonification duration impacts the surface area reduction and simultaneously pore radius will increase up to 38.63% (ZAM1), 44.88% (ZAM2) and 67.25% (ZAM3) due to high intensity ultrasonic liquid processor in operation produced micro bubles in between of primary and secondary flow, this sonification effect can increase the physical properties of materials [3]. Isotherm adsorbtion/desorbtion of ZAM1 and ZAM2 are higher comparing with ZAM3, this condition is related with the phisical properties of samples as shown in figure 7.

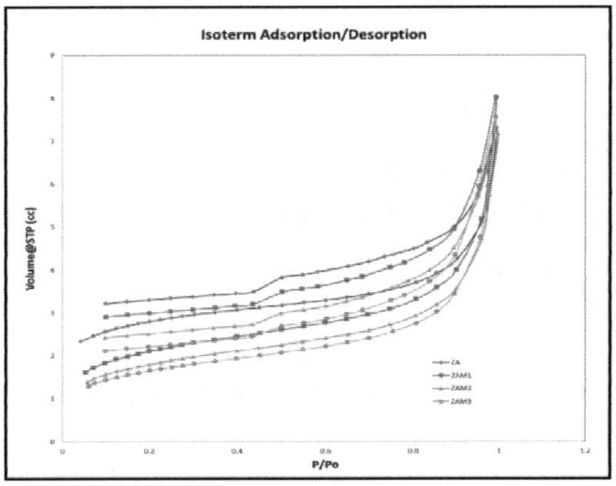

Figure 7. Isotherm adsorption and desorption

4. Conclusion

Natural zeolite (ZA) produced in Bayah mostly consists of mordenite and clinoptilolite phases. Ferric Chloride Hexahydrate addition on Natural zeolite with sonification processes increased Fe element 320.81%wt, besides that sonification effect could increase the physical properties of zeolite material to meet an adsorbent properties-requirements such as increasing 44.03% in surface area and 67.25% pore radius. This result demonstrated that ultrasonic processing on natural zeolite in the ferric chloride hexahydrate solution will produce uniform natural zeolite physical and chemical properties as a candidate of gas adsorbent materials.

Acknowledgments
This research is supported financially by "Hibah Publikasi Internasional Terindeks untuk Tugas Akhir Mahasiswa University of Indonesia" through letter of decree no.1996/UN2.R12/HKP.05.00/2016

References

[1] Trisunaryanti W, Syoufian A and Purwono S 2013 *J. Chem. Chem. Eng.* **7** 176
[2] Al-Jabri M 2008 *Journal standarisasi* **10**(2) 59
[3] Bang J H and Kenneth S S 2010 *Adv. Mater.* **22**(10) 1039–59
[4] Kim W, Choi D and Kim S 2010 *Materials Transactions* **51**(9) 1694-98
[5] Supardi S and Parikin 2010 *Proc. Int. Conf. on Materials Science and Technology* (Jakarta: National Nuclear Energy Agency of Indonesia (BATAN)) p 229
[6] Magdalena and Canovic 2005 *J. Serb. Chem. Soc.* **70**(11) 1335–45
[7] Andac Ö, Tatlıer M, Lu A S and Senatalar 2005 *J. Microporous and Mesoporous Materials* **79** 225–233
[8] Dietmar P 1996 *J. Chemical Mater.* **6**(10) 1605-18
[9] Kamarudin N, Yuan C Y and Hamdan H 2001 *Gas Adsorption Phenomenon in Microporous Zeolite Adsorbent* (Malaysia: University Teknologi Malaysia) p 5
[10] Nilesh P, Parag R C, Vishwas Y G and Aniruddha B D 2001 *J. Ultrasonic Sonochemistry* **8** 23-33
[11] Gong C and Douglas P H 1998 *J. Acoustical Society of America* **104** 2
[12] Thommes M 2010 *Pore Size Analysis by Gas Adsorption* (Florida: Quantachrome Instruments Powder Tech Note 31) pp 1-7

Phase study of SiO_2-ZrO_2 composites prepared from polymorphic combination of starting powders via a ball-milling followed by calcination

Musyarofah[1], R Nurlaila[1], N F Muwwaqor[1], M Saukani[2], A Kuswoyo[3], Triwikantoro[1] and S Pratapa[1]

[1]Department of Physics, Faculty of Mathematics and Natural Sciences, Institute of Technology Sepuluh Nopember, Jalan Arief Rahman Hakim, Surabaya 60111 Indonesia
[2]Department of Mechanical Engineering, Faculty of Engineering, University of Islam Kalimantan Muhammad Arsyad Al-Banjary, South Kalimantan, Indonesia
[3]Department of Mechanical Engineering, Polytechnic Tanah Laut, Pelaihari, South Kalimantan, Indonesia

Email: suminar_pratapa@physics.its.ac.id

Abstract. The effects of SiO_2-ZrO_2 polymorphic combinations as starting powders and calcination temperature on phase composition of the SiO_2-ZrO_2 composites were studied. Stoichiometric (1:1 mol%) mixtures of the SiO_2-ZrO_2 composites were mechanically activated using a ball-milling for 5 h followed by calcinations at 1000, 1100 and 1200 °C for 3 h. The composites used in the present study were *a*-SiO_2+ *a*-ZrO_2, *a*-SiO_2+ *t*-ZrO_2, *c*-SiO_2+ *a*-ZrO_2 and *c*-SiO_2+ *t*-ZrO_2 which were symbolized by AA, AT, CA and CT, respectively. Prefixes *a, t* and *c* denote amorphous, tetragonal and cristobalite, respectively. The phase composition was determined by Rietveld analysis of X-ray diffraction (XRD) data using *Rietica* software. The identified phases for all calcined samples were a combination among *t*-ZrO_2, *c*-SiO_2, *m*-ZrO_2 and zircon ($ZrSiO_4$). Amorphous zirconia formed a transient tetragonal zirconia phase during heating, which reacted with silica to form zircon. The zircon phase was not found to form even at 1200 °C in the AT and CT mixtures and at 1100 °C in the CA mixture. The AA mixture in particular crystallized to form zircon at a lower temperature with more composition fraction than the others, *ca* 82.9 (14) mol%.

1. Introduction

Zircon ($ZrSiO_4$) is a readily available raw material in nature and zircon ceramic has interesting physical and chemical properties for technological uses [1-3] including excellent thermo-mechanical properties such as a very low coefficient of heat conductivity (5.1 W/m°C at room temperature and 3.5 W/m°C at 1000 °C), a low coefficient of thermal expansion (about 4.1×10^{-6}°C^{-1} between 25 and 1400 °C), and good high-temperature mechanical strength even at a temperature higher than 1400 °C [1, 2, 4, 9]. It is also highly inert even when in contact with molten glass or slag which is appropriate for elevated temperature applications with a low chemical attack such as in steel or glass industries [1, 3].

Zircon forms from an equimolar reaction of zirconia (ZrO_2) and (silica) SiO_2. The binary phase diagram of ZrO_2-SiO_2 system has been revealed by Curtis and Sowmann (1953) [10] in a range of 1400-1800 °C, completed by Butterman and Foster (1967) [11] in a range of 400-2800 °C, re-calculated and optimized by Kaiser et al (2008) [12] in a range of 1000-2800 °C. Curtis and Sowman [10] observed the formation of zircon from mixtures of zirconia and cristobalite, zirconia and quartz and zirconia and tridymite. They have determined the temperature range within which the formation of zircon can occur and reported that the lower temperature limit for zircon formation was about 1320 °C. The allotropic (polymorphic) modification silica (SiO_2) that they used in the synthesis had little effect on the final degree of the formation of zircon. Further, Itoh [13] carried out a detailed study of the process from an equimolar mixture of amorphous silica (a-SiO_2) and amorphous zirconia (a-ZrO_2) by means of a high-temperature X-ray diffraction technique. It was reported that zircon ($ZrSiO_4$) formed predominantly from amorphous silica (a-SiO_2) and tetragonal zirconia (t-ZrO_2) at temperatures as low as 1200°C. The formation of zircon dominates the transformation of (a-SiO_2) to cristobalite (c-SiO_2). Tetragonal-to-monoclinic transformation of unreacted zirconia in the reaction may occur during the quenching process. Based on that investigation, Itoh [13] suggested the following four reactions as candidates for zircon formation:

$$a\text{-}SiO_2 + a\text{-}ZrO_2 \rightarrow ZrSiO_4 \quad (1)$$

$$a\text{-}SiO_2 + t\text{-}ZrO_2 \rightarrow ZrSiO_4 \quad (2)$$

$$c\text{-}SiO_2 + a\text{-}ZrO_2 \rightarrow ZrSiO_4 \quad (3)$$

$$c\text{-}SiO_2 + t\text{-}ZrO_2 \rightarrow ZrSiO_4 \quad (4)$$

There has been a lot of research that investigates the methods to produce zircon (solid reaction, hydrothermal, sol-gel routes, chemical reaction like aerosols, reverse micelle process, and micro-emulsion process). Therefore, the equimolar mixtures of polymorphic of silica and zirconia powders in Eq. 1-4 can be used as a basis reaction to form zircon [13]. However, the method based on the mixture of amorphous ZrO_2-SiO_2 revealed the incomplete powders crystallization under heat treatment [14, 15]. Because of its relatively low free energy of zircon formation [16], slow diffusion kinetics, and kinetically high activation energy for the reaction combination between the component oxides (ZrO_2 and SiO_2) [17], the synthesis of pure zircon with reasonably high yields is difficult. This is, in part, due to the large change in the crystallographic environment of Zr between zirconia and zircon structure, and very low solid solubility of silica in zirconia [17].

Due to the very low solubility of SiO_2 in ZrO_2 [16], the addition of material that roles as a binder may increase the wet ability between the powders. Polyvinyl alcohol (PVA) was often used as a binder of solid state reaction between silica and magnesia [18, 19]. The mechanical activation process also has proven to be a useful technique to enhance a solid-state reaction at room temperature [20]. The ''mechanical'' effects of milling, such as the reduction of particle size and mixture homogenization, resulted in active reactants with high surface energy.

This work focuses on the effects of the SiO_2-ZrO_2 polymorphic combination as starting powders and calcination temperature on phase composition of the SiO_2-ZrO_2 composites. Stoichiometric (1:1 mol% ratio) mixtures of a-SiO_2+ a-ZrO_2, a-SiO_2+ t-ZrO_2, c-SiO_2+ a-ZrO_2, and c-SiO_2+ t-ZrO_2, symbolized by AA, AT, CA and CT respectively, were prepared by a mechanical activation process via ball-milling followed by calcination. Prefixes a, t and c denote amorphous, tetragonal and cristobalite, respectively.

2. Experimental

The raw material was zircon natural sand taken from Kalimantan Tengah, Indonesia. The sand was processed to obtain zircon powder with high purity by magnetic separation, soaking with HCl, and reaction with NaOH. This extraction processes produced sodium silicate (Na_2SiO_3) solution and pure zircon ($ZrSiO_4$) powders; both were further processed to obtain zirconia and silica powders. The amorphous zirconia (a-ZrO_2) powder was prepared from the extracted zircon powder by an alkali fusion method with NaOH and followed by a co-precipitation with HCl and NH_4OH. Tetragonal zirconia (t-ZrO_2) was obtained from the amorphous zirconia (a-ZrO_2) powder which was calcined at 700 °C for 3 hours. Meanwhile, a sodium silicate solution was precipitated by HCl to produce the amorphous silica (a-SiO_2) powder. The cristobalite (c-SiO_2) powder was obtained from the amorphous silica (a-SiO_2) powder which was calcined at 950 °C for 8 hours. The zirconia (amorphous and tetragonal) powders and the silica (amorphous and cristobalite) powders with a composition of 1:1 molar ratio and polyvinyl alcohol (PVA) of 3 wt% were mixed and mechanically activated using a planetary ball mill apparatus for 5 hours at a speed of 150 rpm. The milled mixtures were then calcined at various temperatures of 1000, 1100, and 1200 °C for 3 hours. Phase compositions were obtained from X-ray diffraction (XRD) data which were collected using a Philips X'Pert Powder diffractometer with CuKα radiation (λ=1.54060 Å) for all samples. The XRD patterns were recorded with a step size of 0.0170°. The XRD patterns were analyzed using *Match!2* for phase identification and using *Rietica* based on Rietveld method to provide accurate phase compositions of the samples.

3. Results and Discussion

3.1 Qualitative Analysis

Figure 1 shows the XRD patterns of the four mixtures of silica and zirconia which were prepared by a mechanical activation process via ball-milling before calcination. The mixed phases are confirmed by a qualitative analysis of XRD data, i.e. tetragonal-zirconia (COD No. 1-525-706) and cristobalite (SiO_2) (COD No. 1-010-938), while the "amorphous phases" (amorphous zirconia and amorphous silica) are known from the existence of peak hump. Furthermore, the XRD patterns of the calcined mixtures are shown in figure 2. The identified phases for all calcined samples were a combination among tetragonal-zirconia, cristobalite, monoclinic-zirconia (PDF No. 89-426), and zircon (PDF No. 9-582).

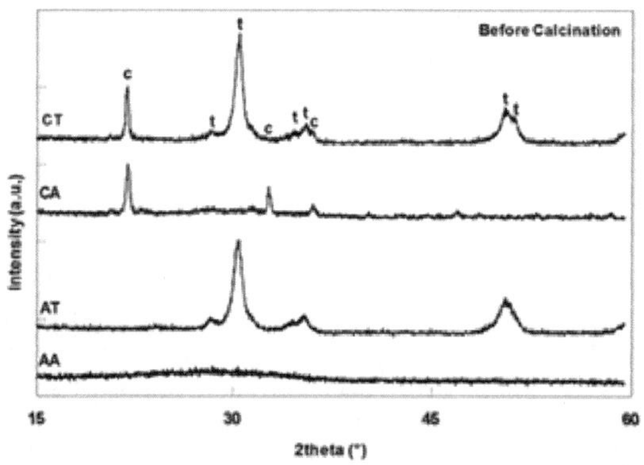

Figure 1. X-ray diffraction patterns (CuKα) of AA, AT, CA, and CT powders before calcinations.

Zircon is not identified in the 1100 °C of the AT and CT composites and in the 1000 °C of the CA composites as seen in figure 2. In contrast, zircon is identified at lower calcination temperature of 1000 °C for sample AA. Formation of zircon by the solid-state reaction usually requires a thermal treatment at 1450-1500 °C in air [21]. In this study, zircon was already formed at a temperature as low as 1000 °C. Zircon is formed at a relatively lower temperature because for the AA sample, the amorphous zirconia and amorphous silica powders have smaller particle size than the other samples. Smaller particles tend to react in an easier way than larger particles [22]. It was reported in a previous study on the mechanism of zircon formation from amorphous powders [13] that zircon formed after a reaction between amorphous-silica and tetragonal-zirconia. It was also found that the formation of zircon dominates over the transformation of amorphous silica to cristobalite. Tetragonal → monoclinic transformation of unreacted zirconia in the reaction mixture occurs during the quenching process.

The zircon formation is not a nucleation process, but, in fact, it is a diffusion of silicon species through a zircon layer, from an external continuous or granular silica phase to the surface of zirconia grains. A model of ZrO_2-SiO_2 reaction mechanism has been proposed by Veytizou et al. [23]. They observed the kinetics of the reaction of a-SiO_2 + t-ZrO_2→ $ZrSiO_4$ by isothermal HTXRD experiments. In a great number of solid–solid reactions, because there were no mobile species like in gas or liquids, it was considered that the growth of product was limited by diffusion. Two models of contracting spheres were proposed: in the first model, the system was described as a series of spherical t-ZrO_2 grains surrounded by a shell of zircon, which is covered by a-SiO_2. In the second model, a fast, superficial diffusion of Si^{4+} and electrons originating from silica grains on the external surface of the zircon layer produced a homogenization of the silicon concentration.

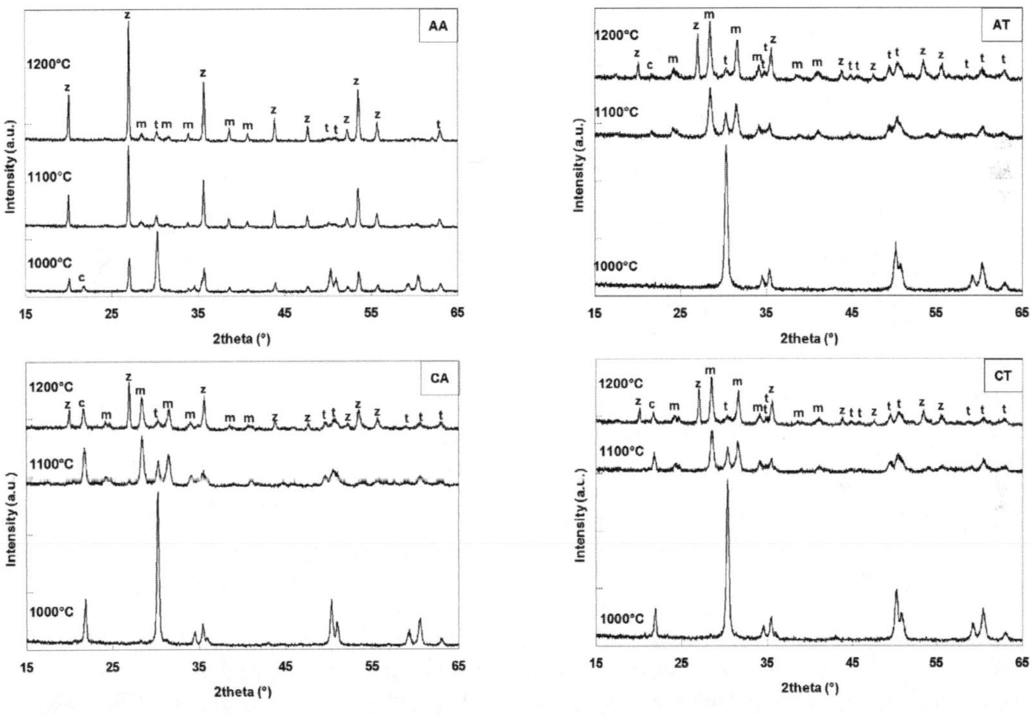

Figure 2. X-ray diffraction patterns (CuKα) of AA (a-SiO_2+a-ZrO_2), AT (a-SiO_2+t-ZrO_2), CA (c-SiO_2+a-ZrO_2), and CT (c-SiO_2+t-ZrO_2) powders after calcined at 1000, 1100, and 1200 °C for 3 hours.

3.2 Quantitative Analysis

Quantitative phase analyses were based on the Rietveld method using *Rietica* software to the XRD data result in the relative mole fraction of the phases in all samples as described in table 1. Its concentration increased with the calcination temperature, i.e. up to 82.9 mole% at 1200 °C for the AA sample. The mechanical activation process via ball-milling played an important role in the lower temperature of zircon formation. The mechanical activation process has proven to be an effective technique to enhance a solid-state reaction at room temperature. The ''mechanical'' effects of milling, such as the reduction of particle size and mixture homogenization, result in active reactants with high surface energy [20].

In a previous study, Pradhan [24] worked on the synthesis of zircon from stoichiometric (1:1 mole%) monoclinic zirconia and amorphous silica powder mixture also via a ball-mill process followed by a calcination. A relatively high-purity zircon was found when a sample was ball-milled for 30 h and calcined at 1200 °C for 1 h, i.e. up to 93 mole% as compared to powder with the same treatment except with milling for 5 min which has only 22 mole%. In our work, milling of AA powder for 5 h followed by calcination at 1200 °C for 3h resulted in 82.9 mole% zircon.

Table 1. Results of quantitative analysis of X-Ray diffraction pattern of samples using *Rietica* by Rietveld method.

Sample	Mole fraction (%)			
	Zircon	t-Zirconia	m-Zirconia	cristobalite
AA 1000 °C	76.4 (54)	2.4 (2)	-	21.2 (64)
AA 1100 °C	77.8 (20)	0.2 (0)	22.0 (16)	-
AA 1200 °C	82.9 (14)	0.1 (0)	17.0 (11)	-
AT 1000 °C	-	100	-	-
AT 1100 °C	-	0.4 (0)	90.7 (25)	8.9 (12)
AT 1200 °C	20.2 (6)	0.2 (0)	75.3 (11)	4.3 (7)
CA 1000 °C	-	1.7 (0)	-	98.3 (2)
CA 1100 °C	4.2 (5)	0.3 (0)	73.6 (22)	21.9 (28)
CA 1200 °C	27.1 (4)	0.3 (0)	37.0 (6)	35.6 (10)
CT 1000 °C	-	5.2 (0)	-	94.8 (1)
CT 1100 °C	-	0.4 (0)	60.6 (1)	39.0 (1)
CT 1200 °C	18.6 (0)	0.1 (0)	49.3 (0)	32.0 (1)

The role of calcinations temperature in the reaction of the powders can also be seen based on the phase composition by quantitative analysis in Table 1. Zircon was already formed at a temperature as low as 1000 °C for AA samples and 1100 °C for CA samples, but for AT and CA samples, zircon was formed at 1200 °C. The composition of zircon increased with increasing calcinations temperature.

4. Summary

The effects of the SiO_2-ZrO_2 polymorphic combination as starting powders and calcination temperature on phase composition of the SiO_2-ZrO_2 composites were studied. Stoichiometric (1:1 mol%) mixtures of *a*-SiO_2 + *a*-ZrO_2, *a*-SiO_2 + *t*-ZrO_2, *c*-SiO_2 + *a*-ZrO_2, and *c*-SiO_2 + *t*-ZrO_2were prepared by a mechanical activation via ball-milling for 5 h followed by calcination at 1000, 1100, and 1200 °C for 3 hours. The phase composition was investigated by Rietveld analysis of X-Ray Diffraction (XRD) data using *Rietica* software. The identified phases for all calcined samples were a combination of *t*-ZrO_2, *c*-SiO_2, *m*-ZrO_2, and zircon ($ZrSiO_4$). Amorphous zirconia formed a transient tetragonal zirconia phase during heating, which reacted with silica to form $ZrSiO_4$. The $ZrSiO_4$ phase was not found to form even at 1200 °C in the AT and CT composites and at 1100 °C in the CA composites. The AA composites in particular crystallized to form $ZrSiO_4$ at a lower temperature with a greater composition fraction than the others, *ca* 82.9 (14) mol%.

5. Acknowledgements

The authors would like to thank Ministry of Research Technology and Higher Education of the Republic of Indonesia and LPPM ITS to support finance through program EPI-Unet 2016 No. 015685/IT2.11/PN.08/2016 given to SP.

References

[1] Rendtorff N M, Grasso S, Hu C, Suarez G, Aglietti E F and Sakka Y 2012 *Ceram. Int.* **38** 1793
[2] Suarez, Acevedo S, Rendtorff N M, Garrido L B and Aglietti E F 2015 *Ceram. Int.* **41** 1015

[3] Terki R, Bertrand G and Aourag H 2005 *Microelectron. Eng.* **81** 514
[4] Rendtorff N M, Grasso S, Hu C, Suarez G, Aglietti E F and Sakka Y 2012 *J. Eur. Ceram. Soc.* **32** 787
[5] Veytizou C, Quinson J F and Jorand Y 2002 *J. Eur. Ceram. Soc.* **22** 2901
[6] Singh R N 1990 *J. Am. Ceram. Soc.* **73** 2399
[7] Mori T, Hoshino H and Ishikawa Y 1991 *J. Ceram. Soc. Jpn.* **99** 227
[8] Shi Y, Huang X X and Yan D S 1994 *J. Eur. Ceram. Soc.* **13** 113
[9] Mori T, Hoshino H, Yamamura H, Kobayashi H and Mitamura T 1990 *J. Ceram. Soc. Jpn.* **98** 1017
[10] Curtis C E and Sowman H G 1953 *J. Am. Ceram. Soc.* **36** 190
[11] Butterman W C and Foster W R 1967 *Am. Miner.* **52** 880
[12] Kaiser A, Lobert M and Telle R 2008 *J. Eur. Ceram. Soc.* **28** 2199
[13] Itoh T 1992 *J. Cryst. Growth* **125** 223
[14] Garvie R C 1979 *J. Mater. Sci.* **31** 6089
[15] Tartaj P, Serna C J, Moya J S, Requena J and Ocana M 1996 *J. Mater. Sci.* **31** 6089
[16] Ellison A J G and Navrotsky A 1992 *J. Am. Ceram. Soc.* **75** 1430
[17] Kanno Y 1989 *J. Mater. Sci.* **24** 2415
[18] Dewa E, Musyarofah U N, Triwikantoro, Firdaus S and Pratapa S 2015 *Adv. Mater. Res.* **1112** 262
[19] Musyarofah U N, Dewa E, Triwikantoro, Mashuri, Firdaus S and Pratapa S 2015 *Adv. Mater. Res.* **1123** 383
[20] Suryanarayana C 2001 *Prog. Mater. Sci.* **46** 1
[21] Ballman A A and Laudise R A 1965 *J. Am. Ceram. Soc.* **48** 130
[22] Jong L 2014 *Ind. Crops. Prod.* **55** 25
[23] Veytizou C, Quinson J F, Valfort O and Thomas G 2001 *Sol. St. Ion.* **139** 315
[24] Pradhan S K and Sinha M 2005 *J. Appl. Cryst.* **38** 951

IOP Conf. Series: Journal of Physics: Conf. Series **817** (2017) 012034 doi:10.1088/1742-6596/817/1/012034

The effect of excitation wavelength on plasma spectrum of metals in vacuum condition

A N Hidayah, M M Suliyanti and Isnaeni
Research Center for Physics, Indonesian Institute of Sciences, Building 442 Puspiptek, South Tangerang, Banten, Indonesia 15314

E-mail: affi001@lipi.go.id

Abstract. Pure metals Aluminum (Al), Copper (Cu) and Brass (CuZn) have been detected using laser-induced plasma spectroscopy technique using nanosecond laser. A Q-switch pulsed Nd:YAG laser operating at infrared (1064 nm), visible (532 nm) and ultraviolet (355 nm) wavelengths have been used. The energy laser used was 12 mJ and the experiment was carried out in vacuum condition (5 Torr). The plasma generated by a focused Nd:YAG laser beam was detected by a spectrometer to identify trace elements quantitatively. The result of quantitative trace element of pure metals at three different wavelengths gave the different results. Al metal excited by 355 nm, 532 nm and 1064 nm wavelengths has 5, 2 and 13 Al emission peaks, respectively. Whereas the analysis of Cu showed 5, 7 and 6 emission peaks upon excitation by 355 nm, 532 nm and 1064 nm wavelengths, respectively. Finally, CuZn metal produced 7 Cu lines and 2 Zn lines upon excitation by 355 nm wavelength, 8 Cu lines and 1 Zn line when excited by 532 nm wavelength, and 8 Cu lines and 3 Zn lines when excited by 1064 nm wavelength.

1. Introduction

When a powerful pulsed laser is focused onto a surface, a tiny amount of the material is vaporized and through further photon absorption, it is heated up until it ionizes and creates a dense area of high energy ions/atoms/electrons which we call as a plasma. This laser-induced plasma is a micro-source of light that can be analyzed by a spectrometer. The obtained spectra consist of lines corresponding to the elements evaporated from the sample surface. This is the principle of laser-induced plasma spectroscopy (LIPS) and it has proven to be a powerful surface analytical technique, capable of performing trace element measurements in any kind of solid material [1-3] as well as in liquids [4-5]. A variety of phenomena in laser interaction with material including heating, melting, evaporation, atomization, excitation and ionization are involved [6].

The efficiency of plasma is described by the ablation rate, which gives the maximum layer thickness ablated during the irradiation with a laser pulse [7]. The ablation rate is strongly influenced not only by the laser beam characteristics (laser spot diameter, laser fluence, laser wavelength, a number of pulses and pulse duration), but also by the optical and thermal properties of the processed material surface (optical absorption, surface reflectivity, thermal conductance and mass density) [8-9]. The influence of laser fluence on ablation rate is more significant when the laser fluence exceeds its threshold value (F_{th}) [10]. Plasma formation requires vaporization of the material surface as a first step. The vaporization of the sample occurs when the energy deposited on the target exceeds the latent heat of vaporization of the target, L_v. Threshold fluence (F_{th}), $F_{th} = \rho\, L_v\, a^{1/2}\, t_e^{1/2}$, where ρ is the sample density, a is its thermal diffusivity ($a = K/\rho C_p$, where K and C_p are the thermal conductivity and specific heat, respectively) and t_e is the laser pulse width. The initiation of plasma over the target surface begins in the hot vapor and is generated by inverse bremsstrahlung absorption during collisions among sampled atoms and ions, electrons and gas species. Theoretical considerations on plasma production and heating by means of laser beams have been discussed by several authors [11].

Here, we experimentally explored the influence of the laser wavelength on the plasma spectrum for material analysis of pure metals. We use three pure metals (Al, Cu and CuZn) that were ablated by three different laser wavelengths from Nd-YAG laser (355 nm, 532 nm and 1064 nm) in vacuum condition.

2. Materials and methods

2.1. Materials

Three pure metal samples used in this research are 99% Aluminum (Al), 99% Copper (Cu) and 99% Brass (CuZn). Samples are high purity metals for research and industry. Samples were cut into 2 × 2 cm dimension.

2.2. Methods

The schematic diagram of the experimental setup used in this research is shown in figure 1. A Nd:YAG laser source (Q-Smart 850, Quantel) with a pulse repetition rate of 10 Hz and pulsed width of 6 ns was used in this experiment. Three optional wavelengths were available, i.e. fundamental (1064 nm), second harmonic generation (SHG) (532 nm) and third harmonic generation (THG) (355 nm) modes. The output of laser beam was focused by a convex lens with a 100 mm focal length onto a copper target mounted on a rotating table in the vacuum chamber. Energy laser used was fixed at 12 mJ/pulse. Plasma emission generated from the laser-sample interaction was collected through a high-resolution fiber optic spectrometer (HR 2000, Ocean Optics). This spectrometer has a detection range from 200 nm to 1050 nm. The experiments were carried out in vacuum condition (5 Torr).

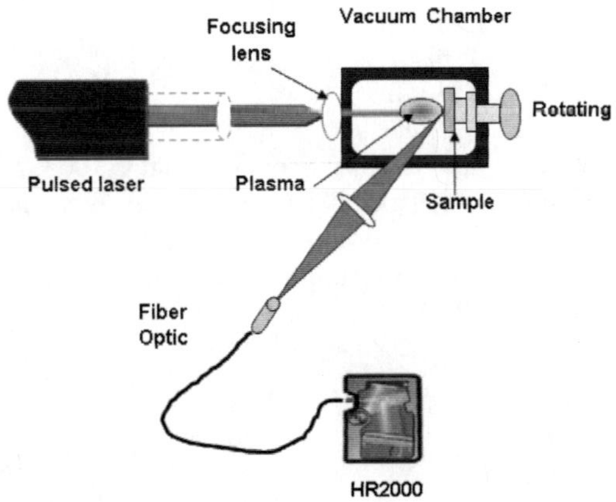

Figure 1. Schematic diagram of the experimental setup.

3. Results and discussion

Our research purpose was to analyze three pure metals (Al, Cu and CuZn) with LIPS technique and to observe the effect of laser wavelength on plasma spectrum for material analysis. We used three different laser wavelengths, i.e., infrared (1064 nm), visible (532 nm) and ultraviolet (355 nm).

3.1. Information about sample properties of Aluminum (Al), Copper (Cu) and Brass (CuZn)

When a laser beam is focused on a solid sample surface, a plasma is formed. The plasma consists of atoms, ions, electrons. When pulsed laser interacts with material, a certain amount of energy is absorbed by metals and converted into heat, causing analyte vaporization as the local temperature exceeds the

Table 1. Thermal and physical constants of the metals used in this work[a]

Element	T_b (K)	T_f (K)	ρ (g cm^{-3})	L_v (J g^{-1})	L_f (J g^{-1})	C_p (J K^{-1} kg^{-1})	K (W m^{-1} K^{-1})	a (cm^2 s^{-1})
Al	2740	933	2,70	10800	388	900	237	0.97
Cu	2840	1356	8.96	4796	205	385	401	1.16
Zn	1180	692	7.14	1748	111	388	116	0.41

[a] T_b, boiling point; T_f, melting point; ρ, density; L_v, latent heat of evaporation; L_f, latent heat of fusion; C_p, specific heat; K, thermal conductivity; a, thermal diffusivity

boiling point of the sample material. The timescale of this interaction is so small that it has been considered for the optical energy is converted into heat instantaneously at the point of absorption [8]. Hence, plasma plume is strongly influenced by the properties of material. The physical properties of Al, Cu and CuZn, especially thermal features such as boiling point, melting point and thermal conductivity are shown in table 1 [12].

3.2. Influence of laser wavelength on analysis of materials

The formation of plasma is also influenced by threshold fluence (F_{th}) and F_{th} strongly influences the ablation rate (Δh). The dependency of the ablation rate and the threshold laser fluence on the laser wavelength arises from three interrelated phenomena: the decrease in the sample's intrinsic absorption, the increase in the surface reflectivity and the enhancement of the absorptivity of the produced plasma with increasing the laser wavelength. The interplay of these three phenomena drives to a larger ablation rate and smaller threshold fluence for the visible than for infrared laser pulses. The enhancement of absorptivity of the produced plasma with increasing the laser wavelength relate with the dependence of the plasma coefficient (α_{IB}) on the wavelength (λ) is showed in equation (1), implies that the attenuation of the incoming laser beam into the ignited plasma-plume via the inverse-Bremssthrahlung effect is significantly stronger in longer wavelengths [11].

$$\alpha_{IB} \approx \lambda^3 \quad (1)$$

In our case, we wanted to observe the effect of laser wavelength (1064, 532 and 355 nm) on plasma spectrum for material analysis of three pure metals: Al, Cu and CuZn. When laser interacts with materials, the plasma is generated and the plasma spectrum are collected and analyzed by HR 2000 spectrometer from Ocean Optics. We used the standard table from MIT to analyze the emission spectrum of metals. The Al analysis result is shown in figure 2, the result of analysis Cu is shown in figure 3 and the result of analysis CuZn is shown in figure 4.

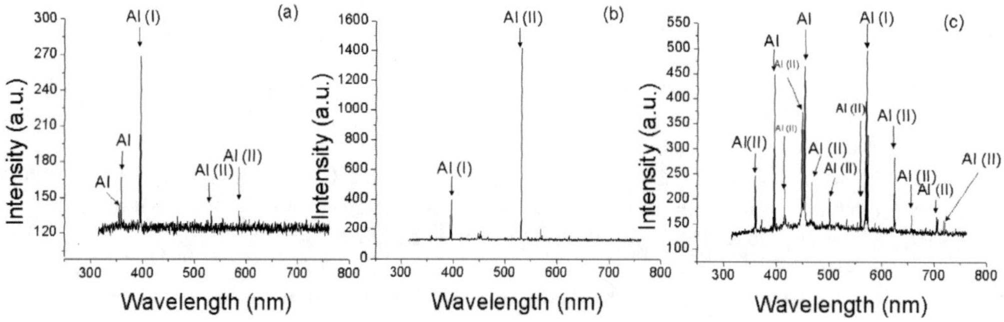

Figure 2. The emission spectrum of Al generated by three different wavelengths (a) 355 nm, (b) 532 nm and (c) 1064 nm, in vacuum condition (5 Torr) and laser energy 12 mJ.

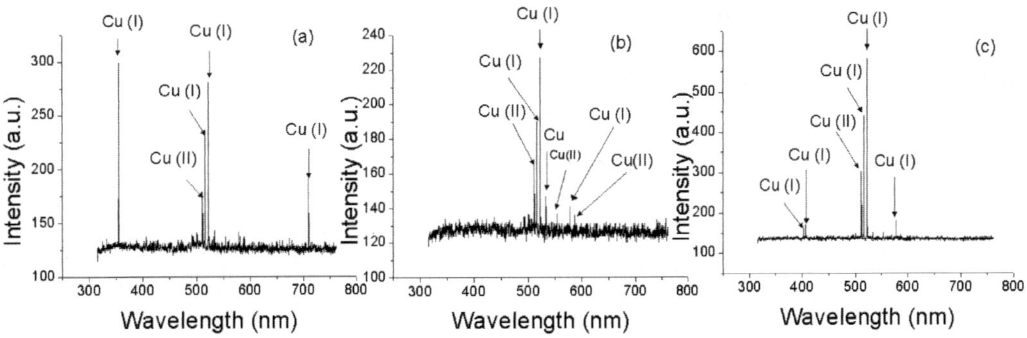

Figure 3. The emission spectrum of Cu generated by three different wavelengths (a) 355 nm, (b) 532 nm and (c) 1064 nm, in vacuum condition (5 Torr) and laser energy 12 mJ.

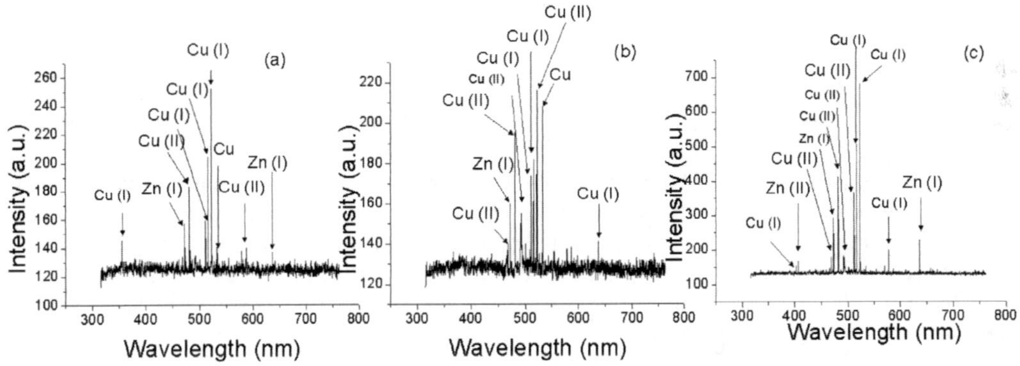

Figure 4. The emission spectrum of CuZn generated by three different wavelengths (a) 355 nm, (b) 532 nm and (c) 1064 nm, in vacuum condition (5 Torr) and laser energy 12 mJ.

Figure 2 shows that different emission spectrum of Al ablated by three different laser wavelengths. The same result is also shown in figure 3 for Cu and in figure 4 for CuZn. Al metal excited by 355 nm, 532 nm and 1064 nm wavelength generated 5, 2 and 13 Al emission peaks, respectively. While the analysis of Cu showed 5, 7 and 6 emission peaks upon excitation by 355 nm, 532 nm and 1064 nm wavelength, respectively. Finally, CuZn metal produced 7 Cu lines and 2 Zn lines upon excitation by 355 nm wavelength, 8 Cu lines and 1 Zn line when excited by 532 nm wavelength, and 8 Cu lines and 3 Zn lines when excited by 1064 nm wavelength. The number of detectable elements peaks depends on the ablation wavelength as predicted by equation 1. Emission spectrum of Al, Cu and CuZn ablated using 1064 nm wavelength yield the most number of element peaks. However, our results do not show cubic dependency between element peaks and wavelength as predicted by equation 1. It is believed that our fiber-optic spectrometer was not sensitive enough to detect and differentiate element peaks due to two reasons. First, some element peaks are similar to the noise level, i.e., high signal-to-noise ratio. Second, the resolution of our spectrometer is limited to differentiate two coinciding peaks. Nevertheless, this results show the trend of wavelength influence.

The emission spectrum of Al, Cu and CuZn generated by three different wavelengths in figure 2, 3 and 4 are the optical emissions from the relaxation of excited species within plasma, which provides information regarding the composition of the material. Therefore, spectroscopic analysis of the optical emission from the excited species is able to identify and reveal pure elements in the material. A spectral line emission occurs when a bound electron in an excited state relaxes to a lower energy state, emitting photons of a specific wavelength in the process. The continuum emission is caused by to free-free transitions (also called 'Brehmstrahlung' radiation) and free-bound transitions (also called recombination radiation).

The influence of wavelength on emission spectrum is briefly described as follow. A previous study revealed that there is a correlation between the fluence of wavelength and plasma threshold in different metals [13]. The initiation of plasma over the target surface begins in hot vapor and the plasma is generated by inverse Bremsstrahlung absorption during collisions among sampled atoms and ions, electrons and gas species. In the case of metal sample, generation of initial electrons, which promotes the plasma process, is produced by a thermal ionization mechanism. It is known that metals have a low ionization potential, high absorption coefficient and high electron number density. Using short laser pulses, which have small thermal diffusion into the material, the generated vapor is rapidly ionized [6]. Consequently, the laser induced breakdown occurs easily. Previous study also revealed that the cascade-like growth of the electron number density is considerably more favorable in the IR than in the UV [13]. Another study reported that the F_{th} in the case of infrared pulse laser is twice as large as F_{th} corresponding to the visible pulse laser [14]. In this work, although we could not show calculation of ablation rate and threshold value, we have clearly showed the influence of spectra emission due to ablation wavelength.

4. Conclusion

The analysis of Al, Cu and CuZn has been done by means of LIPS system. The different ablation wavelengths (1064, 532 and 355 nm) of nanosecond laser were used in this experiment. The results show that wavelength 1064 nm produced the most emission peaks spectra. This phenomenon is influenced by properties of sample material, laser fluence and laser wavelength. Further observations and experiments can be done further if we want to know and learn about the accuracy of the material analysis. If we are able to observe the environmental effects of laser, energy laser and the relationship about the ablation threshold fluence, ablation rate, the impact of the laser spot diameter and with plasma formation, we hope we can explain this phenomenon more detail.

Acknowledgement

The authors would like to thank to Kegiatan Unggulan LIPI and Kegiatan Penguatan Komptensi from Research Center for Physics, Indonesian Institute of Sciences for financial support of this work. The authors would also like to thank to researchers in the Group of Laser, Research Center for Physics for valuable discussions and support during this work.

References

[1] Vander Wal R L, Ticich T M, West J R and Householder P A 1999 *Appl. Spectr.* **53**(10) 1226
[2] Panne U, Haisch C, Clara M, Niessner R 1998 *Spectrochim. Acta B* **53** B(14) 1957-68
[3] Ihlemann J, Scholl A, Schmidt H and Wolff R B 1995 *Appl. Phys. A* **60** 411
[4] Ng C W, Ho W F, Cheung N H 1997 *Appl. Spectr.* **51**(7) 976
[5] Ho W F 1997 *Appl. Spectr.* **51**(1) 87
[6] Weyl G M 1989 Physics of laser-induced breakdown: An Update *Laser-Induced Plasmas and Applications* ed L J Radziemski and D A Cremers (New York: Marcel Dekker) chapter 1 p 1
[7] Stafe M, Vlădoiu I, Negutu C and Popescu I M 2008 *Rom. Rep. Phys.* **60**(3) 789-96
[8] Ready J F 1971 *Effects of High-power Laser Radiation* (New York-London: Academic Press)
[9] Bauerle D 2000 *Laser Processing and Chemistry* (Berlin Heidelberg: Springer-Verlag)
[10] Amoruso A, Bruzzese R, Spinelli N, Velotta R 1999 *J. Phys.* B **32** 131–72
[11] Piepmeier E H 1986 Laser ablation for atomic spectrometry *Analytical Applications of Lasers* ed E H Piepmeier (New York: John Wiley and Sons) p 627
[12] Lide D R and Frederikse H P R 1995-1996 *Handbook of Chemistry and Physics* (Boca Raton Florida: CRC Press)
[13] Calabi´n L M and Laserna J J 1998 *Spectrochim. Acta B* **53** 723-30

[14] Vladoiu I, Stafe M, Negutu C, Popescu I M 2008 *J. Optoelectron. Adv. Mater.* **10**(12) 3177-181

IOP Conf. Series: Journal of Physics: Conf. Series **817** (2017) 012035 doi:10.1088/1742-6596/817/1/012035

An SMS (single mode – multi mode – single mode) fiber structure for vibration sensing

T B Waluyo and D Bayuwati

Research Centre for Physics – Indonesian Institute of Sciences, Kawasan PUSPIPTEK Serpong South Tangerang 15314 Indonesia

E-mail: thom002@lipi.go.id

Abstract. We describe an SMS (single mode - multi mode - single mode) fiber structure to be used in a vibration sensing system. The fiber structure was fabricated by splicing a section (about 300 mm in length) of a step index multi mode fiber between two single mode fibers obtained from a communication grade fiber patchcord. Interference between higher order modes occurs while light from a narrow band light source travels along the multi mode fiber. When the multi mode fiber vibrates, the refractive index profile is changed because of the photo-elastics effect and the amplitude of the interference pattern is changed accordingly. To simulate a vibrating structure we used a loudspeaker to vibrate a wooden table. By using a digital oscilloscope, we recorded and analysed the vibrating signals obtained from the SMS fiber structure as well as from a GS-32CT geophone for referencing. We observed that this SMS fiber structure was potential to be used in a vibration sensing system with a measurement range from 30 to 180 Hz with inherent optical fiber sensor advantages such as light weight, immune to electromagnetic interference, and no electricity in the sensing part.

1. Introduction

To prevent further damage on structures subjected to continuous vibration, monitoring and analysis of vibration that occurs on those structures, such as machines, buildings, and vehicles, are necessary. For this purpose we require sensors to convert physical quantities of vibration (displacement, velocity, and acceleration) into an electrical time-varying signal, hardware to process and record the signal, and software to analyze the signal in time or frequency domain. In this paper, we describe an optical fiber sensor based on an SMS (single mode – multi mode – single mode) fiber structure to be used as a vibration sensor.

Optical fiber sensors have advantages such as immunity to electromagnetic interference, small size, light weight, high sensitivity, and large bandwidth. Fiber interferometers, for example, sense various physical parameters including temperature, strain, pressure, refractive index, and mechanical rotation; it, however, should compete with other already mature technologies in many fields of application [1, 2]. Optical fiber sensors can be placed kilometers away from the monitoring station, because of the intrinsic low optical attenuation of the fiber, and it is also possible to perform multiplexed measurements using arrays of remote point or distributed sensors [3]. Several types of optical fiber sensors such as the Fabry-Perot (FP) cavity, fiber Bragg grating (FBG), and high-birefringence (HB) fiber have been investigated and applied in many areas and, recently, the multi mode interference occurring in an SMS fiber structure has also been studied and developed to act as novel optical devices such as sensors or filters [4]. The advantages of an SMS fiber structure are, among others, high compactness, wide wavelength operation range, low cost, simple to construct, ease of packaging and connection to optical fiber system.

There are several researchers already investigated the use of an SMS structure for vibration measurements. Sun *et al*, for example, employed an SMS structure as a high sensitive optical fiber microphone which can detect human voice within 2 meter range [5]. Wu *et al* used a bent SMS structure for vibration sensing. The bent SMS structure is fixed to a translate stage and a vibration generator. Using an oscilloscope they observed vibration frequencies from 10 Hz up to 1900 Hz [6]. Zhao *et al* demonstrated an SMS structure to measure a vibrating cantilever. Using a lock-in amplifier as a signal acquisition device they obtained sensitivity of 4.741 mV/Hz and measuring range from 2 Hz to 80 Hz [7]. Based on their works, we developed an SMS fiber structure in which the multi mode fiber as the sensing part was coiled on a disc and attached directly on a vibrating structure or platform that would be measured or monitored.

2. Methods

We constructed the SMS fiber structure by splicing a section, about 300 mm long, of a 50/125 step index multi mode fiber (Thorlabs FG050LGA, 0.22 NA) between two 9/125 single mode fibers (Thorlabs SMF-28, 0.14 NA). The multi mode fiber was then coiled on a mini compact disc (diameter: 80 mm) and the whole assembly, acting as the sensing part, was coated with a TFE tape with a total weight is less than 10 grams, see figure 1. As the light source we used a tuneable laser diode (Santec TSL-510, the wavelength is tuned fix at 1560 nm) and for the photodetector we used a Ge photodiode with an adjustable gain amplifier (Thorlabs PDA50B-EC, 0 – 70 dB gain).

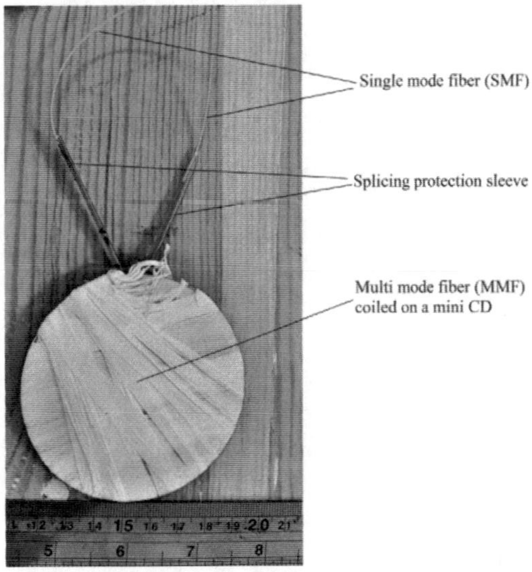

Figure 1. The configuration of an SMS fiber structure for vibration sensing.

When light is coupled from the SMF into the MMF, higher order modes are excited in the MMF and interference between different modes occurs. The relationship between the longitudinal propagation constants for the *m* and *n* order mode can be expressed as [7]:

$$\beta_m - \beta_n = \frac{u_m^2 - u_n^2}{2k_0 a^2 n_{core}} \tag{1}$$

IOP Conf. Series: Journal of Physics: Conf. Series **817** (2017) 012035 doi:10.1088/1742-6596/817/1/012035

where $k_0 = 2\pi/\lambda$, λ: wavelength of light, a: core radius of the MMF, n_{core}: refractive index of the MMF core, u_m and u_n are the normalized propagation constants: $u_m = \pi(m - 1/4)$ and $u_n = \pi(n - 1/4)$.

When the phase difference of two modes is the integer multiple of 2π, interference between these two modes occurs when

$$(\beta_m - \beta_n)L = 2N\pi \tag{2}$$

where L is the length of the MMF, and N is a natural number. The wavelength of the constructive interference can be derived as

$$\lambda = \frac{8(2N+1)n_{core}a^2}{(m-n)[2(m+n)-1]L} \tag{3}$$

When the MMF vibrates, the refractive index profile changes because of the photo-elastics effect and the peak of interference pattern will move so that

$$\Delta\lambda = \lambda_N - \lambda_{N-1} = \frac{16n_{core}a^2}{(m-n)[2(m+n)-1]L} \tag{4}$$

and the output light intensity will change accordingly because the deformation of the MMF will lead to the power losses of all the modes, the higher-order modes experience more power loss than the lower-order modes [7, 8].

In our experiment, to excite vibration we used a 6", 8 Ω, 100 W loudspeaker driven with a low-cost car-audio amplifier. As a signal generator, we used a smartphone with an Android tone generator application. The loudspeaker was used to vibrate a 1 × 1 m^2, 0.5" thick, wooden table. For sensing this vibration we placed the SMS fiber structure nearby the loudspeaker on the table surface side by side with a GS-32CT geophone for referencing. The vibration signals were recorded and observed using a digital oscilloscope (Picoscope 4424™). The set-up of our experiment is shown in figure 2.

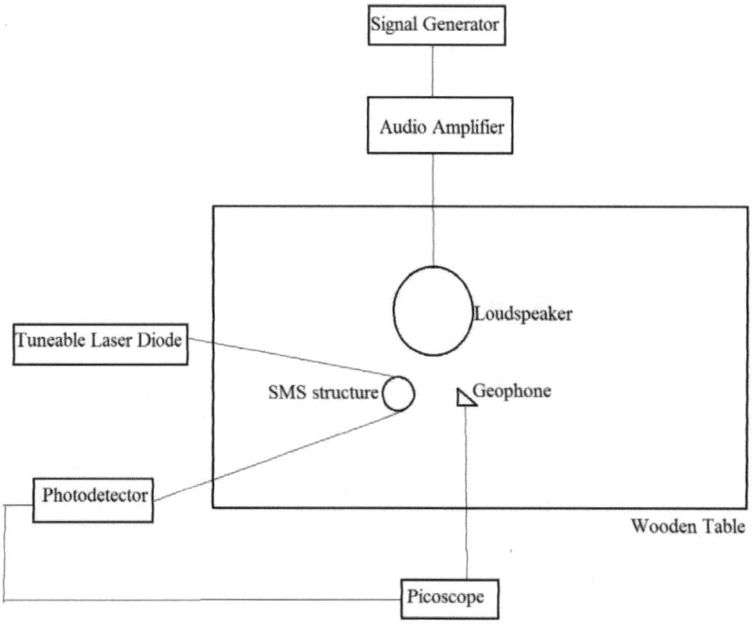

Figure 2. Experimental set-up.

3. Results and discussion

Before using the loudspeaker to excite a forced, linear, and periodic vibration to the wooden table, we conducted an impulse test by tapping slightly the table surface using a rubber hammer. Figure 3 shows the generated vibration sensed by our SMS vibration sensor (top trace, 20 mV/div) and by the GS-32CT geophone (bottom trace, 80 mV/div). It can be seen that the output of our SMS vibration sensor resembled that of the geophone.

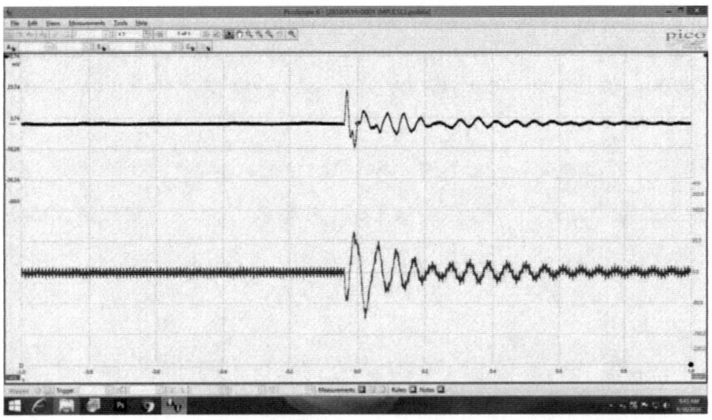

Figure 3. A typical result of the impulse test.

Figure 4 shows the recorded table vibration when we drove the loudspeaker with a 70 Hz sinusoidal signal using the signal generator. It can be seen in this time domain graph that our SMS vibration sensor (top trace, 20 mV/div) well reproduced the driving signal. From the vibration signal reproduced by the geophone (bottom trace, 40 mV/div) we can calculate the displacement of the wooden table caused by this vibration. The sensitivity of the geophone was 0.275 V/cm/s [9], it means that rms vibration velocity of 1 cm/s will produce 275 mV rms signal. The measured rms voltage at this 70 Hz vibration was 14.97 mV so the velocity was (14.97 mV)/(275 mV/cm/s) = 0.054 cm/s or 0.54 mm/s. To convert this velocity (V) value to displacement (X) we used a conversion equation $X = V/(2\pi f)$, valid for a sinusoidal vibration with frequency f [10], so that we obtain the rms displacement at this frequency is 0.0012 mm or 1.2 μm.

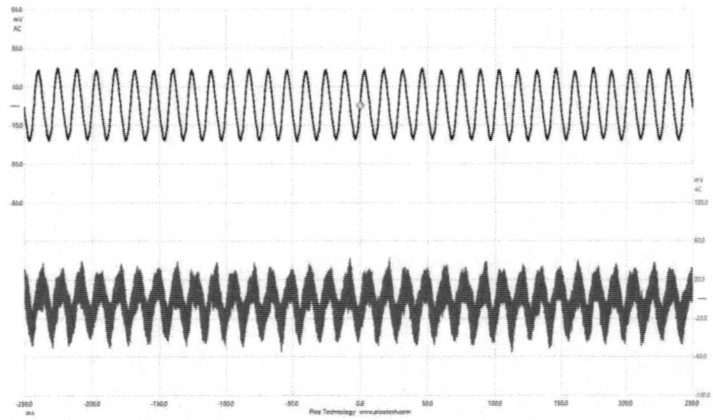

Figure 4. Recorded vibration signal at 70 Hz.

A typical result of the recorded signals from our sensors at the frequency domain is shown in figures 5(a) and 5(b) for vibrating frequency of 30 Hz and 130 Hz respectively. Again it can be seen that the signals from our SMS vibration sensor (top trace, 20 dB/div) resembled that of the geophone (bottom trace, 20 dB/div). Finally, to know the frequency response of our sensor, we drove the loudspeaker with a white noise from the signal generator. The Picoscope™ was used to record the output signal from our sensor in frequency domain when the white noise generator was turned off and then on. The output signal was saved as a CSV file and then, by using MS EXCEL™, we plotted the signal. Figure 6 shows the frequency response in 0 – 200 Hz band when there was no vibration (lower trace) and when the signal generator was turned on (upper trace). It can be seen from this plot that our SMS sensor was potential to be used in a vibration sensing system with a measurement range from 30 to 180 Hz.

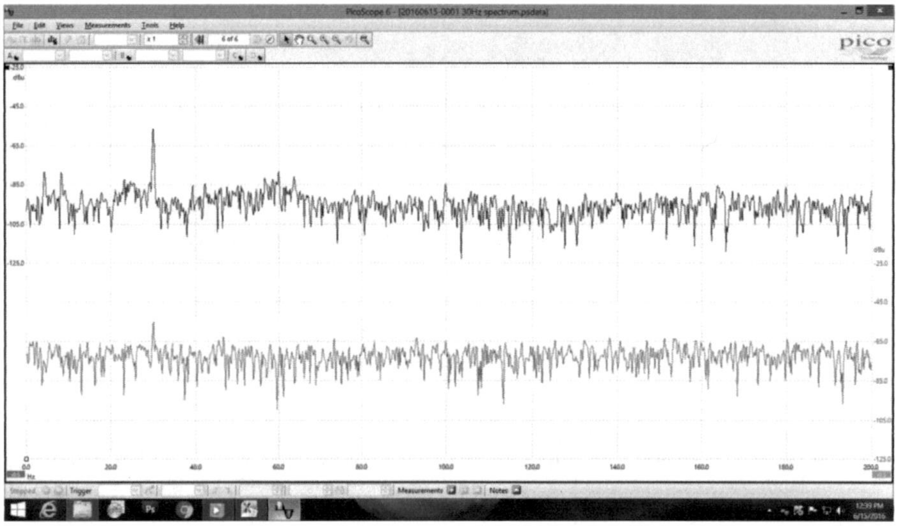

(a)

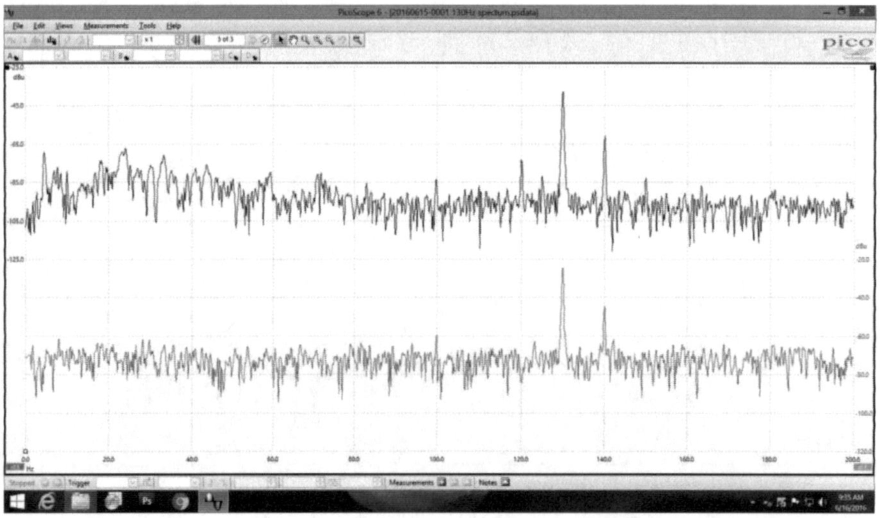

(b)

Figure 5. Typical frequency domain vibration signal at (a) 30 Hz and (b) 130 Hz.

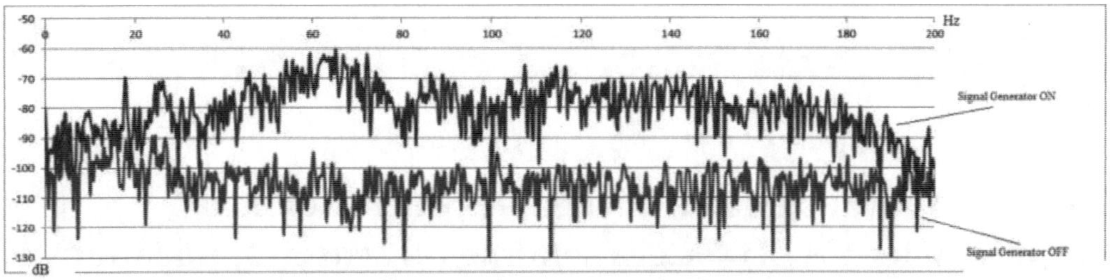

Figure 6. Frequency response of the SMS vibration sensor.

4. Conclusion

We have constructed an SMS fiber structure by splicing a section of 50/125 step index MMF between two 9/125 SMFs. To use this structure as a vibration sensor we coiled the MMF on an 80 mm diameter mini compact disc and coated the whole assembly with a TFE tape. The total weight of the sensing part was less than 10 grams so there was no loading effect when we place the sensor on most vibrating surfaces. Using a 1560 nm narrow line-width laser diode as the light source, we showed that the output of the sensor resembled that of the geophone, was able to detect 1.2 μm rms displacement, and had a good frequency response in 30 – 180 Hz band. This SMS sensor was potential for monitoring vibration in an RF-noisy condition, spark-prone environment, and remote location from the monitoring station due to the inherent optical fiber sensor advantages.

References

[1] Lee B 2003 *Opt. Fiber Technol.* **9** 57
[2] Lee B H, Kim Y H, Park K S, Eom J B, Kim M J, Rho B S and Choi H Y 2012 *Sensors* **12** 2467
[3] Santos J L, Frazao O, Baptista J M, Jorge P A S, Dias I, Araujo F M and Ferreira L A 2009 Optical Fiber Sensing Networks *Proc. Int. Microwave & Optolectronic Conf. 2009 (Belem)* (New York: IEEE) pp 290-298
[4] Wang Q, Farell G and Yan W 2008 *J. Lightwave Technol.* **26** 512
[5] Sun A, Semenova Y, Wu Q, Hatta A and Farrel G 2011 *Microw. Opt. Technol. Lett.* **53** 442
[6] Wu Q, Semenova Y, Wang P and Farrel G 2011 *Proc. SPIE* vol 7753 (Washington: SPIE Publications) 77535G
[7] Zhao Y, Li X G, Meng F C and Zhao Z 2014 *Sensors Actuat. A-Phys* **214** 163
[8] Dong B, Wei L, Zhou D P, Liu W K and Lit J W Y 2009 *Proc. SPIE* vol 7316 (Washington: SPIE Publications) 73160T
[9] OYO Geospace 2004 *GS-32CT Geophone* [http://www.geospacetech.com]
[10] Measurement Specialties, Inc. *Converting Among Acceleration, Displacement, Velocity and Frequency* [http://www.meas-spec.com]

IOP Conf. Series: Journal of Physics: Conf. Series **817** (2017) 012036 doi:10.1088/1742-6596/817/1/012036

Design of optical channel waveguide Mach-Zehnder interferometer (MZI) for environmental sensor applications

D Mahmudin[1], N Huda[2], T T Estu[1], A A Fathnan[1], P Daud[1], S Hardiati[1], L Hasanah[2] and Y N Wijayanto[1]

[1]Research Center for Electronics and Telecommunications (PPET), Indonesian Institute of Sciences (LIPI), Jl. Sangkuriang, Bandung 40135 Indonesia
[2]Faculty of Mathematics and Natural Science Education (FPMIPA), Indonesia University of Education (UPI), Jl. Dr Setiabudi No. 229, Bandung 40154 Indonesia

E-mail: dradrin@gmail.com

Abstract. High sensitivity environmental sensors with a simple and compact structure are required to monitor unwanted pollution such as liquid waste. In this research, a component of environmental sensor based on optical channel waveguide Mach-Zehnder interferometer (MZI) has been designed and also simulated. The materials used in the design were titanium dioxide (TiO_2) as a core and silicon dioxide (SiO_2) with air as cladding. The sensitivity value of the MZI sensor obtained by simulation using computer program was 11 nm/RIU at the angle of 16°. Meanwhile, the obtained fixed length of MZI sensing arm was 4 μm. The proposed design can be used for identifying the existence of liquid material precisely.

1. Introduction

Today, environmental pollution is caused by substances that are extremely dangerous and has reached an alarming level. The environmental damage caused by pollution in the air and water have threatened the sustainability of the living creatures on this earth. Therefore, it required a system for controlling the pollution level. One important component is a sensor with a very high level of sensitivity and selectivity. To meet this need, it is necessary to develop a sensor with the specifications as described above.

Since a few years ago, research and development of sensor components of electronic-based technologies are being replaced by light-based technologies. Optical signals have begun to be applied to the sensors for various needs such as chemical sensors, atmospheric or oceanic with a high level of sensitivity in a compact and simple devices [1-4]. Many types of waveguide sensors have been studied, such as planar waveguide structures, Mach-Zehnder interferometers (MZIs) and silicon micro-ring, for example SMR and MPRR [5-6]. The waveguide ring-resonator-based sensor has several advantages as follow. The shape is small and compact, thus it can be integrated with other components (light source, detector, coupler and so on), and it is easy for mass production [4].

In this paper, the design of the MZI structure for sensor applications environment is reported. The material used were titanium dioxide (TiO_2) as a core, silicon dioxide (SiO_2) and air as cladding with frequency of 180-193 THz. By varying the value of the angle in MZI structure and simulating it using computer program, the optimum value of sensitivity was obtained. The proposed device can be used for identifying difference liquid material precisely.

2. Method and design

The material of waveguide MZI was composed of core TiO_2 (n = 2.4) over a layer of SiO_2 (n = 1.53). The length of the waveguides sensing (*L*) was 4μm [1] with a thickness (*h*) of 1μm. The MZI waveguide structure is shown in figure 1, assuming the values of the air layer around were varied by n = 1.01, 1.1 and 1.2 as well as the variation of angle at 16° and 20°. The frequency range used was about 180 to 193 THz or equaled to 1666-1554 nm. The result from varying refractive index of the cover medium, MZI output power, theoretically is a function of the phase difference ($\Delta\varphi$) which is explained by equation 1:

$$P = \cos^2\left(\frac{\Delta\phi}{2}\right) \tag{1}$$

$$\Delta\phi = \frac{2\pi}{\lambda} \times L \times \Delta n_{eff} \tag{2}$$

where *L* is the length of the sensing pathways. In this study, the length of waveguides was equal and Δn_{eff} was the difference in effective refractive index between the two waveguides, namely the path sensing and reference [4].

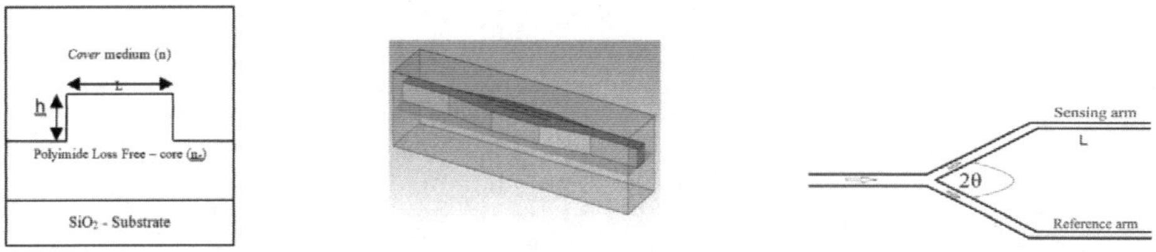

(a) (b) (c)

Figure 1. (a) Cross section of MZI. (b)View of 3-D MZI structure. (c) Scheme of MZI [7].

Field distribution of the waveguide was accomplished by transverse electric (TE) or transverse magnetic (TM) mode which was capable of analyzing 2-D waveguides. Because it was not possible to settle directly using these methods, we used the effective index method (EIM) [8].

In this study, the detection was performed when there was a change of refractive index caused by the change of analyte concentration, which was due to the presence of foreign substances. Such changes resulted in a shift in the frequency response.

Changes in the refractive index around the optical device changed the effective refractive index, which was marked by a shift in the output spectrum when the effective refractive index increased. Therefore, the magnitude of this shift was a function of the sensitivity of a sensor that can be expressed by the Free Spectral Range (FSR). FSR is defined as the maximum frequency difference to each other [9].

3. Result and discussion

Measurement technologies of optical sensors are generally based on changes in the effective refractive index. Refractive index changes due to the change in concentration of the analyte or the detection of foreign substances. Optical mode determines the spectral output of a device in term of resonance spectrum or interference fringe. Changes in the refractive index around the optical device will change the effective refractive index of the optical mode that leads to a shift in the output spectrum. Results displayed by bio-chemical sensors can be characterized by absorption, transmission, fluorescence intensity and polarization or reflection.

Another method of optical sensor is to see a shift in the spectrum of wavelengths or frequencies, which is caused by the changes of the refractive index. Such method is used for sensing the surface because this mechanism is very sensitive to small changes in the refractive index. Related to this shift, it is important to see the steepness of the spectrum peaks or troughs in the view of a previous spectrum or vice versa.

3.1. Design of MZI for value of angle at 16°

Figure 2 shows the graph of transmission loss as a function of wavelength with the value of angle at 16°. The value of the refractive index of the analyte being tested was 1.2 (air). The shift in frequency response occurred for all frequencies.

Figure 3 shows another graph of wavelength versus transmission loss for the value of angle at 16° in which the refractive index value of the analyte being tested was 1.1. The shift was seen more clearly at the beginning of the frequency range, however there were two peak points where the shifts are very small, i.e. at 1600.931 nm and 1584.2 nm.

Figure 4 is a graph of wavelength vs transmission loss for the value of angle at 16° and the refractive index value of the analyte being tested of 1.01. The shift happened more for all frequency ranges. This was due to the refractive index difference of the core material.

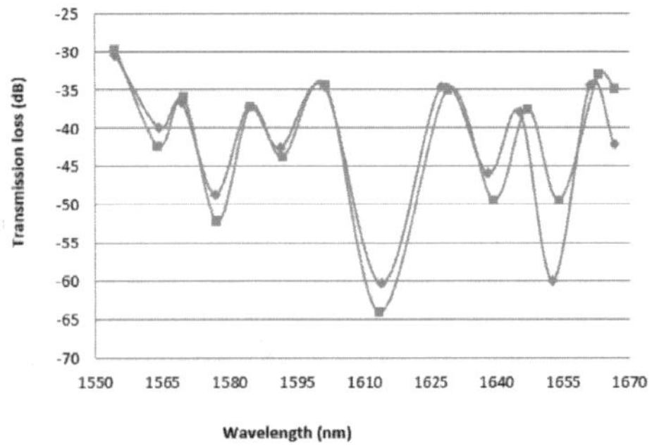

Figure 2. Wavelength versus transmission loss for value of angle 16° and refractive index of TiO_2 1.2.

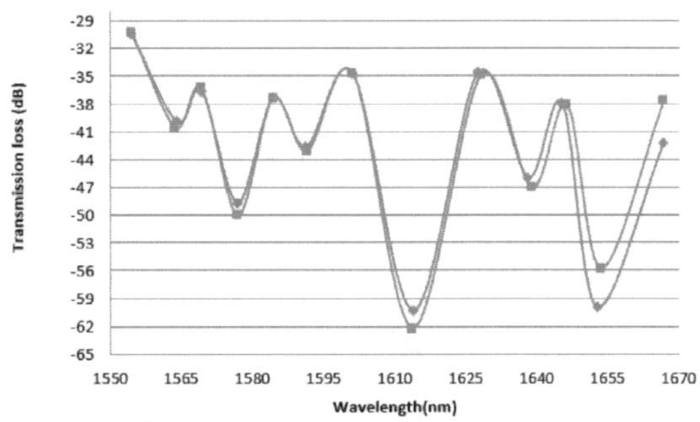

Figure 3. Wavelength versus transmission loss for value of angle 16° and refractive index of TiO_2 1.1.

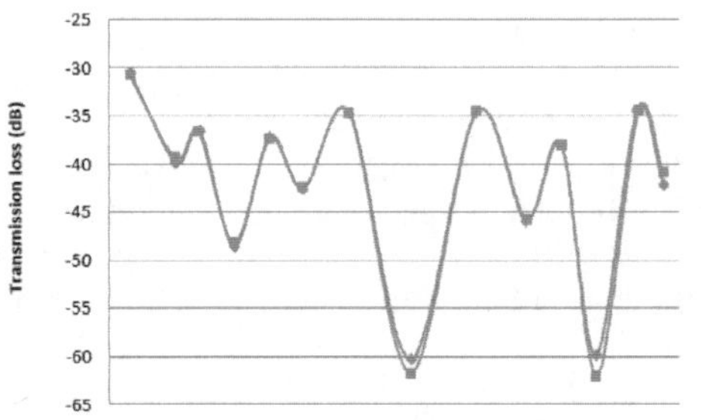

Figure 4. Wavelength versus transmission loss for value of angle 16° and refractive index of TiO_2 1.01.

3.2. Design of MZI for value of angle at 20°

Figure 5 shows the graphs of wavelength against transmission loss for analyte that has refractive index of 1.2 and the angle formed at 20°. The shift occurred across all frequencies and seemed to be greater than that of the angle of 16°.

IOP Conf. Series: Journal of Physics: Conf. Series **817** (2017) 012036 doi:10.1088/1742-6596/817/1/012036

Figure 6 is a graph of the transmission loss as a function of wavelength for the refractive index value of the analyte of 1.1 and the angle formed at 20°. The shift occurred across all frequencies and seemed to be greater than that of the angle at 16°.

Figure 7 is a graph of the wavelength versus transmission loss with a refractive index of the analyte being tested 1.01 on a 20° angle to the core material TiO_2. The shift occurred for all wavelengths. In contrast to figure 6, for the core material and the refractive index of this analyte test, the shift does not occur for all frequencies.

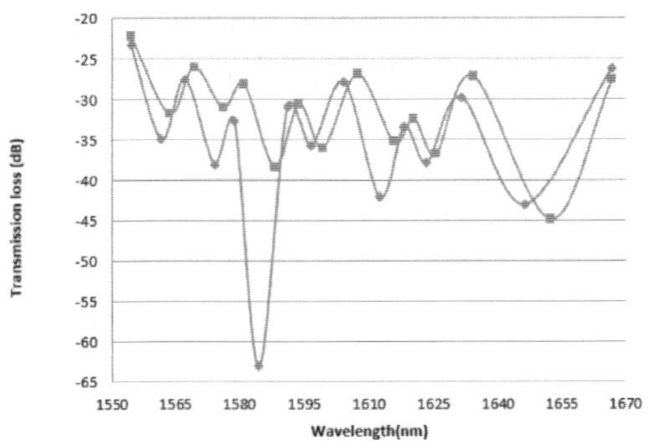

Figure 5. Wavelength against transmission loss for value of angle 20° and refractive index of TiO_2 1.2).

Figure 6. Wavelength against transmission loss for value of angle 20° and refractive index of TiO_2 1.1).

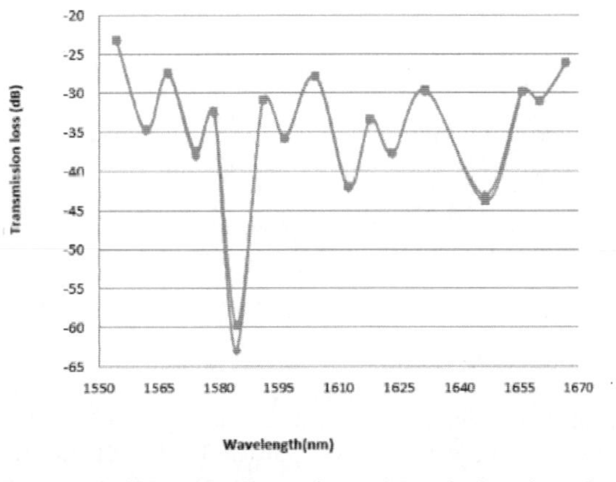

Figure 7. Wavelength against transmission loss for value of angle 20° and refractive index of TiO_2 1.01).

The shift of the frequency response seen in the pictures above, namely the environment (air) and analyte, occurred when the analyte reacted with the cover medium, namely air. The analyte was absorbsed on the surface of the air so that the reduction occurred. This caused electrons in the core material to get difficulties in passing the energy gap.

The nature of the source of the wavelength used was very influential. When light was passed through a medium, the particles absorbed the light. The amount of absorption was determined by the cosine of the angle of light incidence. The greater absorption angle caused greater reflected light.

The simulation done by varying the angle and refractive indices produced different output power. Damping was even greater with enlargement of the angle formed by the second optical path. From the pictures, visible spectrum shift occurred at a greater value of the refractive index of the test. This

IOP Conf. Series: Journal of Physics: Conf. Series **817** (2017) 012036 doi:10.1088/1742-6596/817/1/012036

means that the analyte, interference signal and movement of the membrane of other molecules were detected. Shifting value was determined also by the value of the refractive index core. The refractive index of greater core provided a great shifting spectrum [11]. For a sensor, FSR value can be determined directly by determining the frequency difference between two adjacent peaks [10].

3.3. The values of sensor sensitivity

A sensor is considered good if it has properties of high sensitivity and selectivity. The sensitivity of a sensor can be defined as the ratio between the signal difference and changes in the amount of measured signal. The results of the frequency response shift of MZI structure in earlier tests (figure 2-7) can be used to assess the shift of each peak then to obtain the sensor sensitivity values. The resulting sensor sensitivity can be seen in figure 8 and 9.

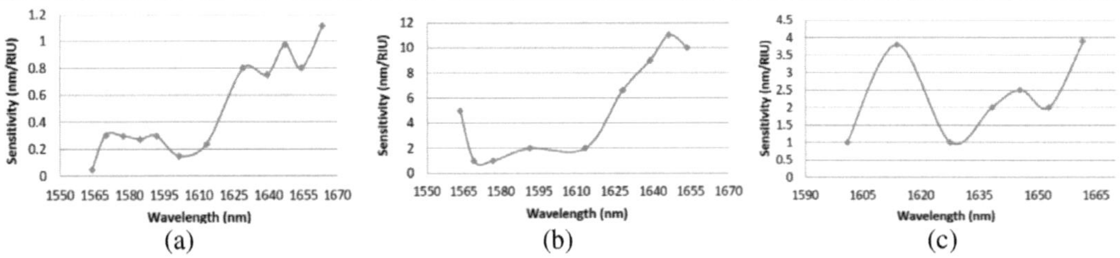

Figure 8. Graphs of MZI sensor sensitivity with the value of angle at 16°. The refractive index of TiO_2 were (a) 1.2, (b) 1.1 and (c) 1.01.

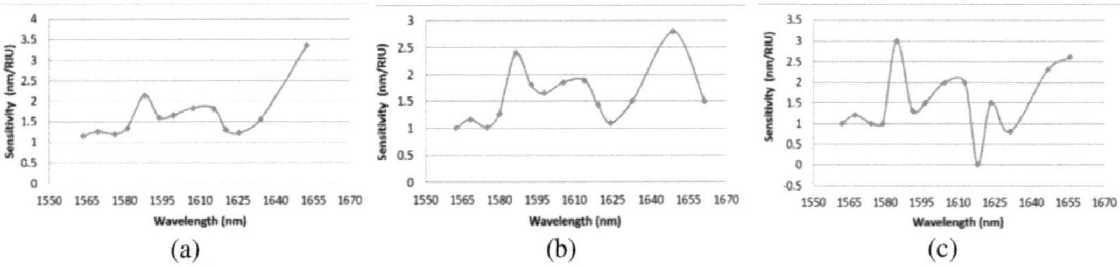

Figure 9. Graphs of MZI sensor sensitivity with the value of angle at 20°. The refractive index of TiO_2 were (a) 1.2, (b) 1.1 and (c) 1.01.

Figure 8 shows the sensitivity of the MZI sensor. The n value of the analytes tested were 1.2, 1.1 and 1.01. The material used as a core was TiO_2 with an angle of MZI structure at 16°. The sensitivity for each MZI structure with the different values of n are 3.9 nm/RIU, 11 nm/RIU and 1.15 nm/RIU. Whereas in figure 9, the sensitivity of MZI sensor structure with an angle value at 20° is shown. The sensitivities value were 3 nm/RIU, 2.8 nm/RIU and 3.36 nm/RIU. From some of these variations, it was concluded that the MZI sensor with 16° angle value has a high value of sensitivity to analyte with $n = 1.1$.

4. Conclusion

The design of the Optical Channel waveguide Mach-Zehnder Interferometer (MZI) was proposed. From the simulation results, we can conclude that Mach-Zehnder Interferometer (MZI) single mode had high sensitivity. Sensor with little attenuation owned by a small angle can be calculated based on the equation 1. Optical path length (L) = 4μm with a very small angle formed optimal small attenuation of FSR values. The proposed device can be used for precisely identifying different liquid material.

Acknowledgements
This research activity was conducted as part of the research project "Design of Smart Sensor Multimode Optical Waveguide Based on Polymer Substrate for Environmental Application" through The Excellent Research Grant from Indonesian Institute of Sciences (LIPI), Indonesia and "Development of Free-Space-Optical Links for Broadband Wireless Communication" through Thematic Program, funded by Indonesian Institute of Sciences (LIPI), Indonesia.

References
[1] Larsen U D, Blankenstein G and Branebjerg J 1997 *Digest of Technical Papers 1997 Int. Conf. on Solid-State Sensors and Actuators (Chicago)* vol 2 (New York: IEEE) p 1319
[2] Chaichuayi C, Preecha P, Yupapin P and Saeung P 2009 *Opt. Appl.* **XXXIX**(1) 175
[3] Fidanboylu K and Efendioglu H S 2010 *Proc. 5th Int. Advanced Technologies Symp. (Karabük)* (Karabük: Karabük University) pp 1-6
[4] La Notte M, Troia B, Muciaccia T, Campanella C E, De Leonardis F and Passaro V M N 2014 *Sensors* **14**(3) 4831
[5] Passaro V M N, de Tullio C, Troia B, La Notte M, Giannoccaro G and De Leonardis F 2012 *Sensors* **12**(11) 15558
[6] Mahmudin D, Estu T T, Hermida I D P, Sugandi G, Daud P, Wijayanto Y N, Menon P S and Shaari S 2015 *Proc. IEEE Regional Symp. on Micro and Nano Electronics (Kuala Terengganu)* (New York: IEEE) p 282
[7] Calo G, Farinola A and Petruzzelli V 2012 *Prog. Electromagn. Res. Lett.* **33** 151
[8] Dattner Y and Yadld-Pecht O 2011 *IEEE Photonics J.* **3**(6) 1123
[9] Fuadi N 2014 *Prosiding Seminar Nasional Geofisika 2014 (Makassar)* (Makassar: Geophysics Study Program, Department of Physics, Hasanuddin University) p 207
[10] La Notte M and Passaro V M N 2012 *Sensors Actuat. B-Chem.* **176** 994
[11] Bruck R and Hainberger R *Proc. SPIE* vol 7138 (Washington: SPIE Publications) 71380N

Design, construction and characterization of fiber extensometer with heart shape structure for detection of displacement

D Bayuwati and T B Waluyo

Research Center for Physics - Indonesian Institute of Sciences (Pusat Penelitian Fisika - LIPI), Kawasan PUSPIPTEK Serpong, South Tangerang 15314 Indonesia

E-mail: dwib001@lipi.go.id

Abstract. This paper discusses the design, construction and characterization of an optical fiber extensometer with heart shape structure. The extensometer was made from a communication grade fiber loop arranged in a housing and pulling mechanism. As the light source, we used 1310 nm LED and a 1325 nm laser diode/LD. To characterize our extensometer, we used a motorized pulling system controlled by a personal computer. From the bending loss characterization using Fujikura fiber as the sensor, we obtained sensitivity of 0.04 dB/mm and measurement range of 28 mm when using the 1310 nm LED, whereas using the 1325 nm LD allowed us to obtain 0.06 dB/mm of sensitivity and 23 mm of measurement range. Using Thorlabs fiber as the sensor, we obtained sensitivity and measurement range of 0.09 dB/mm and 23 mm, respectively, when using the 1310 nm LED. When we used the 1325 nm LD, we obtained sensitivity of 0.03 dB/mm and 31 mm measurement range but with a noted whispering gallery mode/WGM effect. Compromise has to be considered amongst the high sensitivity, measurement range, WGM existence and noise properties to get best data reading for real application.

1. Introduction

Optical fiber sensors offer many advantages when compared to conventional electronic sensors: they are compact, lightweight, immune to electromagnetic interference, noncontact, exhibiting high accuracy, easy to install, non-electrical, explosion-proof and resistant to high temperatures. A number of varieties of fiber sensors have been developed such as for sensing physical properties (temperature, stress, pressure, displacement) and chemical sensors (pH, chemical concentration, gas analysis) [1, 2]. Based on the operating principle, a fiber optic sensor can be classified into several categories: as an intensity, a phase, a frequency, or a polarization sensor. Fiber optics extensometer is a device for measuring land displacement optically, based on the changes in the intensity of the output light through an optical fiber that suffered losses due to bending treatment (due to land displacement). The land displacement will change the transmission of light through the fiber according to the value of land displacement that occurred.

Several types of optical fiber displacement sensors have been developed, for example, fiber micro displacement sensor using fiber-loop structure by Wang et al. [3] and Martinez-Rios et al. [4]. In the first reference, the authors used a tunable laser and an optical spectrum analyzer to explore the influence of bend diameter range of 18~19 mm at a wavelength of 1550 nm and in the later reference, authors examined the displacement range of 3.125 mm by using a 1.280 nm broadband low-power LED source and a single Ge-photodetector in a power transmission sensor.

This paper discusses the design, construction and laboratory characterization of optical fiber extensometer with a heart-shape fiber sensor before undergoing field testing. The basic structure of the heart shape fiber sensor is a fiber loop arranged in a sensor housing equipped by a pulling mechanism. The characterization of the heart sensor fiber was conducted using fiber pigtailed light sources, a PC-controlled pulling mechanism, detector, data acquisition system and PC/reading system. The purpose of the research is to compare the performances of thefiber displacement sensor when using two types of fibers (SMF28 Thorlabs and Fujikura single mode fibers) and two kinds of light sources (LEDat wavelength of 1310 nm and LD at wavelength of 1325 nm; respectively). Previous study by our group on fiber displacement sensor described fiber loop sensor using SMF 28 and LD light source at wavelength 1310 nm and 1550 nm with a manual pulling mechanism [5].

2. Theory

For single mode optical fiber, bending loss (L_s) can be expressed by [6]:

$$L_S = 10\log_{10}(\exp(2\alpha L)) = 8.686\alpha L \quad (1)$$

where 2α is bending loss coefficient, which is determined by fiber structure, bending radius and wavelength of light, L is the length of the bent fiber. D. Marcuse with classic theory of bending loss using coupled mode theory has formulated bending loss as [7, 8]:

$$2\alpha = \left(\pi\delta^2 L_C\left(\frac{2n_1 ka\Delta}{bW}\right)\right)^{1/2} - \Sigma x \exp\left\{-\left[\left(\beta_g - \beta_{1s}\right)\frac{L_C}{2}\right]^2\right\}\frac{J_1^2\left(J_{1s}\frac{a}{b}\right)}{J_0(J_{0s})}\exp\left(\frac{-2a^2}{W^2}\right) \quad (2)$$

For step index fiber, it applies the Bessel function zero and first order (J_0, J_1) and root of Bessel function (J_{01}, J_{1s}) with border condition (J_0 (J_{0s}) = 0, J_1 ($J_1(J_{1s})$ = 0. Practically, an optical fiber consists of not only the core and cladding but also the protection coating. Theoretical formula of bending loss using perturbation theory for step index fiber resulted the following equation [9]:

$$2\alpha' = \frac{2u^2}{\omega' v_{str} V^2 K_1^2(w)}$$

$$x\int_{\zeta=0}^{\infty} \frac{\exp\left[-\left(w^2+\zeta^2 a^2\right)^{1/2}\right] A_i[\chi(0,\zeta)]}{\left(w^2+\zeta^2 a^2\right)^{1/2} B_i[\chi(0,\zeta)]} d\zeta \quad (3)$$

With $\alpha' = w'\alpha/a$, $u = w'\sqrt{n_{co}^2 - \vartheta^2}$ and $V = \omega'^{\sqrt{n_{co}^2 - n_{cl}^2}}$, ζ is angular distance around bend axis of the bent fiber, $K_1^2(W)$ is Bessel function, A_i and B_i are Airy functions, which are the solution of 2 dimensions scalar equation; while ϑ_{str} is the real propagation velocity for straight fiber with the same refraction index. Wang et al. has theoretically and experimentally formulated the bending loss coefficient for optical fiber with more than one layer coating as follows [10]:

$$2\alpha = -2\frac{K^2}{2\pi\beta V^2 K_1^2(a\gamma)} \operatorname{Im}\left(\int_{-\infty}^{\infty} H_1(\zeta) A_i[X_2(0,\zeta)] d\zeta\right) \quad (4)$$

where x is layer thickness, n refractive index, R bending radius, β propagation constant, A_i and B_i are Airy functions and $H(\zeta)$ amplitude function. More details on bending loss fiber phenomenon can be found in references 6-10. The three above equations describe how bending loss was formulated if the

IOP Conf. Series: Journal of Physics: Conf. Series **817** (2017) 012037 doi:10.1088/1742-6596/817/1/012037

fiber consists of only core and cladding (see equation 2), if the fiber is a step index type (see equation 3) and if the fiber consists of core, cladding and multiple coatings (see equation 4).

Meanwhile, the sensor part in the fiber extensometer was formed from one fiber loop, which one of its end was tied into a pulling stick [11]. If the stick is pulled (because of displacement), the loop will be bent and changed into a heart shape form (see figure 1). We do not describe the theoretical calculation of bending loss in heart shape fiber in detail. Equation 1-4 only give brief illustration on what happen inside the fiber if it is bent. Here, we will only investigate the experimental trend between the fiber bending loss and the fiber output intensity.

The optical fiber extensometer consists of light source, optical fiber cable (diameter 3 mm), sensor and reference fibers in a housing system (see figure 1), optical detector, and data acquisition and reading system, which is connected to personal computer. The cable fiber was from flare flex and contains 12 fibers with total diameter 7 mm.

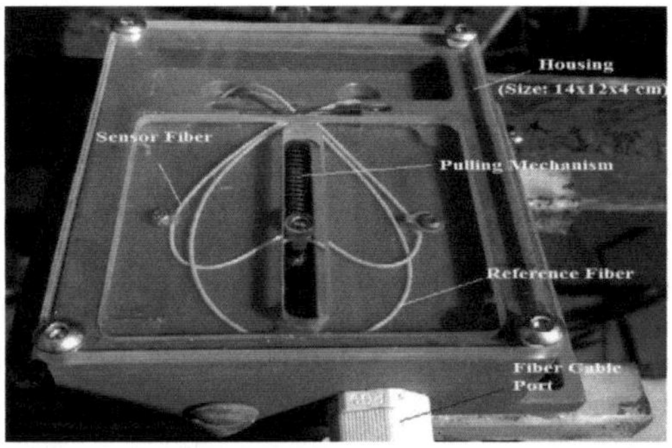

Figure 1. Heart shape optical fiber extensometer.

The sensor and reference fibers are spliced in, each, to one fiber in the fiber cable; and packed in a housing system which was equipped by a pulling stick for displacement acessory. The pulling process here is a simulation of the displacement process at a certain direction. The pulling process will change the fiber sensor loop diameter and consequently the output of light through the fiber.

We investigated the performance of two types of heart shape fiber sensor using SMF 28 Thorlabs and Fujikura fibers. The optical fiber used consisted of the core and cladding (both are glass, 9/125 μm size) and the outer coating, often called the buffer. In this experiment the fiber buffer was uncoated until only acrylic coating left with total fiber diameter 900 nm. Light source was coupled into the input port of the optical fiber loop and the output was detected at the output port. If the fiber sensor loop is pulled in the middle, then the loop will turn into a heart shape and the intensity of light propagating through the fiber will be reduced in accordance with the decreasing of the fiber diameter. The performance of the fiber extensometer depends on the bending properties of the optical fiber therefore bending measurement was conducted for all fibers. The set-up of the bending loss measurement consisted of fiber extensometer, a motorized pulling mechanism, a fiber pigtailed light source (LED or LD) and detector as well as reading system/PC as presented in figure 2.

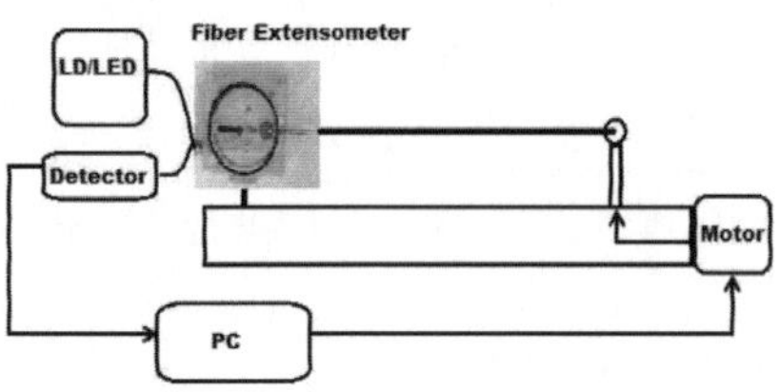

Figure 2. Experimental set up for automatic characterization of fiber extensometer.

Before bending loss measurement, both fibers were checked its full transmission by launching white light source to the fiber and detecting the output at several pulling-distance using optical spectrum analyzer to examine the sensitive bending area in its intensity-wavelength output graphs. The bending loss measurement was then conducted to obtain the intensity versus distance graphs outputs and analyze the data to find the measurement range, sensitivity and resolution of the sensor structure.

3. Experimental results and discussion

The first experiment was the measurement of the full transmission curve for both fibers using *Stabilized White Light Source* SLS201/M from Thorlabs and optical spectrum analyzer ANDO AQ 6312 to check the sensitive bending loss area. Output intensity were measured at three positions of the fiber loop sensor at its housing with pulling distances x = 0 cm, 2 cm and 3 cm for Fujikura fiber. For SMF 28 Thorlabs fiber, the pulling distances were 0 cm, 1 cm and 2 cm. The results are presented in figure 3(a) (for Fujikura fiber) and figure 3(b) (for SMF 28 Thorlabs fiber). We can see that at wavelength around 1310 nm and 1325 nm both fibers showed good respond to bending.

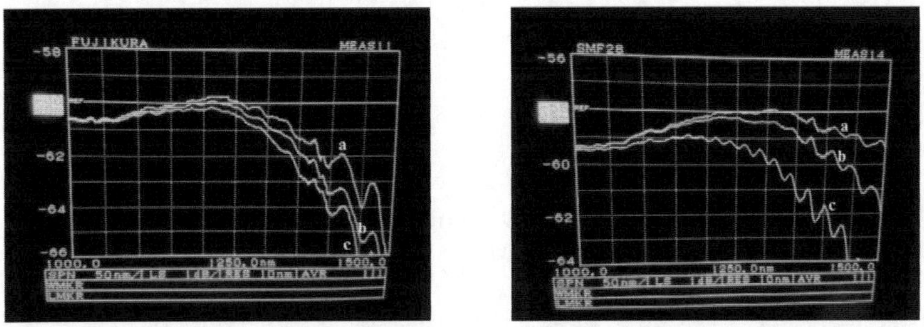

Figure 3. (a) Transmission curve of Fujikura fiber, at pulling distances a = 0 cm, b = 2 cm and c = 3 cm. (b) Transmission curve of SMF28 Thorlabs fiber, at pulling distances a = 0 cm, b = 1 cm and c = 2 cm.

The second experiment was the characterization of bending loss properties of both fiber with experiment set up shown in figure 2. It used two kinds of light sources, LED and LD at 1310 nm and 1325 nm, alternately; to compare its performances. The fiber was pulled automatically from 0-30 mm using a PC controlled pulling mechanism and its output was recorded and read using a PC.

The experiment results were output intensity as a function of distance as shown in figure 4(a) and 4(b) for Fujikura fiber and figure 5(a) and 5(b) for SMF 28 Thorlabs fiber. The data were examined on their linearity and existence of whispering gallery mode (WGM) and further analysed to examine their performances as displacement sensor. The required criteria for good fiber extensometer are free from noise and WGM and possessing high sensitivity, good resolution as well as wide range measurement.

The full data analysis for both fibers with its light source combination is presented in table 1. The combination of the Fujikura fiber with 1310 nm LED, resulted in a quite noisy spectrum but small WGM effect (see figure 4(a) and table 1, first left part) with coefficient linearity R^2 = 0.98511. When it was used as displacement sensor it had 0-28 mm of measurement range (x) with sensitivity of 0.04 dB. Since that the smallest value of the intensity that can be read is 0.05 mV; we can calculate the resolution as 0.05 mV / 0.04 mV/mm = 1.25 mm.

As for comparison, the resulted sensitivities for Fujikura SM fiber were 0.04 dB/mm when using the 1310 nm LED and 0.06 dB/mm when using the 1325 nm LD. The measurement ranges were 28 mm and 23 mm, respectively. In contrary, for SMF28 Thorlabs fiber, the sensitivity value was higher when using the 1310 nm LED compared to that of 1325 nm LD (0.09 dB/mm for 1310 nm LED compare to 0.03 dB/mm for 1325 nm LD), resulting in the resolution of 0.55 mm and 1.60 mm, respectively.

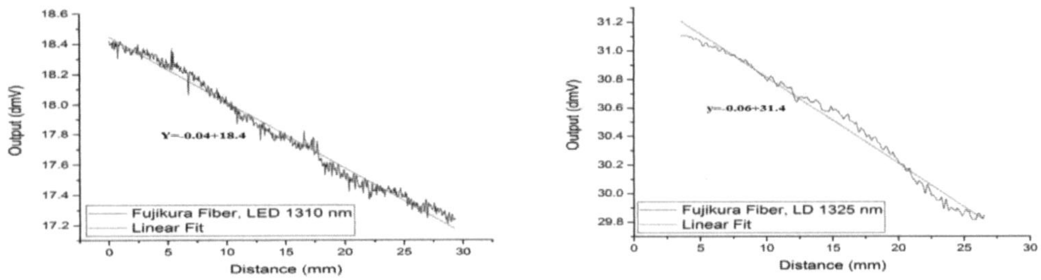

Figure 4. (a) Bending loss curve for Fujikura fiber using LED at 1310 nm, x = 28 mm (0-28 mm). (b) Bending loss curve for Fujikura fiber using LD at 1325 nm, x = 23 mm (3.5-26.5 mm).

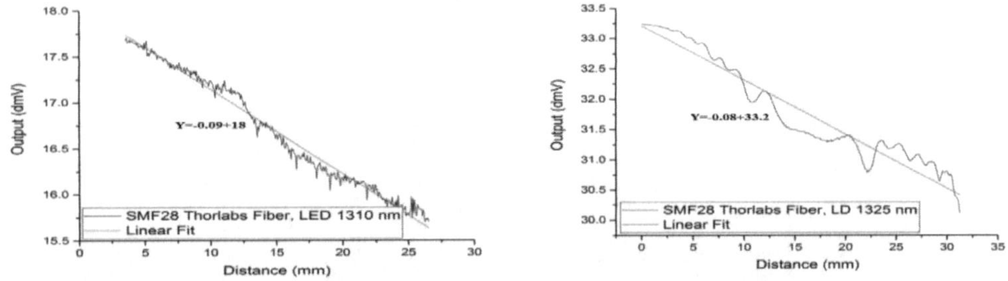

Figure 5. (a) Bending loss curve for SMF28 Thorlabs fiber using LED 1310 nm, x = 23 mm (3.5-26.5 mm). (b) Bending loss curve for SMF28 Thorlabs fiber using LD 1325 nm, x = 31 cm (0-31 mm).

Table 1. Bending loss properties of Fujikura and SMF28 Thorlabs fibers.

	Source wavelength	
Fiber type	LED at 1310 nm	LD at 1325 nm
SM Fujikura Fiber	Spectrum: Noisy WGM effect: small R^2: 0.98511 Measurement range: 28 mm (0-28 mm) Sensitivity: 0.04 dB/mm Resolution: 1.25 mm.	Spectrum: Not noisy WGM effect: small R^2: 0.98549 Measurement range: 23 mm (3.5-26.5 mm). Sensitivity: 0.06 dB/mm Resolution: 0.83 mm
SMF28 Thorlabs Fiber	Spectrum: Noisy WGM effect: small R^2: 0.98389 Measurement range: 23 mm (3.5-26.5 mm) Sensitivity: 0.09 dB/mm Resolution: 0.55 mm.	Spectrum: Not noisy WGM effect: notable R^2: 0.90701 Measurement range: 31 mm (0-31 mm) Sensitivity: 0.03 dB/mm Resolution: 1.60 mm.

The measurement range for SMF 28 Thorlabs fiber is wider when using the 1325 nm LD if compared to when using the 1310 nm LED (31 cm with 1325 nm LD compared to 23 mm with 1310 nm LED); but the WGM effect is notable when using the 1325 nm LD. Oscillation of the bending loss with bending radius has been observed by many researchers and this was explained by the WGM modes [8]. This might be caused by the type of material coating of the Thorlabs fiber which made more back-reflected light in the fiber's bent structure compared to that in the Fujikura fiber. WGM phenomenon should be studied further. The results of this work are comparable to the patented work by Gupta et al. [12] and to our earlier work.

From those results, it can be seen that using a laser diode instead of LED in characterizing bending loss of SMF 28 Thorlabs fiber does not guarantee to have a better performance of the heart shape sensor but there are still other affecting factors. We can get a higher sensitivity of fiber extensometer by using a laser source but it would result in a very narrow measurement range. Furthermore, we have to compromise in choosing a sensor configuration which has a better sensitivity, a wider measurement range, smooth data and acceptable WGM to get the best data reading.

4. Conclusion

We have presented the design, construction and characterization of optical fiber extensometer with heart shape sensor structure for detection of displacement using two types of fiber and light sources. From the bending loss characterization using Fujikura fiber as the sensor, we obtained sensitivity of 0.04 dB/mm and measurement range of 28 mm when using the 1310 nm LED, whereas using 1325 nm LD allowed us to obtain 0.06 dB/mm of sensitivity and and 23 mm of measurement range. Using Thorlabs fiber as the sensor, we obtained sensitivity and measurement range of 0.09 dB/mm and 23 mm, respectively, when using the 1310 nm LED. When we used the 1325 nm LD, we obtained sensitivity of 0.03 dB/mm and 31 mm measurement range but with a noted WGM effect.

Acknowledgments

This research was funded by Program Tematik, Research Center for Physics, Indonesian Institute of Sciences Year 2016. The authors are gratefully acknowledging Mr. Prabowo Puranto, M.T., who have developed the automatic pulling system. The authors also wish to thank Mr. Imam Mulyanto, S.T. for all technical assistances during the experiment.

References

[1] Udd E 2002 Overview of fiber optic sensors *Fiber Optic Sensors* ed F T S Yu and S Yin (New York: Marcel Dekker Inc.) chapter 1
[2] Gholamzadeh B and Nabovati H 2008 *World Acad. Sci. Eng. Techno.* **42** 297
[3] Wang P, Semenova Y, Wu Q and Farrell G 2010 *Microw. Opt. Technol. Lett.* **52**(10) 2231
[4] Martinez-Rios A, Monzon-Hernandez D, Torres-Gomez I and Salceda-Delgado G 2012 *Sensors* **12** 415
[5] Waluyo T B and Bayuwati D 2015 *Proc. SPIE* vol 9444 (Washington: SPIE Publications) S4440V
[6] Harris A J and Castle P F1986 *J. Lightwave Technol.* **4** 34
[7] Marcuse D 1976 *J. Opt. Soc. Am.* **66** 216
[8] Zendehnam A, Mirzaei M, Farashiani A and Horabadi L 2010 *Pramana* **74**(4) 591
[9] Smink R W, de Hon B P and Tijhuis A G 2007 *J. Opt. Soc. Am. B* **24**(10) 1610
[10] Wang Q, Farrel G and Freir T 2005 *Opt. Express* **13**(12) 4476
[11] Bayuwati D, Waluyo T B, Mulyanto I, Adinanta H, Puranto P and Suryadi 2014 Optical Fiber Extensometer with Heart-Shape Configuration Indonesian Registered Patent P00201408136 (in Indonesian)
[12] Gupta N and Nguyen N Q 2013 Fiber Optic Extensometer U.S. Patent US8428400 B2

Development of real-time extensometer based on image processing

H Adinanta, P Puranto and Suryadi

Research Center for Physics - Indonesian Institute of Sciences, Kawasan PUSPIPTEK Serpong, South Tangerang, Indonesia

E-mail: hend019@lipi.go.id

Abstract. An extensometer system was developed by using high definition web camera as main sensor to track object position. The developed system applied digital image processing techniques. The image processing was used to measure the change of object position. The position measurement was done in real-time so that the system can directly showed the actual position in both x and y-axis. In this research, the relation between pixel and object position changes had been characterized. The system was tested by moving the target in a range of 20 cm in interval of 1 mm. To verify the long run performance, the stability and linearity of continuous measurements on both x and y-axis, this measurement had been conducted for 83 hours. The results show that this image processing-based extensometer had both good stability and linearity.

1. Introduction

Extensometer is a device used to measure the displacement of an object [1]. The device is made using various types of sensors, either electrical or optical. The electrical and optical extensometers that are still being developed in Research Center for Physics, Indonesian Institute of Sciences, for example, are fiber-based, potentiometer-based and encoder-based extensometer [2]. Those devices are now in the testing process to be used in ground movement monitoring system [3]. The devices were tested for their sensitivity, resolution, linearity, stability and reliability.

The three extensometers have some advantages and disadvantages over each other, e.g. optical fiber extensometer has high resolution and sensitivity, but its range of measurement is shorter than the others. On the other hand, electrical extensometer that uses a potentiometer has a fairly high linearity, but its reliability is easily influenced by environmental factors such as humidity and temperature. Meanwhile, extensometer using encoder has advantages in stability, linearity, and resolution, but it is more expensive than the others.

In addition, there are other factors to be considered in developing an extensometer that is safe for measurement device. In general, devices that use optical and electrical materials will need a sling tied between the object and the device. The device detects the movement of the sling in order to measure the displacement of the object. However, this measurement method may cause the device to be attracted by the sling. This drawback can be solved by using another technique such as wireless system extensometer.

One of the techniques proposed is webcam-based extensometer. It uses image processing technique on motion-based tracking. This extensometer uses a web camera to detect the movement of the object periodically. The data consists of object positions and the displacement in each interval can be calculated.

2. Methods

To establish webcam-based extensometer system, there were some instruments used. The system employed a web camera with 1080 pixels (full-HD) as a main device, a computer or PC, a servo motor, and a target. This target was designed as a black circle with a diameter of 3 cm [4]. The target was installed in front of a web camera with a distance of 1 meter as shown in figure 1.

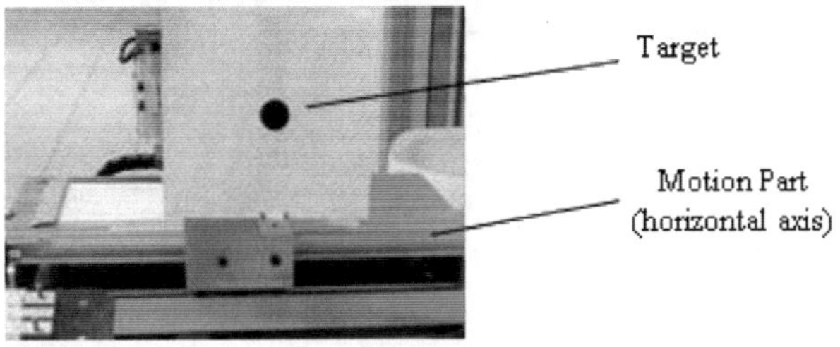

Figure 1. The target object in a circle shape was installed on motion part.

This target can be moved in x and y-axis using a motion part controlled by servo motor controller. This motion part can be moved in a 1 mm resolution using PC interface. In every 1 mm of movement, an image was captured by the web camera at a given position of x and y-axis. Specially developed software was used to control the motion part or servo motor and image capturing by a web camera.

Calibration of the camera's pixel to the actual position was then conducted by performing a stability test. The stability test was used as a parameter to detect the stability of the camera's pixel in capturing the object in 127 Lux of light intensity and 1 meter distance between a web camera and the object for 307830 seconds or approximately 83 hours. The result of the stability test showed the linearity value of camera's pixel and a counter time to find out the capabilities of this system.

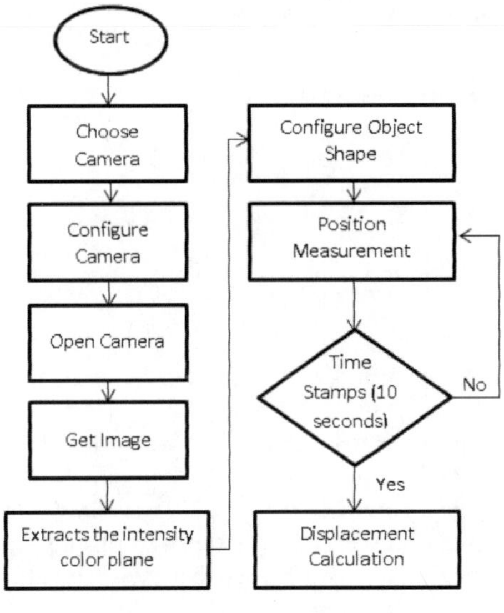

Figure 2. Flowchart of web camera-based extensometer program.

In this system, the sampling time was 10 seconds using a stream mode acquisition. In every 10 seconds, the system calculated the average position of the target to give the accurate value. In order to get the displacement, the system was programmed to calculate the difference of the position from Xt_i to Xt_f, then the displacement value can be calculated using subtraction [5, 6]. Another part of the system was the software, which was developed to get and extract the images then determine the position. The diagram of the software is shown in figure 2.

The arrangement of all parts in this system, which consisted of a PC, a web camera and a motion part, is shown in figure 3. This system measured the position of target in every 10 seconds and it gave a displacement value when the object moved to another point. The displacement measurement was conducted in one single axis, which is x-axis, in ±10 cm range.

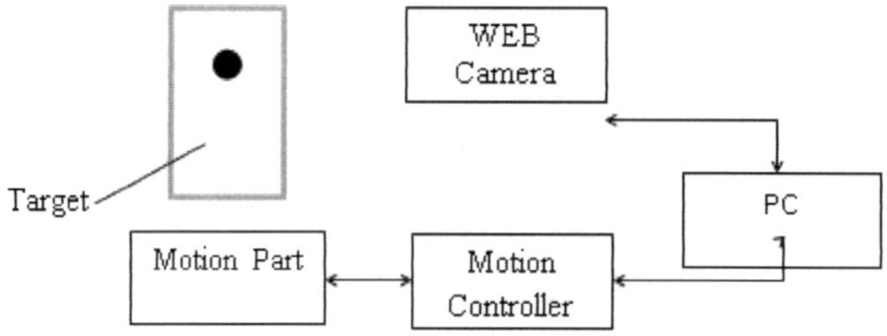

Figure 3. Set up of webcam-based extensometer.

3. Results and discussion

The webcam-based extensometer system was tested in the Research Center for Physics. The test was conducted in two parts. The first test was a stability test and the second was a linearity test. The stability test was conducted by measuring the target for 307830 seconds or approximately 83 hours. This test was done in an indoor area which had a light intensity of 127 Lux and temperature of 28 °C. The result of this test is shown in figure 4.

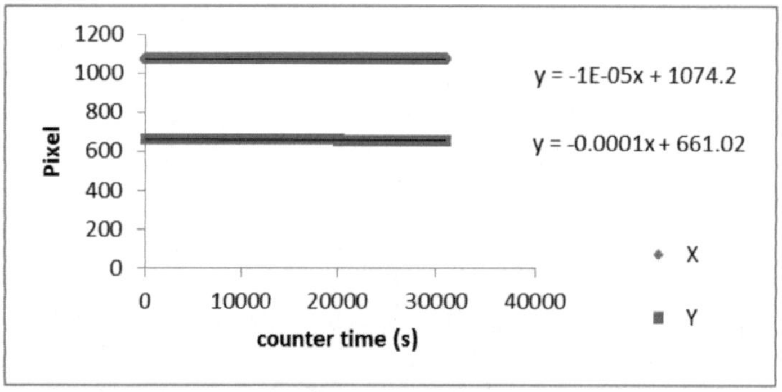

Figure 4. The results of stability test in an indoor area.

Figure 4 shows the results of position stability test in x-axis (blue line) and y-axis (red line). The x-axis of the graph is time counter and the y-axis is camera's pixel. The average position of target in x-axis was 1074 with a least square equation of $y = -1 \times 10^{-5}x + 1074.2$. This means that the equation remained constant. In y-axis, it can be seen that the position of target is at 659 with a least square equation of $y = -0.0001x + 661.02$. This also means that the gradient was close to 0 and it can be concluded that the position still remained constant.

Another test was linearity test using servo motor to move the target along x-axis while the position of the target was measured by web camera-based extensometer. This test required a motor servo controlled by PC to move from the zero point (origin) to 10 cm in the right direction and from zero point (origin) to 10 cm in the left direction. The result can be seen in figure 5.

In figure 5, it can be seen that the x-axis is displacement in mm unit and the y-axis is position in pixel plane of the chart. The displacement referred to the origin in the x-axis of motion part. The target was moved by motor servo to the right direction until it reaches the limiting distance. The limiting distance was 10 cm from the origin.

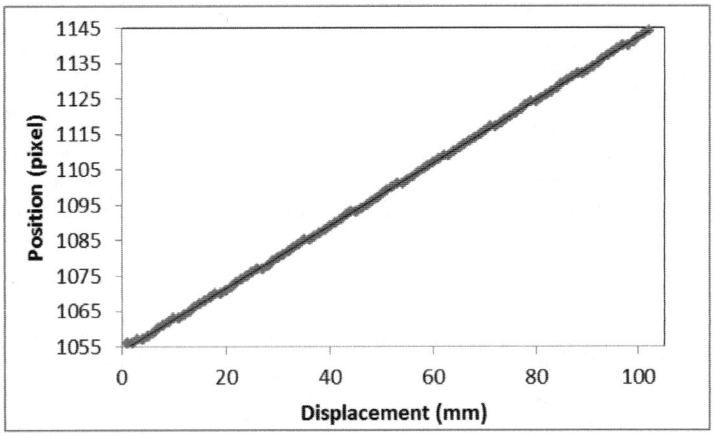

Figure 5. The result of linearity test in positive direction.

To analyze the correlation between displacement and position in the chart area, a least square method was applied. It was used to calculate the linearity of this extensometer. The least square method gave an equation of $y = 0.884x + 1053.7$. The value of 0.884 was also known as the sensitivity and the 1053.7 was the offset. This equation can be used to get the displacement value by inverting the equation. If x is displacement and y is a position in the chart area then it can be converted into $x = y - 1053.7/0.884$.

To complete the test, a measurement was also conducted by moving the target to the left until it reach the limiting distance. The limiting distance was also the same with that in previous test, which was 10 cm to the left from the origin. The result is shown in the figure 6.

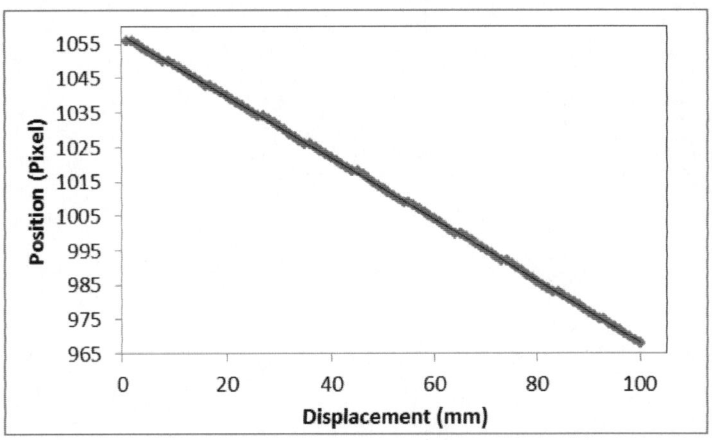

Figure 6. The result of stability test in negative direction.

From the figure 6, it can be seen that there was a correlation between the displacement and the position in the chart shown. This correlation was linear and had a least square formula of $y = 0.88935x + 1067.7$. The displacement can be known by calculation using this formula. If the pixel position can be measured by the webcam-based extensometer, then the displacement of the target can be calculated. The formula for the displacement will be $x = y - 1057.7/0.88935$.

This system was able to give the value of target displacement whenever the target was moved to another position. A web camera captured the image and the software calculated it into a displacement value.

4. Conclusions

A web camera-based extensometer system has been tested to observe its stability and linearity. This system gave a good linearity and stability. Therefore, it can be used to monitor the displacement of object in the lateral area of an object. As a consequence, the web camera also had to be positioned in the lateral area. This system was still able to detect the movement of an object for every 1 mm in 1 meter distance.

Acknowledgments

This research was supported by the Optoelectronic Research Group of Research Center for Physics, Indonesian Institute of Sciences and Mr. Bambang Hermanto who supplied the motor servo to control the movement of the target.

References

[1] Setiono A, Puranto P and Widiyatmoko B 2011 *Prosiding Pertemuan Ilmiah XXV Himpunan Fisika Indonesia Jateng-DIY 2011 (Yogyakarta)* (Yogyakarta: Bagian Penerbitan Himpunan Fisika Indonesia Cabang Jateng-DIY) p 26

[2] Adinanta H, Puranto P and Fauziah K 2012 *Publikasi Ilmiah Annual Meeting on Testing and Quality (South Tangerang)* (South Tangerang: Pusat Penelitian Sistem Mutu dan Teknologi Pengujian, Lembaga Ilmu Pengetahuan Indonesia)

[3] Bayuwati D, Setiono A, Suryadi, Waluyo T B, Mulyanto I, Adinanta H and Puranto P 2015 *Prosiding Seminar Nasional Fisika 2014 (South Tangerang)* (South Tangerang: Lembaga Ilmu Pengetahuan Indonesia, Pusat Penelitian Fisika) pp 78-85

[4] Betke M, Haritaoglu E and Davis L S 2000 *Mach. Vision Appl.* **12**(2) 69

[5] Soetedjo A and Nurcahyo E 2011 *Int. J. Comput. Sci. Netw. Secur.* **11**(2) 109

[6] Gance J, Malet J-P, Dewez T dan Travelletti J 2014 *Eng. Geol.* **172** 26

IOP Conf. Series: Journal of Physics: Conf. Series **817** (2017) 012039 doi:10.1088/1742-6596/817/1/012039

Aggregation effect on absorbance spectrum of laser ablated gold nanoparticles

Isnaeni[1], Irmaniar[2] and Y Herbani[1]
[1]Research Center for Physics, Indonesian Institute of Sciences, Building 442 Puspiptek, South Tangerang, Banten, Indonesia 15314
[2]Department of Physics, Gadjah Mada University, Sekip Utara Bulaksumur, Yogyakarta, Indonesia 55281

E-mail: isnaeni@lipi.go.id

Abstract. Plasmon of gold nanoparticles is one of the hot topics nowadays due to various possible applications. The application is determined by plasmon peak in absorbance spectrum. We have fabricated gold nanoparticles using laser ablation technique and studied the influence of CTAB (Cetyl trimethylammonium bromide) effect on the optical characterization of fabricated gold nanoparticles. We ablated a gold plate using NdYAG pulsed laser at 1064 nm wavelength, 10 Hz pulse frequency at low energy density. We found there are two distinctive plasmon peaks, i.e., primary and secondary peaks, where the secondary peak is the main interests of this work. Our simulation results have revealed that the secondary plasmon peak is affected by random aggregation of gold nanoparticles. Our research leads to good techniques on fabrication of colloidal gold nanoparticles in aqueous solution using laser ablation technique.

1. Introduction

Nanoparticles have been creating a lot of promising advanced nano-technologies. One of potential nanoparticles is gold nanoparticles [1, 2]. There are a lot of researches have been done to study the properties of gold nanoparticles. One unique property of gold nanoparticles is plasmon, which is energy of oscillating electrons inside gold nanoparticles due to external electromagnetic field [1, 2]. In the case, the plasmon is well known as localized surface plasmon, which exists in limited plasmon region surrounding gold particle [2, 3]. The plasmon region only occur few nanometers surrounding the surface of gold nanoparticles. Although the plasmon region is very restricted, the effect of plasmon is very powerful. One of potential applications of plasmon produced by gold nanoparticles is emission enhancement of fluorescent nanoparticles located nearby gold particle [3-5]. However, there are some conditions must be meet to have good enhancement by gold nanoparticles [4, 5]. First, there must be a resonant condition between emitting nanoparticles and plasmon of gold nanoparticles. Second, the distance between emitting nanoparticles and gold nanoparticles must be close enough, approximately 10 nanometers (nm). Based on those conditions, it is difficult to apply specific gold nanoparticle to many emitting nanoparticles. Thererefore, some modification of plasmon region and plasmon frequency is required. Some studies have been conducted to modify plasmon regions and frequencies. Changing diameter of gold nanoparticles is a simple way to vary plasmon frequency. However, this way only changes plasmon frequency up to certain frequency [6]. Fabricating multishell of metallic nanoparticles is one effective way to produce two plasmon frequencies [7]. Many researchers also have proven that gold nanorods produce dual plasmon frequencies due to longitude and transverse waves [8]. Another interesting arrangement of gold nanoparticles is cluster of gold nanoparticles. Plasmon range and distribution significantly change when gold nanoparticles are close to each other. Researches on the change of optical properties due aggregation of gold nanoparticles are very few due the uninteresting application of gold nanoparticle aggregation. Nevertheles, the change of plasmon frequency and distribution is still intersting to study. The aggregation of gold nanoparticles is a well-known knowledge, however, so far only few studies have reported the plasmon effect due to aggregation. In addtion, there several ways to fabricate gold nanoparticles, including chemical process and laser ablation. In this work, we will use laser ablation to fabricate gold nanoparticles. The purpose of this work is to study the change of plasmon frequency due to aggreation of gold nanoparticles fabricated using laser ablation method. In addition, we will use simulation result to explain our experimental findings.

2. Experimental

In order to clarify the effect of aggregation to plasmon frequency and and distribution, we conducted experimental and simulation works. In the experimental work, we fabricated gold nano nanoparticles using laser ablation method as shown in figure 1(a). One gram of gold plate purchased from ANTAM (Indonesian goverment owned general mining company), with purity of 99% was cleaned using methanol and then placed in a glass beaker filled by purified water. Pulsed laser from NdYAG laser with wavelength 532 nm was focused onto gold plate. Detail optical setup used in this expriment is shown in figure 1(b). It took about 1 until 2 hours to ablate gold plate to obtain gold nanoparticles. UV-Vis spectrometer was utlized to analyze fabricated gold nanoparticles. The absorption spectrum was used to characterize plasmon frequency and its other properties. In order to make aggregation and monodispersed gold nanoparticles, we used CTAB (Cetyl trimethylammonium bromide) that was dissolved in purified water solvent prior ablation process. In addition, we also varied the ablation energy of pulsed laser from 12 mJ, 23 mJ, 29 mJ, and 36 mJ.

In the simulation work, we use MPBEM tool box, which was obtained online and run it in MATLAB program [9]. We modified size of gold nanoparticles, space distance in between gold nanopaticles and number of gold nanoparticles. The simulation used boundary element method computational technique to generate extinction cross section, absorption cross section and scattering cross section. We used simulation results to analyze and explain experimental results.

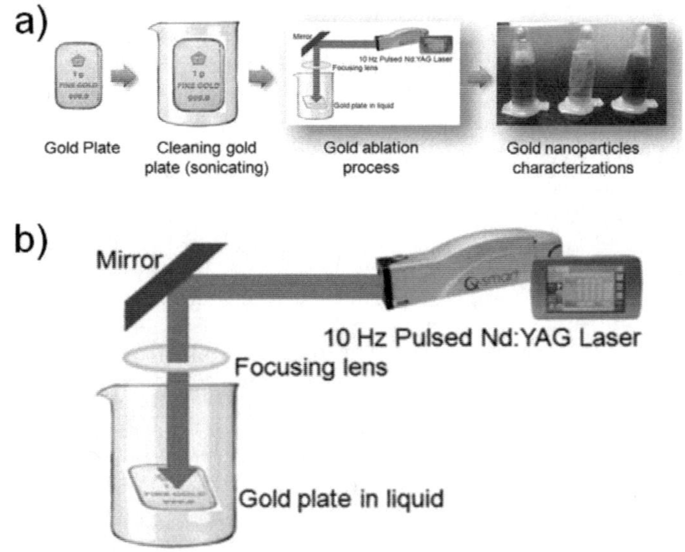

Figure 1. Fabrication process of laser ablation to fabricate gold nanoparticles (a) and laser ablation optical setup used in this work (b).

3. Results and discussion

Laser ablation technique used in this work, has successfully fabricated gold nanoparticles. The direct indication of successfully fabricated gold nanoparticles is the liquid color has changed from colorless into pink and red color as shown in figure 1(a). In general, the color of gold nanoparticle solution determines size, distribution and aggregation of fabricated gold nanoparticles [1, 3]. Pink color usually indicates small size or monodispersed of colloidal gold nanoparticles. On the other hand, purple or dark red color indicates large size or aggregated gold nanoparticles. We have noticed that our fabricated gold nanoparticles were stable until few weeks after fabrication. This good stability gives us enough time to characterize our fabricated gold nanoparticles. Experimentally, we investigated our gold nanoparticles using UV-Vis spectrometer to study the plasmon peak in absorbance spectrum and using TEM (transmittance electron microscope) to study size distribution of gold nanoparticles.

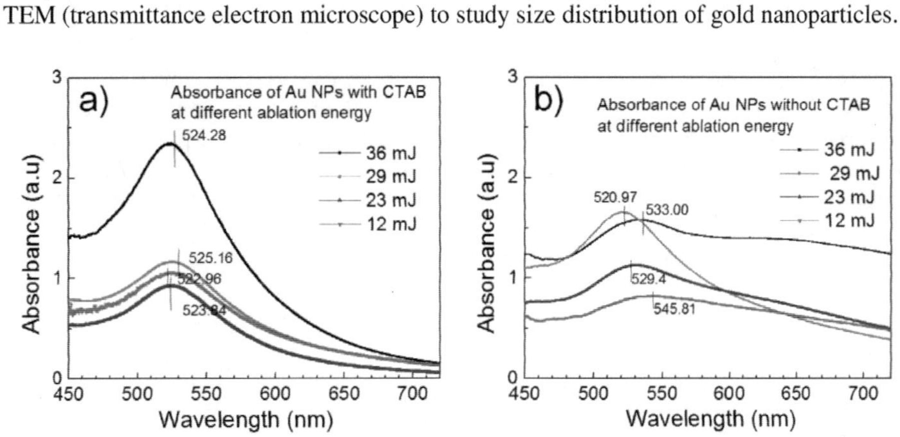

Figure 2. Absorption spectra of fabricated gold nanoparticles with (a) and without (b) CTAB.

We used three different energy of laser to ablate gold plate in aqueous solution. The effect of laser energy is described below. Figures 2(a) and 2(b) describe absorption spectrum of fabricated gold nanoparticles with different ablation laser energy with and without CTAB. Although there is no clear trend, but we can clearly see that higher ablation energy produced higher absorbance values. High absorbance values indicate high concentration of fabricated gold nanoparticles. Visually, we also observed (not shown in the figure) that higher ablation energy produced darker color of colloidal gold nanoparticles. The inconsistence of absorbance value as the function of ablation energy is probably due to aggregation effect. Detail of aggregation effect will be discussed later.

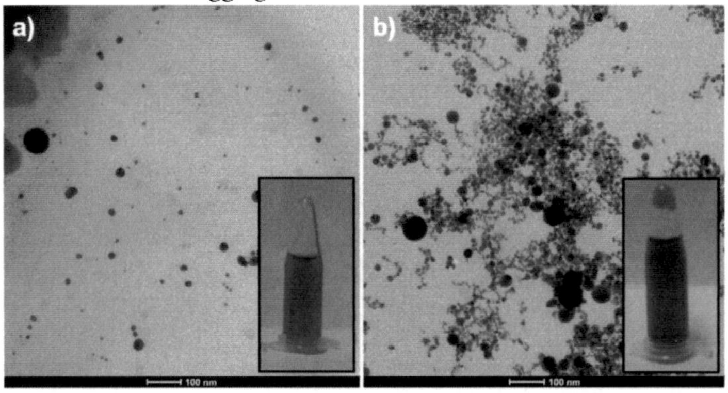

Figure 3. TEM images of gold nanoparticles with (a) and without (b) CTAB. The inset figures are visual appearances of corresponding gold nanoparticles in ambient illumination.

Further effect of aggregation was conducted by investigating fabricated gold nanoparticles without the addition of CTAB. The fabricated gold nanoparticles with CTAB have very distinctive single plasmon peak and absorption spectra as shown in figure 2(a). There is only one plasmon peak occurs for every absorption spectra at around wavelength 523 nm until 524 nm. This indicates that CTAB can protect fabricated gold nanoparticles very well against self-aggregation. The evidence of less aggregation is given by a TEM image as shown in figure 3(a). The fabricated gold nanoparticles is well separated one to another. Visual appearance (inset image in figure 3(a)) also shows that fabricated gold nanoparticles have clear pink color.

On the order hand, without addition of CTAB, the fabricated gold nanoparticles have two distinctive peaks in absorption spectra as shown in figure 2(b). The first peak is the primary plasmon peak. This peak varied from 520 nm until 533 nm. The second peak occurs at around 640 nm. This peak is assumed as secondary plasmon peak that occurs due to aggregation effect of fabricated gold nanoparticles. The evidence of aggregation is shown by TEM images in figure 3(b). The fabricated gold nanoaprticles are aggregated making a group of cluster or a chain group of gold nanoparticles. Actually, the sizes of fabricated gold nanoparticles with and without CTAB are comparable. We found that the size fabricated gold nanoparticles vary from 5 nm until 80 nm. The only different between those samples is only aggregation.

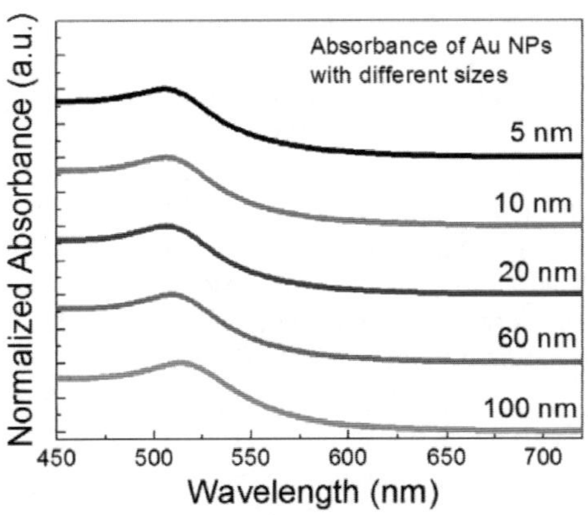

Figure 4. Simulated absorbance spectrum of gold nanoparticles with different sizes.

In order to confirm the origin of aggregation, we conducted simulation work to produce absorption spectra using various gold nanoparticles properties. First, we study the effect of size of gold nanoparticles on absorbance spectra. We simulated absorbance spectra of single gold nanoparticles with different size from 5 nm until 100 nm. The simulated absorbance is shown in figure 4. We found that the plasmon peak slightly shifts to longer wavelength when the size of gold nanoparticles increases. The plasmon peak shifts from 508 nm (for 5 nm gold nanoparticles) to 518 (for 100 nm gold nanoparticles). We found that there is only one distinctive peak in absorbance spectrum. There is no secondary peak occurs at longer wavelength. This simulation agrees with prior researches on basic optical properties of gold nanoparticles [1-3]. Therefore, we can conclude that although fabricated gold nanoparticles have large size distribution, the plasmon peak does not strongly influence by size distribution of gold nanoparticles and there is no secondary plasmon peak that occurs.

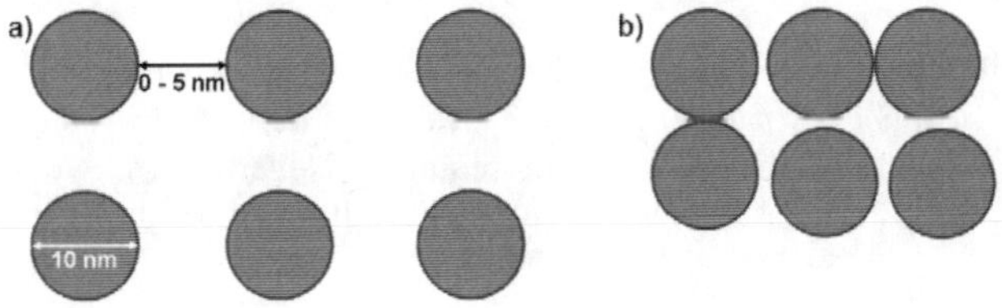

Figure 5. Simulation configuration of gold nanoparticles for space distance effect (a) and random space distance (b).

Furthermore, we simulated aggregation effect of gold nanoparticles. We simulated absorbance using six gold nanoparticles with diameter 10 nm as shown in figure 5(a). First, we varied space distance between gold nanoparticles from 0 nm until 5 nm. We found that the space distance significantly influences absorbance spectrum of gold nanoparticles. The space distance from 0 nm until 1 nm strongly creates secondary peak of absorbance spectrum as shown in figure 6(a). In addition, main plasmon peak shift to longer wavelength by more than 80 nm, compare to plasmon peak of individual gold nanoparticles, when gold nanoparticles are separated by 1 nm. The plasmon peak shifts to shorter wavelength when the space distance between gold nanoparticles increases from 1 nm until 5 nm. It is predicted that the plasmon peak return to its original wavelength when the space distance between gold nanoparticles is more than its diameter. In order to reproduce absorbance spectrum of fabricated gold nanoparticles, we simulated six gold nanoparticles with random space distance between 0 nm until 1 nm as shown in figure 5(b). The simulated absorbance spectrum is shown in the bottom curve of figure 6(a). We compared the simulated the absorbance spectrum of gold nanoparticles having random space distance to the absorbance spectrum of fabricated gold nanoparticles without CTAB as shown in figure 6(b). We clearly see that there are two peaks in absorbance spectrum. The first peak is the primary plasmon peak of gold nanoparticles at around 530 nm. The secondary peak is at wavelength longer than 600 nm. Based on our simulation results, we can confirm that the secondary peak is attributed to aggregation of gold nanoparticles. The wavelength of secondary peak positions of simulated gold nanoparticles is not exactly similar to that of fabricated gold nanoparticles. This is probably due to random distribution of gold nanoparticles and the unknown number of gold nanoparticles in the aggregated cluster. Nevertheless, our simulation results have shown that aggregation plays an important role in secondary plasmon peak. We assume that aggregation of gold nanoparticles constructs overlapped plasmon distribution of individual gold nanoparticle. This overlapping region may generate new plasmon frequency probably due to interference effect in the overlapping region.

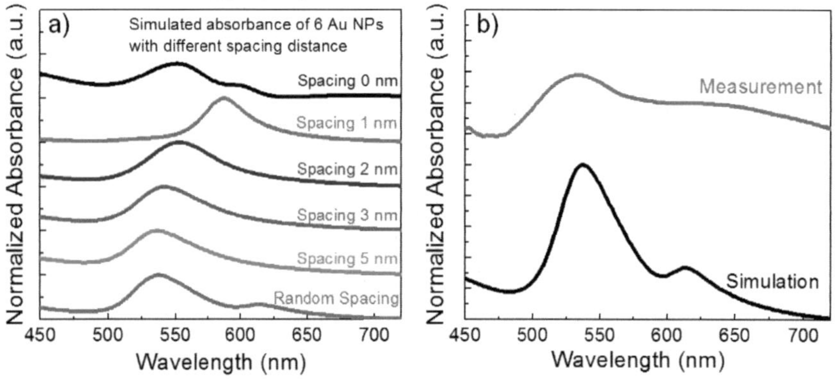

Figure 6. Simulated absorbance spectrum of gold nanoparticles with different space distance (a) and comparison of absorbance spectrum between simulated and fabricated gold nanoparticles (b).

Further work still remains to investigate how powerful the secondary plasmon peak and the possible application of this secondary plasmon peak in the practical technologies. For some application, this secondary plasmon peak is not required. Nevertheless, we still believe that this secondary peak is still useful for certain applications, especially applications in near-infra red region.

4. Conclusions

We have successfully fabricated gold nanoparticles using laser ablation technique with and without CTAB. We have observed that the absence of CTAB strongly affects aggregation of fabricated gold nanoparticles. The aggregation of gold nanoparticles is indicated by secondary plasmon peak in peak We have carefully studied and simulated the effect of gold nanoparticles size and aggregation of gold nanoparticles cluster. We found that size of gold nanoparticles does not generate secondary plasmon peak in absorbance spectrum. Furthermore, random aggregation clearly creates primary plasmon peak around 530 nm and secondary plasmon peak at wavelength longer than 600 nm. The secondary plasmon peak was generated at gold nanoparticles space distances of 0 - 1 nm. These results demonstrate new potential applications in utilizing surface plasmon effect.

Acknowledgements

The authors would like to thanks to Kegiatan Unggulan LIPI and Kegiatan Penguatan Komptensi from Research Center for Physics, Indonesian Institute of Sciences for financial support of this work. The authors also would like to thanks to researchers in the Group of Laser, Research Center for Physics for valuable discussion and support during this work.

References

[1] Rada A G, Abbasi H and Afzali M H 2011 *Physics Procedia* **22** 203 – 8
[2] Phuc T D, Yoshino M, Yamanaka A and Yamamoto T 2013 *Procedia CIRP* **5** 47-52
[3] Zuber A, Purdeya M, Schartner E, Forbes C, Van der Hoek B, Giles D, Abell A, Monro T and Ebendorff-Heidepriem H 2016 *Sens. Actuators, B* **227** 117–27
[4] Hong X, Tan C, Chen J, Xu Z and Zhang H 2015 *Nano Res.* **8**(1) 40-55
[5] Jun H K, Careem M A and Arof A K 2016 *Proceedings 3S* S73-S79
[6] Zhang W, Li Q and Qiu M 2013 *Opt. Express* **21**(1) 172-81
[7] Hong S, Acapulco J A I, Jang H Y and Park S 2014 *Chem. Mater.* **26**(12) 3618–23
[8] Pérez-Juste J, Pastoriza-Santos I, Liz-Marzán L M and Mulvaney P 2005 *Coord. Chem. Rev.* **249**(17-18) 1870–901
[9] Hohenester U and Trügler A 2012 *Comput. Phys. Commun.* **183** 370–81

IOP Conf. Series: Journal of Physics: Conf. Series **817** (2017) 012040 doi:10.1088/1742-6596/817/1/012040

Determining optimum red filter slide distance on creating 3D electron microscope images using anaglyph method

W P Tresna and Isnaeni

Research Center for Physics - Indonesian Institute of Sciences, Bld 442 Puspiptek Area, South Tangerang 15314 Indonesia

Email : wild005@lipi.go.id

Abstract. Scanning Electron Microscope (SEM) is a proven instrument for analyzing material in which a 2D image of an object is produced. However, the optimization of a 3D image in the SEM system is usually difficult and costly. There is a simple method to produce a 3D image by using two light sources with a red and a blue filter combined in a certain angle. In this experiment, the authors conducted a simulation of the 3D image formation using anaglyph method by finding the optimum point of shifting the red and blue filters in an SEM image. The method used in this experiment was an image processing that employed a digital manipulation on a certain deviation distance of the central point of the main object. The simulation result of an SEM image with a magnification of 5000 times showed an optimal 3D effect that was achieved when the red filter was shifted by 1 μm to the right and the blue filter was shifted by 1 μm to the left from the central position. The result of this simulation can be used to understand better the viewing angle and the optimal position of the two light sources, i.e. red and blue filter pairs. The produced 3D image can be clearly seen using 3D glasses.

1. Introduction

An understanding of dimensions will always be associated to field or space. One-dimension (1D) is defined as a point or a line, two-dimension (2D) is defined as the plane of a space and three-dimension (3D) is defined as 3-dimensional space in the vector called x, y, and z. In the case of visual 3D images, it can be produced from modified 2D images with the help of 3D effect software. 3D effect has been used to provide better understanding and view of an image that is seen by eyes. In the world of visualization, the major factor in determining the quality of the 3D image is the logic of the eyes, and this could be pragmatic and subjective from the perspective of the viewers. Although the language of logic in the eye can be translated into real numbers regarding the description of the image, but sometimes tastes and feeling become dominant factors in the industrial market. Simulation of 3D effect with a software is the first step in creating a 3D effect by adjusting and manipulating the position of the light sources [1, 2]. This method can also be applied to microscope, SEM or visual entertainment images.

Some 3D images can be generated using two cameras at a certain position and tilting angle. However, the main problem in producing 3D images from SEM images is to determine the position of proper light sources, which corresponds to the red and blue filter light sources [3, 4]. 3D image produced from SEM images should have illusion effect that shows depth and sharpness of every object and point in the image, so that it will give the impression of parallel horizontal position difference with observers.

The aim of this paper is to suggest a variety of new ways in which anaglyphs can and should be reproduced to create a full-color-stereo-pair image. In this work, we only use single image to produce 3D image. The degree of light source angle in digital manipulation can produce a new perspective of 3D image. Below are a brief description related to 3D image production. Some part of this description will be utilized to generate 3D image from an SEM image.

Digital image capturing process, to produce 3D images, usually requires dual camera rigs. There are two methods regarding dual camera rig, i.e. side-by-side method and mirror rig method. In the side-by-side method, two identical cameras are placed side by side to capture the same object. In the mirror rig method, a camera with a mirror as a splitter is used to produce two shifted images. The problem in side-by-side method is that the interocular or interaxial (distance parallel perspective of two lenses of the two cameras) are not small enough for close-up shots [5]. As a result, the images will be distorted on the sides lengthwise. Meanwhile in the method of mirror rig, it usually tackles the problem above successfully, however it has other drawbacks, such as polarization, reflectance or reflection on the object. This problem can be corrected using a polarizer filter on the lens that contains reflective layer. As a result, the light that enters the camera will change [5, 6].

In computational photography, deconvolution is commonly used to sharpen a "blurry" image. The key concept is that each pixel in the original image, d_i, has a convolved value:

$$d_i = \epsilon_i + \sum p_{ij} u_j \quad (1)$$

where ϵ_i is small level of noise, u_j represents each point in the scene or image, and p_{ij} is fraction contributed to the images that is determined by the point spread function (PSF). A wide range of deconvolution techniques, largely involving analysis in the frequency domain, are mainly use PSF to determine u_j values. It is assumed that the PSF is given as a known input; where it is not specified, it is determined automatically as a part of blind deconvolution [7].

A 3D image, if described in a flat plane, will be like the cardinal directions with the center position as the center of interest. In the most image processing software, front position (front), rear (back), right (right) and left (left) are perspectives that can be changed. While the position of the wind amid such as South East (Southeastern), South West (Southwestern), North West (Northwestern) and North East (Northeastern) are very useful to improve the main perspective, as describe in figure 1 [2, 8-10].

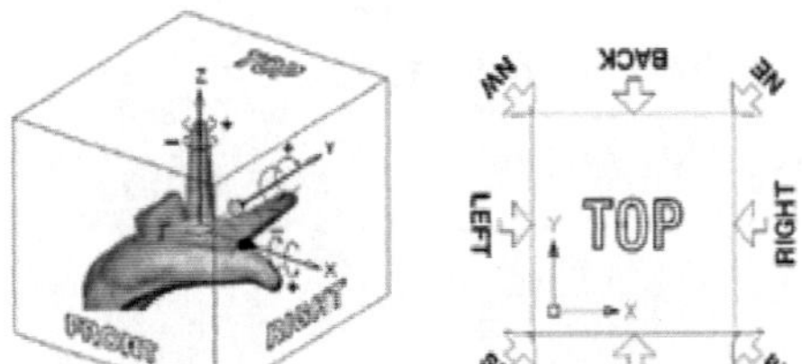

Figure 1. Description of understanding of the 3D effect.

2. Method

In this simulation experiment, the material used was an SEM image of particle sample that contained 72% Fe. It was taken using 5000X magnification in the secondary electron (SE) mode, which focused on the morphological image only. This image was selected because of its clearness, intact and round shape, and there were clear spaces between the main objects (particles) that would make the observation easier. The size scale obtained from 5000X magnification was 10 µm. The SEM image was then processed further using an image processing software. The software used was Photoshop

CS5 portable with friendly features that are quick and simple, therefore it is easy to use even for a beginner.

The image processing step was started by selecting an SEM image which had simple shape and no extreme shadow. The image was selected based on certain criteria to allow better observation using 3D glasses. Firm image boundary which contains not too many color will provide clearer information to the viewer about the depth of field. An example of SEM image used in this work is shown in figure 2.

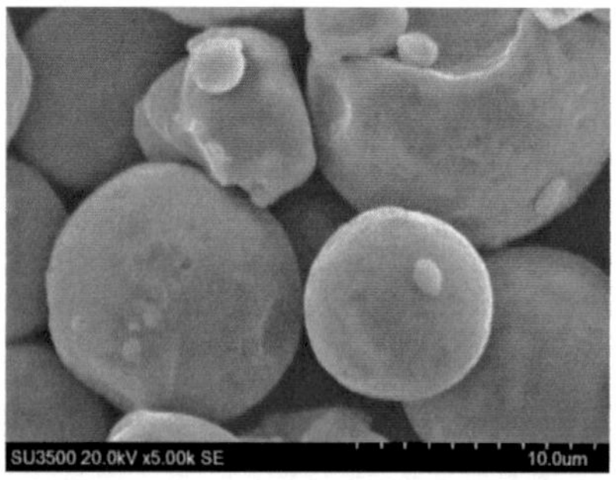

Figure 2. The original image file which is selected as a sample.

The next step was processing the image using the image processing software in which several processes needed to be carried out, such as shown in figure 3. The first was going through Action section to sharpen the left and right sides of the image. In this technique, one important thing to understand is the proportionality of an image so that an image will always look identical when viewed by the left and right eye. Second, the SEM image underwent Blending process, which was the use of color filters to sharpen the contrast, depending on the image. Some colors are usually avoided, such as bright red, orange, pink and cyan. This is in contrast with the colors that can be received in digital processing such as red, maroon, blue, green, purple and brown organic as well as the most dramatic effect, such as metallic color. Insulation colors can also be done on layer style to produce a monochromic infrared image. Finally, a finishing touch was done by combining the working layers together and export it into psd or jpg format.

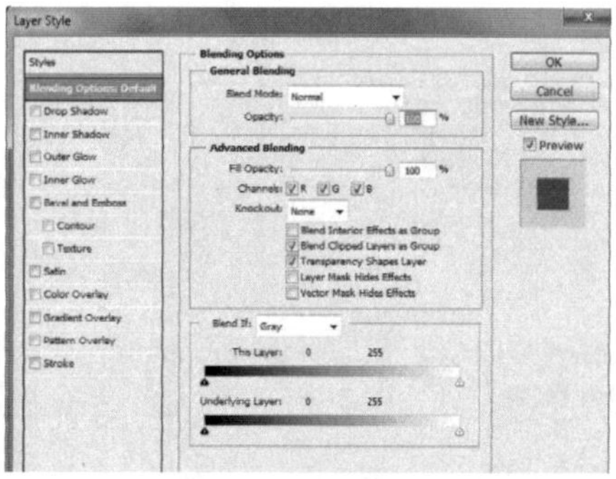

Figure 3. Isolation color can be done with a layer style.

In 3D image simulation using the anaglyph method, one of the important qualitative parameters to be achieved was the convenience of the viewer's eye when observing the 3D image. The convenience factor includes the maximum power of accommodation, the level of alignment with the horizon as well as the brightness of the image.

IOP Conf. Series: Journal of Physics: Conf. Series **817** (2017) 012040 doi:10.1088/1742-6596/817/1/012040

3. Result and discussion

The resulting 3D image from this simulation using anaglyph method was obtained from an original image which was duplicated to give red and blue effect. After that, the separation of the two images (red and blue filtered images) became a major factor in creating the 3D image.

Figure 4 shows the SEM image that had been sharpen. In this step, the process of action increased the sharpness of the image, so that the difference between high and horizontal surfaces were more easily observed. Meanwhile, the process of blending in the figure 5(a) and 5(b) was done by adding red filter and blue filter separately. The addition of the red filter reduced other colors but red as shown in figure 5(a), while the addition of blue filter reduced other colors but blue as shown in figure 5(b). Those two figures were almost similar since there was not much color variation in the original image.

Figure 4. Action in the software process increased the sharpness of the image.

(a) (b)

Figure 5. Blending process was done by adding (a) red filter and (b) blue filter, separately.

Figure 6 shows red and blue anaglyphs of the color stereo pairs produced by reprocessing them using an algorithm based on color analysis. Analysis was also performed on two images with different shifting position, i.e., 0μm, 1μm, 2μm and 3μm. The image with red-blue shifting positing of 0 μm, did not give the impression of 3D anaglyph. Meanwhile, the shifting position of 3 μm generated an extreme image, so that it was uncomfortable for human eyes. Therefore, using anaglyph method, the ideal position was achieved by shifting the red and blue filtered images by 1 μm.

Figure 7 shows the images with shifting distances of 2μm (a) and 3μm (b). The shifting distance between the two images with different filters (i.e., red and blue filters) corresponded to the right eye and the left eye. Largely shifted images made it difficult for human eyes to observe every point in the images as single point. Therefore, creating a 3D image that came from a single image required a digital manipulation, which was done by sliding it towards the opposite side for a few degrees and combining both of the processed images back together.

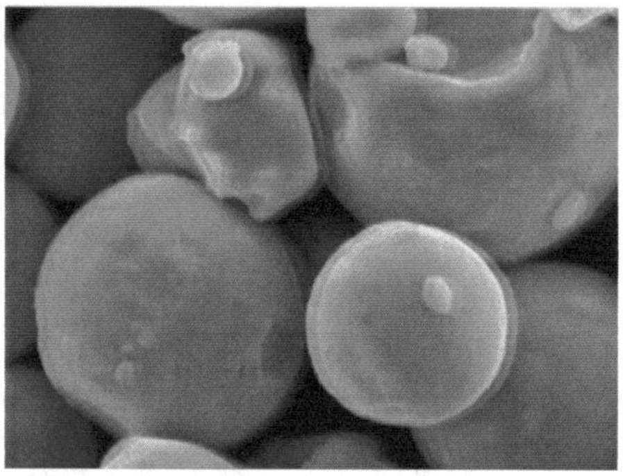

Figure 6. Anaglyph 3D image when two filtered images were shifted by 1 μm.

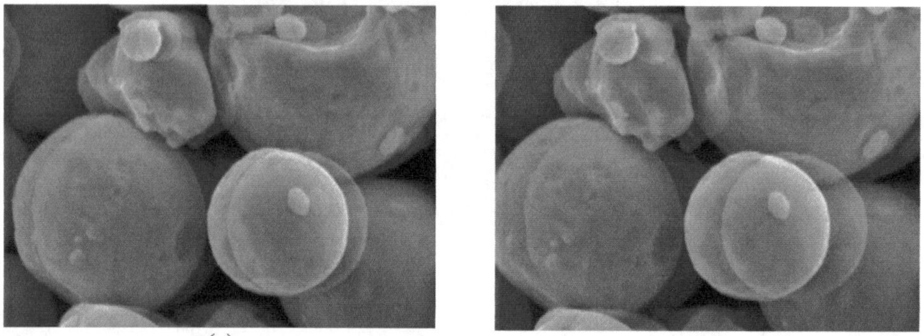

(a) (b)

Figure 7. 3D images with shifting distance of (a) 2 μm and (b) 3 μm.

4. Conclusion

We have managed to create a 3D image using anaglyph method from one original SEM image. The image was duplicated and filtered using red and blue filter. The simulation showed that the optimal 3D effect was achieved when the difference between red and blue filtered images was about 1 μm. The research showed that the 3D image quality was affected by the angle of the filter, the position of the viewer and the polarization. The generated 3D image can be clearly seen using 3D glasses. For further research, it will be useful to apply two light sources (red and blue) on the microscope or SEM to generate anaglyph images.

References

[1] Chang A J, Kim H J, Choi J W and Yu K Y 2008 *Proc. SPIE* vol 6803 (Washington: SPIE Publications) 68031G

[2] Li S, Ma L and Ngan K N 2013 *Signal Process. Image* **28** 597

[3] Dietz H G 2012 *Proc. SPIE* vol 8290 (Washington: SPIE Publications) 82900S

[4] Alkhadour W, Ipson S, Zraqou J, Qahwaji R and Haigh J 2009 *Proc. 4th Int. Conf. on Information Technology (Amman)* (Amman: Al-Zaytoonah University of Jordan) 578

[5] Image Processing Lab The Chinese University of Hong Kong *Anaglyph Image Generation by Matching Color Appearance Attributes* [http://www.ee.cuhk.edu.hk/~snli/anaglyph_database.htm]

[6] Manqing C, Yaowan G, Leisheng Z and Yiqian H 2010 *Proc. 2010 Int. Conf. on Computer Application and System Modelling (Taiyuan)* vol 14 (Chengdu: Institute of Electrical and Electronics Engineers, Inc) pp 562-564

[7] Levin A, Weiss Y, Durand F and Freeman W T 2009 *Proc. 2009 IEEE Conf. on Computer Vision and Pattern Recognition (Miami)* (Washington: IEEE Computer Society) pp 1964-1971
[8] Surman P, Hopf K, Sexton I, Lee W K and Bates R 2008 Solving the 3D problem - the history and development of viable domestic *Three-Dimensional Television* ed Ozaktas H M and Onural L (New York: Springer-Verlag Berlin Heidelberg) pp 471-503
[9] PurePix Images Inc 2010 *TrioScopics 3D* [http://www.trioscopics.com]
[10] TriOviz Company 2010 *TriOviz* [http://www.trioviz.com/]

IOP Conf. Series: Journal of Physics: Conf. Series **817** (2017) 012041 doi:10.1088/1742-6596/817/1/012041

Comparison study between static and dynamic responses of optical fiber weight in motion sensor

D Hanto, R K Ula, A Setiono, P Puranto, H Adinanta, T B Waluyo and B Widiyatmoko

Research Center for Physics, Indonesian Institute of Sciences, South Tangerang, Indonesia

E-mail: dwi.hanto@lipi.go.id

Abstract. This research investigates responses of Optical Fiber Weight in Motion (WIM) sensors when static and dynamic weighing vehicle are applied. The vehicle used in the experiment was a replica of Mitsubishi FE 71 with weight of 23.01 kg. The WIM measurement system consisted of LED 1310 nm, optical fiber based WIM sensor, transimpedance amplified photodetector Ge PDA50 from Thorlabs, data acquisition DT translation 9816 S, speed sensors, and software program for WIM Evaluation. Static weighing was performed when the vehicle stopped. Besides performing static weighing on WIM sensor, measurement was also conducted on Kenko scale as validation. Meanwhile, measurement of dynamic weight was only conducted when the vehicle moved at a constant speed of about 1 km/h through WIM sensor. Experimental results showed two curves corresponded to the front axle load and the rear axle load. During static weighing, both WIM sensor and Kenko scales indicated similar output, which were flat curves. On the other hand, during dynamic weighing, WIM sensor had different output that was a pulse response. This preliminary study resulted in changes of amplitude and output shape. Unlike static weighing, moving vehicles contributed to the output of WIM sensor.

1. Introduction

Monitoring for structures and infrastructures on the transportation system should become government responsibility in order to support economic development. Indonesian government policy concerning transportation appears through Government Regulation or PP No. 43/1993, which regulates the classes of road and vehicle load restrictions [1]. This regulation itself is good, but it is weak in implementation. The reason is due to the lack of instrumentation such as vehicle weighing system.

In the modern countries, measurement of vehicle's weight can be performed by Weight In Motion (WIM) system. WIM is a moving vehicle's weight measurement technology. This concept was introduced a few years ago [2]. The advantages of this technology include more efficient and time saving measurement, especially at a busy traffic [3]. Besides, this technology will support the intelligent transportation systems. Some of these systems have been applied at highways [3-4] which use many kind of WIM technologies.

In the market, there are various kind of sensors used in the WIM system such as load cell, strain gauge, bending plate, piezoelectric quartz sensors, and optical fibers [5, 6]. However, the optical fiber based sensor is a new technology development for WIM system. Optical fiber based sensors can be used as an alternative solution because it has several advantages such as high sensitivity, resistance to electromagnetic interference, high temperature resistance, and corrosion resistance. Thus, optical fiber is suitable for alternative to measure vehicle's weight [2, 7, 8].

Recently, vehicle weight measurement can be predicted from the axle load [9, 10]. The advantage of this method is having small weighing area, which is not as big as the area of the vehicle. Data of vehicle weight is obtained from each wheel load of the vehicle called as axle load. Determination of the total weight of vehicle is calculated from total axle loads. This method can also be applied at WIM system in order to get additional advantages. Those advantages provide simple installation and maintenance.

Indonesia is one of the big countries in the world, which has large area of land and sea transportations. However, there are not many research and implementation of WIM system in Indonesia. So, it is important to start researching and developing WIM system. Moreover, this development will support the existing regulations [1]. The future expectation is to trigger further collaboration with other researchers and to support industry for the implementation.

In this article, authors describe about testing of WIM sensors developed by Research Center for Physics, Indonesian Institute of Sciences. Previous works on WIM sensor have been done by the same research group [11, 12]. The developed WIM Sensors are based on microbending optical fiber. A vehicle replica was used for testing the measurement in both static and dynamic weighing. The purposeof this testing is to compare the performances of static and dynamic weighing so that it can be applied for real vehicle.

2. Method

2.1. WIM system design

Figure 1 shows general block diagram of optical fiber WIM System. The system consisted of a WIM sensor based on microbending of optical fiber, an LED with a wavelength of 1310 nm, a photodiode and amplifier Ge PDA50 for signal conditioning from Thorlabs Inc., an A/D converters data acquisition from DT Translation 9816 S with a sampling rate up to 750 kS/s per channel and 16 bit ADC resolution, and a PC equipped with WIM software evaluation as shown in figure 2. WIM software has some features, e.g. vehicle speed detector and vehicle load graph.

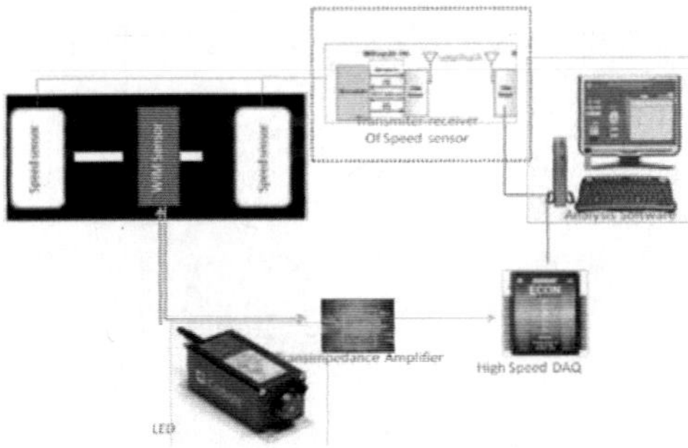

Figure 1. Block diagram of optical fiber WIM system.

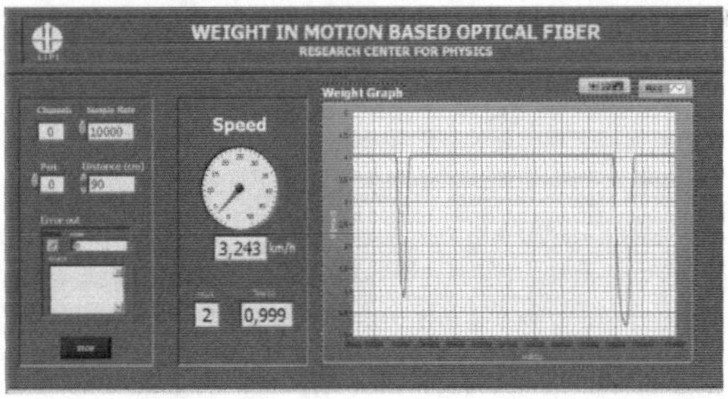

Figure 2. WIM software evaluation.

WIM sensor used for weighing was basically based on microbending optical fiber. When a vehicle passed on this sensor, the WIM sensor generated optical power losses due to microbending effect. The amount of the transmitted optical power after optical power losses was converted into electrical signal (voltage) by the photodiode and amplifier. Then, the voltage data acquired by the A/D converter DT9816-S was processed by WIM software and displayed in a graph as shown in figure 2. In this figure, two curves were appeared at the load graph which indicated that the sensor had been passed by axle loads from vehicle.

2.2. Acquisition system for Kenko scales

Kenko scales was used in this experiment for validation. The Kenko scales had the ability to measure the mass of an object up to 250 kg. It was equipped with RS 232 so that the data can also be acquired by a PC as shown in figure 3. The software system was developed using Labview software based on serial devices.

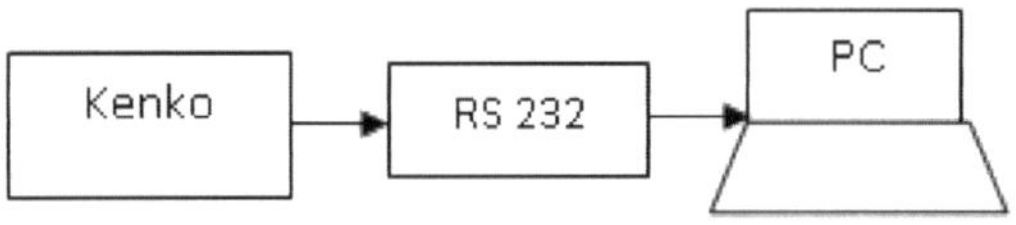

Figure 3. Block diagram of Kenko scales measurement.

2.3. System test

The system test was conducted using replica of Mitsubishi FE 71 with 23.01 kg (empty). The experiment was done in three ways. First, vehicle replica was measured on Kenko scales at a static condition as shown in figure 4. Then the vehicle replica was measured on optical fiber WIM system in static condition as shown in figure 5. Finally, the vehicle replica was moved on the optical WIM sensor with low speed of approximately 1 km/h as shown figure 5.

Figure 4. Measurement using Kenko Scales.

Figure 5. Measurement using Optical Fiber WIM.

The static and dynamic weighing described above were then conducted again by gradually adding 20 to 80 kg of payload into the vehicle replica. The methods on how to place the payloads on the vehicle replica was shown in table 1 and illustrated in figure 6. For instance, in the first experiment, the payload was placed on position 1 in the container of vehicle replica. In the second experiment, the payloads were placed on position 1 and 2, and so on. So, there were 4 experiments in total. The mass of the front and the rear axle load were measured by Kenko scales. The total mass of the vehicle replica was the result of front axle mass and rear axle mass.

Tabel 1. The methods to place the payloads in the vehicle replica.

No	Payload placement	Mass of front axle (kg)	Mass of rear axle (kg)	Total mass (kg)
1	1	16.84	26.23	43.07
2	1 + 2	26.31	36.54	62.85
3	1 + 2 + 3	22.92	60.09	83.01
4	1 + 2 + 3 + 4	28.88	74.61	103.49

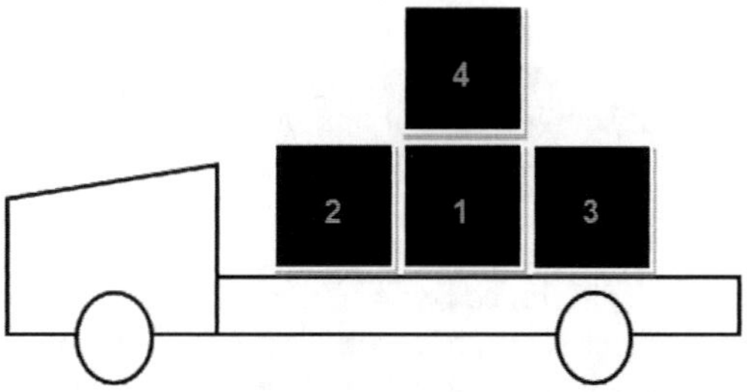

Figure 6. Methods on how to place payloads on vehicle container.

3. Result and discussion

The measurement results of a replica of vehicle using Kenko scales and WIM sensor are shown in figure 7-10. There are three kinds of graph on each figure. Graph (a) of each figure is a result of static weighing using Kenko scales. These graphs were used for measurement reference. The measurement output is showed in kilograms (kg) unit. Meanwhile, graph (b) and (c) of each figure show the performances of optical fiber WIM sensor when vehicle replica was used.

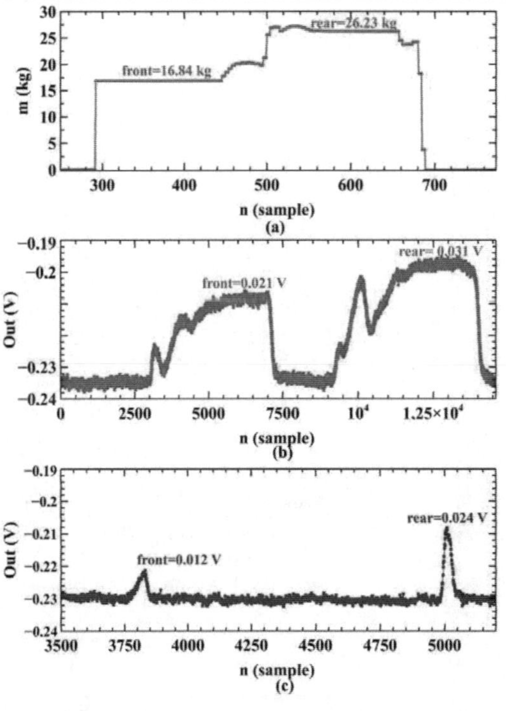

Figure 7. Results of the first experiment, (a) static weighing using Kenko scales, (b) static weighing using optical fiber WIM, (c) dynamic weighing using optical fiber WIM.

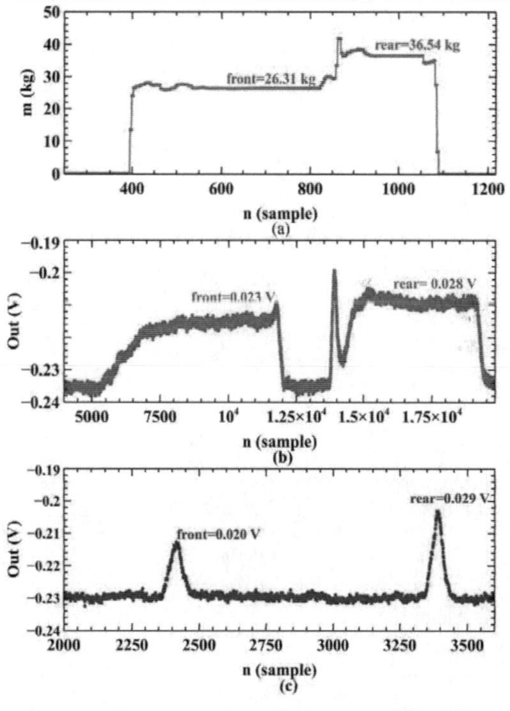

Figure 8. Results of the second experiment, (a) static weighing using Kenko scales, (b) static weighing using optical fiber WIM, (c) dynamic weighing using optical fiber WIM.

Figure 7 shows the response of the vehicle replica when additional payload of 20 kg was given. Figure 7(a) shows the mass of vehicle replica in kg unit by Kenko scales. From this figure, it appeared that the contribution of weight from the front wheels was 16.84 kg and the rear wheels was 26.23 kg, so the total mass became 43.07 kg. In a similar way, optical fiber WIM sensor was used to measure the load of the vehicle replica and the output is shown in figure 7(b). Figure 7(b) shows similar shape of curve to the Kenko scales output. It shows shorter amplitude of curve for front axle load in comparison to the amplitude of curve for the rear axle load. According to figure 7(a), front axle load was shorter than rear axle load. For static weighing using optical fiber WIM, the output appeared in two flat curves. The flat curve of front axle was at 0.021 V and the rear axle was at 0.031 V. The amplitude of the flat area was obtained from the difference between the voltage when flat area was reached and the voltage when there was no load. Figure 7(c) shows the response of the optical fiber WIM sensor when vehicle replica was moving on it with low speed. The resulted curves from this situation were no longer flat but spiky like pulses. Figure 7(c) shows that the amplitude of voltage for front axle and rear axle load were at 0.012 V and at 0.024 V, respectively. This amplitude in dynamic weighing was shorter than that in static weighing.

In the second experiment, another payload of 20 kg was added onto the vehicle replica with position in front of previous payload as shown as figure 6. The results of the second experiment are shown in figure 8. Figure 8(a) shows that the total mass measured by Kenko scales became 62.85 kg. The contribution of the axle loads were 26.31 kg for the front wheels and 36.54 kg for the rear wheels. On the other hand, the result of static weighing using WIM sensor can be seen in figure 8(b). The output of the front axle load was at 0.023 V and the rear axle load was at 0.028 V. When compared to the previous experiment, the amplitude of voltage was also proportional to the vehicle load. Like the dynamic weighing in the first experiment, it showed two pulse responses. The amplitudes of voltage were 0.020 V and 0.029 V. Similar to the first experiment, the responses in dynamic weighing were also shorter than in static weighing.

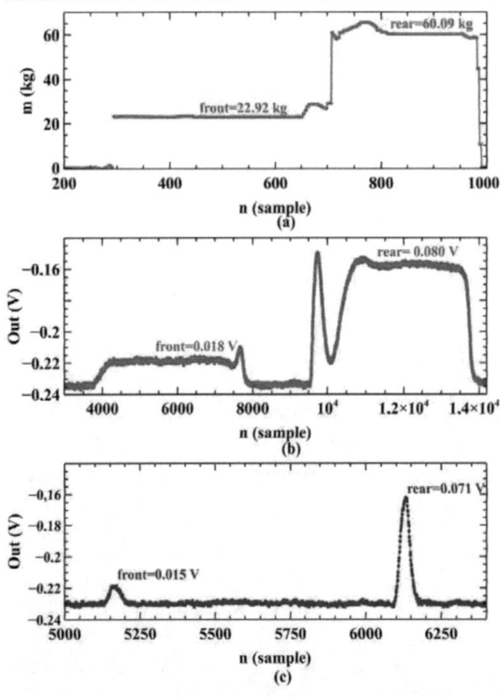

Figure 9. Results of the third experiment, (a) static weighing using Kenko scales, (b) static weighing using optical fiber WIM, (c) dynamic weighing using optical fiber WIM.

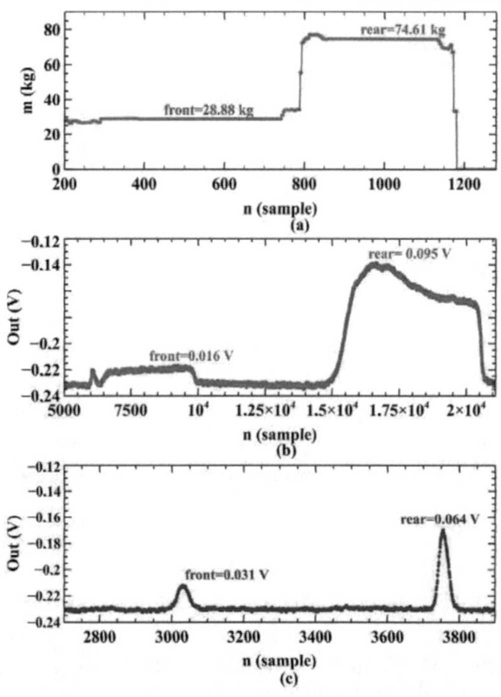

Figure 10. Results of the fourth experiement, (a) static weighing using Kenko scales, (b) static weighing using optical fiber WIM, (c) dynamic weighing using optical fiber WIM.

Figure 9 shows the results of the third experiment. According to figure 9(a), there were interesting results because front axle load was a bit lighter than the second experiment though 20 kg was added. However, the rear axle load got much bigger than the second experiment. This situation was caused by the placement of additional payload exactly above the rear wheels. The response for the static weighing using optical fiber WIM sensor is shown in figure 9(b) which looks similar to figure 9(a). In the case of front axle load, this result indicates that the amplitude of voltage in the third experiment was lighter than that in the second experiment. In the other hand, for rear axle load, the amplitude of voltage in the third experiment was much higher than in the second experiment. Then, figure 9(c) shows similar trends with the first and second experiments. Its shape also shows pulses response and its amplitude of voltage was shorter than the static weighing.

Lastly, the results of the fourth experiment is shown in figure 10. When static weighing on the Kenko scales was carried out, the contribution of mass for the front wheel increased insignificantly, which was from 22.92 kg to 28.88 kg. Otherwise, the mass contribution for the rear axle load increased significanly from 60.09 kg to 74.61 kg. When vehicle replica load was used on the optical fiber WIM sensor, the results was as shown in figure 10(b) for static weighing and in figure 10(c) for dynamic weighing. The results also showed similar trends as previous discussions.

4. Conclusion

This study compared static and dynamic weighing of optical fiber WIM sensor. Kenko scales was also used for performance comparison. Static weighing from vehicle replica using Kenko scales and optical fiber WIM sensor had the similar shape curves. From the experiments, the output of optical fiber WIM sensor was proportional to the mass distribution of axle load. Moreover, the response of dynamic weighing using optical fiber WIM sensor caused decrease in the amplitude of voltage. Study on the effect of speed variation to the WIM sensor is expected in the future research.

Acknowledgments

The work was supported by Riset Unggulan LIPI 2016. The authors also thank Widodo from Indonesia University of Education, Sulastri from Palopo University, then Rayendra, Danisson, and Rizky from Muhammadiyah University of Solo for helping with the experiments.

Reference

[1] Government Regulation 1993 *Peraturan Pemerintah Republik Indonesia Nomor 43 Tahun 1993 tentang Prasarana dan Lalu Lintas Jalan* (Jakarta: The Government of Republic of Indonesia) pp 6-7

[2] Malla R B, Sen A and Garrick N W 2008 *Sensors* **8** 2551

[3] ASTM Standard 1992 *Standard Specification for Highway Weigh-In-Motion (WIM) Systems* E 1318 (Pennsylvania: American Society for Testing and Materials) pp 32-37

[4] Wyman J H 1989 *Proc. 3rd National Conf. on Weigh-In-Motion (St Paul)* (Washington: The National Academies of Sciences, Engineering, and Medicine) pp 6-176

[5] Strathman J G 1998 *The Oregon Department of Transportation Slow-Speed Weigh-In-Motion (SWIM) Project Final Report* OR-RD 99-18

[6] McNulty P and O'Brien E J 2003 *J. Test. Eval.* **31**(6) 497

[7] Bin M and Xinguo Z 2010 *Proc. 2010 Int. Conf. Intelligent Computation Technology and Automation (Changsha)* vol 1 (Los Alamitos: IEEE Computer Society, Conference Publishing Services) pp 458-461

[8] Batenko A, Grakovski A, Kabashkin I, Petersons E and Sikerzhicki Y 2011 *Transport Telecommun.* **12**(4) 27

[9] Kim B H, Park M S, Park H, Bae I H, Kim S J and Soo K 2011 *Dynamics of Civil Structures* vol 4 (Springer: New York) pp 263-270

[10] Liu J and Li L 2012 *Communications in Computer and Information Science* vol 345 (New York: Springer-Verlag Berlin Heidelberg) pp 745-750

[11] Hanto D, Sari D H, Setiono A and Widiyatmoko B 2013 *Prosiding Seminar Nasional 2nd Lontar Physics Forum 2013 (Semarang)* (Semarang: Program Studi Pendidikan Fisika Fakultas Pendidikan Matematika dan Ilmu Pengetahuan Alam IKIP PGRI Semarang) LPF1352

[12] Widiyatmoko B, Hanto D, Setiono A, Adinanta H and Mulyanto I 2013 *Prosiding Seminar Insentif Riset Sinas 2013 (Jakarta)* (Jakarta: Asisten Deputi Relevansi Program Riptek Deputi Relevansi dan Produktivitas Iptek, Kementerian Riset dan Teknologi) pp 167-171

Development of microcontroller-based acquisition and processing unit for fiber optic vibration sensor

Suryadi[1], P Puranto[1], H Adinanta[1], T B Waluyo[1] and P S Priambodo[2]

[1] Research Center for Physics - Indonesian Institute of Sciences
[2] Departement of Electrical Enginering - University of Indonesia

Email: sury011@lipi.go.id

Abstract. Microcontroller based acquisition and processing unit (MAPU) has been developed to measure vibration signal from fiber optic vibration sensor. The MAPU utilizes a 32-bit ARM microcontroller to perform acquisition and processing of the input signal. The input signal is acquired with 12 bit ADC and processed using FFT method to extract frequency information. Stability of MAPU is characterized by supplying a constant input signal at 500 Hz for 29 hours and shows a stable operation. To characterize the frequency response, input signal is swapped from 20 to 1000 Hz with 20 Hz interval. The characterization result shows that MAPU can detect input signal from 20 to 1000 Hz with minimum signal of 4 mV RMS. The experiment has been set that utilizes the MAPU with singlemode-multimode-singlemode (SMS) fiber optic sensor to detect vibration which is induced by a transducer in a wooden platform. The experimental result indicates that vibration signal from 20 to 600 Hz has been successfully detected. Due to the limitation of the vibration source used in the experiment, vibration signal above 600 Hz is undetected.

1. Introduction

Vibration sensor is widely used in many fields of application such as industrial, infrastructure and environmental monitoring to detect and diagnose failure, especially in rotating machine [1]. By adopting vibration sensing mechanism using accelerometer sensor, bridge monitoring system has been developed in order to detect bridge scour in real time [2]. Vibration sensor also has been used for a landslide early warning system [3].

Conventional vibration sensor is commonly based on piezoelectric technology which is sensitive to electromagnetic interference, whereas fiber optic vibration sensor provides more advantages such as immunity to electromagnetic interference, light weight and compact size [4]. Fiber optic vibration sensors can be classified into intensity, interferometers and gratings based sensor [5]. Intensity based vibration sensor have been extensively developed because of its simplicity where vibration can be detected from optical power variation [6]. Interferometric and grating based sensor can provide higher resolution and accuracy, but requires complex construction and interrogation system. Fiber optic vibration sensor can be developed using simple structure of fiber optic called SMS structure where a section of multimode fiber (MMF) spliced to the lead-in single mode fiber (SMF) and lead-out SMF [7].

Fiber optic vibration sensor commonly uses either PC-based Data Acquisition System (PC-DAQ) or Optical Spectrum Analyzer (OSA) to collect vibration signal and to process data [1, 4, 8]. The method that is widely applied to process vibration signal is fast fourier transform (FFT) which is used to extract vibration frequency from signal recorded in time domain [1, 4, 9]. Interrogation of fiber optic vibration sensor using either PC-DAQ or OSA adds more difficulties for implementation in remote area due to its size and complexity.

In this paper, the design of microcontroller-based acquisition and processing unit for fiber optic sensor is developed. The developed device acquires and processes the vibration signal fed from fiber optic sensor. The proposed design gives advantages such as low power, small and compact size, so this design can be implemented for vibration monitoring especially in remote area.

2. Methods

The schematic of MAPU for fiber optic vibration sensor is shown in figure 1. The core of this system is a development board from ST Microelectronics (Nucleo F411RE) which is also used for ST's ARM 32-bit microcontroller, i.e STM32F411RET6. This microcontroller is equipped with some features such as 100 MHz max CPU frequency, 512 KB Flash, 128 KB SRAM, 16 channels 12-bit ADC and built in FPU which provides enough capability as signal processor.

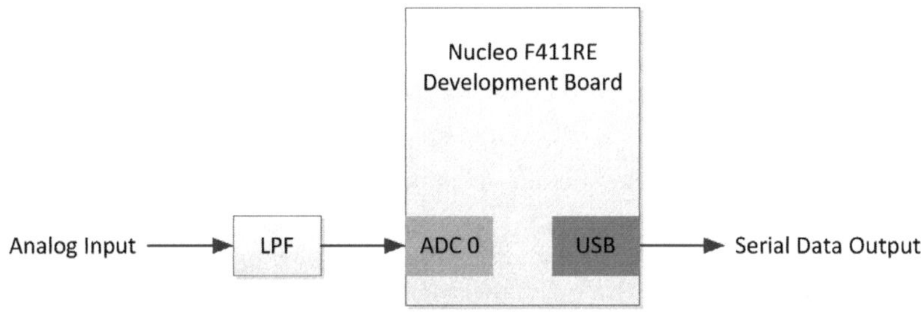

Figure 1. Schematic of acquisition and processing unit for optical vibration sensor.

Analog input signal after filtered using first order low-pass passive filter is fed into channel 0 of MCU built-in Analog to Digital Converter (ADC). This ADC converts an analog signal within 0 - 3.3V range into a 12-bit digital representation. 2048 data are collected in 1 s window and then are processed using FFT method to extract signal feature in frequency domain. With sampling frequency at 2048 Hz, this system can process input signal from 0 to 1024 Hz. The LPF is designed to suppress all input signal beyond this range. The result of signal processing using FFT method is then transferred out serially as a string over USB which is captured by serial terminal software in laptop or pc. The flowchart of vibration measurement using MAPU is shown in figure 2.

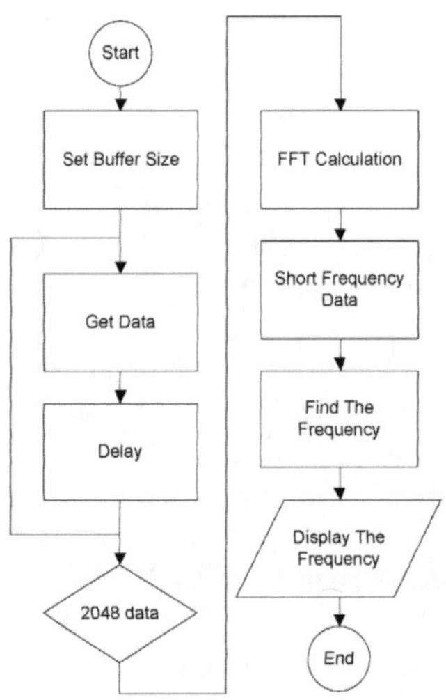

Figure 2. Flowchart of vibration measurement program

3. Result and Discussion

The experiment to characterize MAPU for fiber optic vibration sensor is shown in figure 3. The sinusoidal signal from signal generator is fed to microcontroller, where signal is processed using FFT method to calculate signal frequencies. At the same time, signal amplitude is measured using digital multimeter.

To assess the stability, MAPU is fed with a constant frequency signal at 500 Hz while PC is used to monitor the calculated frequencies. After 29 hours measurement, MAPU shows a stable frequency output at 500 Hz as shown in figure 4. The frequency response of MAPU is examined by swapping input frequency from 20 to 1000 Hz with 20 Hz interval. The measurement is conducted for some level of input signal attenuation, including 0, −10, −20 and −30 dB attenuation.

Figure 5 shows that MAPU can detect input signal between 20 to 1000 Hz range with 0, -10 and -20 dB input signal attenuation. When input signal is attenuated to −30 dB, MAPU fails to detect frequency above 800 Hz. Voltage measurement using digital multimeter falls below 4 mV when input signal is attenuated to -30 dB and its frequency is above 800 Hz. It shows that MAPU can detect input signal in range of 0 to 1000 Hz with minimum amplitude is 4 mV RMS.

Figure 6 shows the experimental setup of MAPU for fiber optic vibration sensor. The vibration sensor is based on fiber optic SMS structure as shown in figure 7. The sensor is attached to a wooden platform which is vibrated by woofer driver loudspeaker. The loudspeaker is driven by sinusoidal signal from signal generator which is amplified using audio amplifier.

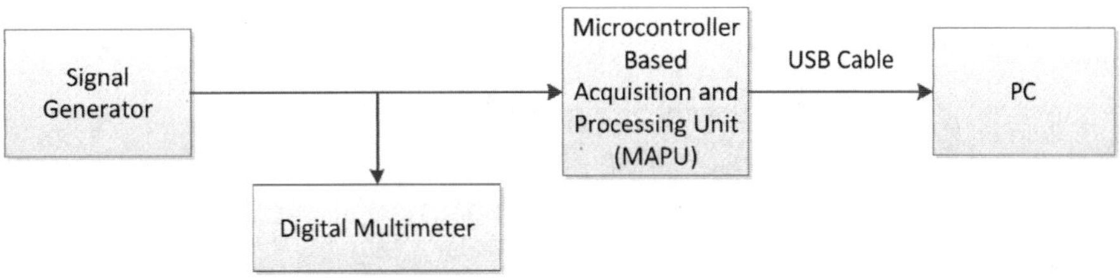

Figure 3. Experiment setup for MAPU characterization.

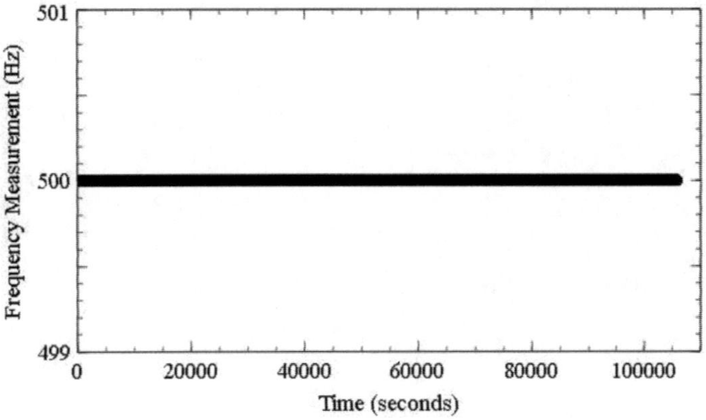

Figure 4. MAPU stability assesment.

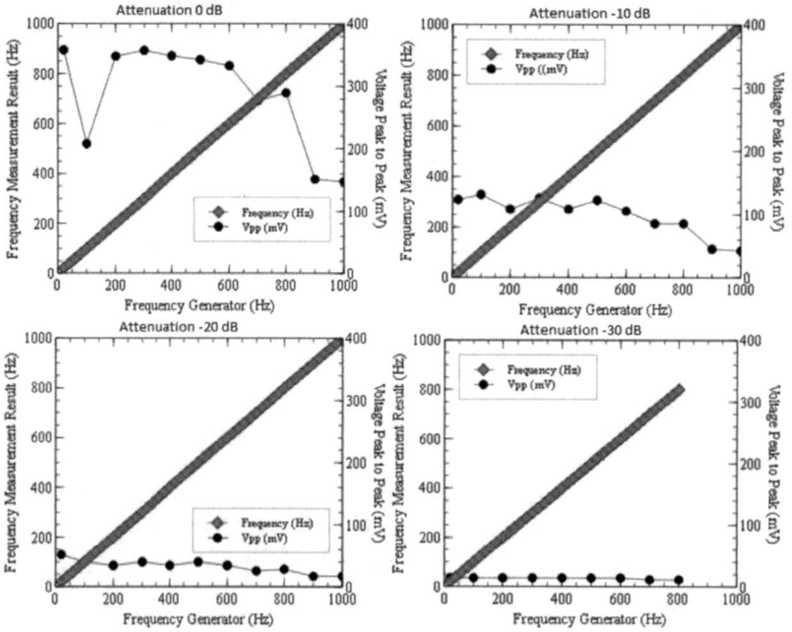

Figure 5. MAPU frequency response

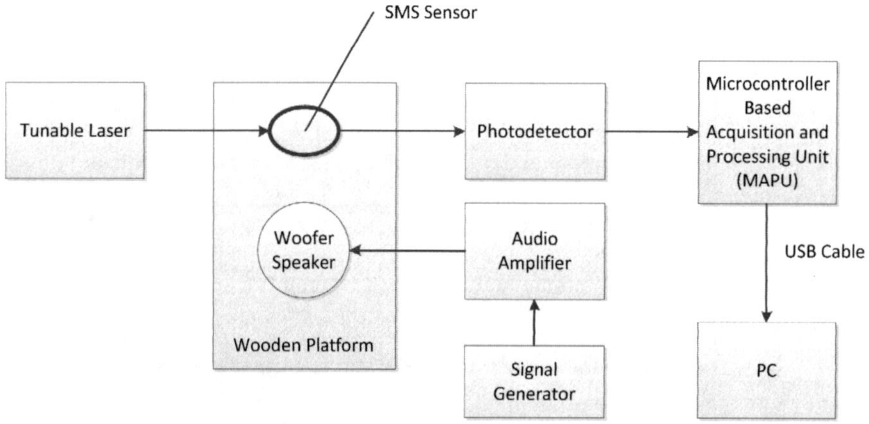

Figure 6. Experiment setup for vibration measurement.

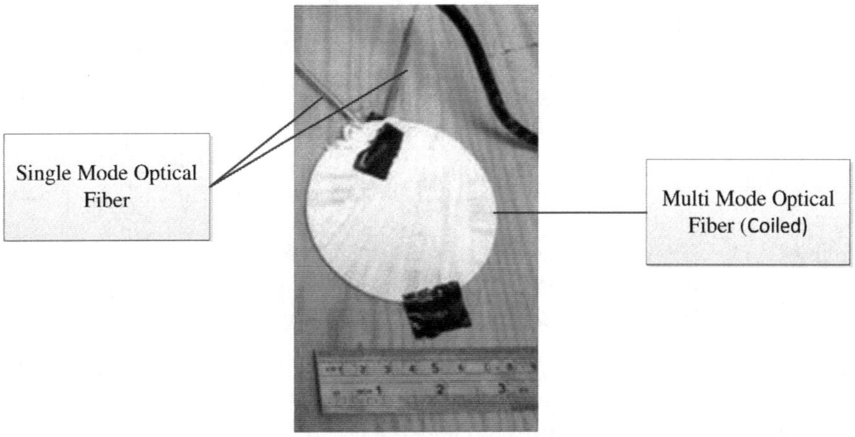

Figure 7. Fiber optic vibration sensor using SMS structure

In this experiment, a tuneable laser source tuned at 1560 nm is used to emit light into SMS fiber sensor. The high-speed photo detector is then adopted to capture output of SMS fiber sensor and convert it into electric signal. Because the transmitted optical power in SMS fiber structure is affected by the multimode interference, the photodetector output will reflect power fluctuation induced by vibration. Photodetector output is then processed by MAPU to extract the vibration frequencies. Figure 8 shows linear correlation data between input signal frequency that drives the loudspeaker and output frequency detected by MAPU. The experimental results show that input signal frequency is similar to the frequency detected by MAPU. Due to limitation of the vibration sources, MAPU cannot detect vibration above 600 Hz.

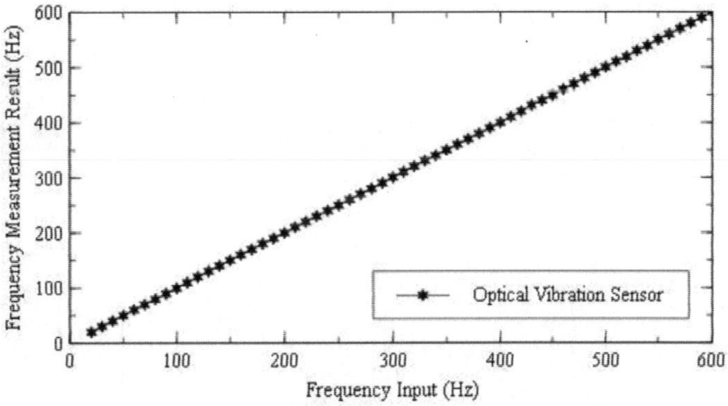

Figure 8. Responsivity test of optical vibration sensor

4. Conclusion

The microcontroller based acquisition and processing unit (MAPU) for optical vibration sensor has been successfully developed. During stability assessment, a stable operation in acquiring and processing input signal has been successfully demonstrated. Characterization shows that MAPU can detect input signal with frequency range from 20 to 1000 Hz and minimum amplitude 4 mV RMS. The experimental results demonstrate that MAPU can detect vibration frequency up to 600 Hz. The reduced detection range is caused by limitation of vibration sources used in the experiment. The development of MAPU will provide a compact and low power interrogation system for fiber optic vibration sensor.

References

[1] Atoui I, Meradi H, Boulkroune R, Saidi R and Grid R 2008 *The 8th Int. Workshop on Systems, Signal Processing and their Applications* (WoSSPA) (USA: IEEE) pp 401-6.
[2] Chen S-Y, Hsieh Y-J, Yang, Wu C-M and Huang C-M 2016 *IEEE Int. Instrumentation and Measurement Technology Conf. Proc.* (USA: IEEE) pp 1 - 6.
[3] Biansoongnern S, Plungkang B and Susuk S 2016 *Energy Procedia* **89** 417 – 20.
[4] Ran Y, Xia L, Han Y, Li W, Rohollahnejad J, Wen Y and Liu D, 2015 *IEEE Photon. J.* **7**(2) 6800607.
[5] Rodrigo S R, Aldaba A L, P´erez-Herrera R A, Bautista M C L, Esteban O, Lo´pez-Amo M, 2016 *J. Lightwave Technol.* **34**(19) 4525 – 30.
[6] Zuo Z-W, Hao Y, Choi S -J, Song M, Kim Y-C and Pan J-K 2017 *IET Sci. Measure. Technol.* **11**(1) 49–56.
[7] Fernandes C S, Giraldi M T M R, de Sousa M J, Costa J C W A, Gouveia C, Jorge P and Franco M A R 2016 *IEEE Trans. Instrum. Meas.* **65**(9) 7485860.
[8] Ali T H, Medjadba H, Simohamed L M and Chemali R 2015 *Proc. IEEE 3rd Int. Conf. on Control, Engineering & Information Technology* (CEIT) (USA: IEEE) pp 1-6
[9] Mohanty S, Gupta K K, Raju K S, Mishra V, Kumar V and Prasad P B 2014 *Twentieth National Conf. on Communications* (NCC) (USA: IEEE) pp 1-5

Improving frequencies range measurement of vibration sensor based on Fiber Bragg Grating (FBG)

Qomaruddin, A Setiono and M I Afandi

Research Center for Physics, Indonesian Institute of Sciences, Kawasan PUSPIPTEK Building 442, Serpong, South Tangerang 15314 Indonesia

Email: qomar_ku@hotmail.com

Abstract. This research aimed to develop a vibration sensor based on Fiber Bragg Grating (FBG). The design was mainly done by attaching FBG at the cantilever. The free-end of the cantilever was tied to a vibration source in order to increase the measurement range of vibration frequencies. The results indicated that the developed sensor was capable of detecting wide range of frequencies (i.e. 10 - 1700 Hz). The results also showed both good stability and repeatability. The measured frequency range was 566 times greater than the range obtained from the previous works.

1. Introduction

There have been many uses of FBG in optical sensor applications, such as vibration sensors [1, 2], temperature [3], load [4], strain [3, 4], and liquid level [5 – 7]. The focus this study is using FGB as a vibration sensor. The working principles of FGB as a vibration sensor are either intensity, phase or wavelength modulation [1, 2]. Generally, FGB as a vibration sensor can work at low frequencies, i.e less than 100 Hz [1, 2, 8 – 10] and using the wavelength modulation technique. This technique utilizes the changes in grating period due to external stimuli resulting wavelength shift Bragg. Therefore, modulation requires a broadband light source and optical spectrum analyzer (OSA) to display the output signal. In this work, we use the intensity modulation based on the light source that has a spectral width that is narrower than the reflection spectrum of FBG. In this case, we use a laser diode as a light source. Fig. 1 shows the working principle of FBG where a dynamic strain on FBG affects the reflected intensity of laser diode before (a) and after (b) vibration.

If the polychromatic or a broadband light source propagate on FBG then there will be a spectrum with a specific wavelength is reflected. When the FBG is not vibrated the laser source spectrum will be fully reflected. If FBG was vibrated there will be the changes in the intensity of the reflected laser source [11]. FBG shows a shift response to the changes in light reflectance on FBG at certain wavelengths. In this system, a laser diode which has a narrow spectrum of wavelength was set to approach Bragg wavelength. Mathematical description of the FBG principle can be written as follows:

$$\lambda_B = 2n_{eff}\Lambda \tag{1}$$

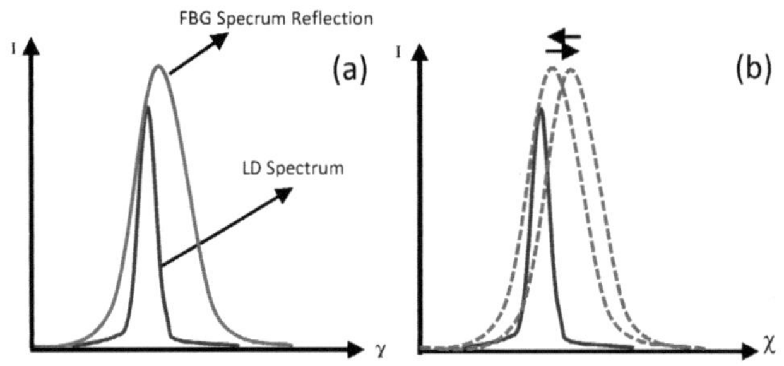

Figure 1. Dynamic strain on FBG affects the reflected intensity of LD (a) before and (b) after vibration

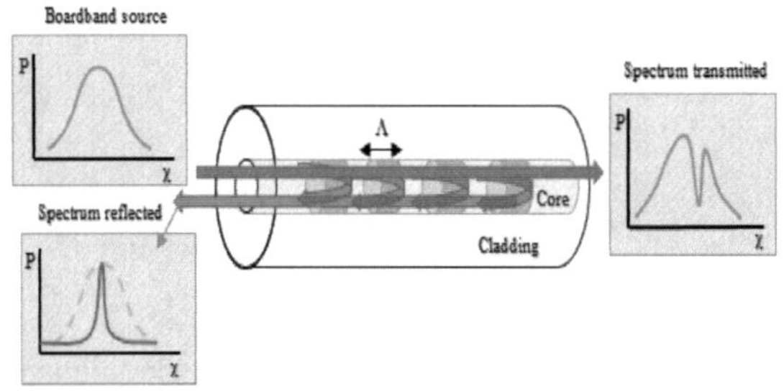

Figure 2. The working principle of FBG that encounter dynamic strain

where λ_B is Bragg wavelength, n_{eff} is effective refractive index and Λ is periode grattings of FBG[8,11-13]. If the grating period FBG is changed due to dynamic strain from outside, the Bragg wavelength also changes. Fig. 2 shows the change in the intensity of the laser diode which is reflected by the dynamic strain of the FBG due to vibration [3, 13].

In this work, the technique used is based on the changes intensity of the reflection by FBG. The aim of this research is to increase the measurement range of the FBG sensor more than 1000 Hz.

2. Methodology

We use a cantilever with a dimension of 60 × 3 × 0.5 mm as a vibration source for FBG. Figure 3 shows the bronze cantilever used in this study. The resonance frequency of the cantilever can be expressed using equation [10, 13, 14]:

$$f = \frac{1}{2}\sqrt{\frac{3EI}{L^3 M}} \quad (2)$$

where f is resonance frequency, E is the Young modulus of bronze, I is a moment of inertia, L is the lenght of cantilever, and M is an effective mass. We found that the cantilever has a resonance frequency of 200 Hz. Moment of inertia, can be expressed using the equation:

$$I = \frac{bd^3}{12} \quad (3)$$

Using equation (3), we found that the cantilever has moment of inertia 3.13×10^{-14} m^4. On the middle of the cantilever, a polyamide FBG with center wavelength λ_B = 1607.7 nm was attached. This FBG has a uniform gratings structure with a length of 1 mm. It also has the bandwidth of 1.6 nm and a reflectance > 50%.

Figure 4 is the experimental set-up of the FBG based vibration sensor. The light source was a tunable laser diode (TSL-510) with a wavelength range of 1560-1680 nm. The laser was propagated through optical sirculator toward the FBG attached on the cantilever. The free-end of the cantilever was then tied to a vibration source (a speaker device) with a frequency controlled by a signal generator. The frequency was tuned in 10 - 2000 Hz range. At a given vibration with a specific frequency input, the intensity-response of the laser source that reflected by FBG was captured by a high-speed photodetector (HSPD). Data acquired from the photodetector was then processed by LabVIEW software. Intensity responses that were recorded and captured were all in the time domain. Representation data in frequency domain were obtained using Fast Fourier Transform (FFT).

To increase the sensor sensitivity, we set the center of Bragg wavelenght at 1604.738 nm and the laser source at 1604.000 nm using an optical spectrum analyzer. Figure 5 shows the Bragg wavelenght and laser source that were set in this work.

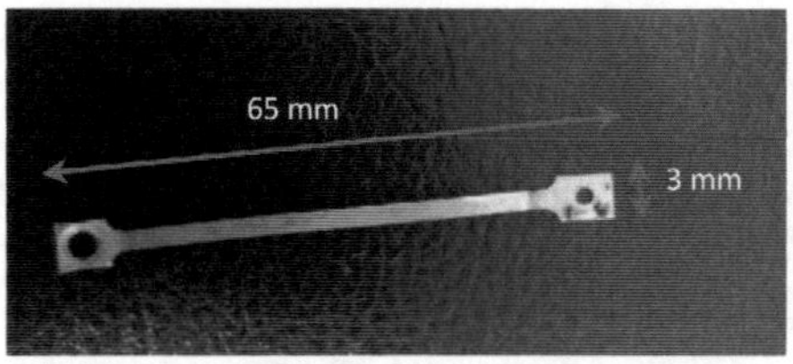

Figure 3. The dimension of cantilever [13]

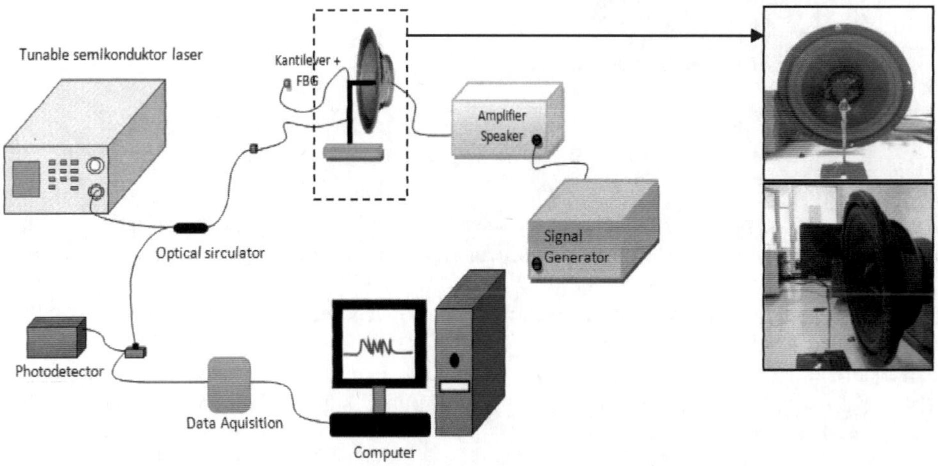

Figure 4. Experimental on set-up of the FBG based vibration sensor. Inset: How the cantilever tied to a vibration source

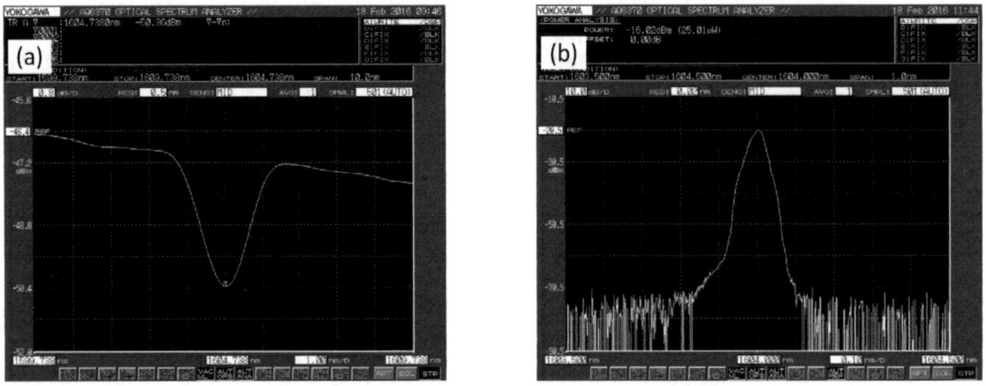

Figure 5. (a) The center of Bragg wavelenght at 1604.738 nm and (b) laser source wavelength at 1604.000 nm which were set by the optical spectrum analyzer.

3. Result and discussion

Intensity responses of laser light due to the vibration of cantilever were captured by a photodetector and processed using a LabVIEW to represent in the time domain. These data were then analyzed using FFT to represent in the frequency domain. Figure 6(a) is one of the raw data from vibration recorded in the time domain for the input frequency of 200 Hz. The raw data was then analyzed directly using FFT to obtain the amplitude and the output frequency. Figure 6(b) is the representation of vibration with a frequency of 200 Hz after analyzed using FFT. Note that there is no shift in the output frequency.

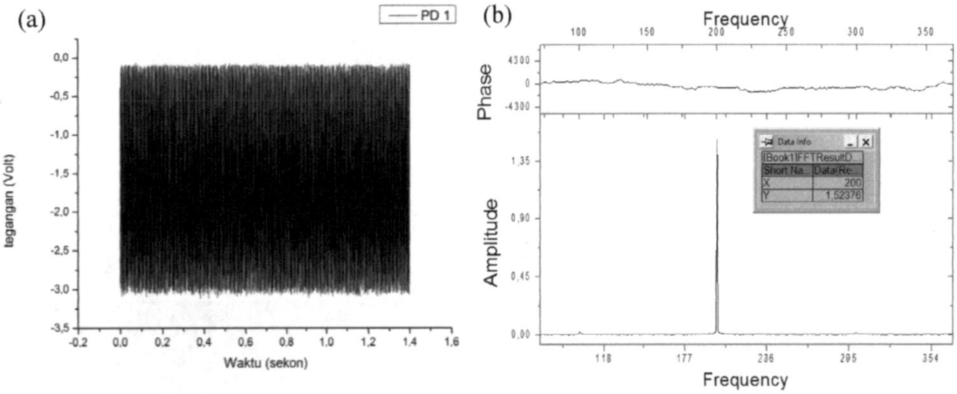

Figure 6. Intensity response of the FBG based vibration sensor for a typical vibration data with an input frequency of 200 Hz (a) in time domain and (b) its representation in frequency domain.

To show the linearity between the frequency of the input vibration and the output frequency, we scanned all the possible frequencies that can be produced by the signal generator. Figure 7 shows the output frequency of 500 and 1000 Hz for the input frequencies at the same values. There was no different in the frequencies between an input signal and output signal. If the input signal increases, the output also increases with the equal value.

Based on the dimension and the type of cantilever's material, we can determine the resonance frequency of cantilever with the equation (2) i.e. 200 Hz. In this experiment, we generated the input frequency of 10 - 2000 Hz and we got the maximum amplitude of 1.5 cm for the resonance frequency of 200 Hz, as shown in figure 8. It is obvious that the maximum amplitude was obtained when the frequency of the input signal was equal with the resonance frequency. This indicated that resonance frequency can make the maximum amplitude in vibration of cantilever. This experiment can detect an amplitude of vibration of frequency up to 1700 Hz. The maximum frequency of the input signal is 1700 Hz to produce amplitude that can be measured by system.When the frequency of the input signal was higher than 1700 Hz, the amplitude of vibration was zero.

We also examined the sensor stability. The stability test showed that the sensor can work properly. There was no interference which could affect the output signal. Figure 9 shows the stability of the sensor in terms of the output frequency. The sensor can operate smootly with frequency up to 1700 Hz. From the lowest frequency until 1700 Hz, stability of sensor still capture the same value of the output frequency.

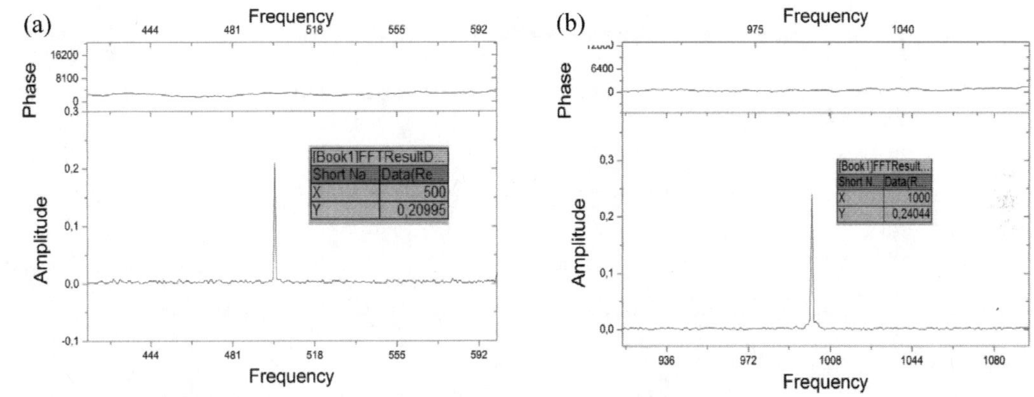

Figure 7. Representation data with frequency 500 Hz (a) and 1000 Hz (b)

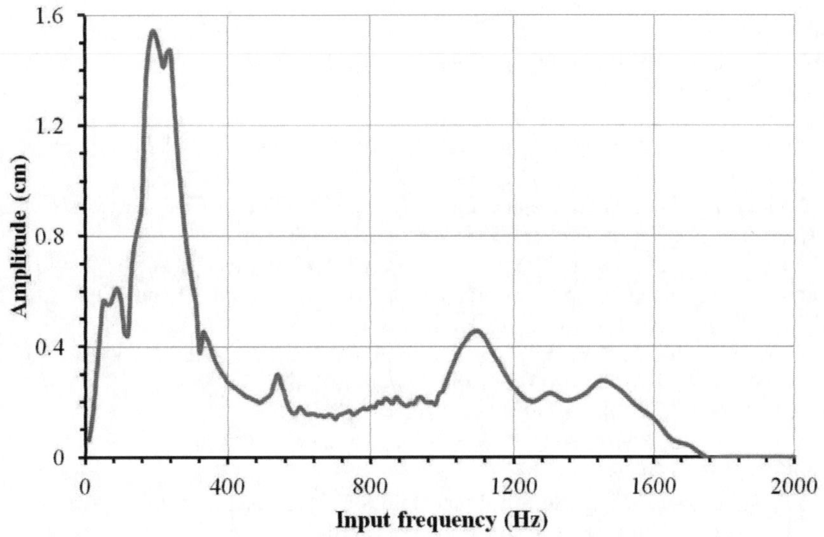

Figure 8. Relationship amplitude and input frequencies that can be produced until 1800 Hz.

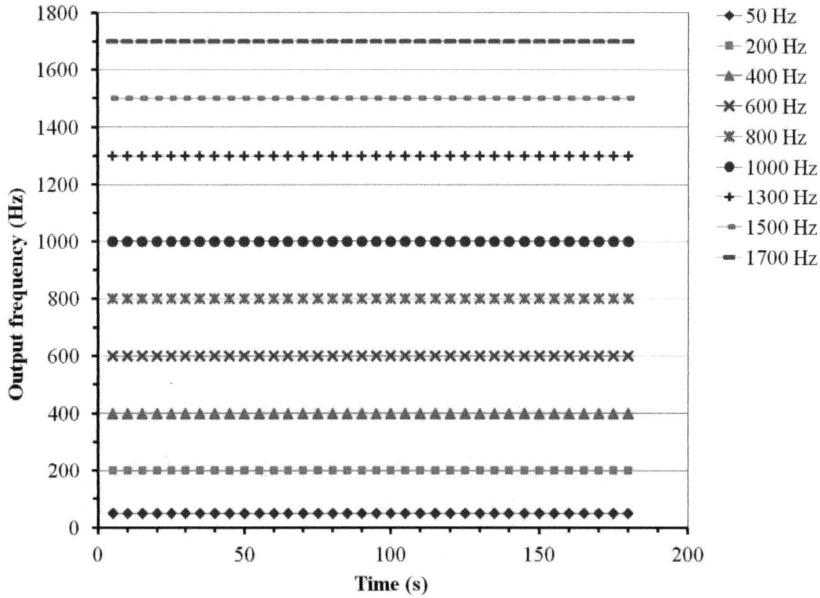

Figure 9. The stability of the sensor for the output frequency produced.

4. Conclusion

Development of FBG based vibration sensor has been done. The sensor was capable of detecting the vibration frequencies at the range of 10 - 1700 Hz with no shift. The results also showed both good stability and repeatability. The measured frequency range was 566 times greater than the range obtained from the previous works.

References

[1] Chang J, Wang Q, Zhang X, Ma L, Wang H, Zhang S, Wang Q, Ni J, and Wu Y 2008 *Laser Phys.* **18**(7) 911–13
[2] Chang J, Wang Q, Zhang X, Huo D, Ma L and Liu X 2009 *Laser Phys.* **19**(1) 134–7
[3] Kadhim S A, Taha S A A and Resen D A 2015 *Int. J. Innov. Res. Sci. Eng. Technol.* **4**(9) 8038–43
[4] Zeng X, Rao Y, Wang Y, Ran Z and Zhu T 2002 *15th Optical Fiber Sensors Conf. Technical Digest. OFS 2002 (Cat. No.02EX533)* vol 1 (USA: IEEE) pp 199–202
[5] M. Shao, X. Qiao, X. Zhao, Y. Zhang, and H. Fu 2016 *IEEE Sens. J.* **16**(8) 2374–379
[6] Marques C A F, Pospori A, Sáez-rodríguez D, Nielsen K, Bang O and Webb D J 2015 *Proc. SPIE* vol 9634 (Washington: SPIE Publishing) 96345V
[7] Lai C C, Au H Y, Liu M S Y, Ho S L and Tam H Y 2016 *IEEE Sens. J.* **16**(16) 6346–50
[8] Sun L, Li H, Ren L and Jin Q 2007 *Sensor. Actuator. A* **136** 572–9
[9] Zeng Y-J and Rao X-K 2001 *Chin. Phys. Lett.* **18** 1617–19
[10] Shan-li W, Yong W, Man-li H. U, Jian-hong W, Cheng G U O and Hai-feng L E I 2013 *J. Basic Appl. Phys.* **2**(1) 20–3
[11] Moyo P, Brownjohn J M W, Suresh R, and Tjin S C 2005 *Eng. Struct.* **27** 1828–34
[12] Gangopadhyay T K 2004 *Sensor. Actuator. A* **113** 20–38
[13] Setiono A, Ula R K, Hanto D and Widiyatmoko B 2016 *AIP Conf. Proc.* vol 1711 (New York: AIP Publishing) p 300031
[14] Weng Y, Qiao X, Guo T, Hu M, Feng Z, Wang R and Zhang J 2012 *IEEE Sens. J.* **12**(4) 800–4

Application of activated zeolite to quantitative analysis of Pb liquid sample using commercial laser-induced breakdown spectroscopy (LIBS)

H Suyanto[1], A P Utomo[1], M Manurung[2] , W G Suharta[1] and Windaryoto[1]

[1]Physics Department, Faculty of Mathematics and Natural Sciences, Udayana University, Kampus Bukit Jimbaran, Badung 80361Bali, Indonesia
[2]Chemistry Department, Faculty of Mathematics and Natural Sciences, Udayana University, Kampus Bukit Jimbaran, Badung 80361Bali, Indonesia

E-mail: hery6@yahoo.com

Abstract. The aim of this research is to figure out an activated zeolite ability to immobilize Pb element in liquid sample. The zeolite was made into powder with size of about 400 μm and was activated by NaOH with different concentrations of 0.5, 1, 1.5 and 2 M. Prior to being analyzed by commercial Laser-Induced Breakdown Spectroscopy (LIBS) technique, 0.05 gram activated zeolite of each concentration were soaked into 10 ml solution containing 100 ppm of Pb and then heated and stirred until become dry powder. The powder was then made into a pellet. After being analyzed by LIBS, activation by 1.0 M NaOH was the optimum condition for zeolite to immobilize Pb in solution. The minimum concentration of Pb in liquid sample which could be immobilized by this method was about 5.86 ppm.

1. Introduction

Zeolite is a very stable solid that is three-dimensional hydrated alumino-silicate crystals constructed by interlinked tetrahedral of alumina and silica (AlO_4-SiO_4) [1]. Because zeolite has porous structure, open structure and being few negatively charged mineral in nature, it can react or accommodate a wide variety of cations, such as alkali or alkaline-Earth: Mg^{2+}, K^+, Na^+, and other metals plus water molecules trapped in the gaps between them to make charged equilibrium [2]. These positive ions are rather loosely held and can readily be exchanged for others in a contact solution. Because of these properties, naturally occurring zeolite is widely used in many applications such as adsorbent, electrons transfer media in voltammetry instrument, and others [3-5]

Natural zeolite is not pure alumino-silicate crystals, but it contains organic and inorganic impurities as mentioned above. However, this zeolite can originally conduct exchange between its positively charged ions with other positively charged ions trapped inside them, although in low concentration. Therefore, it is necessary to increase its performance through an activation process [6]. The activation process is used to remove cations and other impurities for increasing surface area and size of zeolite pore as well as reducing Si/Al ratio [6-10]. There are two activation processes, physical activation method by heating process and chemical activation method by dissolving into base or acid solution [6, 11].

This research used natural zeolite from Penggajawa village, Ende district, East Nusa Tenggara Province, Indonesia that was activated by NaOH in different concentration. This kind of zeolite had already been applied for adsorbing Pb element and was analyzed by using voltammetry method [4]. Therefore, the aim of this work is to find the optimum condition for activated zeolite to adsorb Pb ion in liquid sample and analyzed by using another method, which is laser-induced breakdown spectroscopy (LIBS).

2. Experimental procedure

2.1. Sample preparation

The zeolite sample used in this study was originated from Penggajawa village, Ende district, East Nusa Tenggara Province, Indonesia. Prior to undergoing activation process, the zeolite sample was grounded into fine powder and sieved to approximately 400 μm in diameter. The zeolite powder was then soaked in demineralized water for 24 hours before cleaning. This step was repeated 3 times. For activation process, the cleaned zeolite powder was soaked in various NaOH concentrations of 0.5, 1.0, 1.5, 2 M and stirred for 2 hours with a constant temperature of 70 °C. After that, the mixture between zeolite powder and NaOH solution was left to stay for 24 hours before being washed with demineralized water to neutralize its pH. Finally, the activated zeolite powder was dried in room temperature before being used as adsorbent.

To make a sample, a 0.05 g of activated zeolite powder was dissolved in 10 ml of solution containing Pb in various concentration of 10, 30, 50, and 100 ppm and each samples stayed for 2.5 hours. This solution was then heated up and stirred until become dry powder. Prior to LIBS analysis, the dry powder sample was made into a pellet. The wavelength used to detect Pb element in this experiment was 405.7 nm (Pb I) since it has 1000 times higher intensity compare to the other wavelengths [12].

2.2. Experimental Setup

The commercial LIBS configuration used in this study is shown in figure 1. The system consists of Nd:YAG laser (CRF 200, λ=1064 nm, 7 ns), Ocean Optics HR 2500$^+$ spectrometer, and AddLIBS software for elemental analysis. For this experiment, the laser was operated at 5 Hz repetition rate with laser energy of 120 mJ. The laser beam was focused through biconvex lens onto the surface of the zeolite pellet and the emission lines from the luminous plasma were captured by a spectrometer containing 14,336 CCD pixels in the wavelength range of 200-900 nm with resolution of 0.1 nm. All experiments were conducted with gate delay of 1 μs, accumulation of 3 shots and at 1 atm pressure of surrounding air.

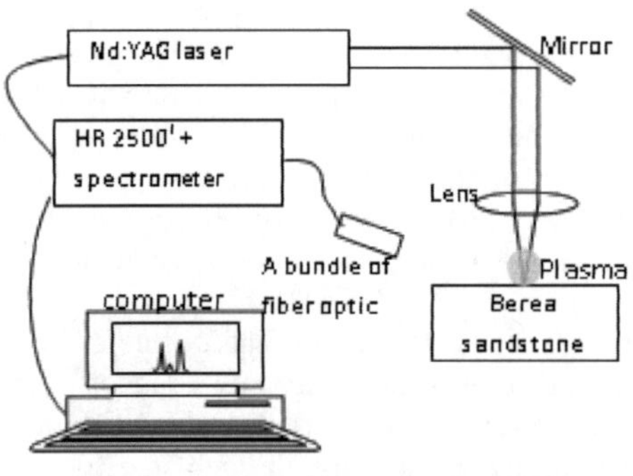

Figure 1. Experimental setup [13].

3. Results and discussion

3.1. Optimum condition for activation

Zeolite is a kind of stone which is usually available in the sea, abundant and invaluable. It is normally able to immobilize some elements in liquid at low ability. The performance of zeolite can be enhanced by doing chemistry activation method. In this work, the zeolite was activated using 1.0 M NaOH. The activated zeolite was then characterized using LIBS and the result is shown in figure 2. The graph exhibits emission lines intensity of element as a function of wavelength on Pb line region. There is no Pb I 405.7 nm line from both activated and inactivated zeolite. Therefore, this zeolite can be applied to detect Pb element in liquid sample. Due to activation process, the emission intensity of K I 404.4 nm reduced to about one-tenth compared to inactivated zeolite and the presences of other impurity elements were also confirmed except for Na, H and O elements. Increases in porosity and surface area of zeolite will enhance its performance to immobilize element in liquid. Besides, it was also already confirmed that before activation process, the zeolite contained some elements such as Al, Ca, Fe, Ba, K, Na, Si, H, and O.

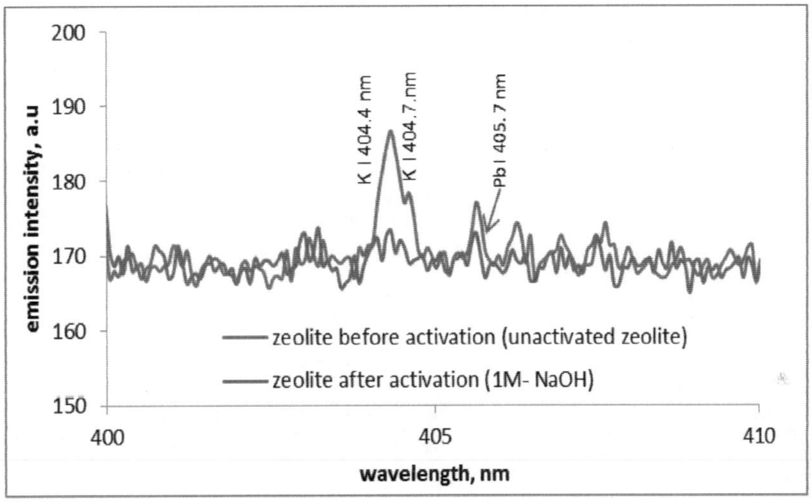

Figure 2. Spectra of activated and inactivated zeolite before being soaked in liquid sample containing Pb element.

In order to apply zeolite, 0.05 gr of activated and inactivated zeolite was each dissolved in 10 ml of standard solution containing 100 ppm of Pb and then heated and stirred until it became dry powder. The result of LIBS analysis is shown in the figure 3. A strong enough emission of Pb I 405.7 nm line was detected in activated zeolite and there was a small signal in the inactivated zeolite. K I 404.4 nm and 404.7 nm signals also increased. It can be predicted that the solution contained K element. The appearance of Pb signal and the rising of K signal indicate that the activated zeolite was more energetic to immobilize some elements in the liquid, whether through adsorption, cation exchange or other processes.

To optimize the analytical zeolite performance in term of immobilizing elements in the solution, it is necessary to vary the NaOH activation concentration of 0.5, 1.0, 1.5, and 2.0 M. 0.05 gr of each activated zeolite was dissolved in 10 ml of standard solution containing 100 ppm of Pb and was then heated and stirred until become dry powder. Figure 4 shows the result after being analyzed by LIBS. The average signal intensity of Pb I 405.7 nm rises with increasing NaoH concentration up to 1 M and then drops in the following concentrations. Decreasing emission intensities in higher NaOH concentration may be caused by broken structure of zeolite and as the result, it did not immobilize Pb element in the solution as much as expected.

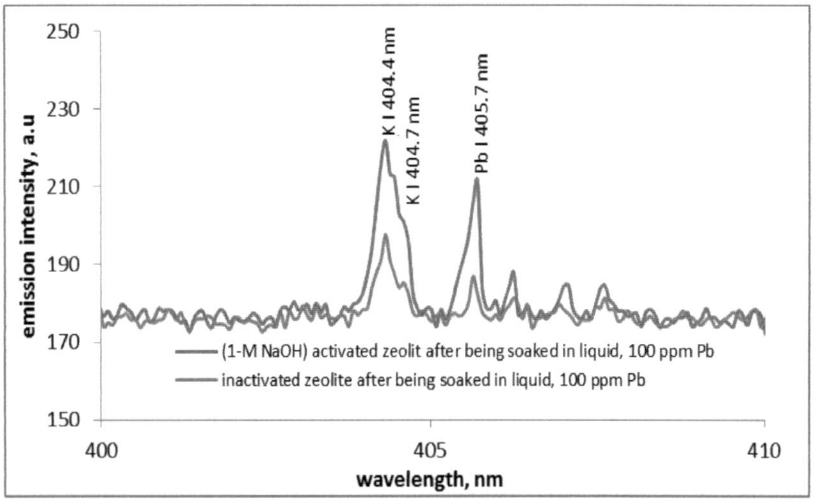

Figure 3. Spectra of 1.0 M NaOH activated zeolite and inactivated zeolite after being soaked in liquid sample containing 100 ppm Pb element.

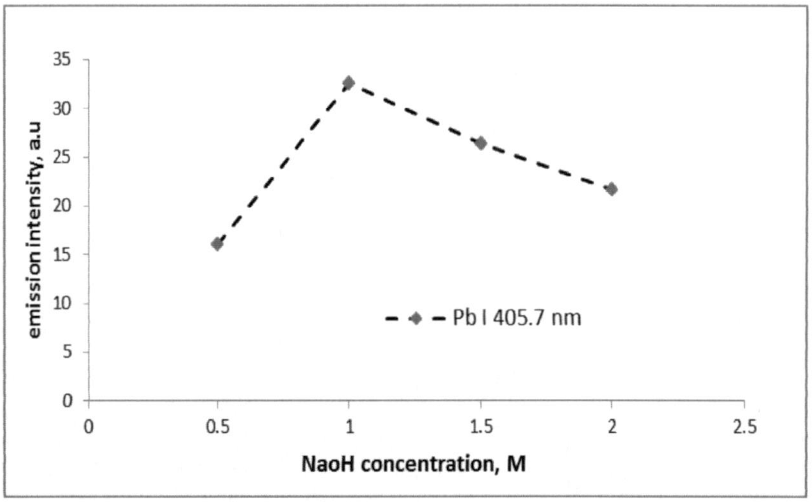

Figure 4. Emission intensity of Pb I 405.7 nm as a function of NaOH activation concentration on zeolite.

3.2. Quantitative analysis

The analytical performance of activated zeolite can be obtained by lowering the quantity or concentration of Pb in liquid sample so that it would be reliably adsorbed and detected by this method. To confirm this statement, a Pb concentration series of 10, 30, 50, and 100 ppm were used in this work and the result is exhibited in figure 5.

Figure 5 depicts the emission intensity of Pb that increases linearly from 10 ppm to 100 ppm in liquid sample with an equation of $y = 0.3274\ x - 1.384$. The limit of detection (LoD) of Pb based on this equation can be calculated to be about 5.86 ppm [14, 15]. This LoD is about three times higher compared to that of voltametry method [4]. However, this method is simpler than voltametry and AAS methods in term of sample preparation and detection. In addition, by using this method, some elements in the sample can be analyze simultaneously.

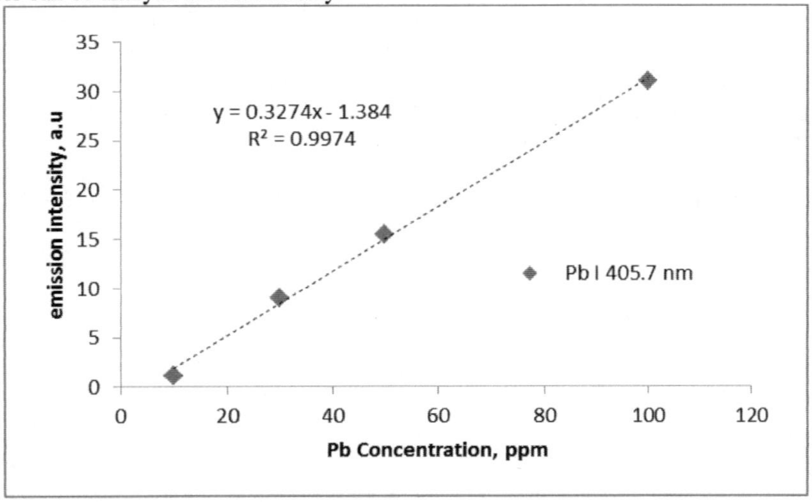

Figure 5. Calibration curve for Pb element in the solution.

4. Conclusion

In conclusion, activated zeolite could increase the ability to immobilize Pb element in solution and to transform it into solid phase. This element can be detected by LIBS technique with following conditions: 120 mJ of laser energy, 1 μs delay time detection and an accumulation of 3 shots. 1 M of NaOH was the maximum condition for zeolite activation so that the LIBS system could detect down to 5.86 ppm of Pb in liquid sample.

References

[1] Xu R, Pang W and Yu J 2007 *Chemistry of Zeolites and Related Porous Materials Synthesis and Structure* (Singapore: John Wiley & Sons)

[2] Newsam J 1992 Zeolites *Solid State Chemistry: Compounds* ed Cheetham A K and Day P (Gloucestershire: Clarendon Press) pp 234-257

[3] Vjunov A, Fulton J L, Huthwelker T, Pin S, Mei D, Schenter G K, Govind N, Camaioni D M, Hu J Z and Lercher J A 2014 *J. Am. Chem. Soc.* **136**(23) 8296

[4] Esati N P, Irdhawati and Suyanto H 2016 *Analysis of Pb, Cd, and Zn Elements with Anodic Stripping Voltammetry Method Using Carbon Paste Electrode Modified by Activated Zeolite* Master's Thesis (Udayana University)

[5] Payra P and Dutta P K 2003 Zeolites: a primer *Handbook of zeolite science and technology* ed Scott M A, Kathleen A C and Prabir K D (New York: Marcel Dekker, Inc) p 16

[6] Ertan A and F Çakicioğlu-OZKAN 2005 *Adsorption* **11**(Suppl 1) 151

[7] Hamdan H 1992 *Introduction to Zeolites: Synthesis, Charcterization, and Modification* (Penang: Universiti Teknologi Malaysia)

[8] Ackley MW, Rege S U and Saxena H 2003 *J. Micropor. and Mesopor. Mat.* **61** 25

[9] Triantafillidis C, Vlesidis A and Evmiridis N 2000 *Ind. Eng. Chem.* **39**(2) 307

[10] Mohamad D, Laeli K, Aprilina P and Setia B S 2010 *Proc. 1st Int. Conf. on Materials Engineering and 3rd AUN/SEED-Net Regional Conf. on Materials (Yogyakarta)* (Yogyakarta: Gadjah Mada University)

[11] Durand G, Plee D and Delcroix D 1993 Process for the Thermal Activation of Zeolites and Resultant Products U.S. Patent US5211929 A
[12] Sansonetti J E, Martin W C and Young S L 2005 *Handbook of Basic Atomic Spectroscopic Data* version 1.1.2 (Gaithersburg: National Institute of Standards and Technology) [http://physics.nist.gov/PhysRefData/Handbook/Tables/leadtable2.htm]
[13] Suyanto H 2016 *AIP Proceeding* (New York: AIP Publishing) vol 1719 030049
[14] Cremers D A and Radziemski L J 2006 Handbook of Laser-Induced Breakdown Spectroscopy (West Sussex: John Wiley & Sons)
[15] Hibberd D B and Gooding J J 2006 *Data Analysis for Chemistry: An Introduction Guide for Students and Laboratory Scientist* (New York: Oxford University Press)

Correlation analysis of solar constant, solar activity and cosmic ray

Y S Utomo

Research Center for Electrical Power and Mechatronics, Indonesian Institute of Sciences, Jl. Sangkuriang, Kompleks LIPI, Bandung 40135, West Java, Indonesia

E-mail: yustomo@gmail.com

Abstract. Actually, solar constant is not constant but fluctuated by ±1.5% of their average value. Solar constant indicates that the value is not constant but varies with time. Such variation is correlated with solar activity and cosmic ray. Correlation analysis shows a strong correlation between solar activity and cosmic ray and between solar activity and solar constant. Solar activity indicates by sunspot number. Correlations between solar constant variations and sunspot number variations were found to be higher than ones between variations in cosmic ray and solar constant. It was also found a positive correlation between solar constant and sunspot number, with correlation coefficient about +0.77/month and +0.95/year. In other hand, negative correlation between solar constant and cosmic ray flux i.e. −0.50/month and −0.62/year were found for monthly and yearly data respectively. A similar result was also found for the relationship between solar activity and cosmic ray flux with a negative correlation, i.e. −0.61/month and −0.69/year. When solar activities decrease, the clouds cover rate increase due to secondary ions produced by cosmic rays. The increase in the cloud cover rate causes the decrease in solar constant value and solar radiation on the earth's surface. Solar constant plays an important role in the planning and technical analysis of equipment utilizing solar energy.

1. Introduction

The sun is the main energy source for the earth and its environment. Through the emitted energy, the sun acted as a climate controller is very important to the air movement (atmospheric circulation) and ocean currents (ocean circulation) as well as to bio-instruments that processes photosynthesis in the biosphere. Therefore, the climate also depends on variations in the flux of solar energy received by the earth's surface. Variation in the solar energy flux is caused by variations in solar activity cycle. Thus the climate is a manifestation of how solar radiation is absorbed, redistributed by the atmosphere, land and oceans, and ultimately radiated back into space. Every variation of solar energy received at the earth's surface and reradiated by the earth into space will have a direct impact on climate change on Earth.

When reaching the upper limit of the atmosphere, the solar radiation is not actually attenuated but when it passes through the layers of the Earth's atmosphere solar radiation it will be scattered and absorbed by molecules and particles of dust clouds. Thus, only a small fraction of the solar radiation reaches the earth. Radiation flux before attenuation by the atmosphere at an average distance of the Sun - Earth (1.5×10^8 km) is called the *solar constant* which is defined as the total amount of solar radiation of all wavelengths, which falls on a unit area of the surface normal to the Sun - Earth line on a clear sky condition [1].

Solar activity was also influenced by variations in solar irradiance received at the earth's surface. Simulation of solar irradiance variability over 113 years (1880-1993) showed a variation of 0.5 %. While satellite observation during one cycle showed a variation of 0.1 % between the minimum and maximum activity [2, 3]. Determination of the solar constant and its variations over time is needed in climate research [4]. Solar constant plays an important role in the planning and technical analysis of equipment utilizing solar energy [5].

The results of this study are expected to use as an input and consideration in the planning and technical analysis of renewable energy, in particular the utilization of solar energy will be intensified and expanded in line with increasing energy needs. The utilization of solar energy is one step in the development of energy diversification. However its field operations should be supported with adequate information about the potential and availability of solar energy at a given location. By knowing the potential and availability of solar energy, then the site selection mistakes can be avoided.

2. Basic Theory

2.1. Solar Constant

Solar constant is defined as the total amount of solar radiation of all wavelengths, which falls on a unit area of the surface normal to the sun - earth line on a clear sky condition [1]. Until now, the absolute value of the solar constant uncertainty is both theoretically and observationally high enough. Determination of the solar constant has experienced a long history, has began by Smithsonian Institute in the early 20th century [6]. Solar constant measurement data obtained at that time were large, but the results did not provide a convincing value of solar constant [7].

The average value of solar constant ranges from 1364.61 to 1371.67 Wm^{-2}. Nimbus-7/ERB measurement is differed by 4.31 to 6.86 Wm^{-2} compared to the three other measurement results. It is caused by the problem of directing the telescope, correction calculation the average distance of the earth-sun and the sensitivity of instruments. Sensor Nimbus-7/ERB has an accuracy of 0.5 Wm^{-2} [8].

2.2. Solar Activity

Fluctuation of turbulence in the surface of the sun is related to the term of what we called as a solar activity. The combination of radiation activity and magnetic activity plays a major role in the cycle of solar activity. One indicator of solar activity is the appearance of black spots on the sun's surface called sunspots. The numbers of sunspots indicate the activity level of the sun. Variations in solar activity can affect the geomagnetic properties of ionosphere and stratosphere layer. Rapid changes in the region of UV radiation and X-rays due to solar activity can affect the ionosphere and the atmosphere through a system of coupling the magnetosphere-ionosphere-atmosphere [9, 10].

Solar activity was also influenced by variations in solar irradiance received at the Earth's surface. Simulation of solar irradiance variability over 113 years (1880-1993) showed a variation of 0.5%, while satellite observation during one cycle showed a variation of 0.1% between the minimum and maximum activity [2, 3]. Determination of the solar constant and its variations over time is needed in climate research [4]. Solar constant plays an important role in the planning and technical analysis of equipment utilizing solar energy [5].

2.3. Cosmic Ray Flux

Cosmic rays are energetic particles originating from space and spread until it reaches the Earth's atmosphere equal in all directions. The composition of cosmic rays consists of protons (90%), a helium nucleus (9%) and electrons (1%). Cosmic rays or the full name of Galactic Cosmic Rays (GCR) has a high energy level in the range of 300 MeV to 10 GeV. The main cause modulation of the cosmic rays is not the activity level of the sun, but the variation in the magnitude of the solar wind. The amount of solar wind is not constant, but varies with changes in solar activity during the 11 annual cycles [11].

Cosmic ray particles that had such high energy enter the Earth's atmosphere at speeds approaching 3×10^8 m.s^{-1}. With this speed and high energy, the cosmic ray particles are possible to collide with molecules of atmospheric when cosmic rays entering Earth's atmosphere. This collision is able to break the composition of the molecules in the atmosphere and induced the formation of secondary ions which serve as condensation nuclei to increase the formation rate of high clouds (at an altitude of 12-15 km). Because cosmic rays have an impact on the formation rate of cloud which further affects the amount of solar radiation to reach the Earth's surface, the contribution of cosmic rays on the Earth's climate cannot be ignored [11].

2.4. Correlation between Solar Constant, Solar Activity and Cosmic Ray Flux

Solar Constant is actually not constant but fluctuated by ±1.5% of their average value [1]. Direct observation via satellite indicates the same that the solar constant value is not constant but varies with time. The variation of solar constant that occurs as a result of changes in the form of solar activity such as sunspots dark or faculae bright in the Sun's surface [4, 10].

Solar activities have close correlation with the solar constant whose inconstant values. Therefore, in addition the sunspots, faculae are also an important source of solar constant variations [10]. The amount of solar energy emitted from the entire surface of the Sun is not the same, but varies with time. Hoyt and Kyle studies (1990) indicated that the solar constant will decrease with an increase in the sunspot area and will increase with the presence of the excellent faculae area on the solar disc. Based on Lean's research (1991) solar constant value at the time of minimum solar activity in 1986 is about 1367±3 Wm^{-2}, and showed a variation of 0.1% [2, 8].

3. Methodology

To determine whether there is a relationship between solar constant and solar activity, in this study the effect of solar activity (with indicators of the sunspot numbers) on the solar constant were analyzed using data with a 39-year period (1978-2016). Daily data for global solar constant were obtained from ftp://ftp.ngdc.noaa.gov/STP/solar_data/solar_irradiance/composite_42_65_1611.dat. Daily data of solar constant were processed into monthly average data, and then compared to monthly average data of sunspot numbers, which were obtained from ftp://ftp.ngdc.noaa.gov/STP/solar_data/sunspot_numbers/ridaily.plt with a 39-year period (1978-2016). These data were used in the analysis of solar activity's influence on the solar constant variation. Monthly Cosmic rays data (Galactic Cosmic Rays - GCR data) were obtained from ftp://ftp.ngdc.noaa.gov/STP/solar_data/cosmic_rays/huancayo.dat with a 39-year period (1978-2016). Furthermore, correlation analysis using a linear regression to determine the degree of influence of solar activity on the solar constant was employed and the effect of cosmic rays on the solar constant variation and its relation to solar activity was studied. In correlation analysis, it can be estimated a sample correlation coefficient which is denoted by r, ranges between −1 and +1. This coefficient quantifies the direction and strength of the linear association between the two variables. The correlation between two variables can be positive or negative. The sign of the correlation coefficient indicates the direction of the association. The magnitude of the correlation coefficient indicates the strength of the association.

4. Analysis and discussion

4.1. Correlation analysis

As already known, that the value of the solar constant is not constant, but varies with time despite minor amendments. The change might be related to solar activity and cosmic rays. Therefore, all three parameters i.e. solar constant, solar activity (which is indicated by the number of sunspots) and cosmic rays) have a relationship or correlation between one parameter with another. Correlation analysis was done by calculating the correlation coefficient and the results are summarized in Table 1.

Table 1 shows the correlation coefficient both per month and year between solar constant, sunspot number and cosmic rays. The values are all above ±0.50 which can be categorized as high value. The high correlation coefficient indicates that cosmic rays and sunspot numbers have strong relationships with the solar constant. There are two types of correlation between the solar constant with both parameters, namely the positive and negative correlations. The correlation coefficient between the solar constant and sunspot numbers are very high with positive number, i.e. +0.77/month and +0.95/year. The correlation coefficient between the solar constant and cosmic rays is high with negative number, i.e. –0.50/month and –0.62/year, while the correlation coefficient between sunspots number and cosmic rays were –0.61/month and –0.69/year. Therefore, if the value of cosmic rays increases then solar activity decreases, indicated by the reduction in sunspot numbers and the solar constant value.

Table 1. Correlation between solar constant, sunspot number and cosmic rays.

Parameter Relation	Correlation coefficient - *r*	
	per month	per year
Solar Constant – Sunspot Number	+0.77	+0.95
Solar Constant – Cosmic Rays	–0.50	–0.62
Sunspot Number – Cosmic Rays	–0.61	–0.69

The influence of solar activity on the solar constant variation has been proposed by previous researcher [2, 3]. A previous researcher simulated the variability of solar irradiance over 113 years (1880-1993) and found the variation of solar constant at 0.5% [2], while another researcher found a variation of 0.1% based on satellite observations during a solar cycle, i.e. the 21th cycle (1976 to 1986) [3]. Determination of the absolute value of the solar constant and its variations over time is needed in climate research. Solar constant plays an important role in the planning and technical analysis of equipment utilizing solar energy. Another researcher reported a correlation between temperature and global cloud coverage with the cosmic rays [12]. The influence of cosmic rays on cloud coverage, could also affect the solar constant [4, 5].

4.2. Influence of solar activity on the solar constant

For purposes of analysis influence of solar activity on the solar constant, the daily data of solar constant and solar activity in the period of November 1978 to November 2016 were used. Solar activity is indicated by the number of sunspots.

Figure 1 shows the pattern of the yearly average value of solar constant is similar with the pattern of the yearly average number of sunspots. The figure also shows the 21th solar cycle (1976-1986) and the 22th cycle (1987-1997). In the 21th solar cycle, the yearly average value of the solar constant reached its maximum value of 1366.56 Wm^{-2} in 1979. This coincided with the peak period of 21thsolar activity cycle which reached the maximum yearly average sunspot number of 155.3. Start from 1980, the value of the solar constant was gradually reduced following the reduction in solar activity. Solar constant value reached minimum when the sun was quiet in 1985-1986 with the average value of 1365.54 Wm^{-2}. The lowest value occurred in 1986, i.e. 1365.54 Wm^{-2} in conjunction with solar activity reached a minimum value. At the beginning of 22th solar cycle (1987–1997), a constant value increases exponentially with the increment of solar activity. The sun reaches its maximum solar constant value in 1989 with the amount of 1365.39 Wm^{-2} in conjunction with solar activity reached its maximum with yearly average sunspot number of 157.8. Solar constant value remained high during the years of 1989-1992 with small variations from year to year ranged from 0.04 to 0.32 Wm^{-2}. When the 23th solar cycle began in 1998, the value of the solar constant grew significantly following the increase in solar activity which was peaked in 2000. The yearly solar constant values were ranging from 1365.43 to 1366.60 Wm^{-2} or varied by 0.09% against the average value of 1366.01 Wm^{-2} [10, 13–15].

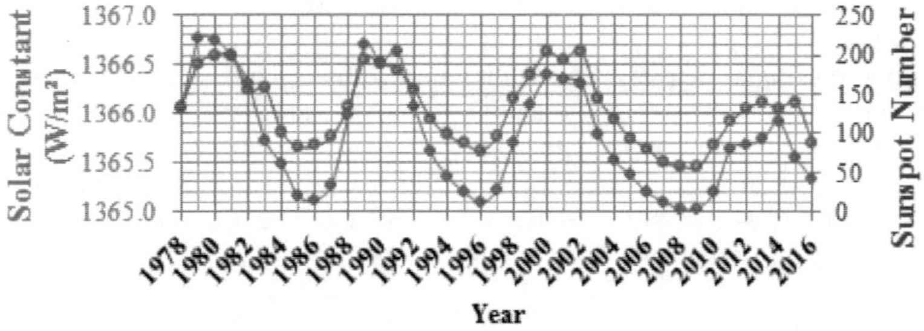

Figure 1. Yearly average of solar constant and sunspot number.

4.3. Influence analysis of cosmic ray on the solar constant

For the influence analysis of cosmic rays on the solar constant, the daily average cosmic ray data of Huancayo, Peru (12° S, 75° W) and solar constant data in the period of November 1978 to November 2016 were used. Huancayo was chosen with consideration that it has the lowest latitude among several other locations on Earth. Figure 2 shows the effect of cosmic rays on the solar constant.

Figure 2 shows yearly averages of solar constant and cosmic rays which has the same pattern but in the opposite direction. Yearly average of solar constant variation cycle was seemed to follow the cycle of the yearly average variation of cosmic rays in the opposite direction. When cosmic rays increased, the solar constant actually declined. As mentioned before, cosmic ray particles that have high energy (10 GeV) enters the Earth's atmosphere at speed approaching to 3×10^8 ms^{-1}. With its high speed and high energy, it is possible to collide with molecules of atmospheric and to break the composition of the molecules in the atmosphere forming the secondary ions which acts as condensation nuclei. The increase in the formation rate of cloud would affect the decrease in the intensity of solar radiation reaching the Earth's surface. The relationship between cosmic rays and solar constant is a "opposite" relationship because of the negative correlation type ($r < 0$). The phenomenon of "opposite" is in a good agreement with the result by Svensmark (1997) who found a correlation between temperature and global cloud coverage with the cosmic rays [12].

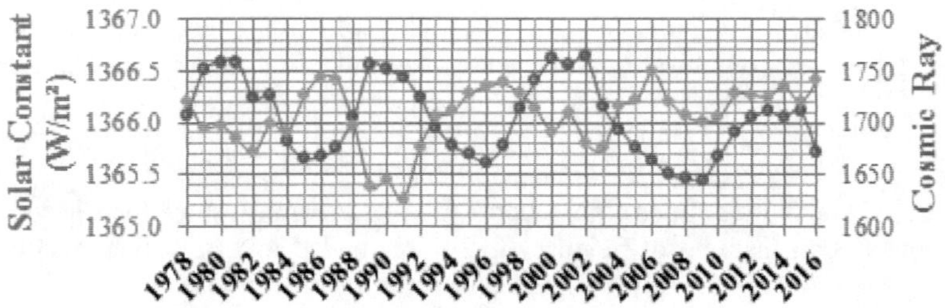

Figure 2. Yearly average of solar constant and cosmic ray.

4.4. Influence analysis of solar activity on the cosmic ray

For the influence analysis of the of solar activity on cosmic rays, a daily average of cosmic rays and sunspot numbers data during the period November 1978 to November 2016 were used. Figure 3 shows the effect of solar activity on cosmic rays.

In figure 3, variation of yearly average cosmic rays is in opposite to the pattern of sunspot numbers. When solar activity is maximum (indicated by the high number of sunspots), cosmic rays is minimum.

This phenomenon of "opposites" is in accordance with what is stated by Svensmark (1997) who found a correlation between temperature and global cloud coverage with the cosmic rays. Physically, particles of cosmic rays when it enters the Earth's upper atmosphere are suffered from modulation caused by two things, namely the solar wind and Earth's magnetic field. The solar wind affects cosmic ray particles to decelerate. According to Yamada (1998), the size of the solar wind is not constant, but varies with changes in solar activity during the 11st annual cycles. The relationship between solar activity and cosmic rays is an "opposite" relationship because of negative correlation type ($r < 0$). The solar activity influenced on cosmic rays has an impact on the cloud formation rate [11, 12].

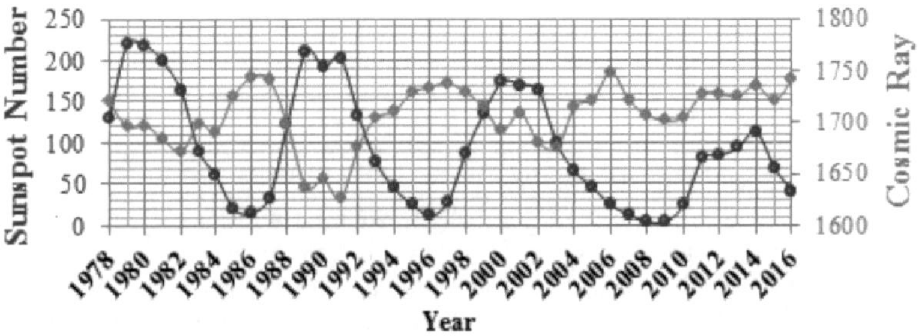

Figure 3. Yearly average of sunspot number and cosmic ray.

5. Conclusion

Correlation analysis showed that there was a relationship between the variation of the solar constant with the variation in solar activity and the flux of cosmic rays. A positive correlation ($r = 0.95$) was found between the solar constant and solar activity, a negative correlation ($r = -0.62$) was for the relationship between the solar constant and cosmic rays, and also a negative correlation ($r = -0.69$) between solar activity and cosmic rays.

The results of this study are expected to be used in the consideration in the work of planning and technical analysis of renewable energy, particularly in the utilization of solar energy that will be intensified and expanded in line with increasing energy needs. The utilization of solar energy is one step in the development of energy diversification, but its application should be supported with adequate information about the potential and availability of solar energy at a given location. By knowing the potential and availability of solar energy, mistake in choosing the site selection can be avoided.

Acknowledgments

The author gratefully acknowledgment the support of the Research Center for Electrical Power and Mechatronics, Indonesian Institute of Science (LIPI) for funded and has given encouragement and opportunity to work in the field of conversion and conservation energy research.

References

[1] Bayong TjHK 2004 *Climatology* (Bandung: ITB Press)
[2] Lean J 1991 *Rev. Geophys.* **29** 505-35
[3] Wilson RC 1997 *Science* **277** 1963-65
[4] Sinambela W 1998 *Warta LAPAN* **56** 1-12
[5] Suwono A 1980 *Proc. Solar Energy Utilization for Rural* (Bandung: LFN-LIPI) chapter 2 pp 1-2
[6] Frohlich C 1990 *NASA Conf. Publication* vol 3086 (USA: NASA) pp 269-77
[7] Frohlich C 1987 *J. Geophys. Res.* **92** 796-800

[8] Hoyt DV and Kyle HL1990 *An alternative derivation of the NIMBUS-7 total solar irradiance variation* ed Schatten K and Arking A *NASA GSFC* (USA: Greenbelt MD 20771) 293-300
[9] Tsurutani BT, Goldstein BE, Smith EJ, Gozales WD, Tang F, Akasofu SI and Anderson RR 1990 *Planet Space Sci.* **38** 109-26
[10] Yatini CL 2004 *J. Fisika HFI* **A4** 04-20
[11] Yamada Y, Yanagita S and Toshida T 1998 *Geophys. Res. Lett.* **25** 2353-56
[12] Svensmark H and Cristensen EF., 1997 *J. Atmospheric and Terrestrial Physics* **59**(11) 1225-32
[13] Lake EG and Drummond AJ 1968 *Science* **161** 888-91
[14] Mecherikunnel AT, Kyle HL and Lee RB 1990 *NASA Conf. Publication* vol 3086 (USA:NASA) pp 309-13
[15] Mihalakakou G, Santamouris M and Asimakopoulos 2000 *Theor. Appl. Climatol.* **66** 185–97

IOP Conf. Series: Journal of Physics: Conf. Series **817** (2017) 012046 doi:10.1088/1742-6596/817/1/012046

Analysis of laser induced plasma of lithium iron phosphate/$LiFePO_4$ using Nd:YAG laser at low pressure

M M Suliyanti, A N Hidayah and Isnaeni

Research Center for Physics, Indonesian Institute of Sciences, Bd. 442 Kawasan Puspiptek Serpong, South Tangerang 15314 Banten, Indonesia

E-mail: mari004@lipi.go.id

Abstract. Preliminary analysis of lithium in Lithium Iron Phosphate ($LiFePO_4$) powder using laser induced plasma spectroscopy at low pressure had been done. Recently, $LiFePO_4$–based batteries are widely used in most electric cars and bikes due to less toxic. However, lithium (Li) element is very difficult to detect since it is a very light element. In this work, we used a Nd:YAG laser (1064 nm wavelength, 5 ns pulse width at 10 Hz repetition rate) that was focused on $LiFePO_4$ sample at low pressure. The main Li peak emission in $LiFePO_4$ powder and sheet can be easily detected using this technique. We report the results of experimental study on Li element emission lines at wavelength 460.18 nm, 610.37 nm and 670.83 nm using 2 mJ and 12 mJ laser irradiation at 5 Torr and 35 Torr air atmosphere. The results of this study showed promising application of laser-induced plasma spectroscopy to detect and analyse Li in various samples.

1. Introduction

Laser induced plasma spectroscopy (LIPS) for spectrochemical analysis has actually been an active field of study for decades. Researchers are drawn to the simplicity of LIPS along with its several other advantages. Apart from eliminating the need of tedious sample preparation, spectrum observation can be carried out without the need of sample transport. This method also allows the implementation of local microanalysis and the feasibility of obtaining real time analytical information. Furthermore, this method is very potential for wide range of applications under various operational conditions by improving its sensitivity and accuracy. LIPS consists of direct analysis and indirect analysis. In the case of direct analysis, atomization and excitation take place in one step. In this case, the analytical signal is obtained from the plasma emission by means of emission spectroscopic detection techniques. In the case of indirect analysis, atomization and excitation processes take place separately. Namely, laser plasma is generated only after the atomization process, while the excitation process is subsequently induced by an auxiliary setup to improve detection sensitivity such as inductively coupled plasma spectroscopy. Recently, the use of lasers for material analysis in laser induced breakdown spectroscopy (LIBS) and LIPS significantly grow for various applications. Many books and reports about these techniques are already available [1-4]. LIBS have been continuously improved and applied in industries and research laboratories [5, 6]. The interaction of high-power laser light with a target material has been an active topic of research not only in plasma physics but also in the field of material science, chemical physics and particularly in analytical chemistry.

A lithium ion battery is a device that can convert chemical energy stored in its active materials directly into electric energy by using electrochemical oxidation reduction reactions [7]. Recently, $LiFePO_4$ has been developed, improved and applied in lithium batteries since it is an excellent material in batteries. Some papers that have been published related to the $LiFePO_4$, studied about sequential $LiFePO_4$ particle charge and discharge by conducting electrochemical measurements [8-11]. The grain sizes of $LiFePO_4$ strongly affect electrode potentials due to the fact that the grain sizes influences electronic contact between $LiFePO_4$ layer and the current collector [12-13]. Therefore, properties of $LiFePO_4$ thin film are very important for battery technologies. However, one main problem in characterization of Li thin film occurs. It is very difficult to detect Li content and concentration in material or sample since Li is one of the light elements that cannot be detected by conventional X-ray diffraction spectroscopy. So far, one powerful method to detect Li element in sample is neutron scattering spectroscopy, which is very expensive due to neutron source production. Therefore, a new and alternative method to detection Li is still required. In this work, we propose LIPS as one of the candidates to detect Li element from $LiFePO_4$ powder sample by utilizing Nd:YAG laser.

2. Experimental setup

Basic experimental setup used in this work is shown in figure 1. The laser source used in this experiment was Nd:YAG laser (Q-Smart 850, 1064 nm wavelength, 5 ns pulse width) operated in the Q-switched mode at a 10 Hz repetition rate. The laser output energy was fixed at approximately 2 mJ and 12 mJ by means of delay switch setting. The laser beam was focused by a lens of focal length f=150 mm onto the sample surface through a quartz window. The diameter of the laser spot on the sample surface was approximately 1.5 mm. The sample was rotated during laser irradiation so that the resulting data represent spatial average of different points on the sample surface. The monitored shot-to-shot fluctuation of the laser was found to be approximately 3-5 %.

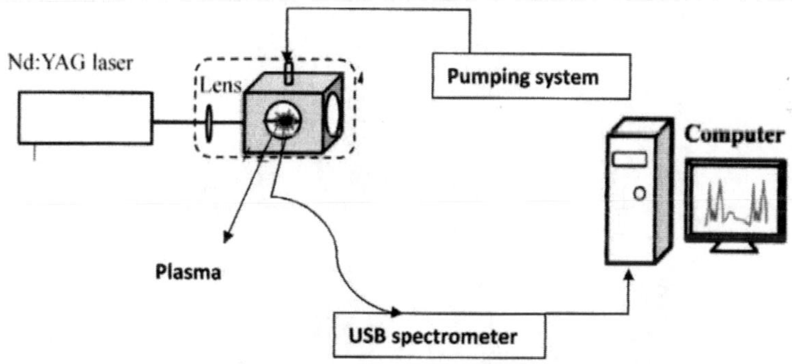

Figure 1. Experimental set-up.

There were two samples used in this work. First, Li sheet of $LiFePO_4$. Second, Li pellet made of $LiFePO_4$ powder using pellet press machine. Furthermore, the samples were placed in a small, vacuum-tight metal chamber (11.5 cm × 11.5 cm × 11.5 cm). The chamber was evacuated using a vacuum pump and then filled again with air until desired pressure was achieved. The gas flow into the chamber was regulated by needle valves in the air line and pumping line. The chamber pressure was monitored by means of a digital Pirani meter.

Plasma emission was detected by means of a mini spectrometer USB2000 from Ocean Optics, which has an optical fiber input. The optical fiber input was placed at a distance of approximately 10 mm from the sample surface and at a position of approximately 100 mm sidewise from the center of the plasma. The optical fiber position was then slightly adjusted to get the best position. At this position, the fiber is expected to collect most of the emitted plasma.

3. Results and discussion

The first LIPS experiment study was conducted on a lithium sheet sample made of $LiFePO_4$. The resulting emission spectrum presented in figure 2 show three typical Li element emission lines of Li(I) at the wavelength of 460.18 nm, Li(I) at the wavelength of 610.37 nm, Li(I) at the wavelength of 670.83 nm. These three peaks can be clearly detected. An unidentified emission spectrum was also observed in the wavelength regions around 395 nm, 450 nm and 570 nm. These unidentified emission peaks are related to impurities in the samples.

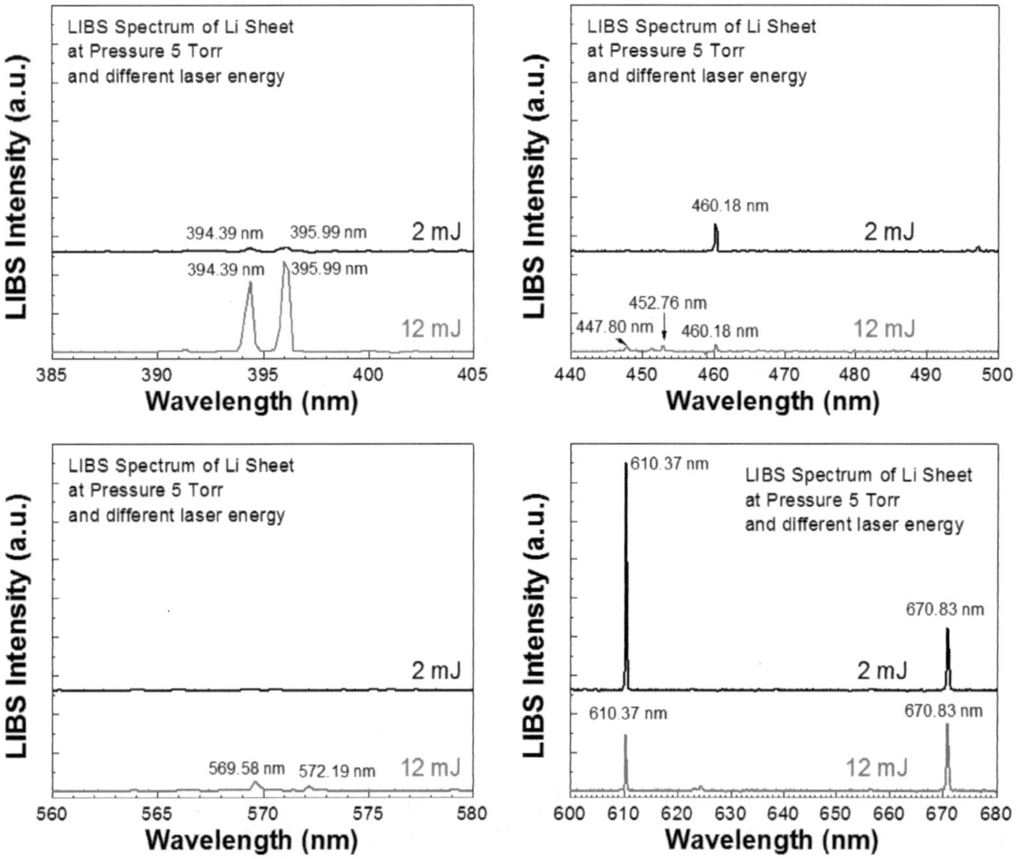

Figure 2. Emission spectrum of $LiFePO_4$ sheets. The laser energies were 2 mJ and 12 mJ and the surrounding air pressure was 5 Torr.

The second experiment was conducted on the Li pellet sample made of $LiFePO_4$ powder. The sample was irradiated using laser Nd-YAG with energy 2 and 12 mJ at surrounding air pressure 5 Torr and 35 Torr. Figure 3 shows the emission spectrum of $LiFePO_4$ pellet.

In the case of $LiFePO_4$ sheet, three typical Li emission lines were detected more clearly than those of $LiFePO_4$ pellet (figure 3). In order to clarify this result, we assume that the atomization process in a thin film sample was easier to proceed than that in the hard bulk or pellet sample [14]. The effect of the sub-target may also occur in this case of $LiFePO_4$ pellet.

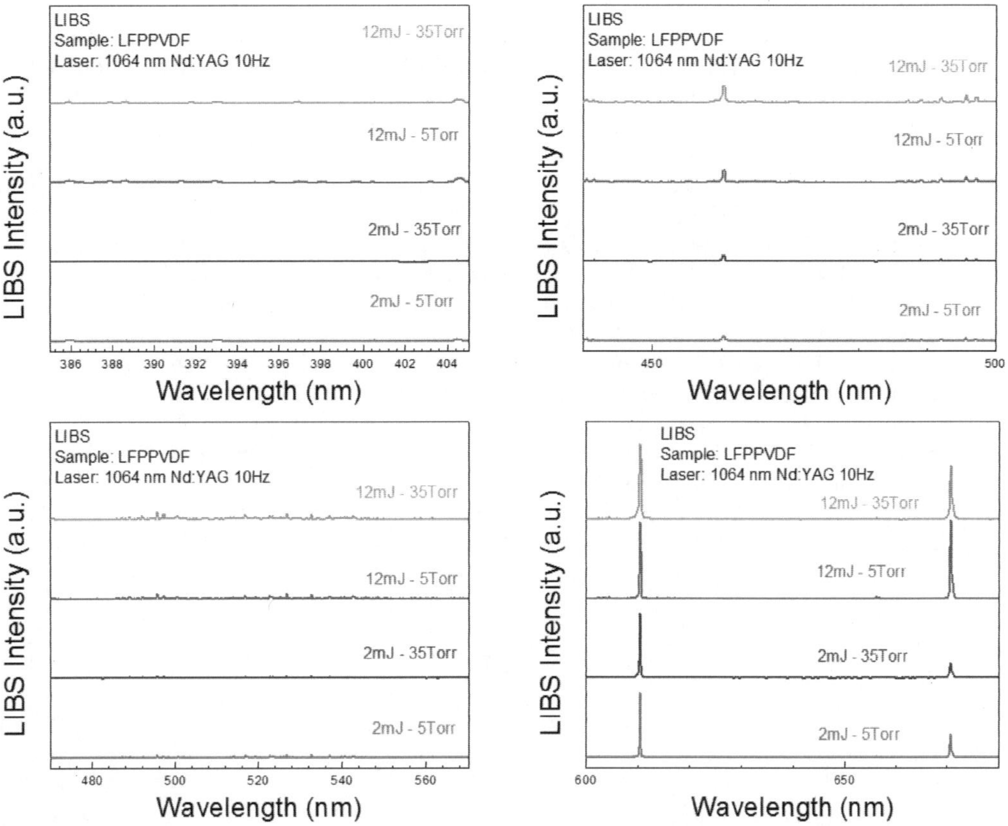

Figure 3. Emission spectrum of $LiFePO_4$ pellet. The laser energies were 2 mJ and 12 mJ and the surrounding air pressures were 5 Torr and 35 Torr.

Figure 3 shows detail comparison of the intensity-versus-wavelength of three Li lines under different laser energies and air pressures. In the case of pellet sample, when laser energy increased, the emission counts increased as well, so that the correlation between energy and intensity emission is proportional.

4. Conclusion

We have shown that three emission lines of Li from $LiFePO_4$ sheet and pellet can be detected using laser induced plasma spectroscopy. In the case of Li sheet sample, the Li emission lines were stronger than those of Li pellet sample even in low energy. It is naturally assumed that in the case of a thin film sample, the atomization will proceed easily, compared to that of a hard bulk sample or pellet. This study can be explored and improved further by varying the number of sample, pressure and energy.

Acknowledgments

The authors express their sincerest thanks to Fadli Rohman for the sample preparation and discussion. We also thank P2F-LIPI for the supporting this research. The authors would like to thank Kegiatan Unggulan LIPI and Kegiatan Penguatan Kompetensi from Research Center for Physics, Indonesian Institute of Sciences for the financial support of this work. The authors would also like to thank the researchers in the Group of Laser, Research Center for Physics for the valuable discussions and supports during this work.

References

[1] Singh J P and Thakur S N eds 2007 *Laser-Induced Breakdown Spectrsocopy* (Oxford: Elsevier)
[2] Cremers D A and Radziemski L J 2006 *Handbook of Laser-Induced Breakdown Spectroscopy* (West Sussex: John Wiley & Sons)
[3] Miziolek A W, Palleschi V and Schechter I eds 2006 *Laser-Induced Breakdown Spectroscopy (LIBS), Fundamentals and Applications* (Cambridge: Cambridge University Press)
[4] Lee Y I, Song K and Sneddon J 2000 *Laser-Induced Breakdown Spectrometry* (New York: Nova Science Publishers)
[5] Wisbrun R, Schechter L, Niessner R, Schroder H and Kompa K 1994 *Anal. Chem.* **66** 2964
[6] Fichet P, Mauchiem P, Wagner J F and Moulin C 2001 *Anal. Chim. Acta* **429** 269
[7] Linden D and Reddy T B 2002 *Handbook of Batteries* 3rd ed (New York: McGraw-Hill)
[8] Sasaki T, Ukyo Y and Novák P 2013 *Nat. Mater.* **12** 569
[9] Dreyer W, Jamnik J, Guhlke C, Huth R, Moskon J and Gaberscek M 2010 *Nat. Mater.* **9** 448
[10] Moskon J, Jamnik J and Gaberscek M 2013 *Solid State Ion.* **238** 24
[11] Roscher M A, Bohlen O and Vetter J 2011 *Int. J. Electrochim.* **2011** 1
[12] Zhu C 2013 *Size Effect on Lithium Storage and Phase Transition in* $LiFePO_4/FePO_4$ *System* Ph.D. Dissertation (University of Stuttgart)
[13] Zhu C, Gu L, Suo L, Popovic J, Li H, Ikuhara Y and Maier J 2014 *Adv. Funct. Mater.* **24** 312
[14] Suliyanti M M, Hedwig R, Kurniawan H and Kagawa K 1998 *Jpn. J. Appl. Phys.* **37** 6628

Analysis of dual peak emission from Rhodamine 6G organic dyes using photoluminescence

I T Sugiarto, Isnaeni and K Y Putri

Research Center for Physics - Indonesian Institute of Sciences
Kawasan PUSPIPTEK Serpong, 442 Building, South Tangerang 15314
Banten, Indonesia

E-mail: iyon001@lipi.go.id

Abstract. Rhodamine 6G (R6G) or Rhodamine 590 is a cationic dye with a strong absorption in the visible and a high fluorescence yield. R6G is commonly used as a tracer dye in water to determine the rate and direction of flow and transport. Some examples of Rhodamine dye applications are fluorescence microscopy, flow cytometry, fluorescence correlation spectroscopy, etc. In this work, the optical properties of R6G with various concentrations in water were observed by mean of photoluminescence (PL) measurement. The experimental results showed that the higher the concentration, the greater the intensity of the photoluminescence. It is observed that R6G dye emission spectrum had two distinctive peaks, at 550 nm and 602 nm wavelengths. R6G dye concentration affected these peaks in a very interesting manner. With the increasing R6G concentration, the emission peak shifted from 550 nm, which is dominated by monomer, to 602 nm, which is dominated by dimer emission. Our findings have given a good clue in optimizing the concentration of R6G dye in order to give the best emission for various applications.

1. Introduction

Rhodamine 6G (R6G) or Rhodamine 590 is a highly fluorescent rhodamine family dye. R6G has a chemical formulas $C_{28}H_{31}N_2O_3Cl$. The structure of the R6G molecule is illustrated by figure 1.

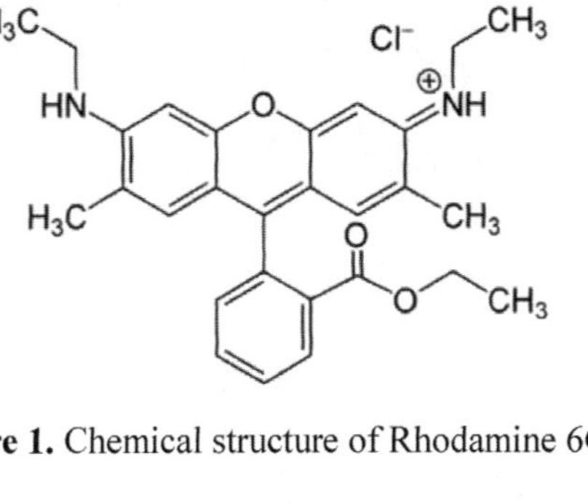

Figure 1. Chemical structure of Rhodamine 6G.

R6G is a cationic dye with a strong absorption in the visible and a high fluorescence yield. The molecule consists of two chromophores, a dibenzopyrene chomophore (xanthene) and a

IOP Conf. Series: Journal of Physics: Conf. Series **817** (2017) 012047 doi:10.1088/1742-6596/817/1/012047

carboxyphenyl group tilted by about 90° with respect to the xanthene ring (see figure 1). The π-systems of the two chromophores of R6G are therefore not conjugated. The strong absorption of R6G in aqueous solution has a maximum of around 530 nm and a vibronic shoulder of around 470 nm [1].

R6G has a molar mass of 479.02 g/mol and 1.26 g/cm^3 in density. Its solubility is 20 g/l in water at 25°C and 400 g/l in methanol. R6G is commonly used as a tracer dye in water to determine the rate and direction of flow and transport. Some applications of rhodamine dye are fluorescence microscopy, flow cytometry, fluorescence correlation spectroscopy, etc. In other application, R6G is frequently used as a gain medium in dye-lasers [2]. The absorption of R6G ranges between 440 nm and 570 nm, with the peak at 530 nm [3, 4]. The resulting emission spectrum vary from about 510 nm to around 710 nm with the peak at 550 nm depending on the solvent and the dye concentration. However, the laser emission range is much narrower, from around 560 nm to 610 nm with the peak wavelength at around 575 nm [3, 5].

The color emission of R6G will change according to the levels of concentration. Molecular aggregates of R6G, in the form of dimers or higher aggregates, act to quench fluorescence and therefore degrade the lasing ability of dye–laser systems. This quenching process is largely due to the transfer of excitation energy between monomers and aggregates, which then decay non-radiatively. Dimers, in particular, are known to have a low quantum yield and so will effectively quench emission [6-8]. Dimerization comes about as the concentration of dye is increased beyond some critical value. In methanol and ethanol, R6G monomers present alone up to concentrations on the order of 0.1 mol/L, whereas in aqueous solutions, R6G begins to form dimers at concentrations as low as 1×10^{-5} mol/L. In aqueous solutions, the R6G dimer consists of two dye molecules bridged by a water molecule. Because of the resulting fluorescence quenching, R6G in water is not able to exhibit lasing for concentrations greater than approximately 1×10^{-4} mol/L [9].

The purpose of this work was to use photoluminescence (PL) to study and explain the color evolution that occurs in R6G due to changes in concentration. In this study, the excitation wavelength used was 532 nm, which was not located at the maximum absorpsion of R6G. This excitation wavelength was close to the 550 nm peak which was alleged as the monomer peak of R6G. Therefore, the shorter wavelength was used to facilitate analysis in determining excitation values.

2. Methodology

The experimental set-up for this work is shown in figure 2. Analysis of dual peak emission from R6G was done by focusing laser beam into a cuvette, from which then the reflected beam was collected and detected by an optical spectrometer. Observation was then made on the optical spectrum produced.

A series of aqueous R6G solutions were prepared with concentrations ranging from 10^{-9} to 10^{-3} g/ml. For this purpose, a concentrated solution was initially prepared and the other samples were prepared by dilution of a small volume of this stock solution in distilled water. Between 10^{-6} and 10^{-4} g/ml, more samples were prepared in smaller increments, i.e.; 2×10^{-6}, 4×10^{-6}, 8×10^{-6}, 1×10^{-5}, 2×10^{-5}, 4×10^{-5} and 8×10^{-5} g/ml to see more detailed changes in their emissions.

For each measurement, a glass cuvette was filled with dye solution and illuminated by a picosecond laser. The laser beam was focused into the cuvette and the reflected beam from the cuvette was then collected using two collimated lenses as shown in figure 2. The emission spectrum was recorded by Ocean Optics SpectraSuite HR2000. The emission scans were performed from 400 to 700 nm.

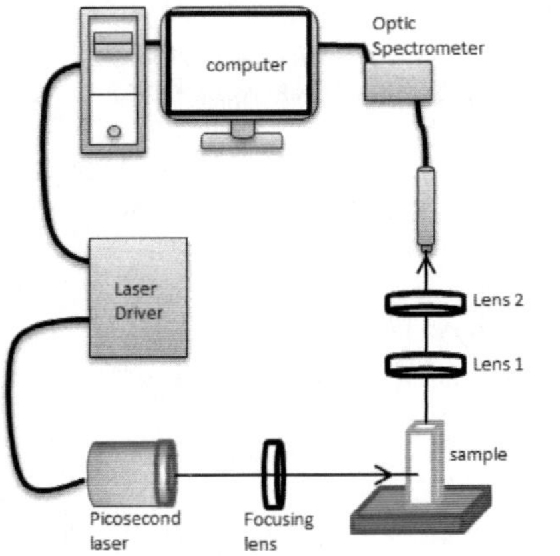

Figure 2. Experimental set-up.

3. Results and Discussion

Figure 3 gives an overview of the emission spectra from the aqueous dye solutions at the concentrations of 10^{-3} and 10^{-9} g/ml.

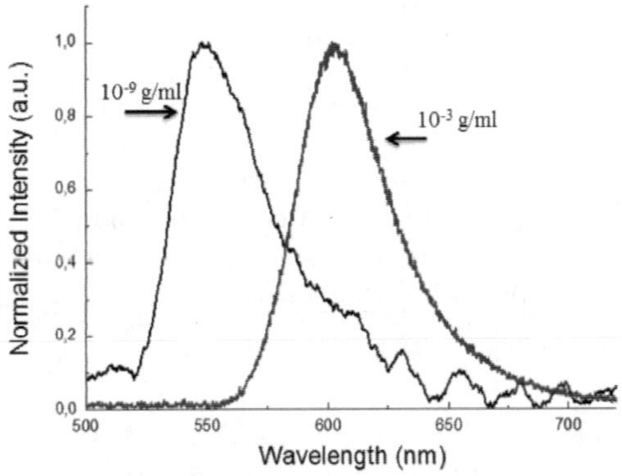

Figure 3. PL spectra for R6G concentration at 10^{-3} and 10^{-9} g/ml.

Based on the graph, it appears that there is a shift in the emission peak of R6G from low to high concentration. The black graph shows the spectra of R6G at a concentration of 10^{-9} g/ml, while the blue graph is the result of R6G spectra at a concentration of 10^{-3} g/ml. There were several peaks appeared in the spectra of 10^{-9} g/ml R6G, in contrast to one-peak-only in R6G spectra at a concentration of 10^{-3} g/ml. In order to prove that there was truly a shift in emission peak, this experiment was conducted on R6G samples with concentrations between 10^{-6} and 10^{-4} g/ml. For a more detailed analysis, the individual spectra for R6G with concentrations ranging from 10^{-6} to 10^{-4} g/ml are displayed in Figure 4. The high concentration solutions emitted light with peak wavelengths beyond 602 nm, while the low concentration emission was at 550 nm. A particularly interesting feature was the change in the band shape of the fluorescence emission. For intermediate concentrations, the spectral shape indicated the existence of two bands overlapping with each other [3]. The two spectral bands observed originated from R6G monomers and dimers. At a high concentration, the R6G molecules arranged themselves in so-called H-type and J-type dimers or larger

IOP Conf. Series: Journal of Physics: Conf. Series **817** (2017) 012047 doi:10.1088/1742-6596/817/1/012047

aggregates, the first of which was energetically favored. This dimer formation influenced the electronic structure and as a consequence, the optical properties result in an emission around 600 nm. At a low dye concentration, most dye molecules were isolated from each other and fully solvated, hence the emission band was around 560 nm.

From figure 4, it appears that a greater concentration, the PL intensity will be even greater as well. It was also observed from the graph that there were changes in spectrum shapes with increasing concentrations, i.e. there were widening of spectrum and it looked like there were two peaks in the spectrum. So, based on this graph it can be assumed that there were monomer and dimer in each spectrum generated as mentioned above [3, 9].

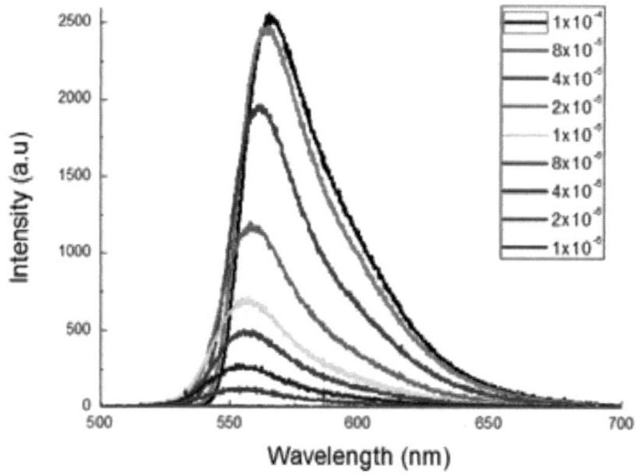

Figure 4. PL spectrum of R6G at various dye concentration.

Figure 5 is a graph of changes in the peak PL intensity and wavelength against the changes in concentration. From the graph, it appears that the effect of concentration changes on the PL intensity produced was not linear. This was because self-absorption would strongly depend on the distance the fluorescence signal traveled inside the dye solution [5].

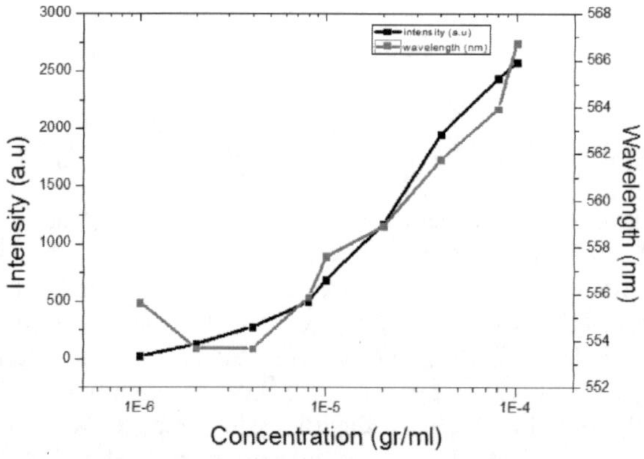

Figure 5. Peak intensity and peak wavelength as function of R6G concentration.

The effect of self-absorption was to rob signal from the emission line. As the number of emitting atoms increased, the likelihood of self-absorption also increased and there was no longer a linear relationship between the number of emitting atoms and the measured intensity. The self-absorption effect occurred in the samples whose absorption and emission spectrum partially overlap, and different emission wavelengths undergo different processes of absorption when propagating in the samples.

In order to study the emission shift of R6G from 550 nm to 602 nm wavelengths, all emission spectra were analysed using photoluminescence to reveal that each spectra consisted of monomer and dimer spectrum. Figure 6 is one of the examples in the monomer-dimer analysis that was assumed in figure 4. Fitting analysis was performed at a wavelength range between 550 nm and 602 nm based on the peak wavelengths generated in figure 3. The green lines were the assumed monomer-dimer presences.

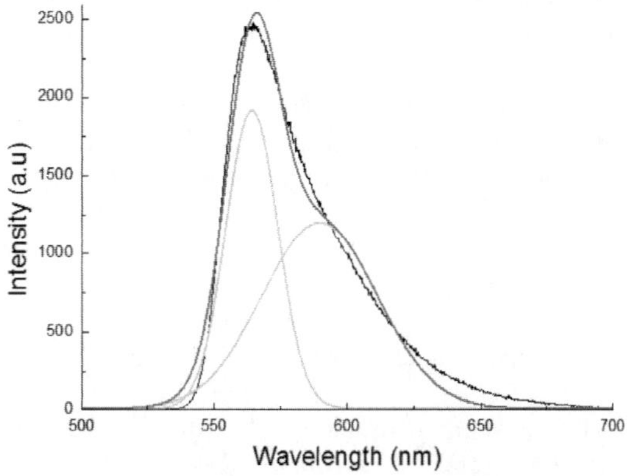

Figure 6. Dual peak emission spectra of R6G.

Figure 7 shows an evidence of monomer-dimer evolution that occurred in any concentration of R6G. From the graph, it appears that the intensity of the monomer dropped with the increase in the concentration of R6G, while the intensity of the dimer went up proportionally to the increase in concentration. The monomer-dimer percentage values should converge at a point close to the concentration of $4x10^{-5}$ g/ml.

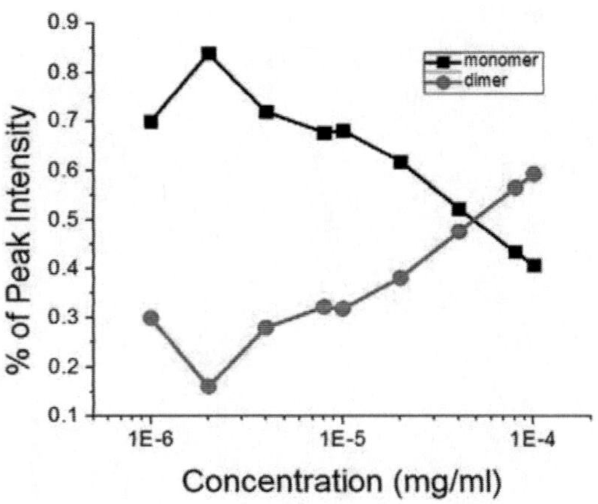

Figure 7. Crossing curve of analysis dual peak monomer-dimmer.

This experiment showed us that the emission of R6G may change in respond to a given concentration and therefore it may be tuned as desired by modifying the dye concentration accordingly. Appropriate concentration should be further studied for certain applications such as bio tagging etc. Regarding applications as source of laser (dye laser) etc., Further investigation on the peak stimulated emission that can be changed should also be carried out.

4. Conclussion

In this work, we studied the concentration effects in the fluorescence emission spectra of Rhodamine 6G. R6G concentrations were tested at concentrations ranging from 10^{-3} to 10^{-9} g/ml. Changing the dye concentration provided tunability between 550 nm at low concentration and 602 nm at high concentration. At a concentration approaching the value of 10^{-4} g/ml, the monomer and dimer curves crossed path and therefore proved the existence of monomer to dimer evolution in the R6G spectrum. However, when highly concentrated dye solutions were used in application such as dye lasers, strong absorption and self-absorption became problems. The results of this experiment is very useful to understand how the evolution monomer-dimer takes place in the concentration changes of R6G and to grasp the approximate value of emissions in accordance with the given concentration for applications such as bio-tagging, etc.

References

[1] Jensen L and Schatz GC 2006 *J. Phys. Chem. A* **110**(18) 5973
[2] Arbeloa FL, Gonzales IL, Ojeda PR and Arbeloa IL 1982 *J. Chem. Soc. Faraday Trans. 2* **78** 989
[3] Zehentbauer FM, Moretto C, Stephen R, Thevar T, Gilchrist JR, Pokrajac D, Richard KL and Kiefer J 2014 *Spectrochim. Acta A Mol. Biomol. Spectrosc.* **121** 147
[4] Brackman U 2000 *Lambdachrome Laser Dyes* 3rd ed (Goettingen: Lambda Physik AG)
[5] Bogdankevich OV, Zverev MM, Krasavina EM, Kryukova IV and Pevtsov VF 1987 *Sov. J. Quantum Electron.* **17** 133
[6] Innocenzi P, Kozuka H and Yoko TJ 1996 *J. Non-Cryst. Solids* **201** 26
[7] Arbeloa FL, Ojeda PR and Arbeloa IL 1988 *J. Chem. Soc. Faraday Trans. 2* **84** 1903
[8] Vogel R, Meredith P, Harvey MD and Rubinsztein-Dunlop H 2004 *Spectrochim. Acta A Mol. Biomol. Spectrosc.* **60**(1) 245
[9] Vogel R, Harvey M, Edwards G, Meredith P, Heckenberg N, Trau M and Rubinsztein-Dunlop H 2002 *Macromolecules* **35**(6) 2063

Silver nanoparticle formation by femtosecond laser induced reduction of ammonia-containing $AgNO_3$ solution

Y Herbani[1], T Nakamura[2] and S Sato[2]

[1] Research Center for Physics, Indonesian Institute of Sciences, Bldg 442 Kawasan Puspiptek Serpong, Banten, INDONESIA.
[2] Institute of Multidisciplinary Research in Advanced Materials (IMRAM), Tohoku University, Katahira Campus, Katahira 1-1-1, Sendai, JAPAN.

E-mail: yuliati.herbani@lipi.go.id

Abstract. This paper reports the synthesis of silver colloids by femtosecond laser ablation of ammonia-containing $AgNO_3$ solution. Effect of ammonia concentration in solution on the production of Ag nanoparticles was discussed. It is found that ammonia rules out significantly to the formation of Ag nanoparticles at which no Ag nanoparticle were formed in the solution without ammonia. Using the solution with the optimum ratio of ammonia to Ag^+ ions, we further investigate the growth process of Ag nanoparticle by monitoring the evolution of its absorption spectra at 402 nm as a function of irradiation time. The result showed that the growth process was fit to the simple exponential function, and confirmed that the addition of ammonia alone to the metal ion system can boost the particle production by femtosecond laser.

1. Introduction

These days the uses of metallic nanoparticles (NPs) in various regions of science and technology are expanding. Many efforts have been devoted to synthesize these materials and to study their optical properties, which strongly depend on the size and shape of the NPs [1]. The surface plasmon resonance (SPR) which consists in a resonance coupling between the electromagnetic field frequency and the free electrons oscillation frequency on the metal NPs is one of the important aspects of the study, at which a strong absorption of the electromagnetic field energy by metallic NPs happened when resonance conditions are matched. On account of these properties, Ag NPs have been generally utilized as antimicrobial agents [2-4], as biolabelling [5], as well as in the application of optical [6] and catalysis science [7]. Different biological [8, 9], chemical [10, 11] and physical [12-14] techniques have been utilized to get Ag NPs, among them is the pulse laser induced photoreduction [15] which is not fully explored.

Here, by using a femtosecond laser induced reduction of metal salts, Ag NPs were synthesized. Differs from pulse laser ablation of metal target in liquid (PLAL) [2, 14], here metal nanoparticles were obtained by photoreduction of metal salts during irradiation by femtosecond laser which is apparently a bottom up process. Photoreduction of metal salts was promoted in the abundantly presence of the hydrated electrons which were produced by Coulomb explosion of the solution due to interaction of intense femtosecond laser pulses with neat solvent [16]. Using this method, several noble metal nanoparticles [17-20] and their alloys [21-22] have been produced in our lab with good stability for months. In the previous report [18], Ag NPs was successfully fabricated at high concentration of Ag^+ ions. The limit concentration of Ag^+ ions was about 0.3 M; none of Ag nanoparticles were observed below this concentration. This fact has the important disadvantage in the production of its alloy with Au, for example, since it will lead to the formation of stable AgCl precipitates even before femtosecond laser irradiation was started. Therefore, we tried to use ammonia (NH_4OH) to modify the chemistry of Ag ions so that the reduction becomes easy. It was reported by D. Goia [23] that redox potential of the Ag ion changes as a result of complex formation. Ag and ammonia, for example, are easy to make a stable complex ion, $Ag(NH_3)^{2+}$, resulting from strong affinity of ammonia to Ag ions and its reduction potential is much higher than that of free ions, $E_0([Ag(NH_3)^{2+}, Ag^+) = +0.38\ V_{NHE}$. Therefore, the ammonia concentration and the nature of the reductant play a major role in the formation of Ag nanoparticles in aqueous solution. In this paper, the result of femtosecond laser irradiation of Ag nitrate solution in the presence of NH_4OH was reported.

2. Methods

Colloidal dispersion of Ag nanoparticles was synthesized by direct irradiation of silver salts using highly intense femtosecond laser pulses at room temperature in the presence of ammonia. All chemicals were used without further purification. Firstly, the solutions of silver ions with the concentration of 2.54×10^{-4} M were prepared by dissolving $AgNO_3$ (Sigma-Aldrich, +99.998%) in extra-pure water containing 0.01wt% of polyvinylpyrrolidone (PVP) as a dispersant. The fresh aqueous liquid ammonia (NH4OH, 25wt%, Sigma-Aldrich, 99.8%) in a variation volume of 1, 3 and 5 μl was added to the Ag^+-PVP solution. In laser irradiation experiment, the ammonia containing solution was introduced into a quarzt glass cuvette and irradiated for 30 minutes by the highly intense femtosecond laser pulses (Spitfire, Spectra-physics Co.) at the fundamental wavelength of 800 nm with the typical pulse width, repetition rate and maximum pulse energy per pulse were 100 fs, 100 Hz and 6 mJ, respectively. The laser beam was directed perpendicularly to the surface of the cuvette into the solutions after tightly focused using an aspheric lens with a focusing length of 8 mm (NA = 0.5). Taking no account for the aberration and considering the diameter of the laser beam before focusing was approximately 3.2 mm, the theoretical laser intensity is about 3.2×10^{18} W/cm^2 at the focal spot considering. This intensity is expected to exceed the optical breakdown threshold of any solvent.

After irradiation, samples were characterized by a UV-visible spectrometer (V630iRM, JASCO Co. Tokyo, Japan) to observe an absorption spectrum in 200 – 700 nm range and then confirm the nanoparticles formation by observing a surface plasmon resonance (SPR) of Ag nanoparticles around 400 nm. TEM experiments were performed using a JEOL2000-EXII (200kV, JEOL Ltd., Tokyo) to obtain the electron micrographs for all samples. Several drops of sample were placed on a carbon coated microgrid and left to dry in room temperature for this TEM observations. The bright field image of each sample was then processed using image analysis software Image J to measure the size of each nanoparticle. At least 500 particles were counted for each sample, taken from three separate areas on the microgrid.

3. Results and Discussion

Figure 1 shows the spectra of colloidal Ag nanoparticle prepared by femtosecond laser irradiation of 2.54×10^{-4} M $AgNO_3$ solution with different concentrations of ammonia for 30 minutes. The irradiation of 3-ml $AgNO_3$ containing 1 μl of 25 wt% ammonia results in the generation of a sharp-symmetric SPR peak at 402 nm which is originated from Ag nanoparticles. It is also confirmed by yellow color of the solution after irradiation. It can also be seen that the increase of ammonia content in solution increases the absorbance before it turns to decrease slightly after the ammonia content reach to 5 μl. This is in contrast with the result that the solution without ammonia which is still transparent after irradiation by femtosecond laser (figure 1 (b)), confirming that no particles were generated in the solution.

The corresponding TEM images of Ag colloids samples presented in figure 1 were shown in figure 2. No particle was found for the sample without ammonia. Using ammonia with the dosage of 1 μl per 3 ml $AgNO_3$ solution, Ag flakes with 100 nm width were generated. If the dosage increases to 3 μl, Ag nanoaparticle were observed and the average diameter was found to be 10.3 ± 8.5 nm with a broad size distribution. With 5 μl ammonia, large aggregates were also found besides small particles with size of 10 nm. This result indicated that the ammonia enhanced the formation rate of Ag nanoparticles, and hence the aggregates were generated due to the absence of dispersant. At this point, it was considered

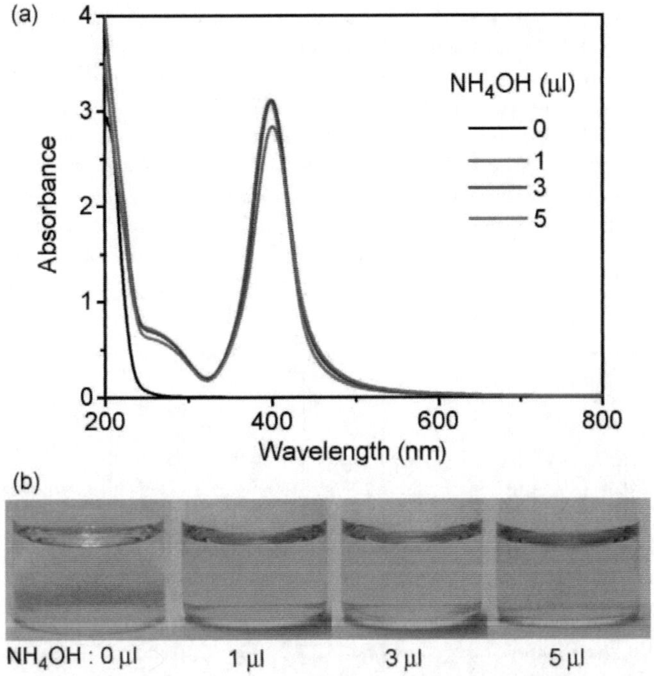

Figure 1. (a) UV-visible absorption spectra of the irradiated Ag ion solution with various concentrations of ammonia, and (b) the corresponding photographs of the solutions.

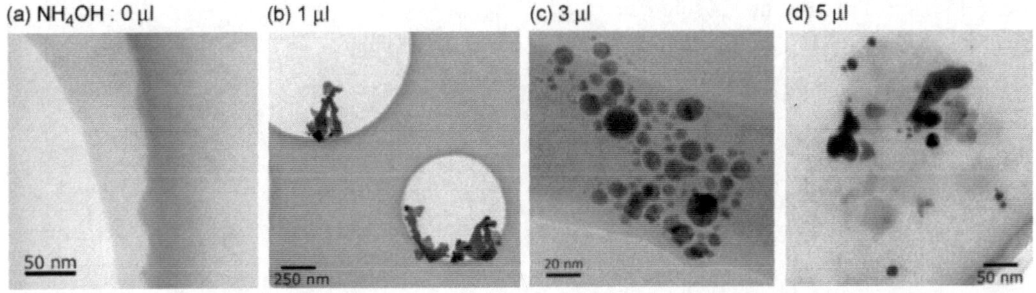

Figure 2. TEM images of the irradiated Ag ion solution for 30 minutes with various content of 25 wt% NH_4OH: (a) 0, (b) 1, (c) 3 and (d) 5 μl in every 3 ml of Ag ions.

that the dosage of 3 μl per 3 ml of $AgNO_3$ was optimum to produce Ag nanoparticles with reliable concentration.

For the monitoring of Ag nanoparticle formation, we followed the evolution of the absorption spectra at 402 nm during laser irradiation. The spectra were collected every 5 minutes for a typical experiment conducted at a solution with 3 μl ammonia. Figure 3(a) shows the UV-vis spectra of the solution every 5 minutes for a total of 30 minutes irradiation. As shown in the figure, the absorbance of Ag nanoparticles started from zero and increased rapidly at the beginning of the reaction without any significant induction time. The black arrow indicates the growth of the SPR at 402 nm, corresponding to the conversion of Ag ions to Ag nanoparticles. If the plasmon feature over the spectral range of 400 – 450 nm is magnified, the maximum plasmon absorbance blue shifts as the reaction upon the continued irradiation and then the absorbance remained stayed at the same absorbance level near the commencement of irradiation. The latter behaviour may be attributed to the inhibited Ag nanoparticle production due to the absorption of laser energy by the produced nanoparticles along the path to the focusing point or to Ag nanoparticle fragmentation by the laser. The maximum plasmon absorbance (peak position) as a function irradiation time is not cubic spline function like the formation of Au nanoparticles [24], but rather fit to the negative exponential function, indicating the formation of Ag nanoparticles goes quickly to completion in the presence of ammonia (figure 3(b)). Within 1 minute, a broad absorption band at 419 nm appeared and it smoothly blue-

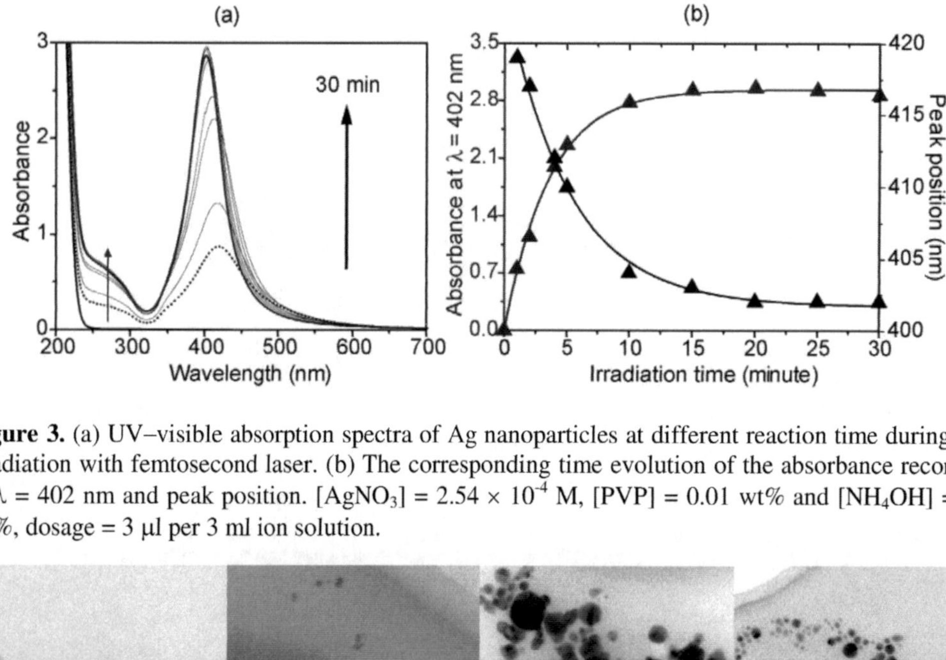

Figure 3. (a) UV–visible absorption spectra of Ag nanoparticles at different reaction time during the irradiation with femtosecond laser. (b) The corresponding time evolution of the absorbance recorded at λ = 402 nm and peak position. $[AgNO_3] = 2.54 \times 10^{-4}$ M, [PVP] = 0.01 wt% and $[NH_4OH]$ = 25 wt%, dosage = 3 μl per 3 ml ion solution.

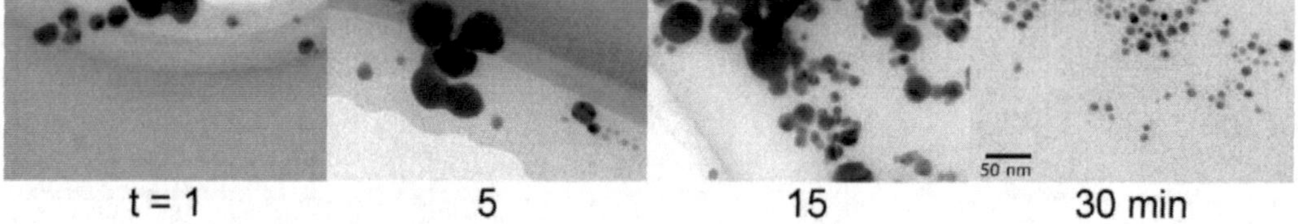

Figure 4. Comparison of TEM image of peak absorbance evolution of Au and Ag nanoparticles during the formation by femtosecond laser for 30 minute irradiation.

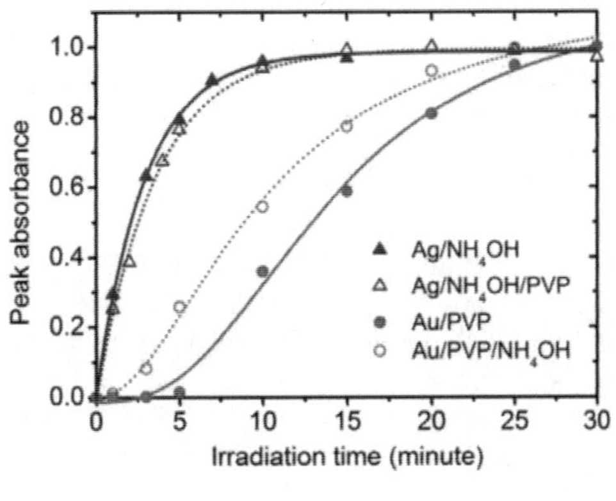

Figure 5. Comparison of peak absorbance evolution of Au and Ag nanoparticles during the formation by femtosecond laser for 30 minute irradiation.

shifted to a constant value of 402 nm with the irradiation time. It suggested that the particles size became smaller as the time increases, as confirmed in figure 4. The absorbance reached maximum approximately after 20 minutes of the irradiation. No significant change in this final value was observed afterward. In addition, the absorbance in the range of 250 – 280 nm increased with the irradiation time (blue arrow in figure 3(a)). The same absorption band has been observed for the reduction of Ag^+ ions in the solution with 2-propanol in alkaline environment [25] as well as during the radiation induced synthesis using synchrotron X-ray [26] and pulsed γ-ray [27]. The cluster Ag_4^{2+} has been identified as the species corresponding to the absorption at 250 – 280 nm [28]. Accordingly, it was assumed in this study that the absorption in the 250 – 280 nm range is due to Ag_4^{2+} clusters.

The evolution of absorption spectra of the Ag^+/NH_4OH without PVP during laser irradiation was also monitored (figure 5). In comparison to the one with PVP, the changes in absorbance with irradiation time for the system were investigated to study the global kinetics behaviour of Ag nanoparticle formation by femtosecond laser irradiation in the absence of additives. A key fact was that the reaction proceeds faster in the absence of PVP. This result confirmed that ammonia alone can boost the formation of Ag nanoparticles without need any dispersant. To test the result, we also conducted the same laser irradiation experiment for ammonia containing $[AuCl_3]^-$ ion solution. Its evolution curves of absorbance shown in figure 4 evinced that Au and Ag have very different reduction and clustering rates under the same reaction condition (blue open triangle for $Ag/NH_4OH/PVP$ and filled green circle for $Au/PVP/NH_4OH$). Even in the presence of PVP or any other additives, the kinetic process that governs the Au nanoparticles formation followed the sigmoidal kinetics, different from the kinetics for Ag nanoparticles which was simply exponential trend. It was also clear that ammonia promoted the formation of both nanoparticles, resulting in fast growth of nanoparticles. The nanoparticles might eventually be aggregated when there is no dispersant used in the system. Nevertheless, the reaction speed is a key factor for multimetallic nanoparticle synthesis, by which alloying process will be assisted, especially for Au and Ag with large difference in reduction potential.

4. Conclusions

We have, for the first time, demonstrated that ammonia addition to the Ag^+ ion solution may boost the production of Ag nanoparticle by femtosecond laser irradiation in the concentration below 0.3 M, at which the mixture of metal ions can be feasibly prepared without introduced any precipitates. The optimum volume is about 3 μl per 3ml of Ag^+ in water. In the presence of ammonia, the growth monitoring of Ag nanoparticle production showed that the process rapidly occurred to fit the expontential function without any significant elapsed time once the irradiation started, even for the system without dispersant. The relative importance of this result is expected to enable greater control over the properties of uncapped Ag nanoparticles using femtosecond laser induced photoreduction, making it possible to use the properties for a variety of application.

References

[1] Zhang J, Malicka J, Gryczynski I and Lakowicz J R 2005 *J. Phys. Chem. B* **109** 167643
[2] Dolgaev S, Simakin A and Voronov V 2002 *Appl. Surf. Sci.* **186** 546
[3] Ravichandran V, Vasanthu S, Shalini S, Shah S A A and Harish R 2016 *Mater. Lett.* **180** 264
[4] Priya Dhas S, Mukerjhee A and Chandrasekaran N 2013 *Int. J. Pharm. Pharm. Sci.* **5** 349
[5] Syrvatka V J, Slyvchuk Y I, Rozgini I I, Gevka I I, Bilyy O I, Osypchuk O S and Zyuzyun A B 2013 *Proc. SPIE – The International Society of Optical Engineering* (San Diego, CA, United States) 8812 88120I
[6] Shao B, Yang Z, Wang Y, Li J, Yang J, Qiu J and Song Z 2015 *ACS Appl. Mater. Interfaces* **7** 25211
[7] Guo J -Z, Cui H, Zhou W, Wang W 2008 *J. Photochem. Photobiol. A Chem.* **193** 89
[8] Dura´n N, Marcato P D, Alves O L, Souza G. IH De and Esposito E 2005 *J. Nanobiotechnol.* **3** (8) 1
[9] Ahmad A, Mukherjee P, Senapati S, Mandal D, Khan M I, Kumar R, Sastry M 2003 *Colloids Surf. B Biointerfaces* **28** 313
[10] Luo Y, Shen S, Luo J, Wang X and Sun R 2015 *Nanoscale* **72** 690
[11] Li X, Shen J, Su A, Zhang Z, Gao G, Yang H, Wu J 2012 *Colloids Surf. A Physicochem. Eng. Aspect* **400** 73
[12] El-Bialy A B, Dawy M, Abd El-Mongy S, Shahin N A M, Eusa W H and Abd El-Aziz Mohamed E 2016 *Res. J. Pharm. Biol. Chem. Sci.* **7** 545
[13] Gu Y, Li N, Gao M, Wang Z, Xiao D, Li Y, Jia H and He H 2015 *Microchim. Acta* **182** 1255
[14] Mohamed H K, Naser M G, Sadrolhosseini Ar, Dehzangi A, Kamalianfar A, Saion E B, Zamiri R, Alhangar H A and Majlis B Y 2014 *Chin. Phys. Lett.* **31** 077803
[15] Nakamura T, Mochidzuki Y and Sato S 2008 *J. Mater. Res.* **23** 968
[16] Pommeret S, Gobert F, Mostafavi M, Lampre L and Mialocq J C 2001 *J. Phys. Chem. A* **105** 11400
[17] Nakamura T, Magara H, Herbani Y and Sato S 2011 *Appl. Phys. A Mater. Sci. Process.* **104** 1021
[18] Sarker M S I, Nakamura T and Sato S 2014 *J. Mater. Res.* **29** 856
[19] Nakamura T, Herbani Y and Sato S 2011 *Proc. Conf. on Lasers & Electro-Optics Europe & 12th European Quantum Electronic Conference CLEO EUROPE/EQEC* (US: IEEE Xplore) 12237041
[20] Chau JLH, Chen CY and Yang CC 2017 *Arab. J. Chem.* **10**(S1) S1395-S1401
[21] Herbani Y, Nakamura T and Sato S 2011 *J. Phys. Chem. C* **115** 21592.
[22] Nakamura T, Herbani Y and Sato S 2012 *J. Nanopart. Res.* **14** 785.
[23] Goia D (2004) *J. Mater. Chem.* **14** 451.
[24] Sarker M S I, Nakamura T, Herbani Y and Sato S 2013 *Appl. Phys. A* **110.**
[25] Tibbets K M, Tangeysh B, Oghner J H and Levis R J 2016 *J. Phys. Chem. A* **120** 3562.
[26] Huang Z Y, Mills G and Hajek B 1993 *J. Phys.Chem.* **97** 11542.
[27] Wang C H, Liu C J, Wang C L, Chien C C, Hwu Y, Liu R S, Yang C S, Je J H, Lin H M and Margarintondo G 2009 *Appl. Phys.* **97** 295.
[28] Henglein A and Tausch-Treml R 1981 *J. Colloid Interface Sci.* **80** 84.

IOP Conf. Series: Journal of Physics: Conf. Series **817** (2017) 012049 doi:10.1088/1742-6596/817/1/012049

Laser-induced photo emission detection: data acquisition based on light intensity counting

N Yulianto, N Yudasari and K Y Putri

Research Center for Physics, Indonesian Institute of Sciences (LIPI), Kawasan PUSPIPTEK Serpong Building 442 Tangerang Selatan, Banten, Indonesia

E-mail: nursidik.yulianto@lipi.go.id

Abstract. Laser Induced Breakdown Detection (LIBD) is one of the quantification techniques for colloids. There are two ways of detection in LIBD: optical detection and acoustic detection. LIBD is based on the detection of plasma emission due to the interaction between particle and laser beam. In this research, the changing of light intensity during plasma formations was detected by a photodiode sensor. A photo emission data acquisition system was built to collect and transform them into digital counts. The real-time system used data acquisition device National Instrument DAQ 6009 and LABVIEW software. The system has been tested on distilled water and tap water samples. The result showed 99.8% accuracy by using counting technique in comparison to the acoustic detection with sample rate of 10 Hz, thus the acquisition system can be applied as an alternative method to the existing LIBD acquisition system.

1. Introduction

Water is a very important element in all aspects of human life, such as for sanitation, drinking, animal husbandry, farming, industry and others. However, the growth of chemical industry has enhanced the water contamination possibility by chemical waste. Some technique has been developed to monitor the water quality by observing micro bacteria, contaminant, and dangerous particulates in water, such as the Photon Correlation Spectroscopy (PCS) technique. The methods could measure colloids whose sizes range between 0.001 and 3μm [1, 2] without requiring calibration. Nowadays, pollutants like pesticides, radioactive elements and viruses are present in groundwater as small colloids; thus need a more sensitive particle counter which can measure the size of particles in nanometer range to detect and prevent the possibility of poisoning [3].

In 2001, T. Bundschuh identified water content by Laser Induced Breakdown detection (LIBD) technique [4]. LIBD is a quantification technique for detecting aquatic colloids of very small sizes (10–1000 nm) at very low concentrations (in the range of 1 ng/L (ppt) to mg/L (ppm)) [4, 5]. The main principle of LIBD technique is the interaction between pulsed laser beam and colloids that causes the occurrence of Photo Acoustic Emission (PAE) from which the optical and acoustic effect in liquid can be measured [6]. Usually, LIBD technique uses CCD camera for image sensor [5, 8] or piezoelectric as a shockwave sensor [7] and they are supported by high speed data acquisition devices. In this paper, we offer a new alternative method for developing a real time data acquisition system to measure the change of the intensity which is caused by plasma formation in the water sample for LIBD application.

IOP Conf. Series: Journal of Physics: Conf. Series **817** (2017) 012049 doi:10.1088/1742-6596/817/1/012049

2. Methods

During LIBD experiment, the interaction between high energy pulsed laser and small-sized colloids can be analyzed to determine the size of colloids. Such interaction results in a breakdown phenomenon from which the transformation of plasma and shockwave can be observed to measure the size of the particle. When the laser energy interacts with the colloidal, first, one atom was ionized by Multiphoton Ionization (MPI). As the result, emerging electrons are subsequently accelerated due to inverse bremsstrahlung in the electric field of high power pulsed laser. When the energy is high enough, other atoms are ionized and the number of electrons release in double number or more. This phenomenon occurs repeatedly and increases the density of the free charge carriers; hence the plasma emission can be obtained. This plasma formation is called breakdown event [8, 9]. Detail experiment set up can be seen in figure 1.

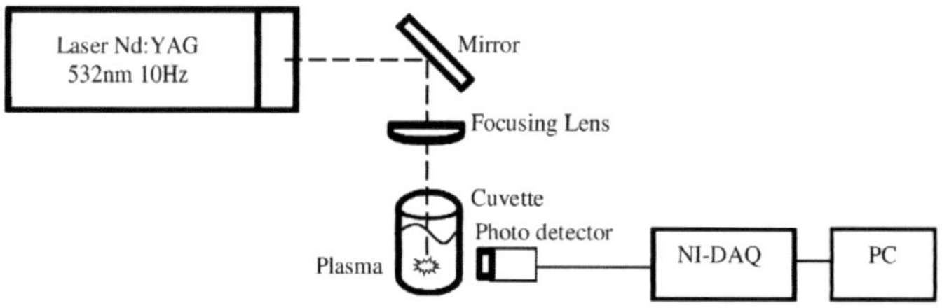

Figure 1. Schematic experimental characterization of nanomaterials in solution using LIBD techniques.

The experiment was conducted by using high energy pulsed Nd:YAG (Neodymium-doped Yttrium Aluminum Garnet) laser with wavelength of 532 nm and pulse frequency of 10 Hz as shown in figure 1. The laser beam was focused into a cuvette, with sample inside, for 10 seconds (1000 pulses). Interaction between laser beam and nanoparticles in the water generated breakdown events in the form of plasma emissions. Enhancement of light intensity was also noticed. The Breakdown Probability (BDP) in this experiment corresponded to the probability of finding plasma within certain laser energy. BDP corresponded to the concentrations of colloids and the size of nanoparticles [9, 10] and the formula was defined as follow:

$$BDP = \frac{Counting\, number of\ breakdown}{Total\, number of\ laser\ shots} \tag{1}$$

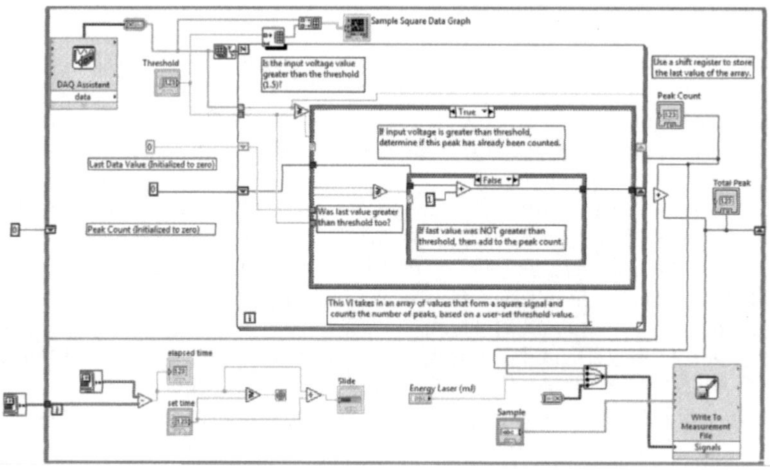

Figure 2. Schematic program photo-counting detection by LABVIEW.

Photodetector captured the optical signal when breakdown event occurred. The light intensity was rising at the moment the plasma occurred and the data acquisition counted it as an analog signal. By using National Instrument device NI DAQ USB-6009 and LABVIEW software, the analog data was converted to a digital signal processed by PC. In this experiment, we built the digital counting program for data acquisition system. The schematic program can be seen in figure 2.

The conversion of analog signal into digital signal by LABVIEW software and NI DAQ 6009 proceeded as following. When voltage increased, which was caused by the change of intensity, the digital signal also increased and reached the peak signal. When the peak reached above the threshold voltage, the signal was counted and saved into PC for every 0.5 second as shown in figure 2. We assumed that each peak signal corresponded to the plasma emission when breakdown occurred.

3. Result and discussion

To check the reliability of the data acquisition system, first, an experiment has been set up to verify its accuracy and stability. The characterization was performed in the dark room and the condition was controlled to resemble LIBD system with 10 Hz light pulse.

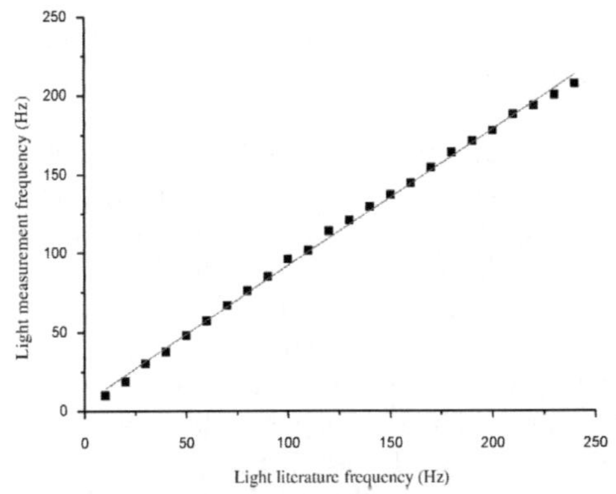

Figure 3. Data acquisition system accuracy for pulsed light with frequency of 1-250 Hz.

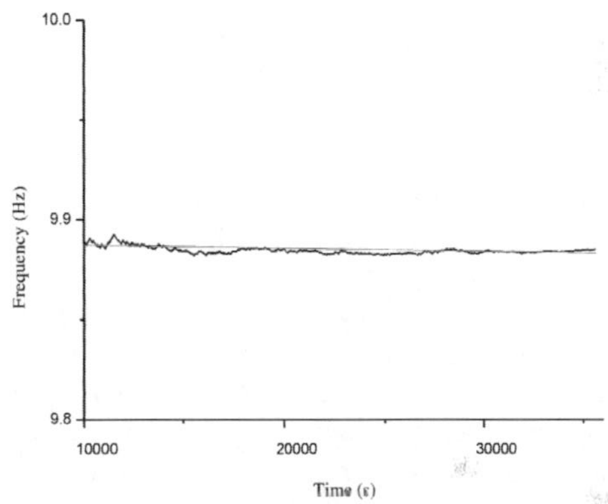

Figure 4. Data acquisition system stability for pulsed light with frequency of10 Hz.

The accuracy of data acquisition system and sensor has been tested by using pulsed light with frequency of 1-250 Hz. As shown in figure 3, the data slope of sensor and data acquisition accuracy is 0.86. The counting stability of data acquisition system using 10 Hz pulsed light was 9.88±0.1 Hz for one hour acquisition, as shown in figure 4. The accuracy and stability were good enough to apply this method into LIBD system. By using schematic in figure 1, we obtained the result application of data acquisition system as shown in figure 5 and figure 6. For the samples, tap water and polystyrene colloids (20 nm) were used as random samples to measure the size of nanoparticles and filtered water were used as control sample (table 1).

Table 1. Data acquisition and sensor result from 3 different samples.

Sample	Threshold Energy (mJ)	Max. BDP at Energy 4.5 mJ
Filtered water	1.5	0.9
Tap water	0.5	1.1
Polystyrene 20nm	<0.5	1.25

(a)

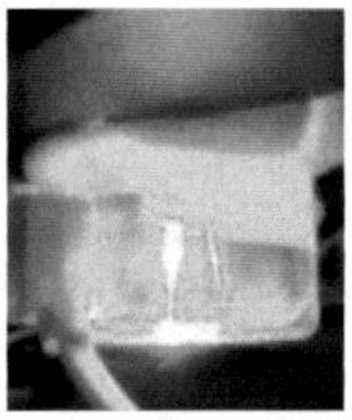

(b)

Figure 5. (a) Breakdown occurred. (b) No breakdown occurred.

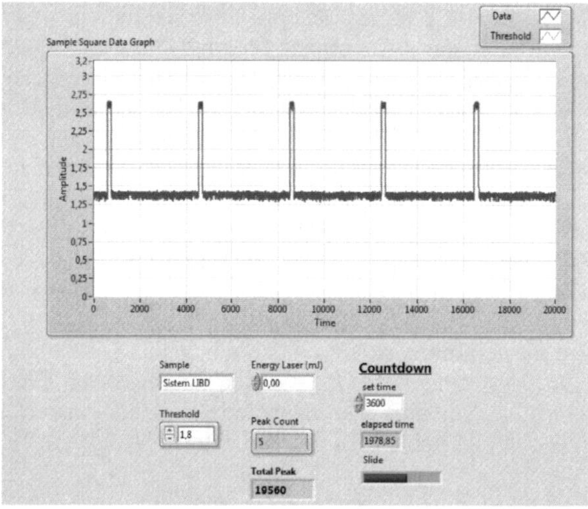

Figure 6. Front panel of LABVIEW program for counting breakdown.

Photodiode responded the increasing intensity when the breakdown occurred and sent the analog signal to ADC in DAQ. The increasing of light intensity as an analog signal was converted into digital signal and read as a peak signal when it was much higher than threshold signal. Then it would be measured as a digital counting as shown in figure 5 and 6. In the condition where there was no breakdown occurred, the peak was not higher than the threshold signal, so the digital count was zero. Voltage background corresponded to the condition of light environment. To ensure that the data acquisition system and sensor could perform well, the experiment of the LIBD data processing based on the light intensity was conducted. The laser frequency was 10 Hz, the number of laser pulses was set to 1000, and the duration of each measurement was 100 seconds. Hence, the range of the signal could be processed for the three samples: tap water, filtered water and 20 nm polystyrene. The final results of data processing in the form of breakdown probability for each variation of pulsed laser energy up to 4.5 mJ can be seen in figure 7.

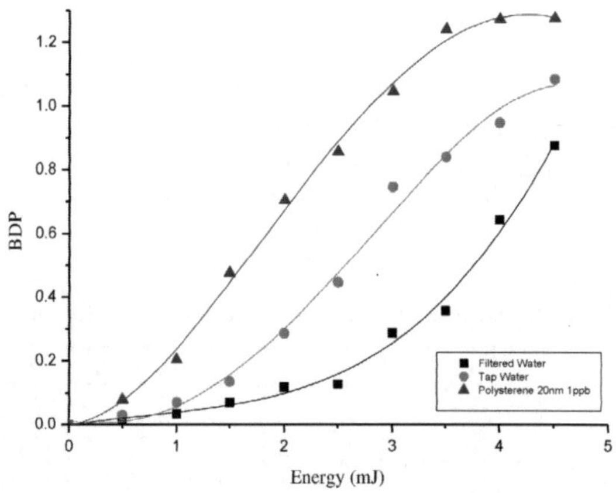

Figure 7. Validation of system data acquisition and sensor in LIBD application.

As seen in figure 7, filtered water had highest breakdown threshold value (when BDP was 0.01) among all samples, followed by colloids in tap water and 20 nm polystyrene, which agreed with the literature [4, 8]. It means that the greater the size of the colloid, the smaller the energy needed to generate breakdown. Colloids in tap water are natural colloids, which generally has average size below 100 nm [11]. As seen in table 1, by using data acquisition and sensor based on light intensity counting, the breakdown probability of 20 nm polystyrene nanoparticle at 4.5 mJ was higher than tap water or filtered water. Therefore, it can be deducted that the size of the tap water colloids are smaller than 20 nm. From table 1 and figure 7, we can also see that the BDP for polystyrene particles or colloids with greater size has BDP more than 1. It happened as the multiple breakdowns occurred or light decay took place in such a long time with intensity higher than the threshold voltage, therefore causing the photodiode to detect it as multiple counting.

4. Conclusion

The data acquisition based on light intensity counting has been built for LIBD application using LABVIEW, DAQ USB NI-6009 and photodiode as the sensor. The light intensity rose as the result of breakdown generation, and then it was counted by photodiode as a digital signal. The frequency characterization gave a slope of 0.86 and counting stability of data acquisition system by 10 Hz pulsed light returned a value of 9.88 ± 0.1 Hz. The counting of breakdown corresponded to the BDP from each sample. From the experiment, it was proven that the size of 20 nm polystyrene colloids was larger than colloids in filtered water. The data acquisition system and the sensor for LIBD application based on light intensity counting can be applied as an alternative technique to the existing LIBD acquisition system.

References

[1] Rees T F 1990 *Water Resour. Res.* **26**(11) 2777
[2] Tscharnuter W 2000 Photon correlation spectroscopy in particle sizing *Encyclopedia of Analytical Chemistry*, ed R A Meyers (Chichester: John Wiley & Sons) p 5469
[3] van Beek C G, de Zwart A H, Balemans M, Kooiman J W, van Rosmalen C, Timmer H, Vandersluys J and Stuyfzand P J 2010 *Water Res.* **44**(3) 868
[4] Bundchuh T, Knopp R and Kim J I 2001 *Colloids Surf. A* **177** 47
[5] Kim J W, Son J A, Yun J-I and Choi J G 2008 *Chem. Phys. Lett.* **462**(1) 75
[6] Knopp R, Scherbaum F J and Kim J I 1996 *TU München Scientific Report* (Germany: Institut für Radiochemie) RCM 00696
[7] Ravishankar S R and Jones B E 2007 Photo-acoustic emission measurements in liquid-based food and aqueous products *NDT.net* **12**(2)
[8] Kim J-I and Walther C 2006 Laser-induced breakdown detection *Environmental Colloids and Particles: Behaviour, Separation and Characterisation* vol 10 eds K J Wilkinson and J R Lead (Chichester: John Wiley & Sons) chapter 12 pp 555-612
[9] Scherbaum F J, Knopp R and Kim J I 1996 *Appl. Phys. B* **63** 299
[10] Walther C, Bitea C, Hauser W, Kim J I and Scherbaum F J 2002 *Nucl. Instr. and Meth. in Phys. Res. B* **195** 374
[11] Kim M A, Panak P J, Yun J-I, Kim J I, Klenze R and Köhler K 2003 *Colloids Surf. A* **216** 97

IOP Conf. Series: Journal of Physics: Conf. Series **817** (2017) 012050 doi:10.1088/1742-6596/817/1/012050

Design of micro bending deformer for optical fiber weight sensor

R K Ula, D Hanto, T B Waluyo, H Adinanta and B Widiyatmoko

Research Center for Physics, Indonesia Institute of Sciences, Building 440, PUSPIPTEK South Tangerang, Indonesia

Email : rini011@lipi.go.id

Abstract. The road damage due to excessive load is one of the causes of accidents on the road. A device to measure weight of the passing vehicles needs to be planted in the road structure. Thus, a weight sensor for the passing vehicles is required. In this study, we designed a weight sensor for a static load based on a power loss due to a micro bending on the optical fiber flanked on a board. The following main components are used i.e. LED 1310 nm as a light source, a multimode fiber optic as a transmission media and a power meter for measuring power loss. This works focuses on obtaining a suitable deformer design for weight sensor. Experimental results show that deformer design with 1.5 mm single side has level of accuracy as 4.32% while the design with 1.5 mm double side has level of accuracy as 98.77%. Increasing deformer length to 2.5 mm gives 71.18% level of accuracy for single side, and 76.94% level of accuracy for double side. Micro bending design with 1.5 mm double side has a high sensitivity and it is also capable of measuring load up to 100 kg. The sensor designed has been tested for measuring the weight of motor cycle, and it can be upgraded for measuring heavy vehicles.

1. Introduction

Many accidents in highway were caused by road damage due to overload vehicles. Although the government has provided weighbridge in some area, the drivers are generally reluctant to measure the weight of their vehicle. Most drivers said that it was time consuming and full of birocracy, so they tend to avoid such a measurement or to cheat whenever they have to get in the weighbridge. Therefore, we need a sensor that can detect a dynamic load with a high sensitivity/fast response so the drivers do not have to stop their vehicles, resistant to water and electromagnetic waves, low cost and capable to be planted in highway structure.

Currently, the weight sensor based optical fiber is one of the alternatives to measure load [1]. Micro bending fiber has been widely used for displacement sensor, pressure and temperature [2]. Such an optics-based sensor has a sufficiently small size, high sensitivity and not interference by electromagnetic waves [3]. Hence, the sensor is quite safe to be planted in highway structures so it will facilitate the process of measurement and avoid a chance for the driver to cheat. Setiono (2012) has developed a weight sensor with optical fiber models with dual side deformer made of wood and LASER 1310 nm as a light source to measure the weight of a motorbike [4]. However, this design has not been developed to a higher load.

Research Center for Physics has developed the weight sensor to detect the load up to 8 tons. In our previous studies, it was found that a 1310 nm LED as a light sources has a high validity to be integrated with a weight sensor and resulted in measuring the load up to 100 kg with 99.1% accuracy[5]. Some important points in the micro bending sensor were a design of deformer, bending period and the type of fiber[6]. Therefore, in this study we used a multimode fiber optic with variation in the deformer i.e. single side and double side (sandwich). The purpose of this research was to get the design of deformer that was suitable to integrate with a weight sensor.

1.1. Optical Fiber and Weight Sensor by Micro bending

Optical fiber is a dielectric waveguide where the core material has a refractive index greater than the cladding. The optical fiber consists of a core to send signal optical data from the light source to the receiver, cladding which is a material surrounding the core that reflects the light into the core, and a buffer coating which is a plastic coating that protects the fiber from damage. Plastic coating around the core and cladding does not only strengthen the fiber core, but also helps the absorption and acts as an extra shield in bending the cable. When the incident angle of light is less than or equal to the acceptance angle, there will be a total reflection in the fiber core, where energy will be transmitted so that any losses larger than an optical power can be avoided.

Micro bending is the bending of an optical fiber in micro-size and affects the optical power loss. This bending is caused by mechanical stress on the optical fiber in the axial direction. If the optical fiber is bent in axial direction of the optical axis, the transmission coefficient of the light will change with a magnitude as the following equation[7]:

$$\Delta T = \frac{\Delta T}{\Delta X} A_p k_f^{-1} \Delta P \quad (1)$$

with $\frac{\Delta T}{\Delta X}$ are coefficient which relates to the change in transmission of light in a deformed fiber, A_p is the deformer area, and ΔP are the changes in pressure applied to the sensor.

Several factors which affect the change of transmission coefficient are the force applied by the vehicle load, bending the period and the characteristics of the optical fiber. Bending period in a step index optical fiber was determined by the equation[1]:

$$SI = \frac{\sqrt{2}\pi R_{core}\, n_1}{NA} \quad (2)$$

The change in the transmission coefficient of the optical wave due to bending period can be analyzed by equation:

$$\Delta T = \frac{y\, \Delta F\, A\, \Lambda^3}{3\pi\, Y\, R_{core}\, \eta} \quad (3)$$

with $y\Delta T$ is the attenuation coefficient of fiber, ΔF is the change in load, A is a cross-sectional area of fiber bent, $\Lambda\Lambda$ is the bending period, Y is modulus Young of core material, R_{core} is radius core fiber, $n_1$is refractive index of core, η is the number of intervals bending, ΔP is the change in pressure due to the load, NA is numerical aperture.

Transmission of light in an optical fiber would result in attenuation. It is caused weakening of energy so that the amplitude of wave would arrive at receiver smaller than those sent by the transmitter [8].

At figure 1, the deformer consists of two grooved plates and the optical fiber which is sandwiched between the plates. When the bend radius exceeds the critical angle, the light begins to leak out of the cladding resulted in intensity modulation. When a small bend is applied on the fiber, a certain portion of the propagation of light in the fiber core are lost and then incorporated in the radiation mode. A

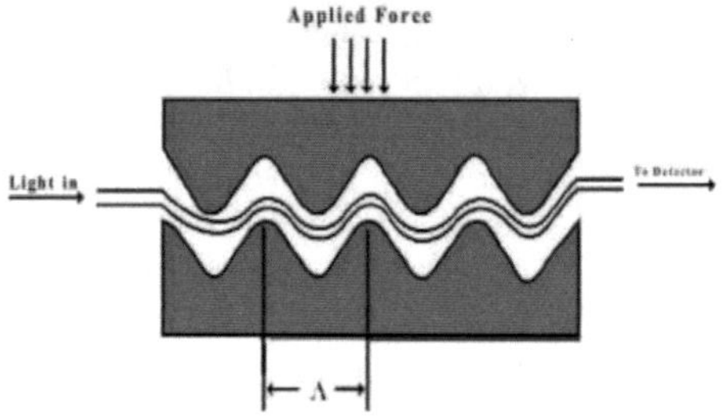

Figure 1. Microbending optical fiber.

merger mode can be achieved by using a corrugated plate that changes the fiber shape into a series of bends. Therefore, micro bending causes the decrease in the light intensity.

Light emitting diode (LED) is a type of diode that can emit light. It is often used as a light source as well as an information carrier. In operation, the LED has a feature that does not require stabilizers for optical circuits and emits light with a little temporal and spatial coherence. That means LED light has a narrow frequency spectrum. LEDs have many advantages for its low power and long lasting.

2. Experimental

In this study, we used a step-index multimode optical fiber with a wavelength range of 850 - 2400 nm, with a radius core of 50μm, a refractive index core of 1.43 and a numerical aperture (NA) of 0.22. The light sensor was an 1310 nm LED. The length of the sensor was 50 cm and an optical power meter was used to detect the transmitted light from the fiber. Using equation (2), we can obtain the bending period of optical fiber. This study aimed to compare the performance of weight sensor with two types of a bending period i.e. (1.5 ± 1) and (2.5 ± 1) mm, both in the single side and the double side configuration as shown in figure 2. The thickness of the deformer plate was ± 1.5 cm.

In a typical experiment, the light from LED propagated on the optical fiber that micro-bended by the deformer, as shown in figure 3. The other end of the fiber was connected to an optical power meter. Initial optical power, P_o, was a power when the sensor in its initial condition with no load and the output optical power was the power when the sensor was being pressed by a load. The difference of both was the optical power loss that we were interested here.

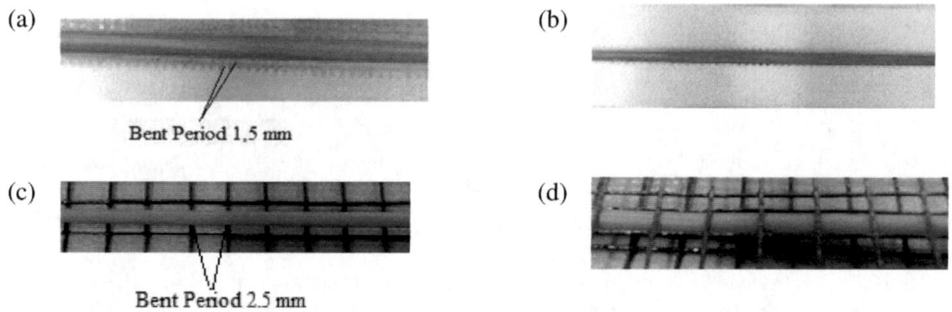

Figure 2. Design of deformer with a bending period of 1.5 (a-b) and 2.5 mm (c-d). Both in the configuration deformer of single (a, c) and double side (b, d).

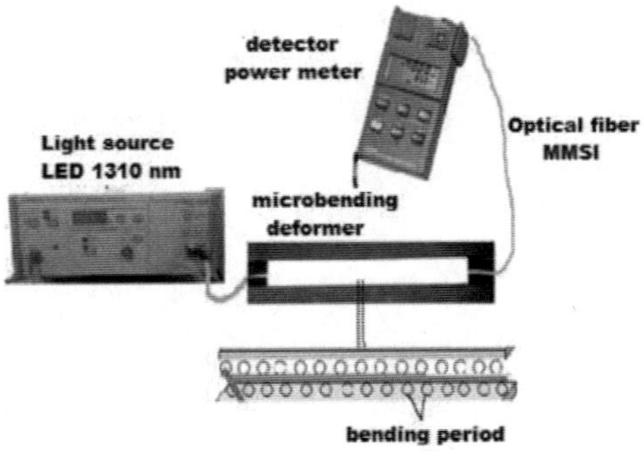

Figure 3. Set up experiment design of micro-bending sensor.

3. Result and Discussion

According to Saleh et.al [9], the change of transmission coefficient can be represented by optical power loss. A good sensor required a linearity correlation between variable control and variable response [10]. As implied by equation (1) and (2), the changes of transmission coefficient for the sensor should be linear to load. Our experimental results for a load of 10 - 100 kg were shown in figure 4.

For the deformer with a bending period of 1.5 mm in the single side configuration, we found that response profile of the optical power loss was not linear but tends to be constant. With an accuracy of 4.3% and the standard error of 95.7%, this type of sensor was not fulfill the criteria as established by the equation (1) and (2), and hence it is not suitable for weight sensor design. In the other hand, for the deformer with a bending period of 1.5 mm in a double side configuration, the linearity was good and the accuracy 98.77% and a standard error of 1.23% were estimated.

Loss of optical power at the sensor with the bending period of 2.5 mm in single side configuration was shown in the graph C at figure 5 for a load of 10 - 100 kg. The linearity was only found when the sensor was pressed by a load of 10-60 kg, but not for a load of 70 - 100 kg. Overload pressure of 70 100 kg resulted in a decrease optical power loss. It might be because of the flattering of the optical fiber after pressed by a load of 10-60 kg. Flattered fiber optic made wave guidance in fiber not perfect, the leakage of optical power in fiber and hence, caused the signal attenuation. The level of accuracy and standard error measurement by were 71.18 and 28.82%, respectively. From these data, it can be stated that the type of weight sensor was only able to detect a load less than 60 kg, not for the load exceeds 60 kg.

Graph D at figure 4 shows the experimental results for the weight sensor using a deformer with a bending period of 2.5 mm in double side configuration. It is clear that the sensor was unstable to measure the load. Loading of 10 - 40 kg and 70 - 100 kg showed a linear increase in power loss, but when it was pressed by a load of 50 - 60 kg, the graph profile became volatile. Level measurement accuracy with double deformer side was 76.94% and the standard error was 23.06% standard error.

It can be concluded that the weight sensor using a deformer with a smaller bending period in a double side configuration was the best design in our case. The reason is that in a single side configuration, the optical fiber was bent only in one side and the produced arch disordered the fiber resulting in a disturbance of light propagation inside the fiber. Hence, only a weakened optical signal with a high power loss was measured. In the contrast, if the optical fiber was bent in a double side configuration, the series of arch were equally formed both side imitated a tooth pattern. In this kind of configuration, the optical signal was repeatedly weakened and strengthened many times during its propagation in the sensor, so the optical power loss due to load pressure was relatively significant. A smaller bending period was preferable, making sure a high number of arches produced in the fiber.

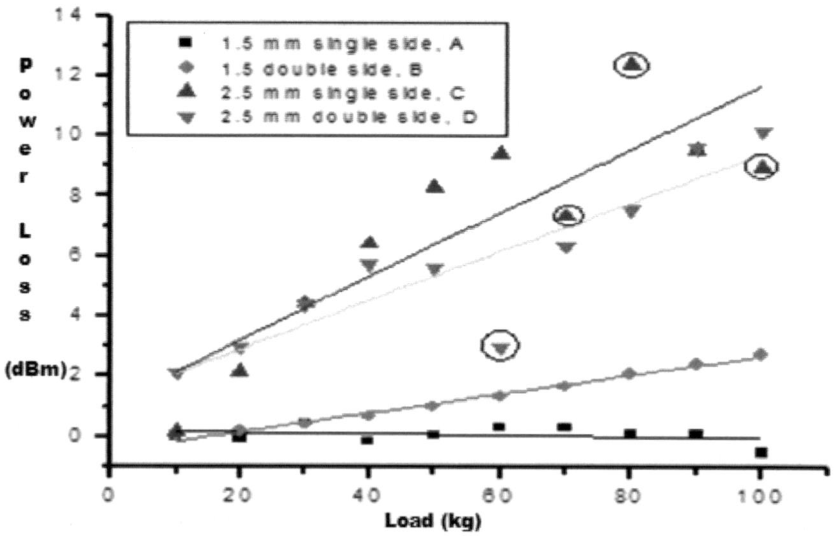

Figure 4. Power loss measured as a function of applied load for all configuration of weight sensor.

Optical fiber multimode has a mode dispersion more than a single mode optical fiber [8], so that the time needed by the light for going out of the fiber would be much longer than the one for a single mode. However, this work showed that the multimode fiber weight sensor with a standard error of 1.23% have been successfully developed to measure the load of 10 – 100 kg if using a deformer with a small bending period in a double side configuration. This design can be prospectively applied as a weight sensor for a two-wheeled motor vehicle with hundreds weight. This research will be developed further in order to measure loads up to 8 ton.

4. Conclusion

We have successfully developed a weight sensor that can measure a static load of 10-100 kg with the accuracy level of 98.77% using a deformer with a bending period of 1.5 mm in double side configuration. The light source used was a low-cost LED with a wavelength of 1310 nm. Further, it is significantly needed to do the next experiment to test the design with a load more than 100 kg.

Acknowledgement

Thanks to our colleagues in group optoelectronics, Research Center for Physics LIPI for feedback and discussion with the authors. We also thank for the support from the Head of Deputy of Science and Engineering LIPI in terms of Kegiatan Unggulan 2016.

Reference

[1] Bin M and Xinguo Z 2010 *Int. Conf. on Intelligent Computation Technology and Attenuation* vol 1 (USA: IEEE) 458-61

[2] Lau D, Chen Z, Member S, Teo J T, Ng S H, Rumpel H, Lian Y, Yang H and Kei P L 2013 IEEE *Transaction on Biomedical Eng.* **60(9)** 2655–62

[3] Puranto P, Hanto D, Setiono A, Ula R K and Widiyatmoko B 2016 *AIP Conf. Proc.* vol 1746 (USA: AIP Publishing) 020009

[4] Setiono, Andi W B 2012 *Telaah* **30**(2) 33–6

[5] Ula R K, Hanto D and Widiyatmoko B 2015 *Prosiding Seminar Nasional Fisika 2015 (Surabaya)* (Surabaya: Program Studi Fisika Fakultas Matematika dan Ilmu Pengetahuan

Alam Universitas Negeri Surabaya) pp 323–326

[6] Ula R K., Hanto D, Regina H Y 2016 *Prosiding Seminar Nasional Fisika dan Pembelajaranya (Malang)* (Malang: Program Studi Fisika Fakultas Matematika dan Ilmu Pengetahuan Alam Universitas Negeri Malang) pp 248–254

[7] Lagakos N, Cole J H and Bucaro J A 1987 *App. Optic.* **26**(11) 2171–80

[8] Keiser G 2000 *Optical Fiber Communication* (Singapore: Mc Graw Hill Int.) pp 51-55

[9] Saleh B E A and Teich M C 1991 *Fundamental Photonics* (USA: Wiley) pp 272-306

[10] Efendioglu H S, Yildirim T, Toker O and Fidanboylu K 2013 *Opt. Fiber Technol.* **19**(4) 279–84

IOP Conf. Series: Journal of Physics: Conf. Series **817** (2017) 012051 doi:10.1088/1742-6596/817/1/012051

Study on acoustic detections of Nd:YAG laser induced breakdown at different wavelengths

K Y Putri, N Yulianto and Y Herbani

Research Center for Physics, Indonesian Institute of Sciences, Bd. 442 Puspiptek Area Serpong, South Tangerang 15314, Banten, Indonesia

Email: kirana.yuniati.putri@lipi.go.id

Abstract. Laser-induced breakdown detection (LIBD) is among many techniques utilizing laser beam to characterize nanoparticles in solution. It is based on the detection of plasma formation resulted from the interaction between particles and laser beam. In acoustic LIBD, the acoustic signal which accompanies the plasma formation is detected by a microphone. In this study, a 1064 nm pulsed laser was used as the laser source for LIBD system instead of the frequently used 532 nm. The presence of colloids in filtered and unfiltered water was investigated using the new system and the result was compared with those from conventional system. The S-curves from 1064 nm system showed similar trend with S-curves from 532 nm system. The presence of larger nanoparticles in unfiltered water resulted in lower breakdown threshold energy. However, the breakdown threshold energy of both filtered and unfiltered water was much higher in the 1064 nm LIBD system. The shifting was attributed to the high absorbance of water at 1064 nm. This research proved the possibility of using 1064 nm pulsed Nd:YAG laser as an alternative wavelength source for LIBD system, nevertheless various setup arrangement might be needed.

1. Introduction

In recent years, nanotechnology is getting a bigger role in science. Smaller nanoparticles are being synthesized. Along with this trend, more and more advance characterization techniques for nanoparticles are being developed. Laser-induced breakdown detection (LIBD) is one of them. Being able to detect and quantify small colloids in low concentration, LIBD is known to be more sensitive compared to other quantification techniques [1,2]. LIBD also needs no special treatment for the preparation step, i.e. a liquid sample can be measured as it is [3].

LIBD based its detection technique on the plasma generation resulted from interaction between a focused, pulsed laser beam and nanoparticle in solution [4-7]. Laser irradiation induces the so-called breakdown of the dielectric properties of a given medium [8-10]. Breakdown threshold, which is the power density needed to induce breakdown, is lowest for solid, higher for liquid, and highest for gases [8,11]. This difference is used to detect solid nanoparticles presences in solution. Breakdown threshold also differs from one particle's size to another. Smaller nanoparticles needs higher power density for a breakdown to occur; therefore, it has a higher breakdown threshold [7]. The presence of more nanoparticles also results in higher breakdown probability, which is the statistical evaluation of breakdown occurrences [7].

Common setup for LIBD involves a 532 nm Nd:YAG laser as a pulsed beam source, a focusing system, a cuvette filled with liquid sample, a detector, and a data acquisition and processing system.

There are two types of breakdown detection, acoustic detection and optical detection [12]. In acoustic detection, a microphone is used to detect the acoustic signal generated along with plasma formation, whereas a charge-coupled device (CCD) camera is used in optical detection to capture the image of the plasma itself. Other types of pulsed laser source have been used in some LIBD experiments, such as a 500 nm excimer-dye laser and a Q-switched 532 nm diode-pumped solid state (DPSS) laser [13]. However, there is still not much information on whether LIBD technique can be performed at different wavelength. An Nd:YAG laser has a 1064 nm wavelength at its fundamental and there are various experiments carried out at this wavelength, e.g. laser-induced breakdown spectroscopy (LIBS) or synthesis of nanoparticles [14,15]. Combining 1064 nm LIBD with other experiments using 1064 pulsed laser will increase the research efficiency.

In this research, LIBD experiment was carried out by the proposed setup using 1064 nm pulsed laser and also the conventional setup using 532 nm pulsed laser. The breakdowns were detected acoustically by a microphone and the resulting breakdown probabilities were compared.

2. Experimental

The experimental setup for acoustic LIBD is shown in figure 1. Pulsed laser beam produced by a nanosecond Nd:YAG laser (Q-Smart, 10 Hz) was focused into the middle part of liquid sample contained in a cuvette. Once pulsed laser beam interacted with colloid, plasma was generated and followed by a breakdown. A microphone connected to a recording system recorded the accompanying acoustic signal. Recording was set to accommodate one measurement in an audio file. Each file was processed and analyzed using a homemade processing program [16]. It is also possible to record several measurements in one audio file since the homemade program provides the option to determine which part of an audio file that to be analyzed. However, it might slow down the data processing since it was heavier and calling such an audio file requires more time.

In this study, two different laser wavelengths were used: (i) the fundamental wavelength of 1064 nm, and (ii) the second harmonics 532 nm. The energy was set in intervals such that it covered from zero breakdown to maximum. Q-Smart is able to provide a maximum energy of 850 mJ at 1064 nm and 430 mJ at 532 nm. Triplicate measurements were carried out, each employed 1000 laser pulses. Water samples were prepared in two conditions: filtered and unfiltered with filtration performed at 0.22 μm. To prevent contamination, a glass cover was placed on the top of the quartz cuvette. The operational pulsed laser energy was measured considering the attenuation by glass cover.

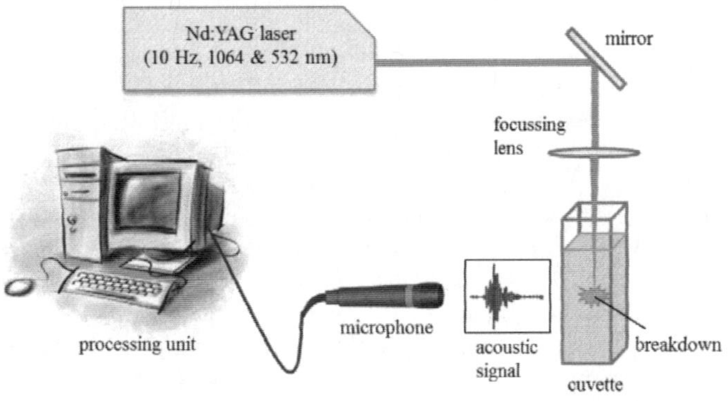

Figure 1. Experimental setup for acoustic LIBD with 1064 and 532 nm Nd:YAG laser.

3. Result and Discussion

Observation during LIBD experiment showed that the interaction between infrared (1064 nm) pulsed laser beam and nanoparticles resulted in the formation of plasma, which was accompanied by acoustic signal emission. This was to be expected since 1064 nm pulsed laser has been frequently used in LIBS experiment to determine the elemental constituents of a sample based on its plasma emission [14]. The acoustic emission has been well detected and recorded by the data acquisition system. Evaluation by the homemade data processing system returned the number of breakdown events. Division by the number of laser shot resulted in the value of breakdown probability [12]. Figure 2(a) showed the S-curve plot of the resulting breakdown probability of filtered and unfiltered water against pulsed laser energy at 1064 nm wavelength.

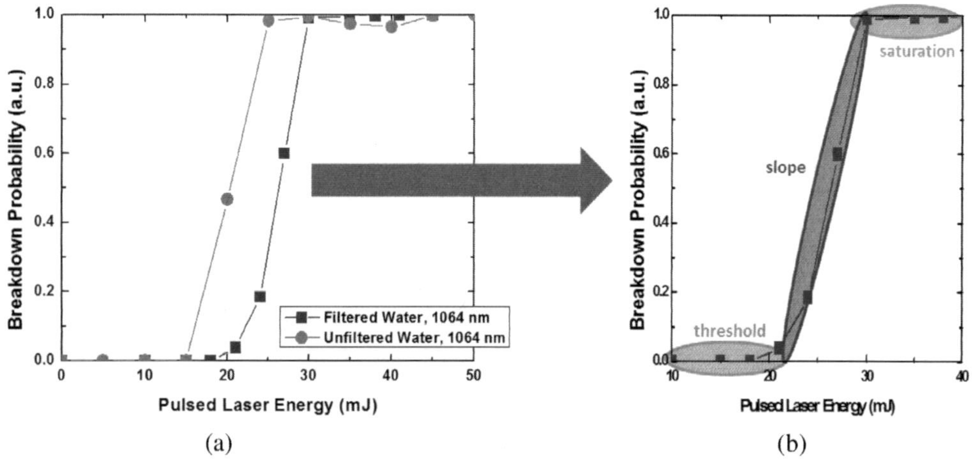

Figure 2. The resulting S-curves of filtered and unfiltered water from the 1064 nm LIBD experiment (a) which can be divided into three regions (b).

The S-curves from 1064 nm LIBD experiment exhibited the trend of a typical acoustic LIBD S-curve [7]. Breakdown probability scales linearly with the effective focal volume, which means more 'target' for laser beam shot in a larger focal volume [12]. Focal volume is proportional to pulsed laser energy [12]. Therefore the breakdown probability increased as the pulsed laser energy was increased. Figure 2(b) displays the three regions of interest: threshold, slope, and saturation region. The threshold region marks the adversity in generating plasma due to low pulsed laser energy. At the end of this region lies the breakdown threshold energy, which is like a 'bottleneck' of breakdown probability. It represents the generation of 'first electron' in the multi photon ionization process where the value depended on the material and the detection process [12].

In the second region, the curve formed a linear slope, which corresponded to the linear range of focal volume as a function of pulsed laser energy [12]. Accordingly, this slope is concentration dependent. The slope of LIBD S-curve can be used to determine the particle concentration in a liquid sample. A higher number of particle has a higher slope and vice versa. The linearity ends near absolute (a probability of 1). At this level of photon flux density, an increase in pulsed laser energy did not result in a higher number of breakdown events and therefore 'saturation' occurred. In this acoustic LIBD, it was assumed that the maximum number of breakdown that might occur due to one laser pulse is 1 [12] and so saturation took place when nearly all of pulsed laser shot had produced breakdown. The acoustic LIBD did not recognize multiple breakdown events (two or more plasma generation by one pulse laser shot).

Figure 2 (a) also implies the ability of 1064 nm LIBD system to differentiate the average size of nanoparticles or colloids presented in the samples. Filtered water sample had a higher value of breakdown threshold energy which showed the smaller colloidal size compared to that of unfiltered water sample. Filtered water samples contained colloids whose average size was smaller than 0.22 μm since the filtration process removed those larger than 0.22 μm. Accordingly, the unfiltered water sample also contained colloids larger than 0.22 μm which gave a rise in the average size value. Smaller colloids require a higher photon flux density to ignite breakdown because the total cross-sections for multi photon ionization is proportional to the particle volume (i.e. the number of valence electrons in the particle) [3,17], hence laser irradiation on filtered water sample ignited breakdown at higher energy and the S-curve of filtered water sample was located on the right side of the unfiltered water sample.

Despite its ability to perform LIBD measurement in the same manner with 532 nm system, 1064 nm LIBD system exhibited a quite different trait in term of breakdown threshold energy. Figure 3 shows the shifting in breakdown threshold energy as a result of changing the 532 nm LIBD system into the 1064 nm LIBD system. The experiment revealed that the 532 nm pulsed laser beam ignited breakdown at relatively low energy, i.e. below 5 mJ; whereas breakdown ignition by 1064 nm pulsed laser occurred at much higher energy, in the order of ten. This phenomenon is attributed to the different light absorbance in water for those two wavelengths. Previous studies on optical characteristics of water [17,18] reported that water has lowest attenuation, i.e. lowest absorbance, at around 500 nm. At 1064 nm, the water attenuation is almost two order of magnitude higher. In another word, a laser beam with 532 nm wavelength is better transmitted by the water than the 1064 nm laser. At 1064 nm, photon were attenuated by the water which resulted in significant energy decrease by the time it interacted with colloids. Additionally, energy per photon also contributes to the shifting of breakdown threshold energy. The energy carried by one photon at 1064 nm is much lower than that carried at 532 nm. Both water attenuation and energy per photon would affect the generation of the 'first electron' in multi photon ionization process. Therefore at 1064 nm, a much higher pulsed laser energy was set in order to compensate the lost.

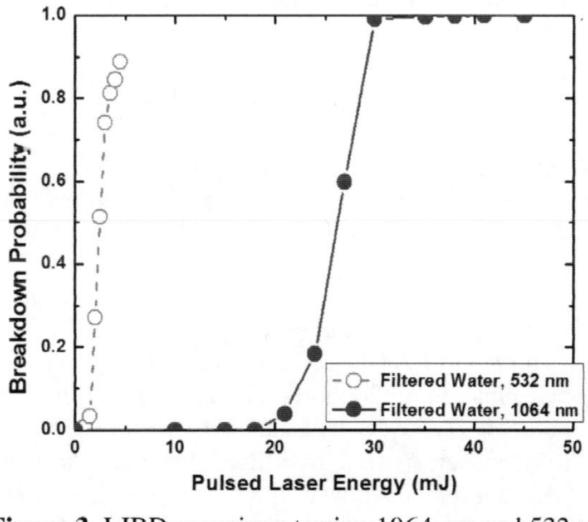

Figure 3. LIBD experiment using 1064 nm and 532 nm pulsed laser resulted in breakdown threshold energy shifting.

There are various experiments that are often performed by 1064 nm Nd:YAG laser. This experiment result showed that an LIBD system might be set-up alongside those experiments and carried out alternately without having to change the laser source and redo the alignment. Furthermore, it may also be possible to integrate the LIBD system into those other experiments, such as laser-induced breakdown spectroscopy (LIBS) and nanoparticle synthesis. For example, an LIBD measurement may be performed during the synthesis of nanoparticle without having to change the laser source. Such an integrated system may be able to provide data on the nanoparticle formation and stability in a more efficient way.

4. Conclusion

It was possible to used 1064 nm pulsed laser for LIBD experiment. The laser was able to ignite plasma emission in the water sample. The number of breakdown occurrences increased as higher pulsed laser energy applied. The trend form an S-curve, a particular trait of LIBD result, was clearly observed. The 1064 nm LIBD system was also able to distinguish sample's particle size from each other based on the difference in breakdown threshold energy. Filtered water, which contained only small size colloids, had higher breakdown threshold energy than unfiltered water, which also contained bigger size particles. However, higher pulsed laser energy was needed in the 1064 nm LIBD experiment, compared to that in the 532 nm. Optical absorbance of water and energy per photon were mostly accounted for this phenomenon. Overall, the LIBD experiment using 1064 nm pulsed laser was a success. However cautions must be taken in such experiment, both for the system tools and users, due to the involvement of higher energy and invisible laser beam.

Acknowledgements

This work was supported by Indonesian Institute of Sciences through the 2015 Excellent Research Program.

References

[1] Thang N M, Knopp R, Geckeis H, Kim J I and Beck P 2000 *Anal. Chem.* **72** 1-5
[2] Bundschuh T, Yun J-I and Knopp R 2001 *Fresenius J. Anal. Chem.* **371** 1063-69
[3] Walther C, Bitea C, Hauser W, Kim J I and Scherbaum F J 2002 *Nucl. Instr. and Meth. in Phys. Res.* B **195** 374-88
[4] Kitamori T, Yokose K, Suzuki K, Sawada T and Gohshi Y 1988 *Jpn. J. Appl. Phys.* **27** L983
[5] Kitamori T, Yokose K, Sakagami M and Sawada T 1989 *Jpn. J. Appl. Phys.* **28** 1195
[6] Fujimori H, Matsui T, Ajiro T, Yokose K, Hsueh Y M and Izumi S 1992 *Jpn. J. Appl. Phys.* **31** 1514
[7] Scherbaum F J, Knopp R and Kim J I 1996 *Appl. Phys.* B **63** 299
[8] Radziemski L J and Cremers D A 1989 *Laser-Induced Plasmas and Applications* (New York: Marcel Dekker) pp 464
[9] Docchio F, Regondi P, Capon M R and Mellerio J 1988 *Appl. Opt.* **27** 3661
[10] Schwebel A H and Ronn A M 1983 *Chem. Phys. Lett.* **100** 178
[11] Bettis J R 1992 *Appl. Spectrosc.* **31** 3448
[12] Kim J I and Walther C 2007 Laser-induced breakdown detection *Environmental Colloids and Particles*, ed K J Wilkinson and J R Lead (IUPAC) chapter 12 555-612
[13] Bundchuh T, Knopp R and Kim J I 2001 *Colloids Surf.* A **177** 47-55
[14] Cremers D A and Radziemski L J 2006 History and fundamental of LIBS *Laser-Induced Breakdown Spectrsocopy (LIBS): Fundamentals and Applications*, ed A Miziolek, V Palleschi and I Schechter (New York: Cambridge University Press) chapter pp 1-39
[15] Mafuné F, Kohno J, Takeda Y and Kondow T 2003 *J. Phys. Chem.* B **107** 4218-223
[16] Putri K Y, Zahar I and Isnaeni 2016 *Prosiding LINOF 2015* (Tangerang Selatan: Pusat Penelitian Fisika LIPI) p 102
[17] Izumida S, Onishi K and Saito M 1998 *Jpn. J. Appl. Phys., Part 1* **37** 2039-42
[18] Irvine W M and Pollack J B 1968 *Icarus* **8** 324-60
[19] Morel A 1974 Optical properties of pure water and sea water *Optical Aspects of Oceanography*, ed N G Jerlov and E Steemann-Nielsen (London: Academic Press) pp 1-24

IOP Conf. Series: Journal of Physics: Conf. Series **817** (2017) 012052 doi:10.1088/1742-6596/817/1/012052

Effect of composition polymeric PVB binder on physical, magnetic properties and microstructure of bonded magnet NdFeB

P Sardjono[1], Muljadi[1], Suprapedi[1], P Sinuaji[2], Ramlan[3] and F Gulo[4]

1)Research Center for Physics-Indonesian Institute of Sciences
2)Departement of Physics-FMIPA, University of North Sumatera, Indoensia
3)Physics Department FMIPA Sriwijaya University,Palembang Indonesia
4)Department of MIPA Sriwijaya University,Palembang Indonesia

Email: priyo_sardjono@yahoo.com

Abstract. The bonded magnet NdFeB has been made by using the hot press method and using Poly Vinyl Butiral (PVB) as a binder. The composition of polymeric binder was varied: 0, 2, 4, 6 and 7 % of weight. Both raw materials are weighed and mixed according to the composition of PVB, then formed by hot press with a pressure 30 MPa, a temperature of 160 ° C and holding time for 30 minutes. The bulk density was measured by using Archimedes method. SEM observation was done to determine the microstructure of bonded magnet NdFeB . The flux magnetic value was measured by using a Gauss meter and the measurement of hysteresis curves was done to know value of remanence Br, coercivity Hc and energy product BHmax by using VSM. According to the characterization results show that the best composition of PVB is 2 of weight. The properties of bonded magnet NdFeB of those compositions are the bulk density around 5.66 g/cm^3. Flux Magnetic value: 1862 Gauss, Br value: 5000 kGauss, Hc value: 8.49 kOe and BHmax value : 5.10 MGOe. According of SEM observation results show that the polymer matrix of PVB appears to have covered on all surface grain and filled grain boundary.

1. Introduction

Permanent magnets are distinguished in two different categories: ferrite and rare earth. Ferrite magnets are classified as oxide magnets and mainly composed of ferric oxide. In the case of rare earth magnets, they retain a higher level of strength than ferrite magnets by approximately 10-fold which are also more expensive [1]. There are two types of rare earth magnets: SmCo-type and NdFeB-type. Currently, more than 98% of rare earth magnets produced are NdFeB ones. There are two types of Nd magnets: sintered and bonded [2]. An Alloy Neodymium Iron Boron (NdFeB) is made from metal elements to form Nd2Fe14B with a tetragonal crystal structure [3]. Bonded magnets are mixtures of permanent magnetic powder and a binder. The powder may be Ferrites, NdFeB, SmCo, AlNiCo and mixtures (hybrides) and the binder will bind the magnetic particles and can produce flexible bonded magnets or rigid bonded magnets. Typical binders for flexible magnets are nitrile rubber and silicon rubber. Binders for rigid magnets include Nylon, Epoxy Resin, Polyphenylene Sulfide, Polyester and Poly Vinyl Butiral[4]. The amount of the NdFeB powder in the bonded magnet is directly correlated with the magnetic and mechanic behaviour of the bonded magnets [5]. A higher content of Nd-Fe-B powder usually results in a higher remanence magnetization (*Mr*) and maximum energy density

(BH)max and therefore, it is desirable from the magnetic perspectives [6]. The applications of bonded magnets are for motors and sensors for the media storage, consumer electronics, automotive, office automation and appliance markets [1,7]. This paper will discuss the synthesis of bonded magnet NdFeB using Poly Vynil Butiral(PVB) as binder. The objective of this study was to determine the bulk density, the magnetic properties (flux magnetic density, coercivity, remanence and energy product) and micro structure by using Scanning Electron Microscope (SEM).

2. Methodology

NdFeB powder type MQP-B was used for preparing bonded magnet NdFeB by using hot press method and Poly Vynil Butiral (PVB) powder was used as binder. The composition of polymeric binder was varied: 0, 2, 4, 6 and 7 % of weight. Both raw materials are weighed and mixed according to the composition of PVB, then formed by hot press with a pressure of 30 MPa, a temperature of 160 ° C and holding time for 30 minutes. The disc samples were measured its bulk density by using the Archimedes method. SEM observation was done to determine the microstructure of bonded magnet NdFeB . The flux magnetic value was measured by using a Gauss meter and the measurement of hysteresis curves was done to know the value of remanence Br, coercivity Hc and energy product BHmax by using Permeagraph.

3. Results and Discussion

The bulk density values of bonded magnet NdFeB with different composition of binder (PVB) are shown in figure 1.

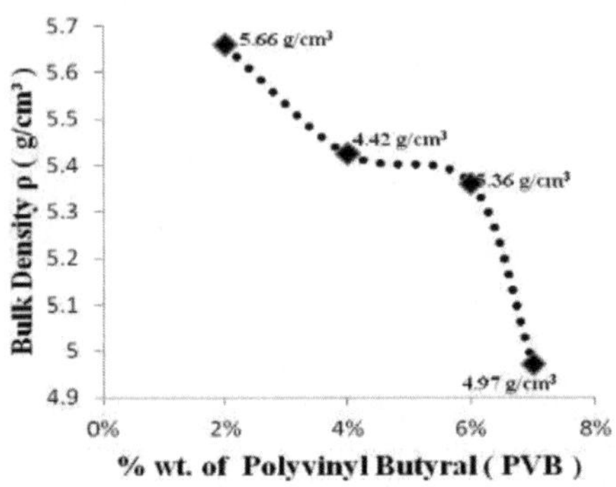

Figure 1. Bulk density value of bonded magnet NdFeB with variation of PVB composition.

The bulk density curve shows that the value of bulk density decreses with increasing of PVB composition. It is because the different density value between NdFeB particle and binder (PVB) particle. The theoretical density of NdFeB particle is higher (about 6 – 7 g/cm^3) compared with the density of PVB (about 1-2 g/cm^3). The highest density value is achieved at 5.66 g/cm^3 (87 % from theoretical density for sample with 2 % PVB) and the lowest density value is achieved at 4.97 g/cm^3 (76.50 % from theoritical density for sample with 7 % PVB). The flux magnetic curve as function of PVB composition is seen in figure 2.

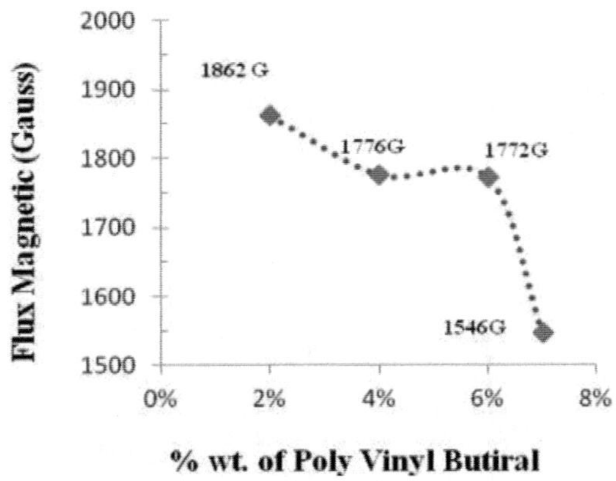

Figure 2. Flux magnetic value of bonded magnet NdFeB with variation of PVB composition.

Figure 2 shows the correlation between flux density with PVB composition. Generally, value of flux magnetic decreases with increasing of PVB concentration, this is due to the properties of both materials, the PVB is nonmagnetic materials. When the amount of nonmagnetic materials in bonded magnet materials, so the magnetic properties of bonded magnet will decrease. According to the application, this material is a magnetic one, so that it becomes a magnetic property is the main consideration. Samples with 2 % and 4 % PVB have the enough high value of flux magnetic, and also they have enough high bulk density as seen in figure 1.

In figure 3, the hysteresis cycles of the samples with 2 % PVB and 4 % PVB are presented. It was found that their shapes are similar, this is due to the small contents of PVB as a binder, hysteresis cycle depends on the content of NdFeB particles, in this case that the content of NdFeB particles are dominant.

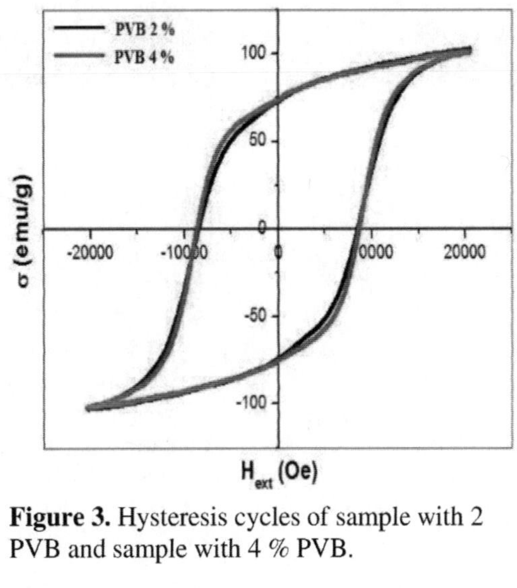

Figure 3. Hysteresis cycles of sample with 2 PVB and sample with 4 % PVB.

Based on figure 3, the value of the remanence (Mr), the magnetic saturation (Ms), the coercivity (Hc) and the maximum energy product (BHmax) can be obtained and the results are shown in table 1. The value of magnetic properties of bonded magnets with 2 % and 4 % PVB are similar.

Table1. Value of Mr,Ms, Hc and BHmax of samples bonded magnet NdFeB

% wt. PVB	**Mr (emu/g)**	**Ms (emu/g)**	**Hc (kOe)**	**BH_{max} (MGOe)**
2%	72.86 [5000 Gauss]	103 [7300 Gauss]	8.490	5.1
4%	73.84 [4000 Gauss]	102 [5700 Gauss]	8.647	5.2

Figure 4 shows structure of a composite bonded magnet with 2 % PVB and 4 % PVB observed in scanning electron microscopy. Their grains are irregularly shaped and occurrence of the small portion of pores was observed in the materials. During the hot pressing at temperature 160 ° C, the binder PVB melted. The melted PVB enveloped surface of gains and bind together with each other. As in the figure 4 (a) and (b) that the PVB binder can be found between grains and it was clear that the polymer binder (PVB) has bonded well the whole NdFeB grain.

(a). Sample with 2 % PVB (b) Sample with 4 % PVB

Figure 4. The SEM microstructure of bonded magnet with 2 % PVB and 4 % PVB.

4. Conclusion

The bonded NdFeB magnet has been made using a Poly Vinyl Butyral (PVB) binder by a hot press method. The composition of PVB binder gives significant effects on the bulk density and flux magnetic value, but it does not influence performance of hysteresis loops and micro structure. The optimum PVB binder is achieved at 2 % wt. of PVB, the bonded magnet, NdFeB with 2 % wt. of PVB has bulk density value about 5.66 g/cm3, flux magnetic value 1862 Gauss, remanence value about 72.86 emu/g [5000Gauss], coercivity about 8.49 kOe and energy product about 5.1 MGOe.

Acknowledgments

The authors would like to thank Joint Research Program UNSRI - PPF-LIPI on "Riset Unggulan Kompetitif Program 2016" for the financial support and thank you for operators of SEM and VSM at Research Center for Physics-LIPI.

IOP Conf. Series: Journal of Physics: Conf. Series **817** (2017) 012052 doi:10.1088/1742-6596/817/1/012052

References

[1] Minowa T 2008 *Resource Geology* **58**(4) 414 – 22
[2] Muljadi, Sardjono P and Sebayang P 2014 *J. Technology Indonesia* **37**(3) 161-66
[3] Shah I A, Abbas T, Ullah Z, Rauf A, Ullah K and Naseem S 2015 *Armenian Journal of Physics* **8**(4) 185-90
[4] Brown D, Ma B M and Chen Z 2002 *J. Magn. Magn. Mater.* **248** 432–40
[5] Kurniawan C, Hutahaean R M, Muljadi 2013 *Advanced Materials Research* **1123** 84-7
[6] Stancu C, Notingher P V, Ionita V, Marinescu V, Panaitescu D 2014 *Electrical Engineering Series* **38** p 98-106
[7] Campbell P, Brown D N, Chen Z M, Guschl P C, Miller D J and Ma B M 2014 *The proceeding of 18th International Workshop on High Performance Magnets andTheir Applications* (France: Annecy) 29 August-2 September 2014

Effect of Te doping on FeSe superconductor synthesized by powder-in-tube

A Imaduddin[1], K Nisa[2], S D Yudanto[1], H Nugraha[1] and B Siswayanti[1]

[1]Research Center for Materials and Metallurgy – LIPI, Gd 470 Kawasan Puspiptek, Serpong, Tangerang, Banten 15314, Indonesia
[2]Airlangga University, Jl. Airlangga No. 4 - 6, Gubeng, Surabaya, Jawa Timur, Indonesia

E-mail: agun001@lipi.go.id

Abstract. FeSe is a superconducting material, which has the simplest crystal structure among the Fe-based superconductors. It has no arsenic element, which is very harmful to the human body. In this study, we analyzed the effects of milling time and Te doping on FeSe superconductors. The synthesis of the samples were carried out using powder-in-tube method in a SS304 stainless steel tube. After the pressing process, followed by the sintering process at 500^0 C for 20 hours, the samples were removed from the tubes. Later, we analyzed its crystal structures, surfaces morphology and the superconductivity properties. δ-FeSe phase (hexagonal, non-superconductor) and β-FeSe (tetragonal, superconductor) were formed in the samples, including minor phases of Fe and Fe_3Se_4. Te doping changed the crystal structure from β-FeSe and δ-FeSe into $FeSe_{0.5}Te_{0.5}$. In addition, the onset critical temperature ($T_{C,\ onset}$) shifted to higher temperature.

1. Introduction

Since the discovery of Fe-based superconducting material in 2008, research on superconductors increase, since Fe which has ferromagnetic properties should not possess superconductivity (generally, superconducting materials are diamagnetic). Due to the advent of Fe-based superconductors, research on superconductors changes its direction and starts to explore this type of superconductor. FeSe is a superconducting material which has the simplest crystal structure among the Fe-based superconductors [1,2]. In addition, it contains no arsenic element, which is very harmful to the human body. The main barrier for FeSe superconductor manufacturing is the emergence of δ-FeSe phase (tetragonal structure with P4/nmm crystal symmetry), which does not have superconducting properties [3,4], during the synthesis of β-FeSe superconductor. Herein, the advanced technique for manufacturing pure β-FeSe superconductor as well as avoiding oxidation is crucial. Move over, special concerns on Se handling is needed to make it inseparable from the powder mixtures due to its low melting and boiling point (220°C and 685°C, respectively).

2. Experiments

The samples were prepared with a stoichiometric ratio (Fe : Se : Te = 1 : 1-x : x) of their elements (Fe, Se, and Te with purity of 90%, 99.99%, and 99.8%, respectively). The powders mixtures were mixed and ground using shaker mill for 2 and 5 hours. The as-milled powder mixtures were then transferred into SS304 stainless steel tubes (with respective outside and inner diameters of 12 mm and 10 mm), and pressed with pressure of 165 MPa at both ends and the center of the tubes to completely remove the air. The sample packs was sintered at 500°C for 20 hours in ambient air with heating rate 10°C/min. The samples were characterized using X-Ray Diffractometer (XRD) and Scanning Electron Microscope (SEM) to examine their crystal structure and surface morphology, respectively. The profiles of temperature dependent resistivity were measured using four-point probe method (Oxford's Cryogenic Magnet System "Teslatron PT"). Table 1 shows the variation of sample preparations. Four samples were prepared with different milling times and Te compositions.

Table 1. Variation of sample preparation.

Sample Code	Milling time (hours)	Values of x in $FeSe_{1-x}Te_x$
Sample 1	2	0.00
Sample 2	5	0.00
Sample 3	5	0.25
Sample 4	5	0.50

3. Results and discussions

Figure 1 shows the results of XRD analyses. There is no oxide phase in all samples, which indicates the suitability of powder-in-tube method for avoiding sample oxidation.

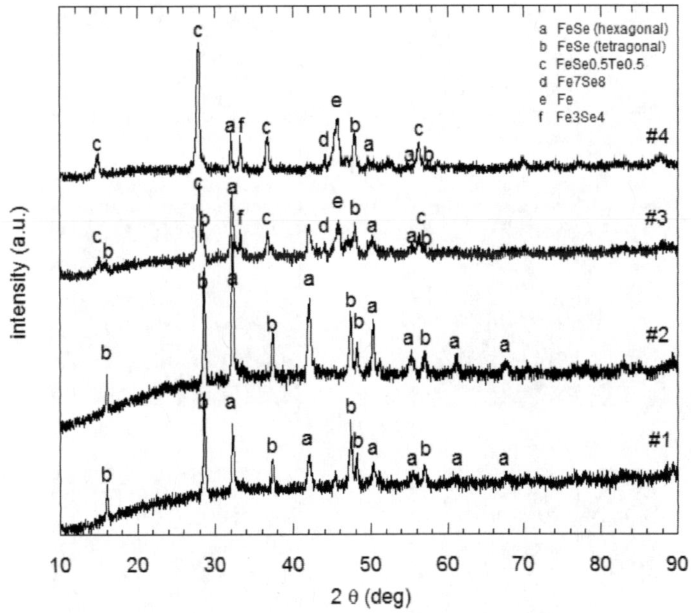

Figure 1. The results of XRD analyses.

Both β-FeSe superconductor and δ-FeSe non-superconductor phase were formed. In sample 1, both peaks have similar high intensity. While in sample 2, the intensity of δ-FeSe peaks is lower than β-FeSe. In sample 3, $FeSe_{0.5}Te_{0.5}$ peaks with a higher intensity appear, while the peak intensity of β-FeSe and δ-FeSe decrease to some extent. In addition, the peaks of Fe_3Se_4, Fe_7Se_8, and Fe phases are begin to form [5]. In comparison with sample 3, sample 4 shows the formation of $FeSe_{0.5}Te_{0.5}$ peaks ("c" peaks) and diminish of β-FeSe peaks ("a" peaks) [6]. Meanwhile, the peaks of Fe, Fe_3Se_4 and Fe_7Se_8 phases begin to emerge [7]. From XRD observation, it is clear that the addition of Te element induces the formation of $FeSe_{0.5}Te_{0.5}$ phase [8,9], while decreases the intensity of β-FeSe and δ-FeSe peaks. Te addition creates more Fe residue and induces the formation of Fe_3Se_4. The addition of Te prevents the formation of hexagonal and tetragonal phases [10, 11]. Further studies using quench method are required to reduce the δ-FeSe phase [12].

Figure 2 shows the results of SEM analyses. Considering these results, the addition of Te promotes particle agglomeration. Comparing sample 1 and 2, the extended milling time promotes particle size reduction and δ-FeSe phase abatement.

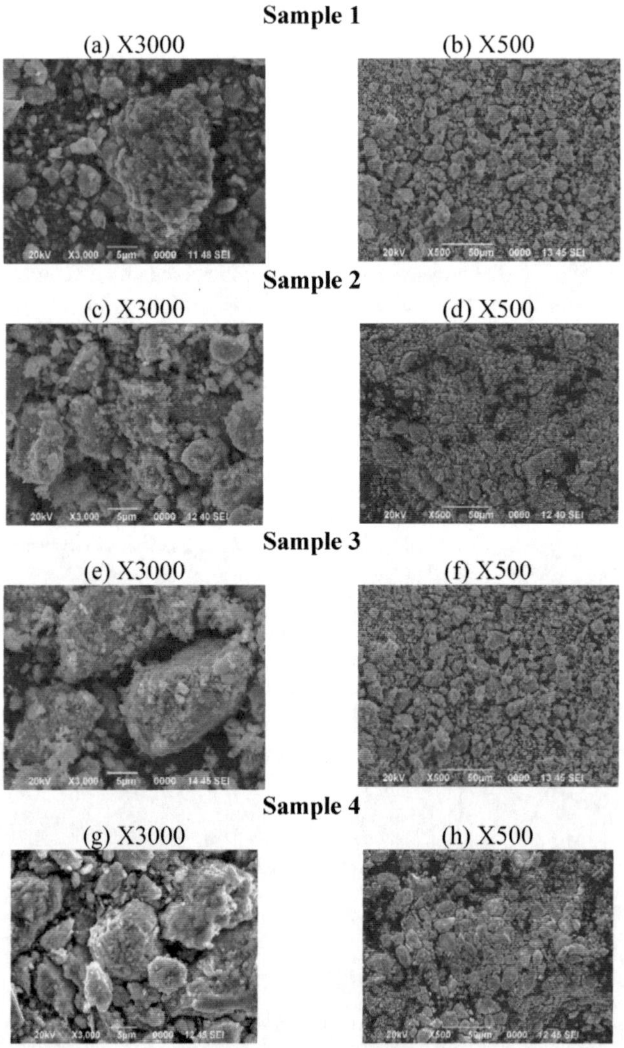

Figure 2. SEM analyses.

Figure 3 depicts the temperature dependence of resistivity of the samples. The measurements were performed from room temperature to around 2 K. As temperatures decreases, Te element (in sample 3 and 4) incites the increase of resistivity value, which indicates the semiconductor property at higher temperature. For sample 2 at temperature above $T_{C,onset}$, the resistivity decreases as the temperature decreases, which indicates the conductor properties than that of sample 1. There are sharp decrease of resistivity in the range of 8 - 16 K for all samples, which indicates the superconductor property. It should be noted that no samples reach zero resistivity until 2 K.

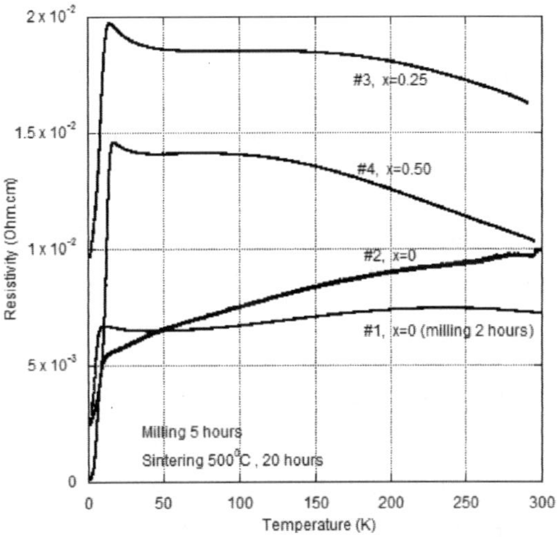

Figure 3. Temperature dependence of resistivity of the samples.

For sample 3 and 4 at higher temperature, resistivity value rises as the temperature decreases, which indicates semiconductor properties. The resistivity value of sample 2 drops as the temperature decreases, which shows the conductor properties.

Figure 4 shows the temperature dependence of resistivity, which is normalized at 18 K. The resistivity is normalized to compare the resistivity change of all samples around critical temperature (T_C). As reported in reference [13], reduction of Fe in the sample increases the $T_{C\ onset}$ value. Further researches on the origin of the increase of $T_{C,\ onset}$ need to be carried out.

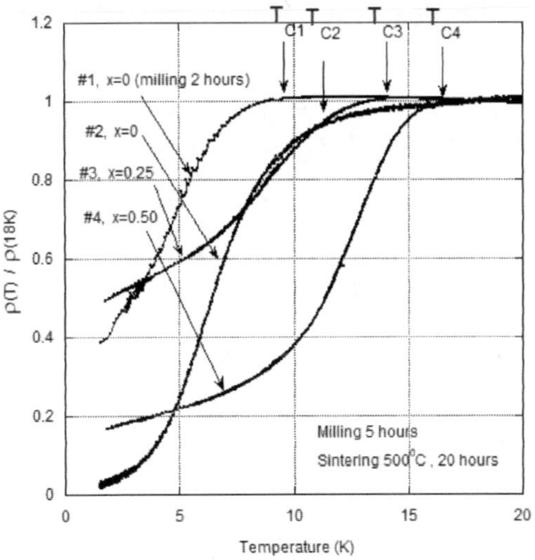

Figure 4. Temperature dependence of normalized resistivity.

The addition of Te element increases the value of $T_{C,onset}$,in the order of 1, 2, 3, and 4, sequentially from 8 K to 17 K. However, below $T_{C,onset}$, no sample reach zero resistivity, which is likely due to the existence of Te element. Considering sample 1 and 2, prolonged milling time increases $T_{C,onset}$, as well as diminishes δ-FeSe phase. While the added Te element instigates the decrease of δ-FeSe and β- FeSe phases, it also increases the possibility to get more Fe, Fe_3Se_4 and $FeSe_{0.5}Te_{0.5}$ phases.

4. Conclusions

$FeSe_{1-x}Te_x$ samples, with variation of milling times and Te compositions, have been synthesized. Based on XRD analyses, β-FeSe and δ-FeSe phases were formed. Prolonged milling time reduces the particle size and diminishes δ-FeSe phase. Te addition incites $FeSe_{0.5}Te_{0.5}$ phase formation and limits the formation of δ-FeSe and β-FeSe phases. Temperature dependence of resistivity analyses at high temperature, show the increase of resistivity as the temperature decreases (sample 3 and 4), which indicates that Te element shifts the sample properties from conductor to semiconductor. At 2 K, the resistivity of the samples with Te element (sample 3 and 4) is higher than that of sample without Te element (sample 2).

References

[1] Tanabe K and Hosono H 2012 *Japan. J. Appl. Phys.* **51** 010005
[2] Ma Y 2015 *Phys. C Supercond. its Appl.* **516** 17–26
[3] Onar K and Yakinci M E 2015 *J. Alloys Compd.* **620** 210–16
[4] Imai T, Ahilan K, Ning F L, McQueen T M and Cava R J 2009 *Phys. Rev. Lett.* **102**(17) 1–4
[5] Nitsche F, Goltz T, Klauss H H, Isaeva A, Müller U, Schnelle W, Simon P, Doert T and Ruck M 2012 *Inorg. Chem.Commun.* **51**(13) 7370–76
[6] Zhang A M, Xia T L, Kong L R, Xiao J H and Zhang Q M 2010 *J. Phys. Condens. Matter.* **22**(24) 245701
[7] Campos C E M, De Lima J C, Grandi T A, Machado K D, Drago V and Pizani P S 2004 *J. Magn. Magn. Mater.* **270**(1–2) 89–98
[8] Caglieris F, Ricci F, Lamura G, Martinelli A, Palenzona A, Pallecchi I, Sala A, Profeta G and Putti M 2012 *Sci. Technol. Adv. Mater.* **13**(5) 054402
[9] Imai Y, Tanaka R, Akiike T, Hanawa M, Tsukada I and Maeda A 2010 *Japan. J. Appl. Phys.* **49**(2) 023101
[10] Shipra R, Takeya H, Hirata K and Sundaresan A 2010 *Phys. C Supercond. its Appl* . **470**(13–14) 528–532
[11] Yadav A K, Sanchela A V and Tomy C V 2015 *Solid State Commun.* 1–5
[12] Okabe H, Takeshita N, Horigane K, Muranaka T, Akimitsu J and Fese P 2010 *Technology*. 1–6
[13] Sala A, Palenzona A, Bernini C, Caglieris F, Cimberle M R, Ferdeghini C, Lamura G, Martinelli A, Pani M and Putti M 2013 *Phys. C Supercond. its Appl.* **494** 69–73

Microstructures, physical and magnetic properties of $BaFe_{12}O_{19}$ permanent magnets with the addition of Al_2O_3-MnO

E A Setiadi, C Kurniawan, P Sebayang and M Ginting

Research Center for Physics, Indonesian Intitute of Sciences (LIPI),
Komplek Puspiptek, Serpong, Tangerang Selatan, Banten 15314 - INDONESIA

E-mail : eko.arief.setiadi@lipi.go.id

Abstract. Barium hexaferrite ($BaFe_{12}O_{19}$) permanent magnets with the addition of Al_2O_3-MnO (0.12, 0.24; and 0.42 wt%) have been prepared by mechanical alloying method. The addition of Al_2O_3-MnO was done in order to improve the density and mechanical properties of the fabricated $BaFe_{12}O_{19}$ permanent magnet. Particle size of the as-mixed powder decreases Al_2O_3-MnO concentration increases. The as-mixed powders was then compacted into pellets and sintered at 1000 and 1100 °C for 2 hours. The crystalline phases and morphology of the pellets were observed using X-ray diffraction (XRD) and Scanning electron microscopy (SEM), respectively. It is found that the increase of the sample hardness has correlation with the bulk density of the samples. While, the flux density decreases linearly as Al_2O_3- MnO additives increases. The analysis of hysteresis curves classified the pellets as a hard magnet.

1. Introduction

Barium hexaferrite ($BaFe_{12}O_{19}$) is classified as a hard magnet with high saturation, coercivity and magnetic anisotropy [1], which is widely used in various fields of science and technology. This type of magnet has unique properties, i.e. high Curie temperature, high durability against corrosion, good chemical stability, and relatively easy to prepare [2]. Due to its unique properties, $BaFe_{12}O_{19}$ can be applied for various applications, such as microwaves absorption, ferrofluid, data storage media, and many more [3-4]. Many methods, such as powder metallurgy, sol-gel [5], hydrothermal, molten salt, and co-precipitation [6],have been utilized to fabricate $BaFe_{12}O_{19}$ with various results.

Several researches have been done to reconstruct the physical and magnetic properties of $BaFe_{12}O_{19}$ by adding or substituting other ions into the crystalline structure. Substitutions can be made by replacing Fe^{3+} ions with other cations, such as Co, Mn, Ru, Zn, Sn, Al, Sc, La, and so on [6]. This study is aimed to investigate the optimum composition of MnO and Al_2O_3 as a additive in $BaFe_{12}O_{19}$ structure, in order to obtain the highest density and mechanical properties. MnO and Al_2O_3 were chosen due their chemical stability, availability in nature, and high melting point (1785°C and 2072°C for MnO and Al_2O_3, respectively). The microstruture, physical and magnetic properties of the doped-$BaFe_{12}O_{19}$ magnet will be analysed to elaborate their correlation with the doping composition.

2. Experimental Method

The raw materials used in this experiment are $BaFe_{12}O_{19}$, Al_2O_3 and MnO powder. Table 1 shows the map of material composition.

Table 1. Composition of samples.

Sample	Composition		Sintering Temperature (°C)
	$BaFe_{12}O_{19}$ (wt%)	Al_2O_3-MnO (wt%)	
A12	99.88	0.12	1000
A24	99.76	0.24	1000
A42	99.68	0.42	1000
B12	99.88	0.12	1100
B24	99.76	0.24	1100
B42	99.68	0.42	1100

Those three materials were wet-milled in a high energy milling (HEM) machine using toluene as a media for two hours, then dried at 100°C for 24 hours. The particle size of the dried powder were estimated using particle size analyzer (PSA - CILAS 1190). The powders were then pressed at 35 kgf/cm^2 to get pelletized samples. The density of the pellets was measured using Archimedes' principle as a green density (ASTM C373-88-2006). After sintering at a specified temperature (Table 1) for two hours, the bulk density of the samples was measured thoroughly. The crystal structure of the samples was analyzed using X-ray Diffraction (XRD Rigaku SmartLab, λ = 1.5418 Å). The microstructure of the samples was observed using Scanning electron microscope (SEM - Hitachi SU 3500). Meanwhile, the hardness of the samples was determined using Micro-hardness Tester (MHT - LECO LM 100AT) at 100 gf load for 15 seconds (Vicker's method). The magnetic properties of the samples were investigated using Gaussmeter and Vibrating Sample Magnetometer (VSM - Electromagnetic VSM250), after magnetization using Impulse Magnetizer (Magnet physics Dr. Steingroever GmbH).

3. Results and Discussion

Figure 1 depicts the average particle size and true density (powder density) of Al_2O_3-MnO-doped $BaFe_{12}O_{19}$. It can be seen clearly, that the average particle size decreases, while the powder density (true density) increases, as Al_2O_3-MnO concentration increases. As the density of the powder becomes smaller, the cavity of the particles should also become smaller.

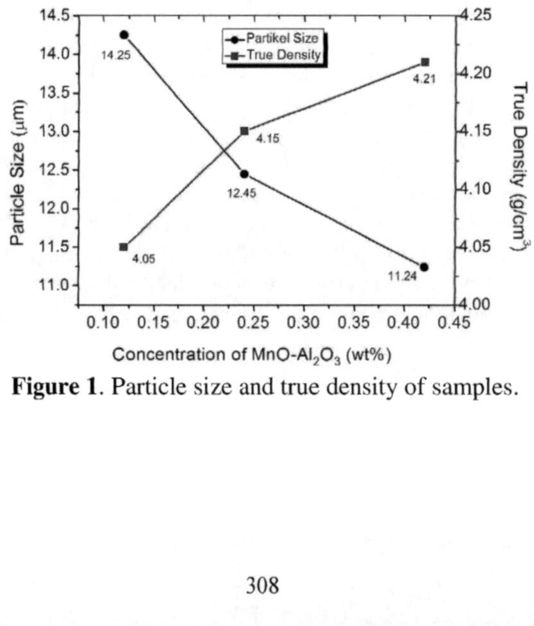

Figure 1. Particle size and true density of samples.

Figure 2 shows the green density and bulk density comparison of the samples (sintered at 1000 and 1100°C for two hours). It can be seen that the green and bulk density become higher as the concentration of the additives increases. The pattern is similar with that of true density, where the sample cavity becomes smaller as the particle size decreases, results to the increase of density value. The sample sintered at 1100°C with additive concentration of 0.42 wt% has lower bulk density than that of sample with additive concentration of 0.24%.This is due to the strong influence of sintering temperature and composition to the bulk density. For the same concentration, effect of sintering temperature can be clearly seen as the process of sample densification. The density of the sample consistently increases for three samples with various additive concentrations and different temperatures (un-sintered and sintered samples). The sintering temperature incites the grain growth and the cavity reduction, so that the density increases as the sintering temperature increases. It should be noted that the sintering temperature and the holding time during the crystals growth are the important factors that influence the crystal size [7]. The optimum density values, i.e. 3.98 g/cm^2, can be obtained with additive concentration of 0.24 wt% and sintering temperature of 1100°C.

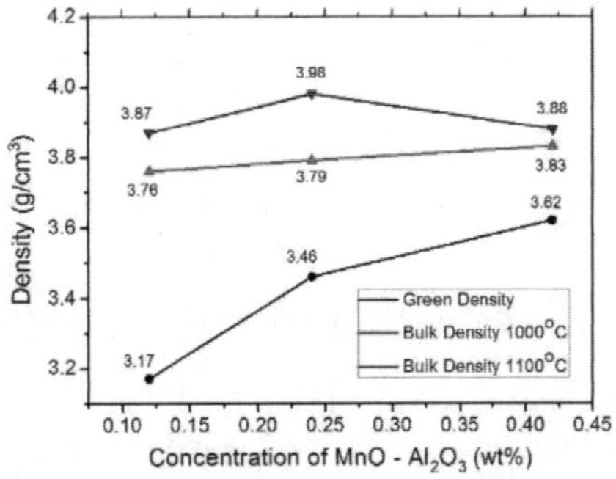

Figure 2. Green density and bulk density of samples.

Figure 3 shows the Vicker hardness plot of the samples sintered at 1000°C and 1100°C . It can be seen that the addition of Al_2O_3-MnO into $BaFe_{12}O_{19}$ tend to increase the hardness of the magnet. At those temperatures, maximum hardness value is 554 HV, which was obtained for additive concentration of 0.42 wt%. While, at 1100°C, maximum hardness value is 816 HV, which was obtained for additive concentration of 0.24 wt%. From those data, it can be seen that the hardness values were influenced by the bulk density values. As the bulk density become higher, the inter-particle density goes higher, while cavities become smaller. In addition, the hardness increases as the increase of the density [8].

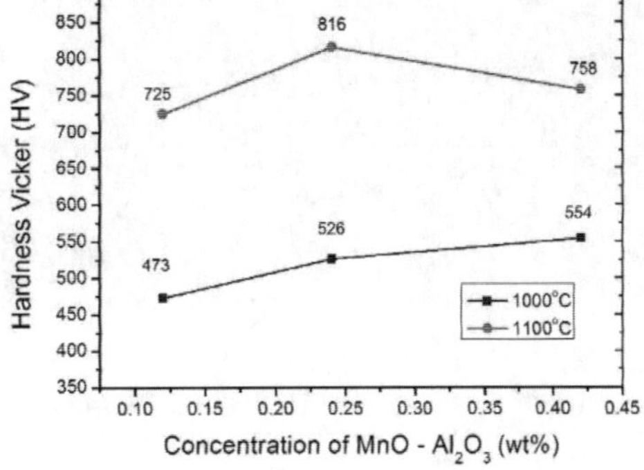

Figure 3. Vicker hardness of the samples.

Based on the above results, 1100°C can be considered as the optimum sintering temperature for the sample. Figure 4 depicts the XRD pattern of doped-$BaFe_{12}O_{19}$ sintered at 1100°C. The samples with additives have single phase of $BaFe_{12}O_{19}$ with hexagonal crystal structure. The additive phases is not detected by XRD measurements, due to the small concentration of additive in the sample. In addition, the additive is likely occur as the ions, where Mn and Al ions replace Fe ions. Nevertheless, the ions substitution phenomena still need to be verified further.

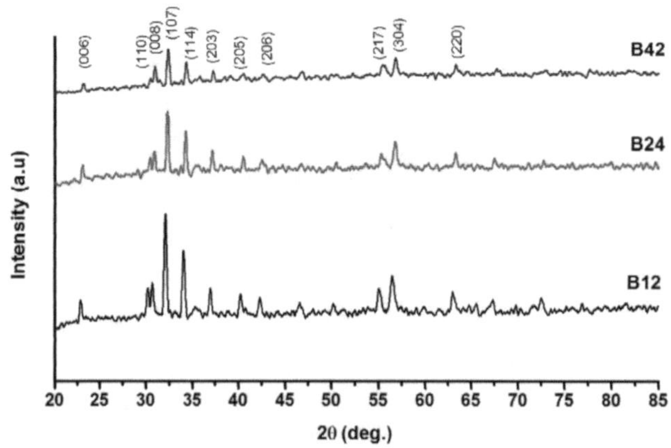

Figure 4. XRD pattern of samples.

Figure 5 and table 2 are results of SEM-EDX analysis of $BaFe_{12}O_{19}$ with Al_2O_3-MnO addition. The morphology analyses confirm, that the size and shape of particles on the surface were relatively nonhomogenous. Furthermore, the sample surfaces seem uneven. Meanwhile, from EDX (energy dispersive x-ray spectrometry) result in table 2, identify some elements, such as Ba, Fe, Mn, Al and O in the sample. The presence of Mn and Al elements confirms the inclusion of the additive into the parent material. Total atomic composition of each element represent the initial stoichiometric value. As the additive concentration becomes higher, the percentage of the atoms also increases.

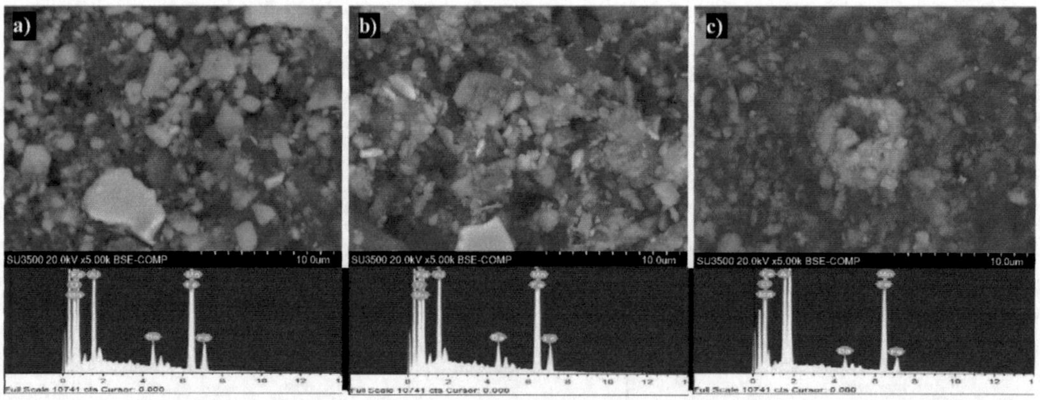

Figure 5. SEM imaging of (a) B12; (b) B24 and (c) B42 samples.

Table 2. EDX measurement result.

	Composition of Samples		
Element	B12 (atomic%)	B24 (atomic%)	B42 (atomic%)
Ba	1.39	1.90	0.60
Fe	21.59	21.26	9.45
Mn	0.03	0.06	0.11
Al	12.15	16.49	26.87
O	64.84	60.29	62.97

Figure 6 shows magnetic flux density of the magnetized sample. The samples were magnetized by an impulse magnetizer. It is indicated for all of the samples, that as the additive concentration goes higher, the magnetic flux becomes smaller, which is happened for the samples sintered at 1000 and 1100°C. Mg and Al elements have paramagnetic behaviour, while $BaFe_{12}O_{19}$ tend to be ferromagnetic. The Al_2O_3-MnO additives change the anisotropy constant of $BaFe_{12}O_{19}$, lead to the decrease of total magnetic flux. The magnetic flux increases, as sintering temperature increases, since the bulk density of the samples sintered at 1100°C sintered is higher than that of the samples sintered at 1000°C. Magnetic properties of material can be influenced by the degree of crystallinity, particle size and any other phase (non-magnetic) [9].

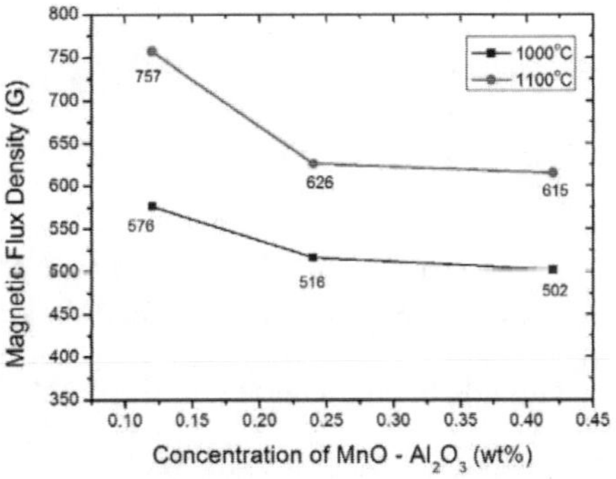

Figure 6. Magnetic flux density of samples.

Figure 7 and table 3 show VSM hysteresis curve of the samples. It can be seen that the magnetic saturation and magnetization remanence decrease as the additive concentration increases, which could be attributed to the change of total magnetic moment. Basically, MgO and Al_2O_3 are classified as non-magnetic material. The addition of those materials should naturally decrease the magnetization values. The intrinsic coercivity measurement reveals, that the higher value of additive concentration increase the coercivity field. Since the smaller particle size increase the density, the coercivity will also increase due to the increase of anisotropy constant. Adeela et al. also reported, that the increase of Mn and Co additives could increase the coercivity value [10]. Considering the above results, the fabricated $BaFe_{12}O_{19}$-(Al_2O_3-MnO) permanent magnets can be classified as hard magnets which can be properly used as magnetic recording media [11].

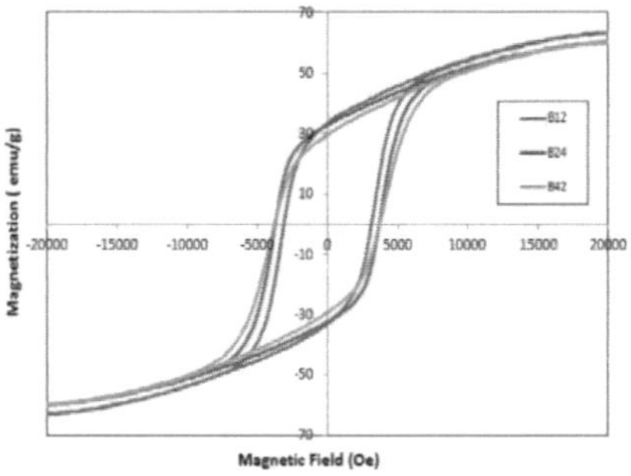

Figure 7. Hysteresis curve of samples.

Table 3. VSM measurement value.

Sample	Coercivity (kOe)	Saturation (emu/g)	Remanence (emu/g)
B12	3.103	63.58	33.10
B24	3.689	60.50	32.64
B42	3.746	60.10	29.25

4. Conclusion

The $BaFe_{12}O_{19}$-(Al_2O_3-MnO) permanent magnets with additive concentration of 0.12; 0.24; and 0.42 wt% have been successfully fabricated. PSA measurements show the linear decreases of $BaFe_{12}O_{19}$-(Al_2O_3-MnO) particle sizes, as the additive concentration increases. After pelletizing, the hardness of the $BaFe_{12}O_{19}$-(Al_2O_3-MnO) pellets increases proportionally, as the increase of bulk density values. While, the decrease of the magnetic flux, saturation, and remanence values of the $BaFe_{12}O_{19}$-(Al_2O_3-MnO) are proportional to that of Al_2O_3-MnO concentration. Moreover, the addition of (Al_2O_3-MnO) could increase the coercivity value. Based on the obtained coercivity values, the fabricated $BaFe_{12}O_{19}$-(Al_2O_3-MnO) permanent magnets can be classified as a hard magnet.

References

[1] Meng Y Y, He M H, Zeng Q, Jiao D L, Shukla S, Ramanujan R V and Liu Z W 2014 *J. Alloys Compd.* **583** 220
[2] Nowosielski R, Babilas R and Wrona J 2007 *J. Achiev. Mater. Manuf. Eng.* **20** 307
[3] Tudorache F, Popa P D, Brinza F and Tascu S 2012 *Proc. Int. Congr. Adv. Appl. Phys. Mater. Sci.* **121** 95
[4] Sharma M, Kashyap S and Gupta H C 2014 *Phys. B :Condens. Matter* **448** 24
[5] Wannawong N, Sridet P, Kaewrawang A, Tonmitr K and Siritaratiwat A 2013 *Proc. Sci. Eng.* p 524
[6] Liu Y, Drew M G B and Liu Y 2011 *J. Magn. Magn. Mater.* **323** 945
[7] Doni W R, Manaf A and Sardjono P 2013 *Int. J. Basic Appl. Sci. IJBAS-IJENS* **13** 65
[8] Birru A K, Shiva K and David S G S 2015 *Mater. Today Proc.* **2** 4402
[9] Suharyadi E, Setiadi E A, Shabrina N, Kato T and Iwata S 2014 *Adv. Mater. Res.* **896** 126
[10] Adeela N, Khan U, Iqbal M, Riaz S, Irfan M, Ali H, Javed K, Bukhtiar I, Maaz K and Naseem S 2016 *J. Alloys Compd.* **686** 1017
[11] Coey J M D 2010 *Magnetism and Magnetic Materials* (New York: Cambridge University Press)

Development of resistance and temperature measurement system for cryogenic equipment

H Nugraha[1], A Imaduddin[1], M A J Mulya[2] and B Hermanto[2]

[1]Research Center for Materials and Metallurgy - LIPI
Gedung 470, Kawasan Puspiptek, Serpong, Tangerang Selatan , Banten, Indonesia
[2]Research Center for Physics - LIPI
Gedung 440, Kawasan Puspiptek Serpong, Tangerang Selatan, Banten, Indonesia

E-mail: heri014@lipi.go.id

Abstract. Cryogenic equipment is used especially for the research on superconductors as the tools to create a condition of low temperature and high magnetic field. However, it needs the measurement system to be used to analyze the material properties. This work developed a control program of resistance measurement system using instruments of current source, nanovoltmeter and temperature controller. We designed and developed a measurement system by building a connection of the instruments and creating a new control program. The created control program can be revised or modified easily when we have to replace or add the instruments. LabVIEW program was used as a control program, and designed for measuring the resistance and controlling the temperature. Four-point probe method was used to measure the resistance value, a sensor to determine the temperature and a heater to control the temperature. The developed measurement system was tested using standard superconductor sample. It clearly showed the graph of the temperature dependence of resistance. And it showed sharp drop of resistance around 98K, indicating its critical temperature.

1. Introduction

Superconductors are the materials that can conduct electricity or transport the electrons with no electric resistance. This means no heat, sound or any other form of energy would be released from the materials when it is below "critical temperature" (Tc), or the temperature at which the materials become superconductors. Unfortunately, most materials must be in an extremely low energy state (very cold) in order to be superconductors [1]. Superconducting materials that mostly used are Nb_3Sn and NbTi (Low Temperature Superconductors, LTS) superconducting wires. Nb_3Sn and NbTi superconducting wires generally are used for high magnetic field equipments. In addition, High Temperature Superconductors (HTS) type superconducting wires have started to be used for many applications in the fields of electricity, such as electrical transport for high voltage power cables [1].

Conducting research on superconductivity requires the resistance measurement system which could be used at low temperature inside cryogenic equipment. The resistance measurement system available on the market is very expensive and using the LabView (Laboratory Virtual Instrument Engineering Workbench) program that cannot be modified or revised. The user only can set and controls the parameters in the front panel. The block diagram of the program cannot be displayed or edited.

IOP Conf. Series: Journal of Physics: Conf. Series **817** (2017) 012055 doi:10.1088/1742-6596/817/1/012055

So that, the aim of this study is to develop the resistance measurement system controlled by LabVIEW programs [2, 3]. The created LabVIEW program has source code or block diagram that can be modified easily.

As a general-purpose measurement instrument, a nanovoltmeter (Keithley 2700) was used to measure the voltage of the samples, and a current source (Keithley 6221) was used to generate electrical current. A programmable temperature controller (Oxford Mercury iTC) was employed to control and measure the sample temperatures. The system was controlled by a set of dedicated LabVIEW applications running on computers. The communication between computers and instruments was done via GPIB bus, TCP/IP and USB. LabVIEW is a system-design platform and development environment for a visual programming language from National Instruments [3]. The measurement system attached in existing equipment was the built-in programs and instruments. Thus, the original programs can't be revised when we need to add or replace the instruments.

The cryogenic magnet equipment that we used can create super low temperature (as low as 1.5 Kelvin) and high magnetic field (up to 8 Tesla). This study developed the measurement system of resistance and temperature. We developed this measurement system that also can be used as current density measurement system to analyze the performance of superconducting wires. This resistivity measurement system is using the state of the art programmable temperature control (Mercury iTC), nanovolt meter (Keithley 2700) and current source (Keithley 6220), thus we can measure low noise and precise resistivity value. We are combining this measurement system with the using of cryogenic magnet that made this measurement system is the first and the only programmable resistivity measurement system for cryogenic system that developed in Indonesia. We used $Bi_2Sr_2CaCu_2O_x$ (or Bi-2212) standard superconducting film to test the developed measurement system.

2. Development Methods

To develop a measurement system for cryogenic equipment, it requires a control or processing unit that can handle several tasks such as measuring and controlling temperature, storing, transmitting and analyzing data. These tasks can be performed by using LabVIEW program [4-6]. The applications of the LabVIEW programs (VI) have two components, which are the user interface and underlying code. The user interface take place at front panel and the underlying code is located at the block diagram.

Each VI has a front panel that can be used to design the user interface. We designed user interface by placing controls and indicators on the front panels. The controls are used to supply inputs, and the indicators are used to see the results. The indicators are typically in the form of graphs, charts, LED's, and status strings [7-8].

3. Results and discussions

3.1. Measurement system

The resistance measurement system was conducted using four-point probe method. The sample was put into cryogenic equipment to be measured under low temperature. Electrical resistance is frequently measured using 4-terminal-potentiometric DC methods. This method involves the connection of two current leads and two potential leads to the sample [7]. The scheme and the image of the four-point probe method were shown in figure 1. As shown in figure 1(a), resistance value is calculated and temperature value is collected using the following loop program in figure 2.

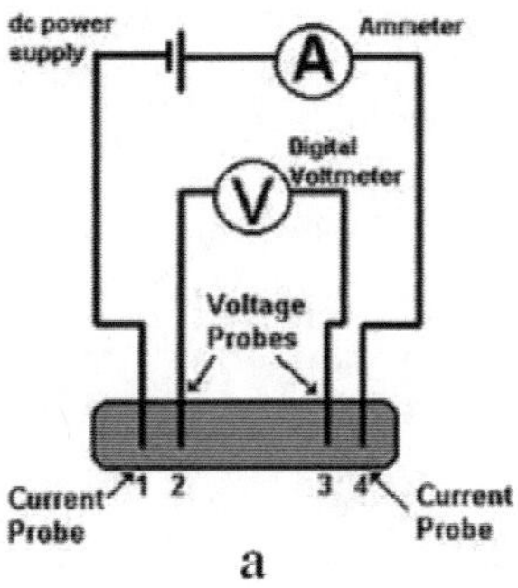

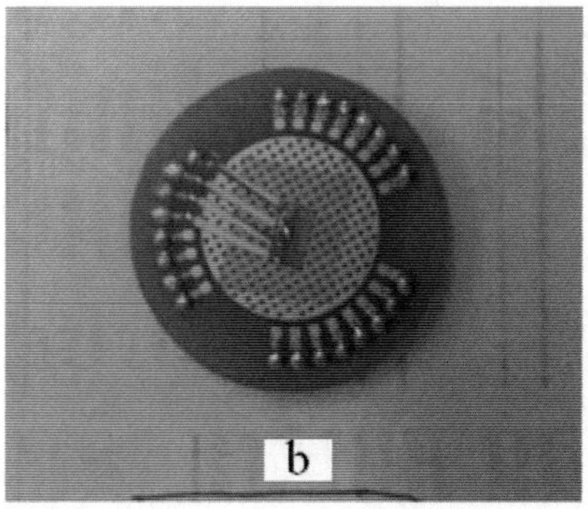

Figure 1. The scheme (a) and the image (b) of sample holder of the four-point probe method (provided by manufacturer)

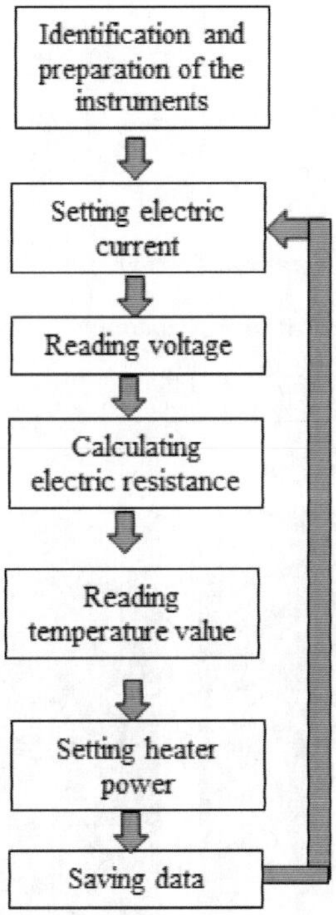

Figure 2. Loop program of the resistance measurement system

Standard sample $Bi_2Sr_2CaCu_2O_x$ (or Bi-2212) superconductor was installed on the sample holder as shown in figure 1(b). The sample holders were included as complementary of cryogenic equipment.

Sample holder has 24-pins connector that used to connect the sample and measurement instruments. Pin #1 to pin #8 were connected to the scanner card (Keithley 7709) at back of Keithley 2700 instrument [9-11]. The sample holder was mounted on the bottom of the sample rod and then inserted into cryogenic equipment. The voltage was measured using nanovoltmeter (Keithley 2700) controlled with LabVIEW program. Superconductor sample was tested from room temperature to low temperature. The measurement system was performed using nanovoltmeter, current source, and temperature controller instrument.

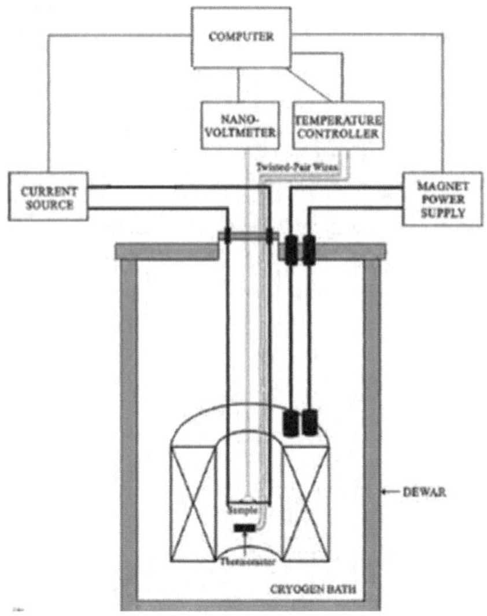

Figure 3. The scheme of the resistance measurement system and cryogenic magnet equipment.

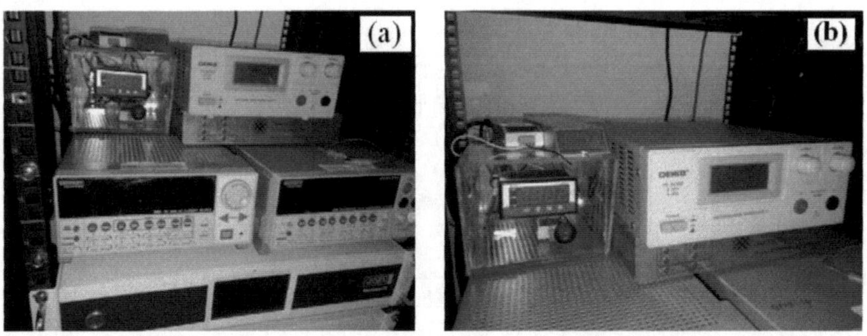

Figure 4. (a) All measurement instruments. (b) An addition instrument to measure the current density

Figure 4(a) shows all measurement instruments which consists of Current Source (Keithley 6221), Nanovoltmeter (Keithley 2700), and Temperature Control (Oxford Mercury iTC) that were connected to the computer. Figure 4(b) shows additional instruments which consists of a rectifier and an ampere meter for measuring current density. We developed the program that could identify and control these instruments. Since the original program provided by the manufacturer is an executing program and cannot be modified, we have to develop our program that can be revised and modified. The revision and modification are needed when we have some trouble with the program and when we have to replace and add the instruments.

Keithley 2700 is a 61/2 –digit high performance multi-meter/data acquisition system with Keithley 7709 module inside. It measured the sample voltage, which was used to calculate the resistance of the sample. The measurement system was using Keithley 6220 to generate current source. Oxford Mercury iTC was used as a temperature controller for controlling the heater and monitoring the temperature of the sample. All of these instruments were controlled remotely using a computer via GPIB and TCP/IP communications.

Table 1. Remote addresses of the instruments

Instrument name	Address	Function
Keithley 2700	GPIB0::16::INSTR	Voltmeter
Keithley 6221	GPIB0::30::INSTR	Current source
Keithley 7709	GPIB0::8::INSTR	Scanner card
Oxford Mercury iTC	tcip::192.168.001.247::7020::socket	Temperature Controller

Table 1 shows that each instrument has its own address to be connected with the computer program. Three instruments from Keithley connected with GPIB communication and Oxford Mercury iTC is connected with TCP/IP communication.

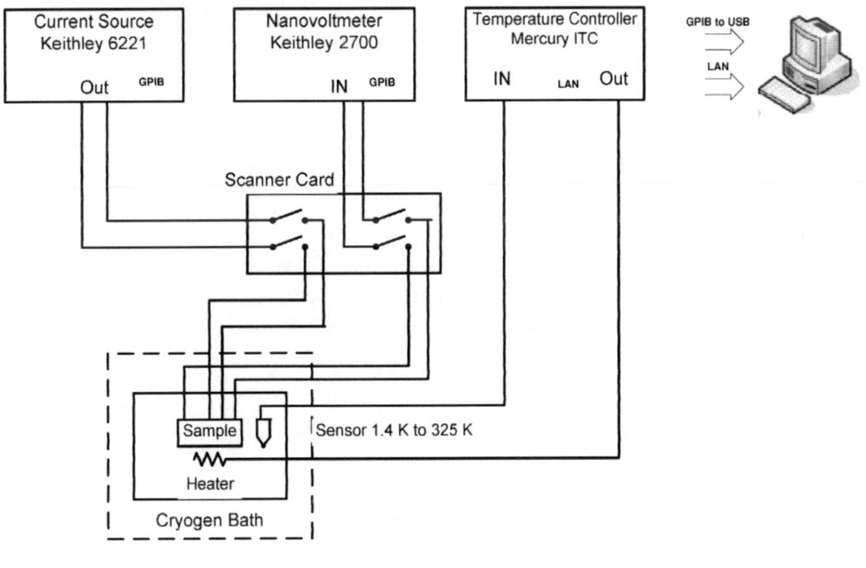

Figure 5. Layout of the measurement system

Figure 5 shows the layout of the measuring system. Scanner card was inserted inside Keithley 2700. Scanner card worked to close and open terminal/ channel of the probes of the current and the voltage of the sample. Scanner card also was used to measure two samples instead of one sample. Temperature of the sample was controlled by temperature controller using a temperature sensor and a heater located near the sample. Current source, nanovoltmeter, temperature controller and scanner card were controlled by LabVIEW programs developed in this study. LabVIEW program was controlling sample temperature from room temperature to as low as 1.5 K, at the same time measuring the resistance of the sample and then saving the data into a file.

3.2. Controller system

Figure 6(a) shows the user interface of the LabVIEW program from built-in program and figure 6(b) shows the program developed in this study. The built-in program is an executing program that cannot be upgraded, modified or changed. The developed program can be upgraded, modified and changed easily. The developed program has block diagrams that can be opened and modified. The program was running during the measurement process. It continuously measured the resistance, controlled the temperature, collected data and saved the data into a file.

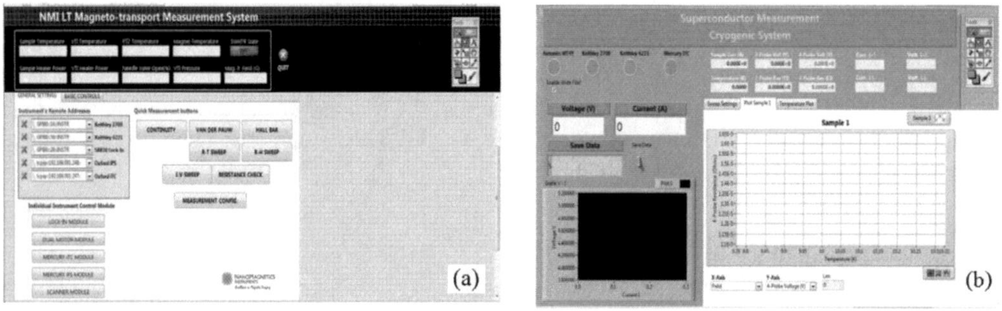

Figure 6. (a) Front panel of the original built-in software. (b) Front panel of the developed software

Figure 7 shows the block diagram of the LabVIEW program developed in this study. LabVIEW is a graphical language program that uses graphical items and wires as its source code. It can also display indicators and graphs. The graph of temperature dependence of resistance was set to be shown so that we can analyze the results during measurement process. The program was controlling the temperature using a heater. Measurement parameters, calculated resistance data and temperature data were saved on a file during the measurement process.

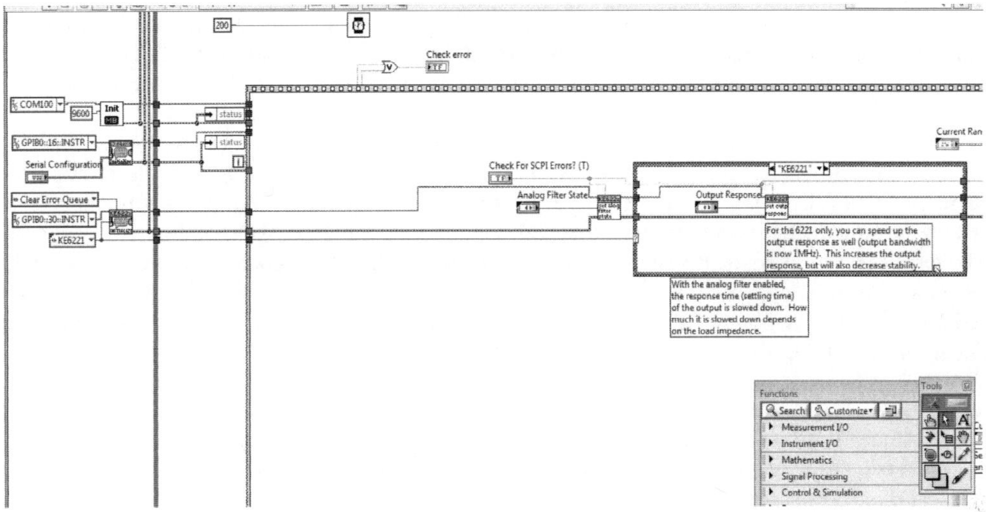

Figure 7. A part of block diagram measurement system

3.3. Testing result

Figure 8 shows the result of the measurement test using the standard superconducting sample. The measurement was conducted from room temperature to low temperature using the LabVIEW program developed in this study. It showed a sharp drop of resistance at around 98K and the zero resistance below 98K, indicating its critical temperature.

The measurement result was clear and had a very small noise. It indicated that the LabVIEW program developed in this study had been successfully performed measuring temperature and resistance of the sample.

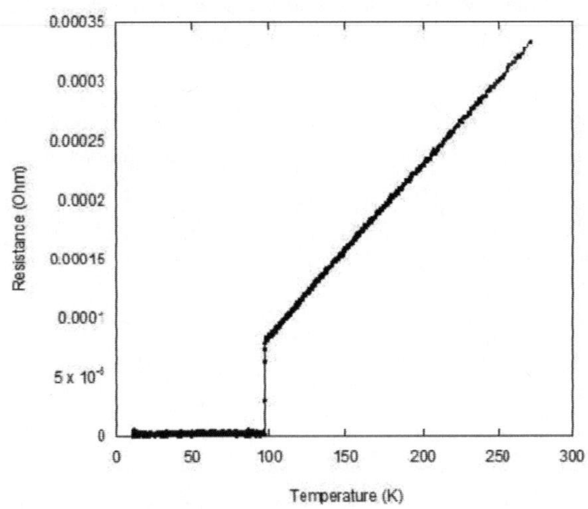

Figure 8. Bi-2212 standard sample measurement result.

4. Conclusions

The development of the resistance measurement system for cryogenic equipment controlled by LabVIEW program has been done. The developed LabVIEW program can be modified and revised. The instruments can also be replaced and added easily. When we have to replace or add some instruments, we only have to modify the block diagram of the LabVIEW program. We also can modify the program when we have to develop another measurement system, such as specific heat, etc. It was using several instruments to measure voltage, generate electrical current, and determine/ control the temperature of the sample. Samples were measured by four-point probe method to calculate the resistance value and a temperature sensor to determine the temperature of the sample.

The developed measurement system was tested using standard superconducting sample. It clearly showed its temperature dependence of resistance, and it showed that critical temperature is at around 98K, indicating that the measurement system was successfully developed. The measurement system has its advantages to be able to measure current density too.

References

[1] van Duzer T 1990 *Cryiogenic* **30** 980
[2] Madaro L, Rijllart A, Saban R, Walckiers L and Bottura L 1996 *Proc. European Part. Accel. Conf. (Barcelona)* vol 1-3 (New York: INSPIRE HEP)
[3] Dhanalakshmi K and Avinash A 2010 *Int. J. Smart Sensing Intell. Syst.* **3**(2) 156
[4] Kohl M and Harison J P1991 *Cryogenic* **31** 369
[5] Sarwinski R E 1977 *Cryogenics* **17** 671
[6] Ferguson G 1988 *Cryogenic* **28** 688
[7] Honsberg C and Bowden S 2016 *Four Point Probe Resistivity Measurements* [http://www.pveducation.org]
[8] Stark S, Rossi A A and Palmieri V 2012 *SPES Facility and Radioactive Ion Beams Annual Report* p 185
[9] Song W J, Fang X Y, Fang J, Wei B, Hou J Z, Liu L F, Lu K K and Li S 2015 *Physica C Supercond.* **518** 111
[10] Kotarba A and Jurkiewicz P 2011 *Proc. Int. Part. Accel. Conf. (San Sebastián)* (USA: Physical Review Special Topics - Accelerators and Beams) p 1756
[11] Afsarmanesh N and Ahmed PZ 2011 *Int. J. Smart Sensing Intell. Syst.* **4**(4) 57

IOP Conf. Series: Journal of Physics: Conf. Series **817** (2017) 012056 doi:10.1088/1742-6596/817/1/012056

Effect of additive Na_2O on sintering temperature, crystal structure and magnetic properties of $BaFe_{12}O_{19}$ magnet

Suprapedi[1], P Sardjono[1], Muljadi[1], N Rusnaeni [1] and S Humaidi[2]

[1] Research Center for Physics-Indonesian Institute of Sciences

[2]Departement of Physics-FMIPA, University of North Sumatera, Indonesia

E-mail: mulj004@lipi.go.id

Abstract. The $BaFe_{12}O_{19}$ magnet has been made at several sintering temperature (1100, 1150, 1200, 1250 and 1300°C) and various additive Na_2O (0, 1, 2 and 3 % of weight) by using Powder Metallurgy. The $BaFe_{12}O_{19}$ powder and $NaHCO_3$ powder as Na_2O source were mixed with ethanol until homogeneous by using magnetic stirrer, dried at 80°C and calcined at a temperature 1000 °C for 1 hour. The calcined powders were milled by using a ball mill in dry condition. Then, the powders were mixed with 3 % PVA as a binder, then, the powder is put into a dies mold and pressed with a pressure of 30 MPa. The disc samples were sintered by using furnace Thermolyne. The characterization of the sintered sample was done such as measurement of density and porosity, analysis of crystal structure and measurement of magnetic properties. The characterization results show that the optimum sintering temperature is 1200 °C with additive 1% of Na_2O, if more than 1 %, the physical and magnetic properties decrease. The composition of additive N_2O at the sintering temperature 1200°C does not influence the crystal structure.

1. Introduction

Barium ferrites are well known hard magnetic materials, which are based on iron oxide. They are also called as ferrite magnets and could not be easily replaced by any other magnets [1]. Hexagonal barium ferrite having the chemical formula of $BaFe_{12}O_{19}$ are widely used in magnetic recording media, microwave devices, and electromagnetic shielding fields [1]. The barium ferrite is attractive materials for permanent magnet due to their large uniaxial anisotropy, high chemical stability and corrosion resistance [2]. The Magnetic saturation of Barium hexaferrite is 380 emu/cm^3 or 72 emu/g at room temperature and the Curie point is 450^{0}C [3]. Many methods of synthesis have been developed to obtain a low production cost of powder particles of barium ferrite. There are many methods such as powder metallurgy method and chemical routes (sol-gel or co-precipitation). The powder metallurgy method is preparation process by using solid-solid mixing, and continued with sintering process. Sintering is a heat treatment applied to a powder compact in order to impart strength and integrity. The temperature used for sintering is below the melting point of the major constituent of the Powder Metallurgy material [4]. The properties of final product can be influenced by impurity of raw materials, mole ratio BaO to Fe_2O_3 and particle size or grain size and sintering temperature [4,5]. The sintering temperature of Ba-Ferrite is high enough about 1300°C [5], and during the sintering process, the grain will be growth. The development of grain size after sintering process can influence of magnetic properties, where the grain size becomes bigger, the magnetic properties will reduce [6]. To prevent grain size does not grow rapidly after the sintering process it is necessary to add ingredients additive. Some additives may be used such as silica (SiO_2), Al_2O_3, Na_2O, Fe_2O_3 [7]. Not only

IOP Conf. Series: Journal of Physics: Conf. Series **817** (2017) 012056 doi:10.1088/1742-6596/817/1/012056

dampen the grain growth, but those additives can also serve lower the sintering temperatures [7]. The aim of this paper is to know the effect of Na_2O additive on sintering temperature, the crystal structure and magnetic properties of $BaFe_{12}O_{19}$ prepared by Powder Metallurgy.

2. Methodology

The raw materials used in this work were barium hexa ferrite $BaFe_{12}O_{19}$ powder (purity 99.0%) and $NaHCO_3$ powder (purity 99.90 %) as a Na_2O source . The sintering aid Na_2O was varied as 0, 1, 2 and 3 wt% in the raw material mixture. The raw materials were weighed in calculated amount and mixed with ethanol until homogeneous by using magnetic stirrer, then they are dried at 80^oC and calcined at a temperature 1000^oC with holding time 1 hour by using Thermolyne Furnace. The calcined powders were milled by using a ball mill in dry condition for 24 hours. The fine powders were mixed with 3 % PVA as a binder, then, they are put into a dies mold and compacted with a pressure of 30 MPa. The disc samples were sintered by using Thermolyne Furnace with heating rate about 8^oC/minute from room temperature to 400^oC, and hold for 1 hour, then the temperature was increased from 400^oC to sintering temperatures (1100, 1150, 1200, 1250 and 1300^oC) with heating rate of 10^oC/minute , and at those temperature were hold for 1 hour. The sintered samples were measured for density and porosity by using Archimedes method. The phase identification study of sintered samples was carried out by X-ray diffractometer. Magnetic properties of the barium hexaferrite samples were measured by using Gauss meter and vibrating sample magnetometer (VSM).

3. Results and Discussion

The XRD results of the samples that have been sintered at a temperature of 1200^oC are shown in figure 1.

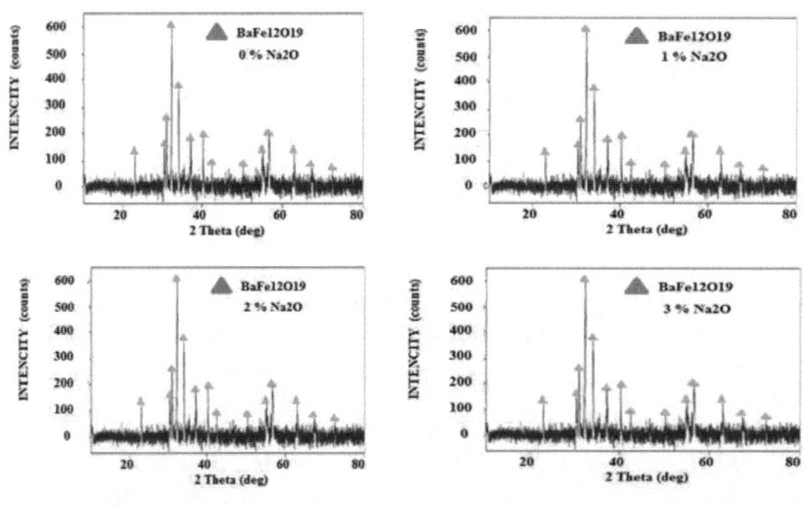

Figure 1. XRD- patterns of 1200^oC sintered samples with variation of Na_2O additives.

The XRD analysis is performed only on samples using additives and without additives and had been sintered at a temperature of 1200°C. The results of X-ray diffraction pattern show that all the samples have the same diffraction pattern, as presented in figure 1. Based on ICDD card No. 00-027-1029 that peaks that formed an $BaO6Fe_2O_3$ phase with a hexagonal crystal structure and have lattice parameters a = b = 5,892 Å, c = 23 198 Å, cell volume = 697.44 Å³. The addition of additives to 3% by weight Na_2O does not give effect to changes in the crystal structure.

The measurement result of bulk density of sintered pellets is shown in figure 2. The curve shows that the value of bulk density of sintered sample with 0 % wt.Na_2O increases sharply from sintering temperature 1100°C to 1300 °C. However the bulk density values of samples with Na_2O additives (1% and 2 %) also tend to increase the sintering temperature range between 1100 to 1200 °C, and tend to be slightly down between 1200 to 1300°C. But sample with 3 % Na_2O tends to increase the sintering temperature range between 1100 to 1200 °C and tends to decline sharply at the sintering temperature range between 1200 to 1300 °C.

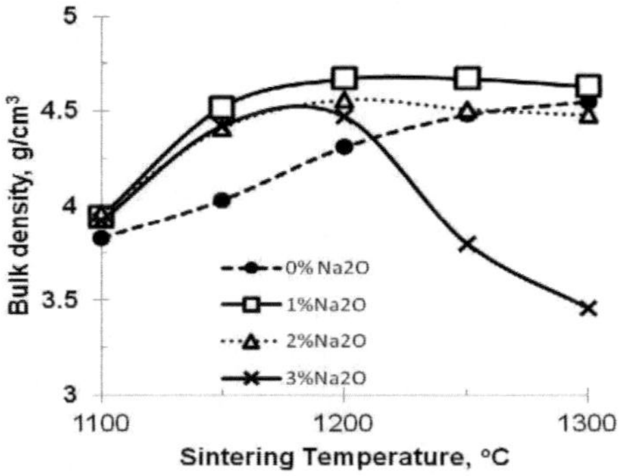

Figure 2. The bulk density of sintered sample as function of sintering temperature.

Based on the results shown in figure 2, the densification process samples without additive has been required high temperatures up to 1300°C to obtain maximum bulk density value (4.55 g/cm³). Meanwhile, the maximum bulk density of samples with addition of Na_2O (1% to 3 % Na_2O) can be achieved at sintering temperature 1200°C. In this case that the Na_2O additive has a function as flux materials so as to lower the melting temperature of a materials, which consequently, the sintering temperatures can be reduced.

Figure 3 shows a curve of porosity of sintered samples as function of sintering temperature. The value of porosity tends to decrease for all sample, it is due to a mechanism of sintering process.

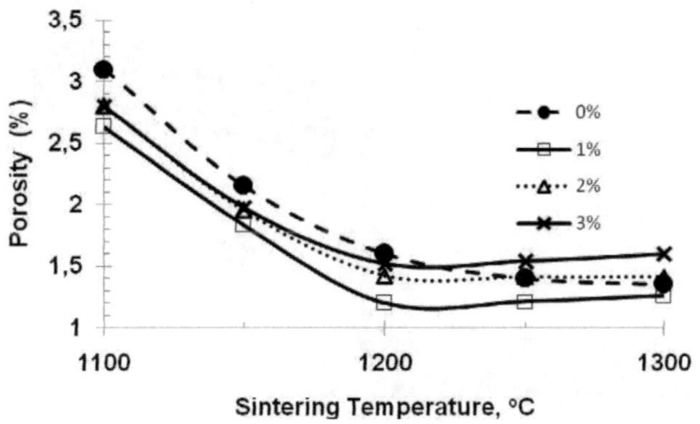

Figure 3. The curve of porosity sintered sample with variation of Na_2O additives.

Densification and grain growth occur during a sintering process at high temperature, this can cause shrinkage and pore reduction [8]. The porosity value of sample with Na_2O additives is lower compared to sample without Na_2O additive. The lowest porosity value (1.12 %) is achieved on sample with 1% Na_2O at a temperature sintering 1200 ^oC, but the porosity values of samples with Na_2O additive greater than sample1% tends to increase again. The value of porosity of samples with Na_2O additive tends to be slightly down at sintering temperature more than 1200^oC. Na_2O additive capable of depressing the value of the porosity and improve densification to a temperature of 1200^oC, but to a temperature above 1200^oC tends to diminish the level of densification, because the grain growth during the sintering process to a temperature of 1200^oC can be prevented. And grain will grow up again at sintering temperatures above 1200^oC.

The measurement results of magnetic properties of sintered samples show at figure 4 and figure 5. Figure 4 shows a flux magnetic (FM) value with variations of sintering temperature. The flux magnetic is a magnetic strength on surface material or amount of magnetic flux in an area taken perpendicular to the magnetic flux's direction and it is measured by using Gauss meter.

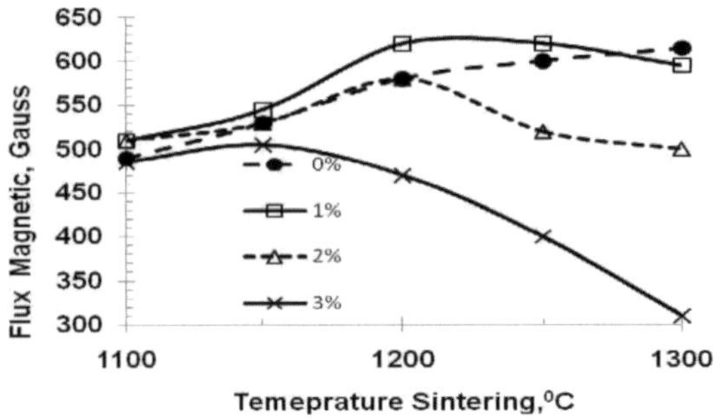

Figure. 4. Flux magnetic curves for sintered samples with variation of sintering temperature.

Where value of FM for samples with 0 wt% Na_2O tends to increase with increasing of sintering temperature, the maximum value of FM is 610 Gauss at sintering temperature 1300^oC. The value of FM of samples with Na_2O additive tends to increase from temperature sintering 1100^oC until 1200^oC. Because all samples with the additive at this condition have maximum densification, and the grain growth can be prevented by using the Na2O additive. But the FM value of samples with Na_2O additive tend to decrease at temperature sintering more than 1200^oC, it is due to lower bulk density value. The highest value of FM is about 620 Gauss for the sample with 1 wt% Na_2O and at sintering temperature 1200^oC. Figure 5 shows hysteresis curve of 1200^oC sintered sample with the variation of Na_2O additives.

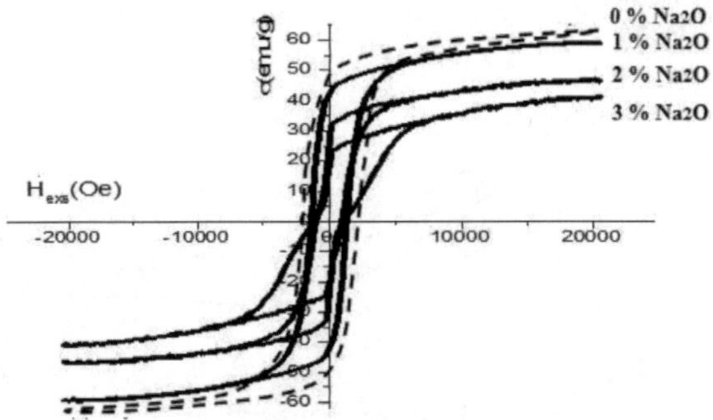

Figure 5. Hysteresis curve of 1200^oC sintered sample with variation of Na_2O additives.

The hysteresis curve as seen in figure 5 shows that the Na_2O additive composition can influence the hysteresis loops. The area of hysteresis loops trends smaller and it indicates that magnetic properties such as magnetic saturation (ms), remanence (mr) and coercivity (Hc) become lower. According to the measurement results of bulk density and porosity, the Na_2O additive can significantly influence the densification until sintering temperature 1200°C. It means that there are not grain growth until temperature 1200°C. According to the research results of J. Topfer et.al. [7], the magnetic properties tend to increase as the grain size decreases. Further, they reported a decrease in saturation magnetization with decreased grain size [7]. It can be proved that based on table 1, samples without additives Na_2O have magnetic saturation value (ms) is higher than the ms value of the sample with additives Na_2O. That means grain growth does not occur in accordance with the level of densification tends to increase.

The Na_2O additive value gives significantly effect on the value of ms and mr, and coercivity value (Hc) is nearly constant. Table 1 shows the value of parameter-parameter magnetic properties according to results of hysteresis loop graph. The sample without Na_2O additive has the highest value of mr (48.50 emu/g) compared to other samples, but value Hc, ms and energy product BHmax are near constant.

Table 1. Value of Remanence (**mr**), coercivity (**Hc**), magnetic saturation (**ms**) and energy product **(BH)max** of 1200°C sintered sample with variation of Na_2O additives.

Sample	mr (emu/g)	Hc (kOe)	ms (emu/g)	BHmax (MGOe)
$BaFe_{12}O_{19}$	48.50	1.322	61.07	0.144
$BaFe_{12}O_{19}$ + Na_2O 1%	42.15	1.314	52.18	0.135
$BaFe_{12}O_{19}$ + Na_2O 2%	25.72	1.282	47.24	0.129
$BaFe_{12}O_{19}$ + Na_2O 3%	18.14	1.252	41.28	0.119

4. Conclusion

In this study, the effect of Na_2O additive has been investigated on the densification, magnetic properties and crystal structure of barium ferrite by solid state sintering method. It was observed that Na_2O additive can influence significantly on densification and magnetic properties, but it does not give effect on the crystal structure. The optimum percentage of additive is 1 % wt.of Na_2O, and the sintering temperature becomes lower at this condition. Permanent magnet $BaFe_{12}O_{19}$ with 1 % wt. of Na_2O and sintered at 1200°C has properties such as : bulk density = 4.67 g.cm^{3}, porosity = 1.21 %, flux magnetic = 620 Gauss, remanence =42.15 emu/g, coercivity =1.314 kOe and energy product =0.135 MGOe.

Acknowledgments

The authors would like to thank on "Riset Mandiri Program PPF-LIPI 2016" for financial support and thank you for operators XRD and VSM at Research Center for Physics-LIPI.

References

[1] Nowosielski R, Babilas R, Dercz G, Paj L, Wrona J 2007 *Arch. Mater. Sci. Eng.* **28** 735-42
[2] Molaei M J, Ataie A, Raygan S j, Picken, Mendes E, Tichelaar F D 2012 Powder Technol **221** 292–5
[3] Sadullahoslua G, Ertusb B, Gökçeb H, Altuncevahirc B, Öztürkd M, Topkayad R, Akdosand N, Öveçoslub M L and Addemirb O 2015 *Acta Phys. Pol.* **128**(3) 377-82
[4] Vidyawathi S S, Amaresh R and Satapathy I N 2012 *Bull. Mater. Sci.* **25**(6) 569–72
[5] Zhang H W, Li J, Su H, Zhou T C, Long Y, and Zheng Z L 2013 *Chin. Phys. B* **22**(11) 1-15
[6] Mahgoob A and Hudeish A Y 2012 *Acad. Schol. Res. J.* **4**(5) 112-20
[7] T Li J, Zhang H W, Li Y X, Liu Y L and Ma Y B 2012 *Chin. Phys. B* **21**(1) 1-4

Preparation and characterization of Fe-Mn-doped Barium Hexaferrite permanent magnet

C Kurniawan[1], M M Nainggolan[2], K Sebayang[2], M Ginting[1] and P Sebayang[1]

[1] Research Center for Physics – LIPI, South Tangerang, Banten, Indonesia
[2]Department of Physics, Universitas Sumatera Utara, Medan, North Sumatera, Indonesia

Email : pard003@lipi.go.id

Abstract. Barium Hexaferrite-based permanent magnets ($BaFe_{12}O_{19}$) was known for its high magnetic anisotropy and suitability in broad applications. Some dopants and atomic substitutions have been utilized to improve its properties for special purposes. In this paper, the Fe-Mn system was used as a dopant for preparing Fe-Mn-doped barium hexaferrite permanent magnet using mechanical alloying method. The physical properties of the samples, such as bulk density, and porosity were examined to study the effect of the dopant. In addition, the crystal structure and magnetic properties of the samples were analyzed using X-Ray Diffractometer (XRD) and Vibrating Sample Magnetometer (VSM), respectively. It is found that the addition of Fe-Mn into barium hexaferrite contributes on the appearance of minor phases such as iron oxide-based magnetite and hematite. In addition, the XRD peak shifted to smaller angle which is likely due to Mn ion substitution and lattice strain within the hexaferrite crystal. It is also observed that the magnetic properties of Fe-Mn-doped barium hexaferrite was inferior to that of the undoped samples. It means that the formation of magnetite and hematite from Fe-Mn dopant during the sintering process is dominant and results to the reduction of hard magnetic properties of the samples.

1. Introduction

Barium hexaferrite-based ferromagnetic materials ($BaFe_{12}O_{19}$) have been investigated for many applications such as permanent magnet, microwave absorber, and magnetic recording media [1,2]. Barium hexaferrite is a long lasting research area for nearly a century due to its high magnetocrystalline anisotropy, chemical stability, excellent frequency, and Curie temperature [3]. In recent years, the study focused on the effect of elemental substitution to M-hexaferrite structure in correlation with its magnetic and electrical properties. The addition of rare-earth element e.g. Samarium contributes to the decrease of remanence, with the slight increase of coercivity of M-hexaferrite structure [4]. In other studies, the addition of Cerium into M-hexaferrite structure produces an enhancement of magnetization and coercivity, which makes it suitable as a microwave absorbing material [5,6]. The common addition of Ti-based ion generally increase the electromagnetic properties such as permeability and permittivity [7,8].

On the other hand, the development of nanocomposite-based magnets is also interesting due to its useful magnetic characteristics, which can be achieved by employing the exchange-spring interaction between hard and soft magnetic structure [9,10]. As an example, the combination of ferromagnetic and anti-ferromagnetic oxide have revealed a new property which is important for magnetic recording and permanent magnet applications [11].

In this paper, we investigated the effect of Fe-Mn alloy addition to M-type structure of barium ferrite on the physical and magnetic properties of the barium hexaferrite-based magnet.

2. Experimental Method

The starting materials used for this experiment were commercial barium hexaferrite and Ferromanganese (Fe-Mn) powders. Based on the XRD data, the barium hexaferrite powder has major phase of $BaFe_{12}O_{19}$ and minor phase of Fe_2O_3. Whereas, the ferromanganese powder is composed by α-Fe and MnO_2. At first, barium hexaferrite and ferromanganese powder were processed separately using dry-ball mill for 24 hours and wet-high energy milling method for 1 hour, respectively, to refine the particle size. The as-milled powders were then blended in a shaker mill for 15 minutes in order to obtain the homogeneous mixture. The as-blended powder was compacted in a magnetic field press machine under 40 kgf/cm^2 pressure and 3.6 kOe parallel magnetic field. The magnetic field was utilized in the compaction process to produce magnetically-anisotropic samples and increase the magnetization. The as-compacted samples were sintered in a furnace at 1100, 1150, 1200, and 1250 °C, respectively, for 2 hours.

The particle size distribution was determined using optical microscope with the assistance of imageJ software [12], by measuring the particle sizes and estimating them to get the size distribution. The crystal structure of the mixed powders were analyzed using X-Ray Diffratometer (Rigaku Smartlab). The crystallite size of different phases were calculated using the Debye-Scherer equation [13]:

$$d = \frac{0.9\lambda}{\beta \cos\theta} \quad (1)$$

Where d is the crystal size, λ is the X-ray wavelength, β is the Full Width at Half Maximum (FWHM) value of the related diffraction peak, and θ is the diffraction angle.

The magnetic properties were analyzed using Vibrating Sample Magnetometer (VSM-250) under external magnetic field of 2 Tesla. The physical characteristics were measured using Archimedes density and porosity measurement method according to ASTM C-373.

3. Results and Discussions

The particle sizes of the as-milled powders were determined using optical microscope which is equipped with CCD camera Pax-Cam (40x objective lens magnification). Size analyses were performed using ImageJ analysis software to produce size distribution data as displayed in figure 1. The particle size of doped and undoped-barium hexaferrite shows a Gaussian distribution with average value of 126, 121, and 120 nm for undoped, 3 wt% Fe-Mn, and 7 wt% Fe-Mn, respectively. Interestingly, the size distribution of milled Fe-Mn alloy shows irregular shape with average size value of 93 nm.

After compacted and sintered, the bulk density and porosity of the-doped barium hexaferrite samples were characterized using Archimedes method. By measuring the ratio of dry and wet mass (ASTM C373), the density and porosity of the samples can be depicted in Figure 2. The figure shows that the addition of Fe-Mn alloy into hexaferrite structure gives the decrease of density as well as the increase of porosity. The optimum value of density and porosity are 4.81 g/cm^3 and 20.7 %, respectively, which was achieved by 3 wt% Fe-Mn doped sample, sintered at 1200 °C. Based on figure. 2, the density and porosity values are affected by sintering temperature. It shows that, in general, the densification processes are inversely proportional with the pore formation during sintering step. The effective sintering process is needed to obtain higher sintered density and minimize porosity in order to increase the magnetic properties [14].

IOP Conf. Series: Journal of Physics: Conf. Series **817** (2017) 012057 doi:10.1088/1742-6596/817/1/012057

The XRD analyses of the sintered samples were performed to characterize the microstructural effect of Fe-Mn addition. The polycrystalline structure was observed, with $BaFe_{12}O_{19}$ as the main phase. Figure 3 shows the diffraction profiles, which concludes that the hexaferrite structure is still preserved after the addition of Fe-Mn alloy. The Fe-Mn alloy itself is composed by α-Fe and MnO_2 structure, which are marked with star (*) and hole-circle (o), respectively, in the diffraction profile. Meanwhile, the starting barium hexaferrite powder is composed by $BaFe_{12}O_{19}$ structure and Fe_2O_3 as a minor phase (~ 13%). Figure 3 reveals the increase of iron-oxide formation as Fe-Mn composition increases. The magnetite (Fe_3O_4) and hematite (Fe_2O_3) peak was observed for 3 and 7 wt% Fe-Mn addition, respectively. The α-Fe phase from Fe-Mn alloy was expected to be oxidized during sintering process and form the more stable oxide structure.

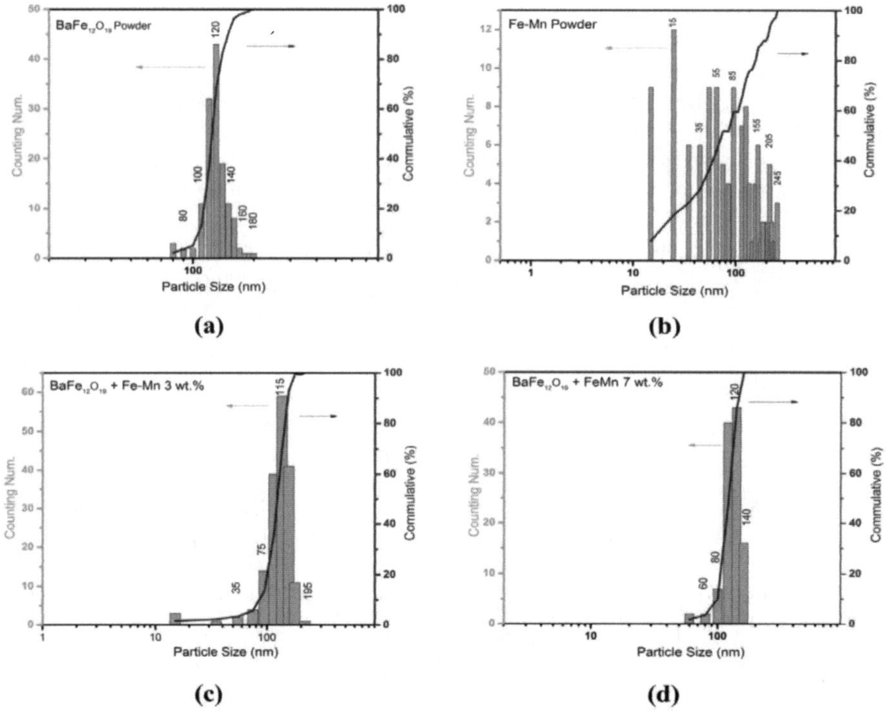

Figure 1. Particle size distribution of (a) undoped, (c) 3wt% Fe-Mn, and (d) 7 wt% Fe-Mn barium hexaferrite, compared to (b) the milled Fe-Mn alloy.

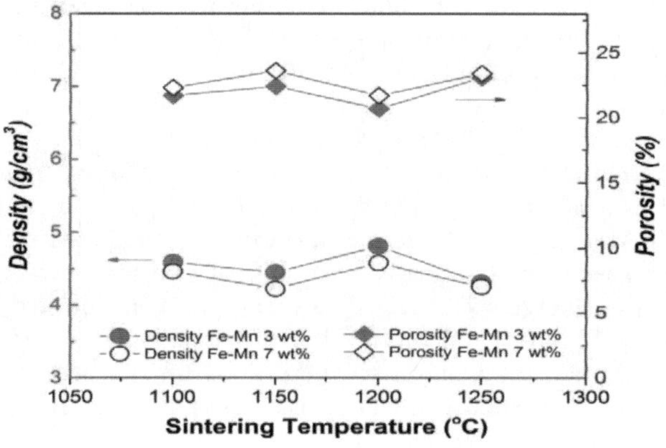

IOP Conf. Series: Journal of Physics: Conf. Series **817** (2017) 012057 doi:10.1088/1742-6596/817/1/012057

Figure 2. Bulk density and porosity values of 3 wt% (red mark) and 7 wt% (blue mark) Fe-Mn-doped barium hexaferrite.

Based on equation (1), the crystallite size of all samples are in nanometer scale, ranged from 33.5 to 44.4 nm. The average values of hexaferrite crystallite size for 4 major peaks at (110), (008), (107) and (114) planes are 40.28 nm, 40.27 nm, and 39.68 nm for undoped, 3 wt% and 7 wt% Fe-Mn, respectively. It means that the addition of Fe-Mn alloy into hexaferrite structure slightly reduces the crystalline size, which become significant as the Fe-Mn addition increased.

MnO_2 phase, which is originated from the starting Fe-Mn alloy, does not appear in the diffraction profile for 3 and 7 wt% addition. However, the peak shift is detected for this addition value. The inset (right profile) of figure 3 shows that the (110), (008), (107) and (114) planes shift to smaller 2θ as Fe-Mn composition increases. The lattice parameters (a and c) of hexagonal ferrite crystals can be determined for (110) and (008) planes using following equations:

$$a = \frac{\lambda}{\sqrt{3}\sin\theta}\sqrt{h^2 + hk + k^2} \tag{2}$$

and

$$c = \frac{\lambda}{2\sin\theta} l \tag{3}$$

where λ is x-ray wavelength and θ is selected diffraction peak position for specified ($h\ k\ l$) planes. The calculated lattice constant a (and c) for undoped, 3 % and 7 % Fe-Mn-doped barium hexaferrite are 5.854 Å (22.941 Å), 5.881 Å (23.158 Å), and 5.88 Å (23.180 Å), respectively. It means that the addition of Fe-Mn alloy also increases the lattice constant (both a and c) as the Fe-Mn addition increases. The Mn^{2+} ions might partly substitute the Fe^{2+} ions in hexagonal structure since the ionic radii of Mn^{2+} (0.66 Å) is slightly larger than Fe^{2+} (0.63 Å). Therefore, we believe that these substitutions contribute to the increase of lattice constant as the Fe-Mn addition increases.

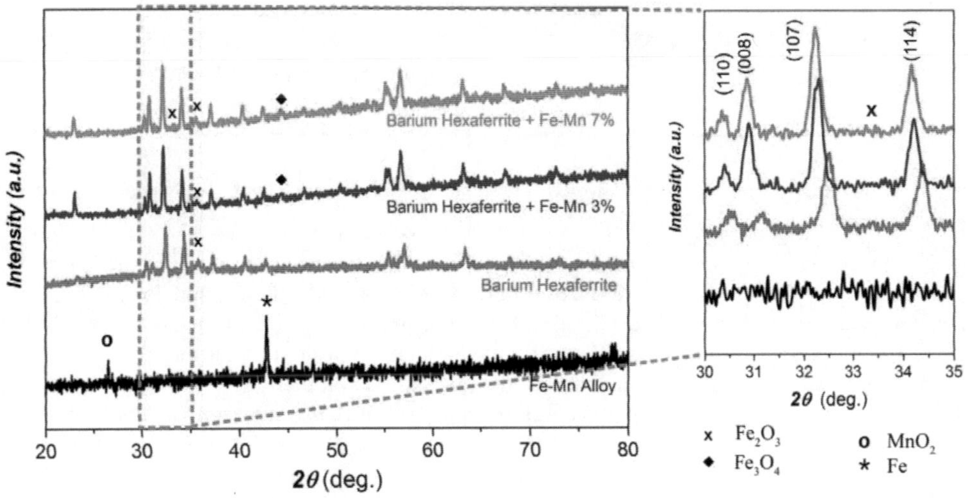

Figure 3. XRD profile of the Fe-Mn-doped barium hexaferrite samples. Inset (right profile) shows peak shift of the doped and undoped samples.

The magnetic properties of sintered Fe-Mn-doped barium hexaferrite samples were analyzed using vibrating sample magnetometer as displayed in figure 4. The hysteresis graph shows that Fe-Mn-doped barium hexaferrite is a permanent magnet. The effects of Fe-Mn addition are described as the change in the magnetic characteristics, which are magnetic saturation (σ_s), remanence (σ_r), and coercivity field (H_{ci}). Figure 4 shows that the magnetic properties slightly decrease as the Fe-Mn composition increases. The reduction of magnetic characteristics is correlated with the crystal structure formed during addition and sintering process. The iron-oxide formations with soft magnetic characteristics contribute to the magnetic properties reduction of Fe-Mn-doped hexaferrite. The increase of iron-oxide, such as in the 7 wt% Fe-Mn-doped barium hexaferrite, lowers the remanence and coercivity. From the inset, the remanence (coercivity) of samples have a decreasing trend with the value of 33.8(3.8), 31.4 (3.0) and 29.3 emu/g (2.73 kOe) for undoped, 3, and 7 wt% Fe-Mn-doped barium hexaferrite, respectively.

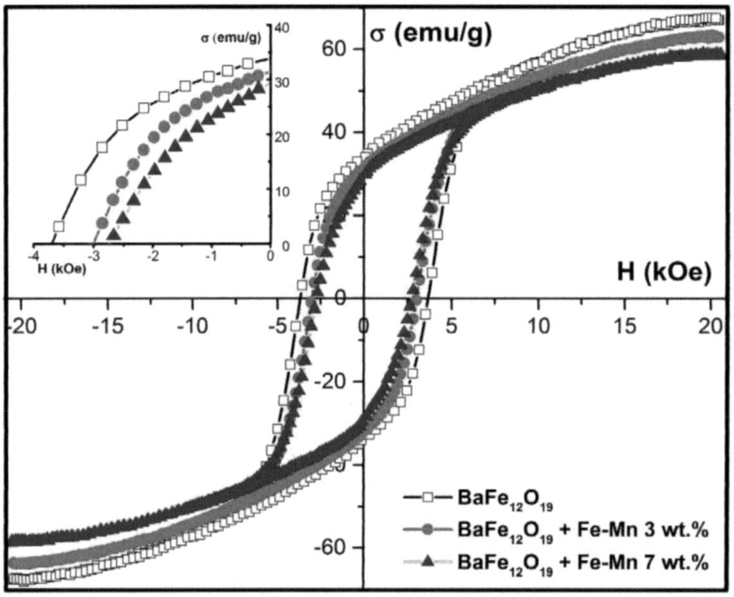

Figure 4. Magnetic properties of undoped (rectangle-black curve) and Fe-Mn-doped barium hexaferrite samples.

Based on the above data, we find that the addition of Fe-Mn alloy into barium hexaferrite structure contributes to the magnetic properties reduction due to the iron-oxide (Fe_2O_3 and Fe_3O_4) formation during sintering process [15,16]. We expect that the nanocomposite formation of Fe-Mn and barium hexaferrite should be free from the humid environment. Thus, the iron as a ferromagnetic intergrain phase will raise the exchange interaction and enhance its magnetic properties. The Mn ion also play an important role in the lattice structure modification and the increase of magnetic anisotropy. We propose to consider this study for the development of high properties composite permanent magnet.

4. Conclusion

In this experiment, the Fe-Mn-doped barium hexaferrite permanent magnet has been prepared by mechanical alloying method. The as-milled powder of commercial barium hexaferrite and ferromanganese (Fe-Mn alloy) was employed in the study. Based on the microstructure analyses, the addition of Fe-Mn into barium hexaferrite contributes to the appearance of minor phase such as iron

IOP Conf. Series: Journal of Physics: Conf. Series **817** (2017) 012057 doi:10.1088/1742-6596/817/1/012057

oxide-based magnetite and hematite. In addition, the Mn ion shifts the XRD peak into smaller angle which is correlated with lattice strain within hexaferrite crystal lattice. The magnetic properties of Fe-Mn-doped barium hexaferrite were smaller than that of the undoped sample. This means that the iron oxide formation during sintering process plays the important role in the reduction of hard magnetic properties of the samples. Moreover, the environmental condition need to be considered for the development of high properties composite permanent magnets.

References

[1] Robert C P 2012 *Prog. Mater. Sci.* **57** 1191
[2] Vincent G H, Anton G, Yajie C, Soack D Y, Mingzhong W, Aria Y, Zhaohui C, Peng H, Patanjali V P, Xu Z, Carl E P, Manasori A, Olivier A and Carmine V 2009 *J. Magn. Magn. Mater.* **321** 2035
[3] Naini D, Mansi C, Inderjeet S S, Jaspreet S J, Shivani M and Rajeev A 2016 *Cogent Phys.* **3** 1208450
[4] Kush R, Preeti T, Atul T, Monika T, Vinay G, Jean L M and Patrick 2016 *Ceram. Int.* **42** 8413
[5] Ram A P, Desai S S, Tamboli Q Y, Sagar E S and Sunil M P 2015 *J. Magn. Magn. Mater.* **378** 59
[6] Zahra M, Parviz K, Poorbaferani A, Mojtaba R and Salamati H 2016 *J. Magn. Magn. Mater.* **397** 101
[7] Charanjeet S, Shalini B N, Hudiara I S, Bai Y and Marina K 2009 *Mater. Lett.* **63** 1921
[8] Yuan C L and Tuo Y S 2013 *J. Magn. Magn. Mater.* **342** 47
[9] Pal M, Bid S, Pradhan S, Nath B, Das D and Chakravorty D 2004 *J. Magn. Magn. Mater.* **269** 42
[10] Roy D, Shivakumara C and Anil Kumar P S 2009 *J. Magn. Magn. Mater.* **321** L11
[11] Kodama R H and Edelstein A S1999 *J. Appl. Phys.* **85** 4316
[12] Schneider C A, Rasband W S and Eliceiri K W 2012 *Nat. Methods* **9** 671
[13] Cullity B D 1978 *Elements of X-Ray Diffraction* 2d ed Addison-Wesley Pub
[14] Liu Y, Li Y, Zhang H, Chen D and Wen Q 2010 *J. Appl. Phys.* **107** 09A507
[15] Nowosielski R, Babilas R, Dercz , Pajak L and Wrona J 2007 *Arch. Mater. Sci. Eng.* **28** 735
[16] Molaei M J, Ataie A and Raygan S 2011 *J. Phys. Conf. Ser.* **292** 012016

IOP Conf. Series: Journal of Physics: Conf. Series **817** (2017) 012058 doi:10.1088/1742-6596/817/1/012058

Fabrication of visible light-triggered photocatalytic materials from the coupling of *n*-type zinc oxide and *p*-type copper oxide

A B Gorospe and M U Herrera

Institute of Mathematical Sciences and Physics, University of the Philippines Los Banos, College Laguna, 4031Philippines

Email: abgorospe4@up.edu.ph

Abstract. Coupling of copper oxide (CuO) and zinc oxide (ZnO) was done by chemical precipitation method. In this method, copper sulfate pentahydrate and zinc sulfate heptahydrate salt precursors were separately dissolved in distilled water; then were mixed together. The copper sulfate-zinc sulfate solution was then combined with a sodium hydroxide solution. The precipitates were collected and washed in distilled water and ethanol several times, then filtered and dried. The dried sample was grounded, and then undergone heat treatment. After heating, the sample was grounded again. Zinc oxide powder and copper oxide powder were also fabricated using chemical precipitation method. X-Ray Diffraction measurements of the coupled CuO/ZnO powder showed the presence of CuO and ZnO in the fabricated sample. Furthermore, other peaks shown by XRD were also identified corresponding to copper, copper (II) oxide, copper sulfate and zinc sulfate. Results of the photocatalytic activity investigation show that the sample exhibited superior photocatalytic degradation of methyl orange under visible light illumination compared to copper oxide powder and zinc oxide powder. This may be attributed to the lower energy gap at the copper oxide-zinc oxide interface, compared to zinc oxide, allowing visible light to trigger its photocatalytic activity.

1. Introduction

Photocatalysis using oxide semiconducting materials is used for environmental cleaning [1-10]. When light, with energy greater than or equal to the band gap of the semiconductor, is incident on the material, it can excite the valence electrons to the conduction band, leading to the generation of excitons (electron-hole pairs). The generated holes can react with water molecules to form highly reactive hydroxyl radicals (OH•). The hydroxyl radical has the ability to decompose organic molecules (e.g., pollutants) since it has a high oxidation potential. Therefore, there is a need to increase the production of these hydroxyl radicals to improve the photocatalytic degradation ability of a material.

Commonly used metal oxide photocatalytic materials such as ZnO and TiO_2, which are n-type semiconductors, have wide band gaps: 3.2-3.7 eV for ZnO [11], 3.2 eV for anatase TiO_2 and 3.0 eV for rutile TiO_2 [12]; thus, they are ultraviolet light-triggered. But most of the abundant light available, such as indoor light and sunlight, are primarily in the visible light region. Because of this, efforts have been made to enhance the photocatalytic response of metal oxide materials under visible light illumination. Some of these methods include doping and coupling. On one hand, doping produces energy levels within the band gap of the material. However, the process of doping demands complicated procedures. On the other hand coupling results to a reduced energy gap at the interface leading to a lower energy is needed to generate excitons, allowing visible light to trigger its photocatalytic activity. It also increases the lifetime of charge carriers. Since coupling is merely a contact between two semiconductors, its fabrication method is relatively easy.

Copper oxide is a narrow band gapped semiconductor (1.7 eV), which corresponds to visible light. Although it is known to have low photocatalytic activities since it is intrisically a *p*-type semiconductor, it is a good candidate for coupling with that of ZnO and TiO_2. The energy diagrams of coupled CuO and ZnO are shown in different literatures [13-19]. Coupling would result to the formation of an energy barrier that leads to the formation of *p-n* heterojunction. The existence of this interface may lower the energy needed (compared to that of ZnO) to trigger the formation of excitons that is crucial in the activation of photocatalysis.

This research aims to fabricate photocatalytic material that can be triggered using visible light. This was done by coupling wide band gapped *n*-type zinc oxide and narrow band gapped *p*-type copper oxide by precipitation method. It is one of the simplest methods involving low-cost materials and equipment. In precipitation method, copper and zinc metal salts were used as precursors. These metal salts can easily be dissolved in water where it dissociates into metal ions. By using a precipitating agent, in this study NaOH, metal hydroxides are formed. These are the collected precipitates after the reaction. The hydroxides are known to be metastable and can easily oxidize into metal oxide.

2. Methodology

2.1 Fabrication of CuO/ZnO mixture via chemical precipitation

Coupled copper oxide and zinc oxide powder was prepared by precipitation method. In this method, 0.2 M $CuSO_4 \cdot 5H_2O$ metal salt and 0.2 M $ZnSO_4 \cdot 5H_2O$ metal salt were used as precursors. Both reagents were completely dissolved in distilled water separately and mixed thoroughly. Then the two solutions were combined into one solution, and poured into an aqueous solution of 0.2 M NaOH, and mixed vigorously. The formed precipitates were collected, washed with distilled water and 90% ethanol solution several times, then filtered and dried overnight. The dried samples were grounded, and then heated at 500°C for 1 hr. Finally, the heat treated samples were grounded again. Zinc oxide and copper oxide powders were also fabricated using precipitation method.

2.2 Characterization

X-ray diffraction (XRD) spectroscopy was used detect the different solid phases present in the sample. The XRD patterns were recorded using a Shimadzu XRD 7000, using CuKα X-rays at a scanning rate of $2^{o}min^{-1}$ between 3°-90°. The tube voltage and current used were 40.0 kV and 30.0 mA, respectively.

2.3 Test of Photocatalytic activity

The photocatalytic activities of the samples were determined through their ability to degrade methyl orange dye under visible light illumination. About 38 mM of methyl orange solution was placed in a vial. Then, 250 mg of the photocatalyst was added into the methyl orange solution and was shaken in the dark. The mixture was allowed to stand in the dark for 30 mins, and then illuminated. A light source with a UV cut-off filter was used. The degradation of methyl orange was determined by measuring the transmission of light through the sample. The intensity of the transmitted light was measured every 30 mins, for 4 hours. The photocatalytic degradation for each specific time was calculated.

3. Results and Discussion

Three powder samples were fabricated: (a) "CuO/ZnO powder", (b) "ZnO powder" and (c) "CuO powder". The sample labelled "CuO/ZnO powder" was fabricated using copper and zinc metal salt precursors. The second sample, labelled "ZnO powder" made use of zinc metal salt precursor, while the last sample labelled "CuO powder" made use of copper metal salt precursor. The oxidation process of the "CuO/ZnO powder" sample involved color change from blue to light brown. On the other hand, the oxidation of "ZnO powder" sample had no color change, while the "CuO powder" sample involved color change from blue to brown.

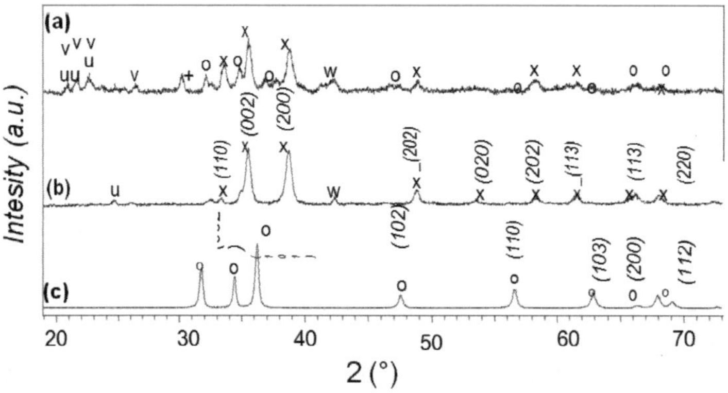

Figure 1. XRD patterns of the prepared (a) "CuO/ZnO powder", (b) "CuO powder", and (c) "ZnO powder" samples. Markers correspond to: **x**-CuO, **o**-ZnO, **w**-Cu, +-Cu_2O, **u**-$CuSO_4$, **v**-$ZnSO_4$.

XRD shows that the three highest peaks of the CuO/ZnO powder (figure 1(a)) correspond to CuO and ZnO. The peaks corresponding to CuO occurred at 2θ~35.54^{o} and 2θ~38.88^{o}, while the peak corresponding to ZnO occurred at (2θ~34.76^{o}). Other components present in the powder are Cu, Cu_2O, $CuSO_4$ and $ZnSO_4$. For the CuO powder sample (figure 1(b)), the highest peak occurring at (2θ~35.54^{o}) correspond to CuO, however, other components were also detected including $CuSO_4$ and Cu. XRD also shows that all the peaks detected in the ZnO powder sample (figure 1(c)) correspond to ZnO.

The photocatalytic activities of the samples were evaluated based on their ability to photodegrade methyl orange under visible light. Figure 2 shows the graph of percent degradation as a function of illumination time for: (a) "CuO/ZnO powder", (b) "CuO powder" and (c) "ZnO powder". The degradation of methyl orange through time suggests the presence of photocatalytic activity of the samples. The highest percent degradation at 4 hours was found out to be 71.134% by the "CuO/ZnO powder" sample. This was followed by the "CuO powder" sample with 48.283% then the "ZnO powder" sample with 32.983%. Figure 2(a) shows that the highest photocatalytic activity was exhibited by the "CuO/ZnO powder" sample. It can be seen that the percent degradation from the initial time to 1 hour changes drastically. At 1 hour, the percent degradation reaches to 46.59%. From that point on, the percent degradation seems to saturate until it reaches the maximum at 4 hours. For the CuO powder sample, the percent degradation is very low for the first hour then increases with time at an approximately linear trend (figure 2(b)).

After 3 hours, the percent degradation reaches 30.32% and continues to increase until the fourth hour. The ZnO powder sample seems to have little to no percent degradation for the first 2 hours and 30 mins (figure 2(c)). It then increases to reach its maximum at 4 hours.

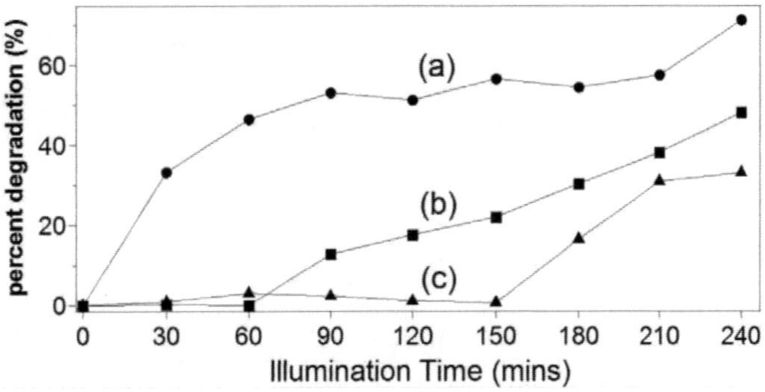

Figure 2. The figure shows the plot of illumination time versus the percent degradation of methyl orange using the fabricated samples: (a) "CuO/ZnO powder", (b) "CuO powder", and (c) "ZnO powder".

The improved photocatalytic activity of the fabricated CuO/ZnO powder compared with ZnO and CuO powder is attributed to the creation of a p-n heterojunction/interface. Figure 3 shows the energy band diagram of the coupled CuO and ZnO powder. The bending of the conduction band (E_C) and valence band (E_V) at the interface results to equalization of the Fermi energy level (E_F) of the two materials. A positively charged depletion layer is created by the upward bending of bands in the n-type ZnO (n-ZnO) material while a negatively charged depletion layer is created by the downward bending of bands in the p-type CuO (p-CuO) material. Likewise, because of the difference of the band gaps of the two semiconductors, the difference between the valence and conduction band at the interface is lower than that of zinc oxide. Thus, the generation of excitons at the interface is triggered by visible light. It is expected that the more copper oxide-zinc oxide interfaces, the better visible light photocatalytic activity.

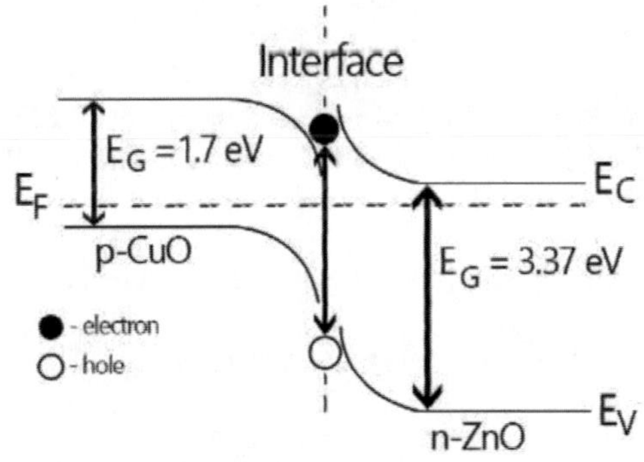

Figure 3. Energy band diagram of coupled CuO and ZnO powders. E_C is the conduction band, E_V is the valence band and E_F is the Fermi energy level, and E_G is the band gap of the semiconductor.

The visible light photocatalytic activity of CuO powder is lower compared to CuO coupled with ZnO, but higher than that of ZnO powder. It is known that CuO has low photocatalytic activity. However, this may be enhanced by coupling it with copper metal (as detected by XRD measurements) forming a Schottky-type junction/interface. Such interface/junction may work synonymously like that of the p-n heterojunction. On the other hand the low photocatalytic activity of the ZnO sample may be associated to its wide band gap, requiring high energy light source to generate excitons which will react with water to produce OH radicals.

4. Conclusion

The researchers have successfully fabricated coupled CuO/ZnO powder samples for visible-light photocatalytic degradation of methyl orange. The CuO/ZnO powder was fabricated via chemical precipitation method which involved copper and zinc metal salts in aqueous solution as precursors and NaOH as the precipitating agent. The precipitates were grounded and heated to form the final powder samples, then were dispersed in methyl orange solution. Its photocatalytic activity was evaluated by its ability to degrade methyl orange under visible light illumination. The fabricated CuO/ZnO powder sample exhibited superior percent degradation compared to the CuO and ZnO powder samples. This is attributed to the formation of a p-n heterojunction.

Acknowledgments

We are grateful for the funding support given by the University of the Philippines Los Baños Basic Research Program.

References

[1] Ahmad M, Ahmed E, Hong Z L, Jiao X L, Abbas T and Khalid N R 2013 *Appl. Surf. Sci.* **285** 702-12
[2] Almanza D L, Garcia M C T, Prodigalidad A D and Herrera M U 2016 *Key Eng. Mat.* **705** 385-89
[3] Han F, Li H, Fu L, Yang J, and Liu Z 2016 *Chem. Phys. Lett.* **651** 183-87
[4] Li B and Wang Y 2010 *Supperlatice. Microst.* **47** 615-23
[5] Liu Z-L, Deng J-C, Deng J-J, and Li F-F 2008 *Mat. Sci. Eng.* B **150** 99-104
[6] Mageshwari K, Nataraj D, Pal T, Sathyamoorthy R, and Park J 2015 *J. Alloy Compd.* **625** 362-70
[7] Sathishkumar P, Sweena R, Wu J J, and Anandan S 2011 *Chem. Eng. J.* **171** 136-40
[8] Shirzadi A and Nezamzadeh-Ejhieh A 2016 *J. Mol. Catal. A* **411** 222-29
[9] Yang C, Cao X, Wang S, Zhang L, Xiao F, Su X, and Wang J 2015 *Ceram. Int.* **41** 1749-56
[10] Yi S, Cui J, Li S, Zhang L, Wang D, and Li Y 2014 *Appl. Surf. Sci.* **319** 230-36
[11] Lee K M, Lai C W, Ngai K S, and Juan J C 2016 *Water Res.* **88** 428-48
[12] Lee S-Y and Park S-J 2013 *J. Ind. Eng. Chem.* **19** 1761-69
[13] Ghosh A and Mondal A 2016 *Mater. Lett.* **164** 221-24
[14] Katoch A, Choi S-W, Kim J-H, Lee J H, and Lee J-S 2015 *Sensor. Actuat. B-Chem.* 111-16
[15] Kim J-H, Katoch A, and Kim S S 2016 *Sensor. Actuat. B-Chem.* **222** 249-56
[16] Liu H, Wu X, Li X, Wang J, and Fan X 2014 *Chinese J. Catal.* **35** 1997-2005
[17] Xu Q, Ju D, Zhang Z, Yuan S, Zhang J, Xu H, and Cao B 2016 *Sensor. Actuat. B-Chem.* **225** 16-23
[18] Yanga L, Xiea C, Zhanga G, Zhaoa J, Yua X, Zenga D, and Zhanga S 2014 *Sensor. Actuat. B-Chem.* **195** 500-8
[19] Zhang Y-B, Yin J, Li L, Zhang L-X, and Bie L-J 2014 *Sensor. Actuat. B-Chem.* **202** 500-7

IOP Conf. Series: Journal of Physics: Conf. Series **817** (2017) 012059 doi:10.1088/1742-6596/817/1/012059

Crystal structure and phase transformation of $BiFeO_3$ multiferroics on the temperature variation

U Nuraini and S Suasmoro

Departement of Physics, Faculty of Mathematics and Sciences
Institute of Technology Sepuluh Nopember, Surabaya 60111 Indonesia

E-mail: suasm@its.ac.id

Abstract. In this study, multiferroic of $BiFeO_3$ was successfully synthesized using solid state methods. The materials was calcined by variation temperature at 800, 840, 860, and 880°C for 4hours in air atmosphere. Structure analysis and phase transformation were carried out using X-ray diffraction. The results show that not only $BiFeO_3$ phase is formed but also other phases as secondary phases on all variation temperatures. The secondary phases,i.e. $Bi_{25}FeO_{40}$ and $Bi_2Fe_4O_9$, have different molar or weight percentage at that variation. $BiFeO_3$ has a perovskite structure and considering as*R3c* space group, for all secondary phases consist of cubic structure. The results showed that increasing calcination temperaturedecrease the amount of $BiFeO_3$ phases and increase the secondary phases. The highest percentage of $BiFeO_3$ phases (91.69%) is obtained at 800°C.

1. Introduction

Multiferroics are materials that can behave as ferroelectric, ferromagnetic, and often ferroelastic on single or composite structure. This material is getting more attention due to those properties and their potential for various applications. Multiferroics are the combination of ferroelectric and ferromagnetic, the interaction between magnetic and electric fields correspond the magneto electric (ME) effect in their materials [1]. ME effect provides the possibility of controlling magnetic properties using electrical fields (E) and electrical properties using magnetic fields (H). Some practical applications of multiferroic materials are transducers, magnetic fields sensor, memory device, and photovoltaic devices.

$BiFeO_3$ (BFO) is one of the materials that can exhibit as ferroelectric and antiferomagnetic properties at room temperature in the same phase [2]. BFO, a rhombohedra ferroelectric with perovskite structure (ABO_3, has high ferroelectric Curie temperature (Tc ~ 830°C) and G-type antiferomagnetic below Néel temperature (T_N~370°C) [3]. BFO is very difficult to synthesizein a pure single phase; it is often on polycrystalline condition and other secondary phases, such as $Bi_2Fe_4O_9$ and $Bi_{25}FeO_{39}$.Liu et al reported on their paper that secondary phase is formed in all their variation temperatures from 650– 850°C[4]. Golić et al by using different method also confirmed that during temperature 620– 770°C the secondary phase is appears together with BFO phases[5]. The synthesis process of BFO can be done by various methods, such as solid solution [6], sol gel [7], and auto-combustion methods [8].

In this work, we studied the phases during the heating process at various temperatures in air atmosphere. The change of phase's percentage are further examined and presented, thus it may provide some useful experimental guidance for the preparation of pure phase of BFO.

2. Experimental

In this work, Bi_2O_3 (99%) and Fe_2O_3 (99%) were used to prepare BFO's precursor by conventional mixed oxide methods. The chemical reaction for this synthesis is given as follow:

$$Bi_2O_3 + Fe_2O_3 \rightarrow 2BiFeO_3 \qquad (1)$$

The mixtures of powder are milled by planetary ball mill with speed of 150 rpm for 5 hours. The process is carried out in ethanol media using ZrO_2 balls, then dried and homogenized using evaporator. The calcination process is completed with various temperatures, 800, 840, 860, and 880°C with holding time of 4 hours. The temperature variation is selected based on the phase diagram of Bi_2O_3 - Fe_2O_3 systems that stable of BFO in the stoichiometric of Bi:Fe = 1:1 at the temperature range of 800-900°C, at higher than 900ºC it will disappears [9]. Furthermore Cheng reported by in-situ XRD, the BFO formation located at temperature higher than 600°C [10].

The calcined powders are characterized by XRD (Philips X'Pert MPD) using CuKα1 radiation then analyzed using Match program for phase identification. Crystal structure and percentage of phases are characterized using Rietveld methods by Rietica software.

3. Result and discussion

X-ray diffraction of BFO samples with various temperatures are presented in figure 1. It shows that all samples not only distinguished the presence of BFO, but also other secondary phases. The secondary phases, $Bi_2Fe_4O_9$ and $Bi_{25}FeO_{39}$, appear with different intensity and compositions. The peak positions on XRD pattern of all various samples can be considered similar. The increasing of temperatures induces the changes on the number of phases. The BFO phases became decrease while the secondary phases increase as the calcination temperature increase.

The lattice parameter and percentage of phases of samples were quantitatively calculated by using Rietica with Rietveld methods. The results are summarized in table 1. All samples variation and phases indicated having similar lattice parameters. It means that the increasing temperatures, within 800 – 880°C, not significantly influence into the lattice parameter of materials.

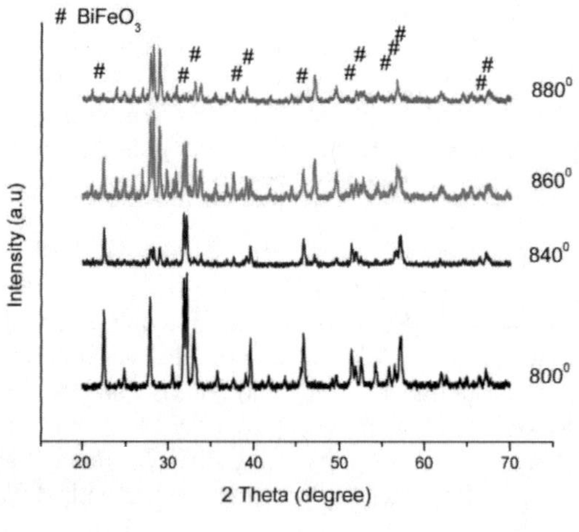

Figure 1. XRD pattern of BFO at various temperature; 800, 840, 860, and 880 °C.

Table 1 Lattice parameter of BFO samples.

Phase	Calcination temperature (°C)			
	800	840	860	880
BFO	a = 5.581 (7) c = 13.876(3) α = 90° β = 120°	a = 5.574(4) c = 13.856(2) α = 90° β = 120°	a = 5.574(1) c = 13.858(6) α = 90° β = 120°	a = 5.552(8) c = 13.939(6) α = 90° β = 120°
$Bi_2Fe_4O_9$	a = 7.891(4) b = 8.482(9) c = 5.988(4) α = β = γ = 90°	a = 7.956(4) b = 8.441(8) c = 5.996(7) α = β =γ = 90°	a = 7.953(6) b = 8.434(6) c = 5.993(4) α = β = γ = 90°	a = 7.972(1) b = 8.455(4) c = 6.039(1) α = β = γ = 90°
$Bi_{25}FeO_{39}$	a = 10.172(4) α = β=γ= 90°	a = 10.173(6) α = β=γ= 90°	a = 10.117(3) α = β=γ= 90°	a = 10.148(3) α = β=γ= 90°

The percentage of BFO and the secondary phases, $Bi_2Fe_4O_9$ and $Bi_{25}FeO_{39}$,, are shown on figure 2 and 3. From figure 2, during temperature of 800 – 880°C, the graph simultaneously showed the decreasing of molar percentage of BFO phase and increasing of molar percentage of $Bi_2Fe_4O_9$. Figure 3, the graph showed some different result; in the similar temperature condition both molar percentage of BFO and $Bi_{25}FeO_{39}$ are decrease. In here, the decreasing of $Bi_{25}FeO_{39}$ phases shows significantly lower than other phases.

The chemical reaction of bismuth – ferrite is ideally formed as indicated in equation (1), where it should be only BFO single phase. However, XRD pattern showed not only the BFO phase but also other phases, i.e.$Bi_2Fe_4O_9$ and $Bi_{25}FeO_{39}$ phases, as qualitatively and quantitatively identified. However, other possibilities of reaction also can be considered, as follow:

$$Bi_2O_3 + Fe_2O_3 \rightarrow Bi_2Fe_4O_9 \quad (2)$$

$$25Bi_2O_3 + Fe_2O_3 \rightarrow 2Bi_{25}FeO_{39} \quad (3)$$

As shown in figure 2 and 3, the increasing temperature effects on the decreasing of BFO and $Bi_{25}FeO_{39}$ phases, and on the contrary increasing $Bi_2Fe_4O_9$ phase. These phenomena were also reported by Cheng, et al. [10]: at above 840°C the secondary phase, $Bi_2Fe_4O_9$, is more stable than $Bi_{25}FeO_{39}$. Furthermore, both those secondary phases, $Bi_{25}FeO_{39}$ and $Bi_2Fe_4O_9$, actually still can be further transformed become pure BFO by addition of Bi_2O_3 and Fe_2O_3, due to the following equations:

$$Bi_2Fe_4O_9 + Bi_2O_3 \rightarrow 4BiFeO_3 \quad (4)$$

$$2Bi_{25}FeO_{39} + 24Fe_2O_3 \rightarrow 50BiFeO_3 \quad (5)$$

Indeed, the results of this study were not in line with as indicated in the phase diagram, in which BFO phase will become better formed at higher temperature. This can be explained due to the phenomena of the volatility of Bismuth (Bi) [11]. So that, the Bi ions became deficit to react with Fe and resulting on fewer of pure BFO phase is formed. As reported on [11], the volatility of Bi can be minimized by an addition of bismuth up to 5-10 wt.%.

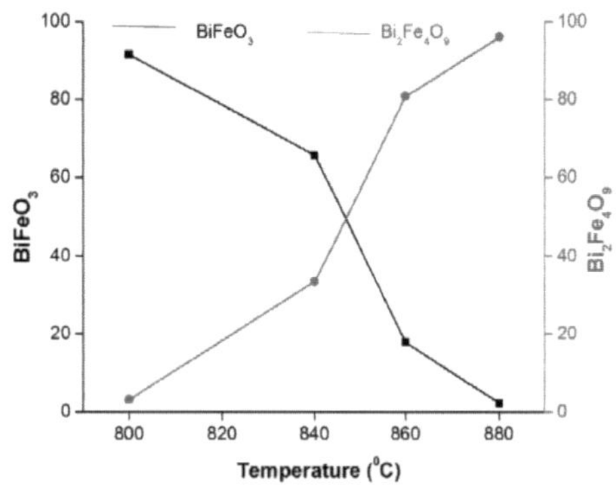

Figure 2. The percentage of BFO and $Bi_2Fe_4O_9$.

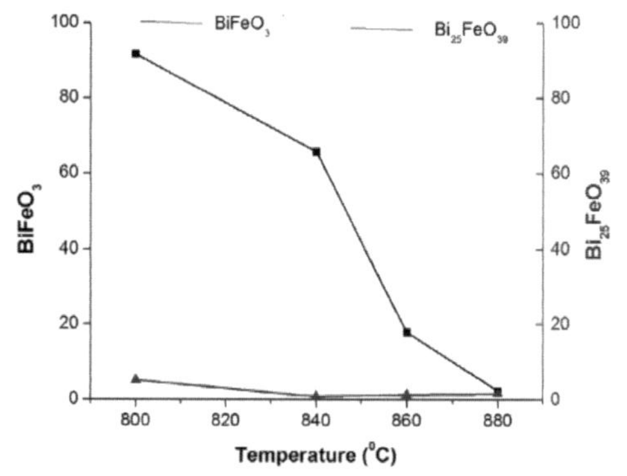

Figure 3. The procentage of BFO and $Bi_{25}FeO_{39}$.

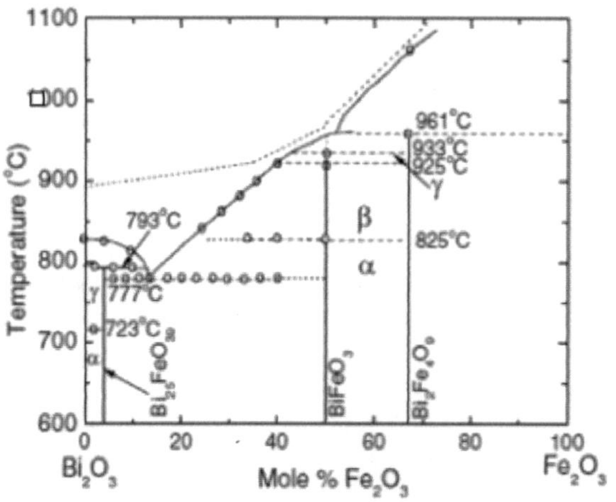

Figure 4. Phase diagram of Bi_2O_3- Fe_2O_3 systems [9].

4. Conclusion

$BiFeO_3$ (BFO) is formed as polymorph and perovskite structure at range of temperature of 800-880 °C, including $Bi_2Fe_4O_9$ and $Bi_{25}FeO_{39}$ as secondary phases. The molar percentage of BFO decreased and $Bi_2Fe_4O_9$ increases. This is due to the volatility of Bismuth (Bi) that evaporates at increasing temperature. Decreasing of Bi atoms is causing greater of Fe numbers, as a consequent $Bi_2Fe_4O_9$ dominantly to be formed.

Acknowledgement
The first author acknowledges Department of Physics, Institute of Technology Sepuluh Nopember (ITS) to support this work.

References

[1] Lakshmi B D, Caltun O F, Dumitru I, Pratap K, Rao B P and Rao P S V S 2015 *Mater. Today. Proc.* **2** (6) 3806–12
[2] Graf M, Sepliarsky M, Machado R and Stachiotti M G 2015 *Solid State Commun.* **218** 10–3
[3] Ilić N I, Bobić J D, Stojadinović B S, Džunuzović A S, Petrović M M V, Dohčević-Mitrović Z D and Stojanović B D 2016 *Mater. Res. Bull.* **77** 60–9
[4] Liu Z, Liang S, Li S,Zhu Y and Zhu X 2015 *Ceram. Int.* **41** S19-25
[5] Golić D L, Radojković A, Ćirković J, Dapčević A, Pajić D, Tasić N, Savić S M, Počuča-Nešić M, Marković S, Branković G, Stanojević Z M and Branković Z 2016 *J. Eur. Ceram. Soc.* **36**(7) 1623–31
[6] García-Zaleta D S, Torres-Huerta A M, Domínguez-Crespo M A, Matutes-Aquino J A, González A M and Villafuerte-Castrejón M E 2014 *Ceram. Int.* **40**(7) Part A 9225–33
[7] Sharma S, Saravanan P, Pandey O P, Vinod V T P, Černík M and Sharma P 2016 *J. Magn. Magn. Mater.* **401** 180–7
[8] Tripathy S N, Mishra B G, Shirolkar M M, Sen S, Das S R, Janes D B and Pradhan D K 2013 *Mater. Chem. Phys.* **141**(1) 423–31
[9] Palai R, Katiyar R S, Schmid H and Scott J F 2008 *Phys. Rev. B: Condens. Matter.* **77**(1) 0141101-1
[10] Cheng G F, Ruan Y J, Liu W and Wu X S 2015 *Mater. Lett.* **143** 330–2
[11] Yan J, Gomi M, Hattori T, Yokota T and Song H 2013 *Thin Solid Films* **542** 150–4

IOP Conf. Series: Journal of Physics: Conf. Series **817** (2017) 012060 doi:10.1088/1742-6596/817/1/012060

Material processing of oil palm empty fruit bunches for use as raw material of conductive carbon paper

F Destyorini and N Indayaningsih

Research Center for Physics, Indonesian Institute of Sciences, Kawasan Puspiptek, Tangerang Selatan, Banten, Indonesia

Email: fred007@lipi.go.id; fredinadestyorini@gmail.com

Abstract. Empty fruit bunches of oil palm is a by-product of the palm oil industry that contains a high element of carbon. This by-product can be processed into a conductive carbon paper that could be applied as fuel cell electrodes. Carbon paper for this application must be conductive, porous, and hydrophobic. Utilization of oil palm empty fruit bunches begins with the carbonization process at a temperature of 500°C that produced charcoal. It is followed by heating at temperature of 900°C and 1300°C. To obtain the carbon paper, powdered charcoal with polymer binder (PEG and EVA) were mixed in solvent and molded using tape casting method. This process successfully produced carbon paper with dimensions of ±(20x20) cm^2 and a thickness of 0.1-0.3 mm. Properties of carbon paper were characterized and analyzed in terms of electrical conductivity, porosity, hydrophobic property, and microstructure. Polytetrafluoroethylene (PTFE), a hydrophobic agent, was treated on carbon paper to enhance the hydrophobicity of the carbon paper. PTFE coating on the surface of the carbon paper could change their physical properties. Carbon paper shows excellent properties in terms of porosity and hydrophobicity. Whereas, its electrical property needs to be improved further by increasing the pyrolysis temperature. But overall, this might show a potential GDL material for PEMFC.

1. Introduction

A proton-exchange membrane fuel cell (PEMFC) is an electrochemical cell that is fed with hydrogen and oxygen. The hydrogen is oxidized at the anode; while the oxygen is reduced at the chatode. The protons released during the oxidation reaction of hydrogen and passed through proton ex-change membrane to the cathode side, while the electrons travel through an external circuit to provide an electric current [1]. Electrode is one of the main components of PEMFC, it consists of a gas diffusion layer (GDL) and the catalyst. GDL acts as gas diffusion media, provides mechanical support, an electrical pathway, and also helps to remove by-produced water outside of the catalyst layer. The GDL must be porous, high conductivity, appropriate hydrophobic, high air permeability and sufficient mechanical strength. GDLs are typically carbon-based materials, and usually in cloth or paper form [2-3]. Generally, GDL are composed of commercial carbon materials, which is produced from non-renewable resources such as carbon nano-tube (CNT) [4], polyacrylonitrile (PAN)-based carbon fibre [5], carbon black [6], etc. It could raise the sustainability issues.

Carbon material can be obtained by a simple carbonization process of natural fibers. Natural fibers are renewable resources and abundant. These resources consist of complex organic compounds namely hemicellulose, cellulose and lignin. Each of these compounds decomposes at different temperatures and produces charcoal as a residue with high carbon content (± 50% to 90%) and porous [7-10]. Hemicellulose is polymers of several monosaccharides such as pentosan and heksosan. The earliest decomposed at temperatures of 200°C to 260°C, followed by decomposition of cellulose at temperatures of 240°C to 350°C, and lignin decomposed at 280°C to 500°C [11]. Oil palm empty fruit bunches (EFB) is one example of natural fibers that can be used. It is a waste of the palm oil industry that has not been utilized optimally. Pyrolysis process at a temperature of 1300°C produced carbon material with electrical conductivity of (1.4x10) S/m to (1.7x10) S/m [12]. Thus, if the EFB can be utilized as a base material for GDL, it will be able to resolve the issue of sustainability.

The present study describes the preparation of conductive carbon paper with the carbon material from EFB. And the objective of this study is to produce conductive carbon paper by using EFB carbon. Carbon material from EFB was used to replace commercial carbon, which has been used as raw materials of carbon paper production. Conductive carbon paper could be applied as GDL for PEMFC electrode. In addition, by using natural fiber materials, it can be resolved the sustainability issues due to the renewable nature of oil palm.

2. Material and methode

2.1. Material

Carbon materials were prepared from pyrolysis products of oil palm empty fruit bunches (EFB) fibers. Ethylene vinyl acetate (EVA) and Polyethylene glycol (PEG) obtained from Aldrich Chemical Co., Inc. (St. Lois, MO, USA) were used as binder. Xylene obtained from Brataco Chemika was used as solvent. Teflon emulsion PTFE 30 obtained from Fuel Cell Earth LLC was used as hydrophobic agent of carbon paper. Teflon emulsion PTFE 30 should be diluted to a concentration of 10% before treated on the surface of the carbon paper. For comparison, a commercially available carbon paper, TGP-H-090 (5% wet-proofed) Toray carbon paper, was obtained from Fuel Cell Earth and was used as received.

2.2. Preparation of carbon materials

The carbon materials were prepared from EFB fibers by a two stage process: carbonization at 500°C and pyrolysis at 900°C followed 1300°C. In the carbonization process, the EFB fibers was heated at 500◦C for an hour under N_2 atmosphere and cooled down to room temperature. The carbonization process produced charcoal from EFB with high carbon content. Then the EFB charcoal was subjected to pyrolysis at 900°C then 1300°C. In the pyrolysis process, the EFB charcoal was heated at 900◦C for an hour under N_2 atmosphere and cooled down to room temperature. It was followed by the pyrolysis process with the same conditions at a temperature of 1300°C. These process produced EFB carbon materials. Pyrolysis processes were carried out to improve the electrical conductivity, eliminate impurities, and improve other properties of carbon.

2.3. Preparation of carbon paper

To prepare the conductive carbon paper, EFB carbon was used as raw materials. First, polymer EVA and PEG were dissolved in xylene as solvents. The solution was stirred and heated at 90°C for 20 minutes. After the polymer dissolved perfectly, then EFB carbon was added. The mixture consisted of EFB carbon, EVA, and PEG with a mass ratio of 80:14:6 in 46 ml solvent. The mixture was continuously stirred and heated at 90°C for 80 minutes to make slurry, and then casted on glass mold with dimension of 20 cm x 20 cm. The carbon paper sheet was dried at room temperature for 24 h to evaporate the solvent. The carbon paper thickness ranged from 0.2 to 0.4 mm.

Hydrophobic coating was made to enhance the hydrophobic property of carbon paper. It was immersed in 10 % PTFE suspension for 10 minutes and dried at room temperature for 24 hours. Then, it was heated at 150°C in an oven for 30 minutes to evaporate the remaining solvent, and then heated at 350°C for 30 minutes under N_2 atmosphere.

2.4. Characterization of composite carbon paper

Characterization was carried out on carbon paper before and after PTFE coating, as well as commercial carbon paper as a comparison. Through-plane electrical conductivity was measured using HIOKI 3522-50 HITESTER LCR-meter. Porosity was determined by the kerosene density method using Archimedes principle in accordance with BS 1902: Part 1A standard. Hydrophobic properties are determined by the value of the contact angle using sessile drop test. For each measurement, a 50 μL water droplet was placed on the carbon paper surfaces by placing the tip of the syringe close to the sample surface, and images were captured. Furthermore, the droplet shape was analyzed using Bashforth and Adams tables [13] to determine the contact angle of the samples. Scanning electron microscope (JEOL JSM-6390 Series) was used to observe the morphology of the carbon paper before and after PTFE coating. Energy-dispersive spectroscopy (EDS) mapping was carried out to analyze the PTFE distribution in the carbon paper.

3. Result and discussion

3.1. Electrical conductivity

Carbon materials in powder form that had been prepared from EFB at different carbonization temperatures were characterized for their electrical conductivity. As shown in figure 1, with increase in the carbonization temperatures from 500°C to 1300°C, the electrical conductivity of the carbon materials increased from a value of 0.8×10^{-5} to 0.8×10^{1} S/cm. The increase in carbonization temperature leads to an increase in the crystallinity degree of carbon materials that derived from natural fibers [12]. In other words, higher carbonization temperatures enhance amount of the conductive phase in the carbonized natural fibers [14]. The property of EFB carbon changed from an insulating material into conductive material. Therefore, EFB carbon was treated under 1300°C used as raw materials of carbon paper production. Figure 2 shows the electrical conductivity of differentially prepared carbon paper in the through-plane direction.

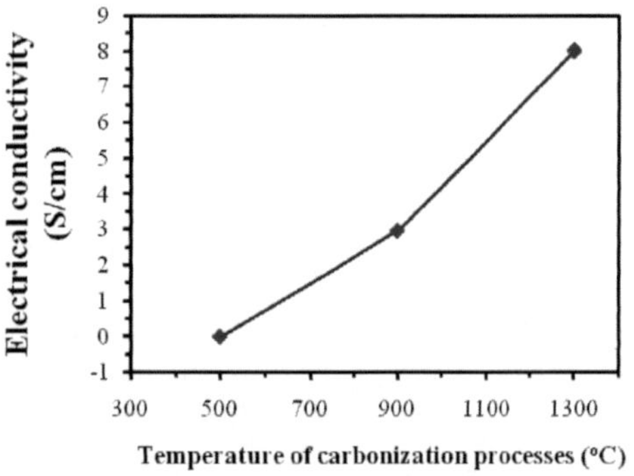

Figure 1. Electrical conductivity of carbon powder from EFB as the function of carbonization temperature

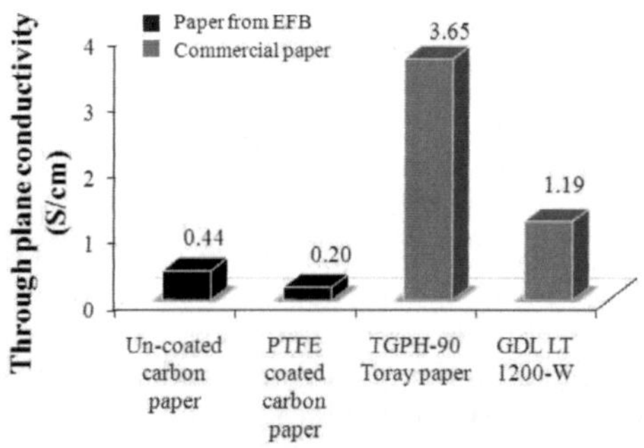

Figure 2. Electrical conductivity of un-treated carbon paper, PTFE-treated carbon paper and commercial carbon paper

The conductivity value of un-treated carbon paper decreased approximately 94.5% compared to that of EFB carbon as its raw material. It was due to the presence of non-conductive polymer (EVA and PEG) as binder which may interrupt the formation of the conductive network in carbon paper. Moreover, the presence of non-conductive PTFE particles in the carbon paper resulted further decrease in the electrical conductivity up to 50% compared to the un-treated carbon paper. And, it was also found that the electrical conductivity of the 10% PTFE-treated carbon paper from EFB carbon was still lower than commercial carbon paper (TGP-H-090 Toray and GDL LT 1200-W carbon paper). However, this electrical property could be improved further by increasing the pyrolysis temperature in the production of carbon materials from EFB to enhance the amount of conductive phase.

3.2. Porosity

The porosity of the composite carbon paper from EFB was measured and compared to commercial carbon paper. As shown in figure 3, the porosity of carbon paper decreased from 73.01% to 68.73% with the wet-proofing treatment of 10% PTFE. The decrease in porosity is mainly due to the blockage and narrowing of the pores by PTFE particles. PTFE particles deposited onto the surface and filled the open pores of carbon paper, as a consequence, reduces the overall porosity. Even though the porosity of carbon paper from EFB is slightly lower than the commercial carbon paper, this value still within the allowable range of porosity values as GDL of PEMFC. As GDL, carbon paper has functions to allow the reactant gases to reach the reaction zones and the product water movesout. To perform these functions effectively, carbon paper should have high porosity from 50 to 90% [15].

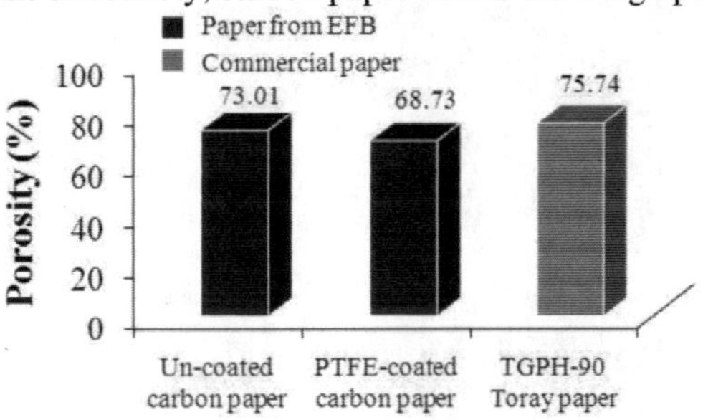

Figure 3. Porosity of un-treated carbon paper, PTFE treated carbon paper and commercial carbon paper

3.3. Hydrophobic property

Hydrophobic property is required by carbon paper as the GDL for the removal of product water from catalyst-layer area to flow-field channels in the PEMFC stack. Hence, carbon paper was water-proofed by introducing PTFE in this work. Furthermore, water contact angles on carbon paper were measured to determine the hydrophobic properties. The photographs and contact angles of water droplet on the surface of each carbon paper sample are presented in figure 4. As shown, the untreated carbon paper did not exhibit the hydrophobicity with low value of contact angle (less than 90°). When the carbon paper was treated with PTFE, the contat angle greatly increased to 117° and displayed the hydrophobic property. The contact angle value of both carbon paper were greater than 90° indicating that the PTFE treated carbon paper from EFB are resistant to wetting as well as the commercial products of Toray TGP-H-090.

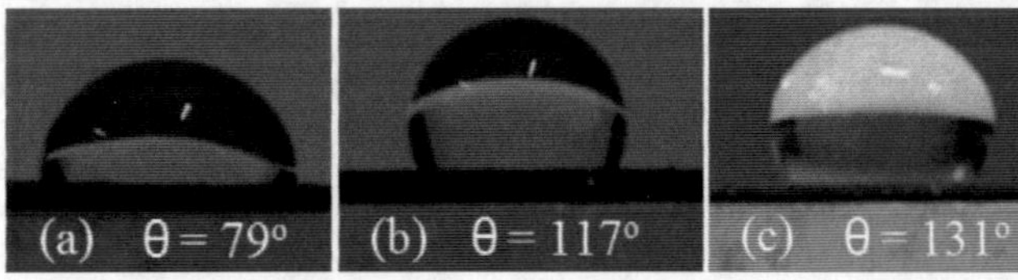

Figure 4. Water drops photographs and contact angle of (a) un-treated carbon paper, (b) PTFE treated carbon paper, and (c) TGP-H-090 Toray carbon paper

3.4. Morphologies and elemental analysis

The surface and cross section morphologies of the untreated, PTFE treated and commercial carbon paper used in thiswork are shown in figure 5(a) to figure 5(f). The gray-colored regions and dark regions are assigned to the the carbon materials and the pores, respectively. As can be seen in figure 5(a) and figure 5(b), the pores are spread evenly on the untreated carbon paper surface and that of the PTFE treated. As pointed out previously, PTFE particles were deposited onto the untreated carbon paper surface and this reduces the overall porosity. This phenomenon is reflected in the SEM micrographs. Meanwhile, the SEM micrographs in figure 5(d) and figure 5(e) show the cross-sections of the untreated and PTFE treated carbon paper, respectively. It can be observed the porous morphology of carbon paper with thickness approximately from 0.40 to 0.50 mm. Whereas the surface and cross-section morphologies of commercial carbon paper are slightly different compared to carbon

IOP Conf. Series: Journal of Physics: Conf. Series **817** (2017) 012060 doi:10.1088/1742-6596/817/1/012060

paper from EFB. As shown in figure 5(c) and figure 5(f), this is composed of carbon fibers arranged randomly.

Figure 6(a) and 6(b) show the elemental F distribution (represented by pink dots) of cross section of PTFE-treated and commercial carbon paper, respectively. It is clearly observed that the PTFE particles successfully deposited onto the surface of both carbon paper, and some of them penetrated through the carbon paper.

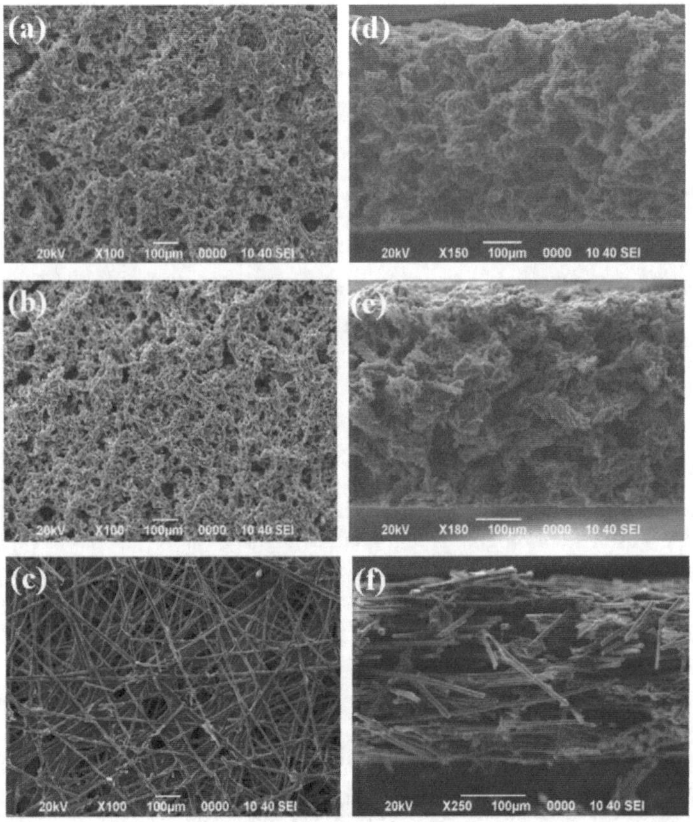

Figure 5. Morphology of the carbon papers: surface morphology of (a) untreated carbon paper, (b) PTFE treated carbon paper, (c) TGP-H-090 Toray carbon paper;and cross-sectional morphology of (d) untreated carbon paper, (e) PTFE treated carbon paper, (f) TGP-H-090 Toray carbon paper

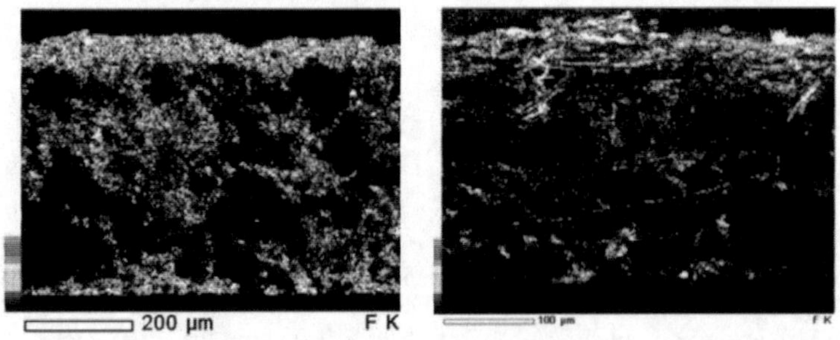

Figure 6. EDX mapping image for the element F (represented by pink dots) of cross section of (a) PTFE-treated carbon paper, and (b) TGP-H-090 (5% wet-proofed) Toray carbon paper

4. Conclusion

In the present work, conductive carbon paper was successfully prepared by utilizing oil palm empty fruit bunches through the process of carbonization, pyrolysis, tape casting, and hydrophobic treatment. Carbon paper was shown to have excellent properties in terms of porosity and hydrophobicity. Whereas, its electrical property needs to be improved further by increasing the pyrolysis temperature. But overall, this can prove as a potential GDL material for PEMFC.

Acknowledgments
The authors gratefully acknowledge the financial support from Indonesian Institute of Sciences and the facilities provided by the Research Center for Physics—Indonesian Institute of Sciences.

Reference

[1] Litster S and McLean C 2004 *J. Power Sources* **130** 61
[2] Mathur V K and Crawford J 2007 *Recent Trends in Fuel Cell Science and Technology*, ed S Basu (New York: Springer) pp 116-128
[3] Kandlikar S G, Garofalo M L and Lu Z 2011 *Fuel Cells* **11**(6) 814
[4] Gao Y, Sun G Q, Wang S L and Zhu S 2010 *Energy* **34** 1455
[5] Mathur R B, Maheshwari P H, Dami T L, Sharma R K and Sharma C P 2006 *J. Power Sources* **161** 790
[6] Shim J, Han C, Sun H, Park G, Lee J and Lee H 2012 *Trans. of the Korean Hydrogen and New Energy Society* **23**(1) 34
[7] Abdul Rani N H, Mohamad N F, Matali S and Kadir S A S A 2014 *Key Eng. Mater.* **594-595** 44
[8] Isahak W N R W, Hisham M W M and Yarmo M A 2010 *J. Chem.* **2013** Article ID 620346 6 pages
[9] Indayaningsih N, Priadi D, Zulfia A and Suprapedi 2011 *Key Engineering Materials* **462-463** 937
[10] Indayaningsih N, Zulfia A, Priadi D and Kartini E *2010 Proc. of the 12th Asian Conference on Solid State Ionics* (China: Wuhan University of Technology Press) p 1045-1-50
[11] Browne FL 2000 *Theories of the Combustion of Wood and its Control* (United States: Forest Products Laboratory) p 69
[12] Indayaningsih N, Zulfia A, Priadi D and Hendrana S 2011 *Advanced Materials Research* **277** 137
[13] Bashforth F and Adam J C 1883 *An attempt to test the theories of capillary action by comparing the theoretical and measured forms of drops of fluid with an explanation of the method of integration employed in constructing the tables which give the theoretical forms of such drops* (Cambridge: Cambridge University Press) p 158
[14] Kwon J H, Park S B, Ayrilmis N, Oh S W and Kim N H 2013 *Composites: Part B* **46** 102
[15] Yoon G and Park Y 2013 *Int. J. Precis. Eng. Manuf.* **13**(7) 1153

IOP Conf. Series: Journal of Physics: Conf. Series **817** (2017) 012061 doi:10.1088/1742-6596/817/1/012061

Elemental analysis and radiation hazards parameters of bauxite located in Saudi Arabia

S Alashrah[1] and A E Taher[2]

[1] Physics Department, college of Science, Qassim University, Buraydah 51452, KSA
[2] Physics Department, college of Science, Al-Azhar University, Assuit 71452, Egypt

E-mail: salashra@gmail.com; ashrh@qu.edu.sa

Abstract. Since Bauxite has been widely used in industry and in scientific investigations for producing Aluminum, it is important to measure the radionuclides concentrations to determine the health effect. The Bauxite mine is located in Az Zabirah city in Saudi Arabia. The concentrations of the radionuclides in the bauxite samples were measured using γ-ray spectrometer NaI (Tl). The average and range values of the concentrations of ^{226}Ra, ^{232}Th and ^{40}K were 102.2 (141.1-62.7), 156.3 (202.8-102.8) and 116.8 (191.7- 48.9) Bq/kg respectively. These results were compared with the reported ranges in the literature from other locations around the world. The radiation hazard parameters; radium equivalent activity, annual dose, external hazard were also calculated and compared with the recommended levels by International Commission on Radiological Protection (ICRP-60) and united nations scientific committee on the effects of atomic radiation UNSCEAR reports. There are no studies for the natural radioactivity in the bauxite mine in Az Zabirah city, so these results are a start to establishing a database in this location.

1. Introduction

Natural occurring radionuclides are present in many natural resources. Human activities may enhance the concentrations of radionuclides and/or enhance the potential of exposure to naturally occurring radioactive material (NORM). The industrial residues containing radionuclides have been receiving considerable global attention, because of the large amounts of NORM containing wastes and the potential long-term risks of long-lived radionuclides [1]. They have become concentrated and exposed not only to the environment but also to human workers in manufacturers, water suppliers, or mines. Actually there are numbers of authors studied the radiological impact resulting from phosphate production and oil or gas production [2]. However Aluminum productions have been widely used in industry and in scientific investigations (bauxite is a rock consisting of aluminum oxide) but the number of studies on the radiological impact resulting from bauxite or red mud (the waste and tail material from primary aluminum production) were low [3]. Table 1 shows the range of radioactivity concentrations in the bauxite. The activity concentration is high because of ^{238}U and ^{232}Th decay series, depending on the geological composition. O'Connor [4, 5] measured natural activity concentration in the bauxite, solid residues and red mud resulting from alumina production in Western Australia. ^{238}U activity concentration results were 120 – 350 (bauxite) 5 – 200 (soil residues) and 150 – 600 (red mud). Furthermore, the activity concentrations of ^{232}Th were (450 – 1050), (300 - 800) and (1000 –

1900) for bauxite, soil residues and red mud respectively. The radionuclide concentration of red mud increases three times compared to the original bauxite mineral due to refine bauxite and produce alumina. The most concentration of uranium or thorium radionuclides is transferred to the solid waste. However, little, if none, of the radionuclides from them are found in the alumina. To reduce the occupational exposure, many studies reported that a percentage of red mud can be mixed with concrete or cement [6]. The absorbed dose of the mixed red mud should not exceed the limit determined by UNSCEAR (2000) [7]. Therefore, manufacturing operation reduces the radiation hazard parameters. Furthermore, Righi et al. [8] measured the natural radioactivity in bauxite. Their results showed that the activity concentrations of ^{238}U were between to 500 and 600 Bq/ kg and from 400 to 450 Bq/ kg for ^{232}Th.

Table 1. Content of radionuclides in bauxite and red soil [9].

Radionuclide	Specific activity (Bq/Kg)	
	Bauxite	Red soil
^{238}U-^{226}Ra decay series	10-900	100-3000
^{232}Th decay series	35-1400	100-300
^{40}K	10-600	10-100

Ademola et al. [10] studied the activity concentration of ^{226}Ra, ^{232}Th and ^{40}K using bauxite ore in Nigeria country. They used NaI(Tl) detector for measurements. They concluded that the activity concentrations were between 134 ± 21 and 47 ± 14. Using High pure germanium and neutron activation methods for bauxite measurement, the results were 37±12 Bq.kg^{-1} for ^{238}U and 154±16 Bq.kg^{-1} for ^{232}Th [1]. The activity concentrations of the bauxite samples in the Gulf of Corinth, Greece country were determined using direct γ-ray spectroscopy by Papatheodorou et al.[11]. They results showed that the activity concentrations of ^{226}Ra, ^{232}Th and ^{40}K were 150, 205 and 38 respectively. The average world activity concentrations of ^{226}Ra, ^{232}Th and ^{40}K were 50, 50 and 500 Bq/Kg respectively.

The present work aims to i) measure the natural radioactivity level of (^{226}Ra, ^{232}Th and ^{40}K) in samples of bauxite in Az Zabirah mine, Saudi Arabia by using γ-ray spectrometry. ii) calculate the radiological parameters such as radium equivalent activity Raeq, external hazard index Hex and absorbed dose rate which is related to the external γ-dose rate. iii) Compare the results of activity concentrations and radiation hazards with similar studies carried out in other countries and UNSCEAR data.

2. Materials and methods

2.1 Study area

Az Zabirah (N27056', E43043') is located on the northeast of Buraidah in Qassim region. It is 180 km from Buraydah. It has a bauxite mine. It is indicated as laterite profile because of rich in iron and aluminum. It is over a distance of 105 km in the scarp face of an early Cretaceous feature. The estimated total reserves are 101.8 million tons. Of course, the laterite development could be older than Oligocene. The thickness of bauxite is approximately 8.5 m described by Vincent [12]. Az Zabirah is in the cuesta region of central Saudi Arabia.

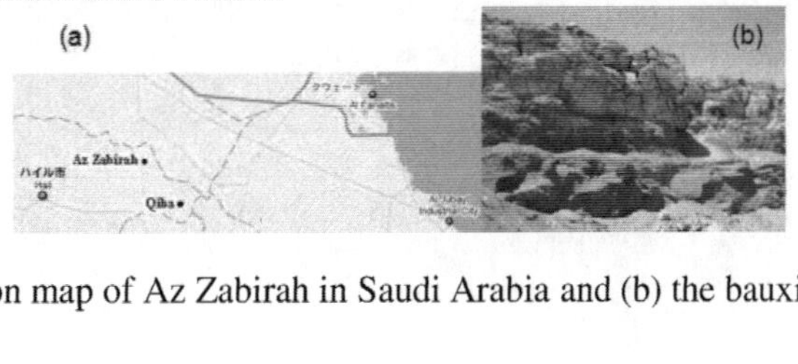

Figure 1. (a) The location map of Az Zabirah in Saudi Arabia and (b) the bauxite rocks in the location of the study area

2.2 Sampling and sample preparation
A total of 24 samples of the bauxite were collected for investigation from Az Ziberah in Qassim region. The samples were ground, homogenized and sieved using a standard set of sieves. Weighed samples were put in polyethylene bottles, of 350 cm^3 volume. The bottles were completely sealed for more than one month to allow radioactive equilibrium to be reached. This step was necessary to ensure that radon was removed from the volume [13, 14, 15]. Each powdered sample was homogenized using an electric shaker [16].

2.3 The γ-ray spectrometer (NaI(Tl)
NaI(Tl) detector was used to measure the natural radionuclides activity (count rate in the environmental samples). It was calibrated using known source such as 60Co and 137Cs point sources. In order to calculate the radionuclide activity concentration (activity per unit mass) for each gamma-ray photo-peak rely on the secular equilibrium between parents and daughters in the samples, the equation1 was used.

$$A = \frac{N_P \times 100}{e \times \eta \times m} \tag{1}$$

where:
N_p is the net count rate (cps)
e is the abundance of the γ-peak in a radionuclide
m is the mass of sample
η is the measured efficiency for each gamma-ray peak observed for the same number of channels.

The count rate was subtracted from an empty polystyrene container using the same manner as measured samples to obtain the net count rate. The measuring time for gamma-ray spectra was 43200 s [14, 17, 18, 19, 20, 21].

3. Results and discussion

3.1 Activity concentrations of ^{226}Ra, ^{232}Th and ^{40}K
The activity concentrations of ^{226}Ra , ^{232}Th and ^{40}K in the bauxite samples in Az Zabirah were shown in table 2. According to the recommended reference level by [7], the activity concentration should be less than 50, 50 and 500 Bq/kg for ^{226}Ra,^{232}Th and ^{40}K respectively [22]. The obtained results of ^{226}Ra and ^{232}Th in the bauxite samples were higher than the recommended reference level (table 2). However, the activity concentrations of ^{40}K in the study were lower than the recommended reference level by [7]. Table 3 shows the comparison between the activity concentrations of ^{226}Ra , ^{232}Th and ^{40}K in the bauxite samples in this study with those by other investigations in different locations in the world. The average value of ^{226}Ra in this study was higher than the average value of ^{226}Ra measured in India, Guinean, Greece Gulf of Itea and Brazil. However, our result was lower than Greece, Antikyra Bay, Hungary, China and Turkey. Regarding to the result of ^{232}Th, our result was lower than all countries except Brazil. However, ^{40}K was the highest value. Figure 2 illustrates the correlations between ^{226}Ra, ^{232}Th and ^{40}K. It is noted that a good correlation between ^{226}Ra and ^{232}Th and between ^{226}Ra and ^{40}K with correlation coefficients of 0.875 and 0.647, respectively.

Table 2. Activity concentrations (^{226}Ra, ^{232}Th and ^{40}K) in the bauxite samples in the study area.

Sample No	Activity concentrations		
	^{226}Ra	^{232}Th	^{40}k
1	65.9	109.7	59.6
2	62.7	105.8	59.9
3	62.9	102.8	48.9
4	75.4	119.4	69.7
5	88.1	142.1	93.2
6	81.5	127.0	89.4
7	94.7	149.1	101.7
8	91.8	147.1	99.8
9	78.9	131.0	117.4
10	127.5	149.8	191.7
11	107.6	171.5	104.7
12	115.5	193.4	141.2
13	107.3	174.9	154.8
14	106.1	171.4	134.0
16	89.6	144.0	124.82
17	102.0	163.7	168.03
18	99.2	157.3	63.11
19	130.6	180.8	142.07
20	119.3	168.5	117.24
21	137.4	202.8	141.34
22	127.4	180.7	120.95
23	141.1	200.1	184.42
24	138.5	201.0	158.07
Average	102.2	156.3	116.8

Table 3. Comparison of natural radioactivity concentration (Bqkg^{-1}) in the bauxite for present study with previous study reported from different countries of the world.

Country	Description	Mean activity concentration (Bqkg^{-1})			References
		^{226}Ra	^{232}Th	^{40}K	
Saudi Arabia, Az Zabirah	bauxite	102.2	156.3	116.8	Present study
Guinean	Bauxite	51.25	227	-----	[23]
India	Bauxite	70.2	499	----	[23]
Brazil	Bauxite	64	154	9.4	[1]
Greece, Gulf of Itea	Bauxite	72.8	185	43	[24]
Greece, Antikyra Bay	Bauxite	150	205	28.3	[11]
Hungary	Bauxite	419.0	256.0	47.0	[25]
China	Bauxite	370.0	400.0	63.0	[8]
Turkey	Bauxite	235.3	243.4	74.3	[26]

IOP Conf. Series: Journal of Physics: Conf. Series **817** (2017) 012061 doi:10.1088/1742-6596/817/1/012061

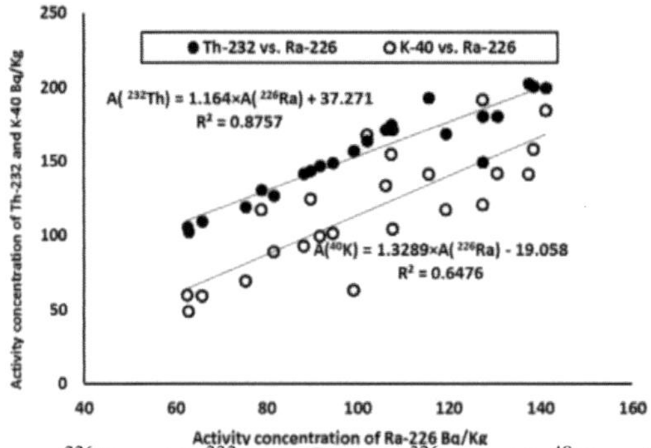

Figure 2. Correlations between ^{226}Ra with ^{232}Th (●) and ^{226}Ra with ^{40}K and (○) in the bauxite samples in Qassim province Saudi Arabia.

The radiological risk can be estimated using indices such as radium equivalent activity, the absorbed dose and the annual effective dose. The distribution of ^{226}Ra, ^{232}Th and ^{40}K in samples are not uniform. To obtain the same gamma dose rate, the activity concentration from the three radionuclides assuming to be 370 Bq·kg^{-1} from ^{226}R, 259 Bq·kg^{-1} from ^{232}Th and 4810 Bq·kg^{-1} from ^{40}K. This is the definition of radium equivalent and is given as equation 2:

$$Ra_{eq}=A_{Ra} +1.43\times A_{Th}+0.077\times A_k \tag{2}$$

where A_{Ra}, A_{Th} and A_K are the activity concentrations of ^{226}Ra, ^{232}Th and ^{40}K in Bq·kg^{-1}, respectively. Regarding to radiological hazard, the maximum Ra_{eq} in the bauxite samples must be less than 370 Bq kg^{-1} for safe use. The radium equivalent results ranged from 213.7 ± 10 to 464.6 ± 34 Bq·kg^{-1} and mean value was 340.2. The average value was in the safety recommended limit. These results were higher than the results obtained from bauxite in Nigeria (189-262 Bq·kg^{-1}) [10]. Furthermore, the absorbed dose rate in air at 1 m above the ground surface was calculated at different sampling points using equation 3.

$$D_R\ (nG\ h^{-1}) = 0.427ARa + 0.623ATh +0.043AK \tag{3}$$

The values varied from 97.0 to 213.7 nGy/h and the mean value was 155.5 nGy/h in the Az Zabirah bauxite mine. The obtained average value based on Equation (3) was higher than the worldwide average value of 60.00 nGy/h [7]. The calculated effective dose to which population are exposed to gamma rays in the Az Zabirah mine ranged from 0.119 to 0.262 mSv/year with a mean of 0.19 mSv/year. The obtained average values in bauxite mine was lower than the safe limit of 1.00 mSv/year recommended by [7]. The calculation value of the absorbed dose rate in air at 1 m above the ground level and the annual effective dose to which population may likely be exposed at different sampling points in this study sites are shown in figure 3. The obtained values of external index (H_{ex}) ranged from 0.60 to 1.3 with an average of 0.9. The internal hazard index (H_{in}) values in the bauxite mine in Az Zabirah varied from 0.7 to 1.6 with a mean of 1.2. Comparing the obtained average values of both external and internal hazard indices in both areas, it can be seen that the obtained average values of the external hazard parameters in Az Zabirah was lower than the safe limits values except for the internal hazard index value was slightly higher than the recommended average values by [7].

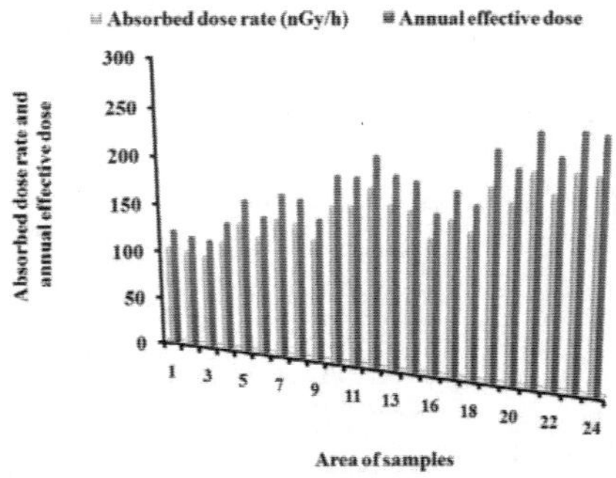

Figure 3. Air- absorbed dose rates and annual effective doses calculated the bauxite mine samples collected from Az Zabirah in Qassim province.

4. Conclusion

It is important to determine background radiation level in order to evaluate the health hazards resulting from Az Zabirah bauxite mine. The method of gamma spectrometry has been used to measure the radioactivity concentration of 24 samples collected from Az Zabirah bauxite mine in Qassim Province, Saudi Arabia. From this study, the range and average values of the concentrations of ^{226}Ra, ^{232}Th and ^{40}K were 102.2 (141.1-62.7), 156.3 (202.8-102.8) and 116.8 (191.7- 48.9) Bq/kg respectively. Overall, the study showed that the measured values for ^{226}Ra and ^{232}Th were higher than the values in the worldwide soil. The mean values of radium equivalent and total absorbed dose rate were 340.2 Bq/kg and 155.5 nGy.h^{-1} respectively. The world average values of radium equivalent and absorbed dose are 370 Bq.kg^{-1} and 65 nGyh^{-1}. The annual effective gamma doses was lower than the world's average. The external hazard parameters in Az Zabirah were lower than the safe limits values. The internal hazard index value was slightly higher than the recommended average values by UNSCEAR. The study area is in the zones of high radiation level, and it is possible to produce radiation hazards to the environment as well as the human health. However, this data may provide a general background level for the area studied and may serve as a guideline for future measurements and assessments of possible radiological risks to human health in this province.

References

[1] Cuccia V, De Oliveira A H and Rocha Z 2011 *Proc. Int. Nuclear Atlantic Conf. (Brazil)* (Brazil: Associação Brasileira de Energia Nuclear (ABEN)) p 24
[2] Abbady A G, Uosif M and El-Taher A 2005 *J. Environ. Radioac.* **84** 65
[3] Jobbágy V, Somlai J, Kovács J, Szeiler G and Kovács T 2009 *J. Hazard. Mater.* **172** 1258
[4] Cooper M B 2005 *Report for to the Radiation Health and Safety Advisory Council* (Australia: EnviroRad Services Pty. Ltd.) ERS-006
[5] O'connor B H, Donoghue A M, Manning T J and Chesson B J 2013 *Ann. Occup Hyg.* **57** 63
[6] Yang Qu and Bin Lian 2013 *J. Bioresour. Technol.* **136** 16
[7] United Nations Scientific Committee on the Effects of Atomic Radiation (UNSCEAR) 2000 *Sources and Effects of Ionizing Radiation: United Nations Scientific Committee on the Effects of Atomic Radiation 2000 Report* (Vienna: UNSCEAR) pp 453-87
[8] Righi S, Verità S, Albertazzi A, Rossi P L and Bruzzi L 2009 *J. Environ. Radioac.* **100** 540
[9] International Atomic Energy Agency (IAEA) 2003 *Techical Report Series* No. 419 (Vienna: IAEA)
[10] Janet A A and Adekunle M O 2013 *World J. Nucl. Sci. Technol.* **3** 150
[11] Papatheodorou G, Papaefthymiou H, Maratou A and Ferentinos G 2005 *Radiopro.* **40** S549
[12] Vincent P 2008 *Saudi Arabia: An Environmental Overview* (UK: Taylor and Francis)
[13] El-Taher A and Alashrah S 2015 *AIP Conf. Proc.* vol 1674 Issue 1 (USA: AIP Publishing) 020007

IOP Conf. Series: Journal of Physics: Conf. Series **817** (2017) 012061 doi:10.1088/1742-6596/817/1/012061

[14] Alashrah S and El-Taher A 2016 *J. Environ. Sci. Technol* **9** 238
[15] Alashrah S 2016 *J. Radiat. Res.Appl. Sci.* **9** 363
[16] El-Taher A and Alharbi A 2013 *Appl. Radiat. Isot.* **82** 67
[17] El-Taher A and Al-Zahrani J H 2015 R *Indian J. Pure Appl. Phys.* **52** 147
[18] El-Taher A and Abdelhalim M A K 2014 *J. Radioanal. Nucl. Chem.* **299** 1949
[19] El-Taher A and Abdelhalim M A K 2014 *J. Radioanal. Nucl. Chem.* **300** 431
[20] El-Taher A and Madkour H 2014 *J. Isot. Environ. Health Stud.* **50** 120
[21] Alharbi A and El-Taher A 2014 *Life Sci. J.* **11** 829
[22] United Nations Scientific Committee on the Effects of Atomic Radiation (UNSCEAR) 1988 *Sources, Effects and Risks of Ionizing Radiation: United Nations Scientific Committee on the Effects of Atomic Radiation 1988 Report* (Vienna: UNSCEAR) pp 565-71
[23] Abbady A G and El-Arabi A 2006 *J. Radiol. Prot.* **26** 415
[24] Karagiannidi T, Papaefthymiou H and Papatheodorou G 2009 *J. Radioanal. Nucl. Chem.* **279** 923
[25] Somlai J, Jobbagy V, Kovacs J, Tarjan S and Kovacs T 2008 *J. Hazard. Mater.* **150** 541
[26] Turhan S, Arıkan I, Demirel H and Güngör N 2011 *Radiat. Phys. Chem.* **80** 620-5

IOP Conf. Series: Journal of Physics: Conf. Series **817** (2017) 012062 doi:10.1088/1742-6596/817/1/012062

Powder metallurgy preparation of Mg-Ca alloy for biodegradable implant application

D Annur[1], A Suhardi[2], M I Amal[1], M S Anwar[1] and I Kartika[1]

[1]Research Center for Metallurgy and Materials - Indonesian Institute of Sciences Gedung 470, Kawasan Puspiptek, Serpong, Tangerang Selatan, 15314, Indonesia
[2]Universitas Sultan Ageng Tirtayasa, Banten, Indonesia

E-mail: dhya001@lipi.go.id

Abstract. Magnesium and its alloys is a promising candidate for implant application especially due to its biodegradability. In this study, Mg-7Ca alloys (in weight %) were processed by powder metallurgy from pure magnesium powder and calcium granule. Milling process was done in a shaker mill using stainless steel balls in various milling time (3, 5, and 8 hours) followed by compaction and sintering process. Different sintering temperatures were used (450°C and 550°C) to examine the effect of sintering temperature on mechanical properties and corrosion resistance. Microstructure evaluation was characterized by X-ray diffraction, scanning electron microscope and energy dispersive X-ray spectroscopy. Mechanical properties and corrosion behavior were examined through hardness testing and electrochemical testing in Hank's solution (simulation body fluid). In this report, a prolonged milling time reduced particle size and later affected mechanical properties of Mg alloy. Meanwhile, the phase analysis showed that α Mg, Mg_2Ca, MgO phases were formed after the sintering process. Further, this study showed that Mg-Ca alloy with different powder metallurgy process would have different corrosion rate although there were no difference of Ca content in the alloy.

1. Introduction

Owing biodegradable properties, magnesium alloys have been intensively researched in these recent years [1-5]. Magnesium alloys can be used as a temporary implant which will be dissolved without the need for further surgery due to its biodegradation properties [6]. Moreover, magnesium alloys are beneficial due to its elastic modulus which is close enough to the human femur bone (15-20 GPa) [6]. Hence, it is assumed that magnesium alloys can minimize the risk of stress- shield phenomena which currently exist in other metallic implants such as stainless steel or titanium alloy [7, 8].

Though magnesium alloys has shown great potential to be used as bone implant material, more development is needed. The high corrosion rate of magnesium alloy is also unfavorable since magnesium alloy act as supporting device during bone recovery. Accordingly, it is important to ensure the alloying element of magnesium alloy can increase the degradation properties of magnesium alloy. Furthermore, alloying elements must show biocompatibility properties and also give no toxicity to the bone. Previous studies had been conducted for developing Mg- Ca based alloys [2, 9, 10]. Calcium, as the main composition in human bone, showed a great performance in improving bone healing process, and having low density just like magnesium, is a desirable alloying element for biomedical application [2, 10].

Those previous studies had shown the possibility of processing Mg alloy into biomedical devices. Further, it should be noted that the chosen processing method will affect the applicability of magnesium alloy. Until now, limited efforts have been made to magnesium-based alloys through powder metallurgy method. Meanwhile, powder metallurgy is one of a solid- state method that can be chosen in processing magnesium alloy into a near net shape. Powder metallurgy also allows flexibility of alloy design with different elemental or master alloy powder in order to get better mechanical properties. Basically, powder metallurgy is competitive enough in comparison with conventional casting, forging, or machining route. However, processing magnesium through powder metallurgy was not as easy as it sounds.

Wolff et al. observed the difficulty process of sintering magnesium because sintering magnesium powder was strongly inhibited by a stable oxide layer [11] that would form immediately after exposure to air. It was concluded that calcium content would improve sinterability because it would protect the magnesium against additional oxide layer. This is based on the lower formation enthalpy of CaO (-635 kJ/mol) in comparison with MgO (-601 kJ/mol). It was also shown that the gas atomized master alloy Mg-7Ca turned out to be a promising candidate to sinter magnesium [11].

In the present work, mixing parameter and temperature sintering were varied in order to achieve the most optimum condition for preparing Mg-7Ca master alloy by means of powder metallurgy method. Moreover, structure, phase, and corrosion behavior of master Mg-7Ca master alloy were studied.

2. Materials and method

High purity magnesium powder (Merck, particle size: 0.06-0.3 mm) and calcium granules were used to prepare Mg-7Ca alloy. Calcium granules were previously crushed by mortar followed by mixing process with Mg powder in a shaker mill for 3, 5, and 8 hours. The final milled powder of Mg-7Ca alloy was then cold compacted under the pressure of 260 MPa . The resulted green compacts had a diameter of around 26.1 mm followed by sintering process. These green compacts were sintered in a different heat treatment profile which is 450°C and 550°C for an hour under Argon atmosphere. Heating rate used in this process was 5°C/minute. Each specimen was indexed based on its treatment, as shown in table 1.

Table 1. List of specimens.

Specimen Code	Milling Time (hour)	Sintering Temperature (°C)
3M-450C	3	450°C
5M-450C	5	450°C
8M-450C	8	450°C
3M-550C	3	550°C
5M-550C	5	550°C
8M-550C	8	550°C

The microstructure of as sintered alloys were observed by scanning electron microscope (SEM) and phase formation were analyzed by X-ray diffraction (XRD). Mechanical properties of these alloys were examined through hardness testing of each specimen. To investigate corrosion behavior in vitro, electrochemical testing had been done in Hank's solution [12] at pH 7.4 and 37°C under air atmosphere. The counter electrode material was graphite, the specimen was working electrode, and the reference electrode was saturated calomel electrode (SCE). The potential scanning rate was fixed at 1.5V/s. Potential corrosion (E_{corr}) and corrosion rates (Corr rate) were observed.

3. Results and discussion

Figure 1 shows the powder condition before and after milling process. It could be seen from figure 1, miling process would reduce particle size after 3 hours, but in a prolonged duration, milling process would not reduce the particle size (for 5 hours and 8 hours). Time of milling during powder metallurgy process should be optimized due to its importance to create a homogeneous green compact. But it should be noted that the level of contamination might be increased and unfavorable phases might be form if the duration of milling were prolonged [13]. Further, in a prolonged duration of milling aglomeration can occured. Therefore, the longest time of milling might not always be the best milling time. The reduction process is related to the collision between ball mill and the particles. As for mechanical alloying, it is defined by a high energy ball milling process which deformation, fracture, and welding are repeatedly occurred in between constituent powders and grinding media to form a homogeneous alloy microstructure [13]. During milling process, the particles break up into smaller particles making it easier to be dispersed in the alloy.

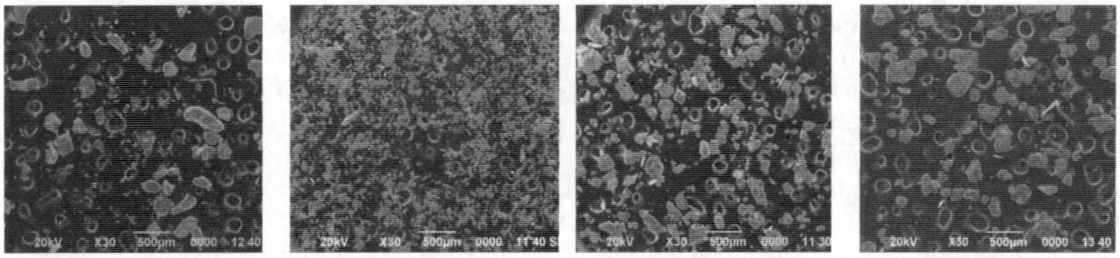

(a) (b) (c) (d)

Figure 1. SEM images of Mg-7Ca alloy. (a) Before milling and after milling for (b) 3 hrs, (c) 5 hrs and (d) 8 hrs.

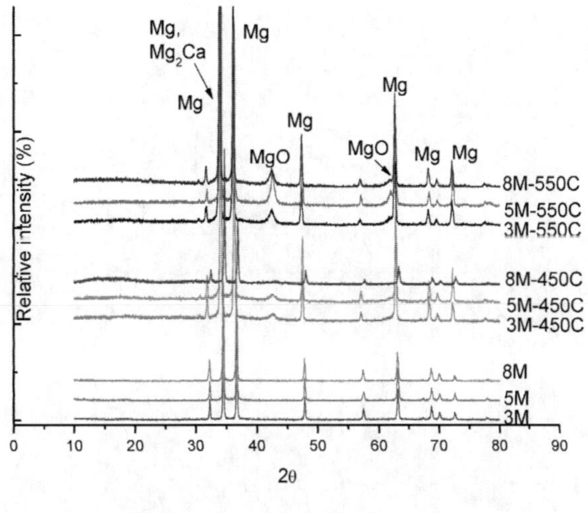

Figure 2. XRD spectra for as- milled powder before sintering and after sintering process.

Further analyses were done through XRD examination. Figure 2 shows XRD spectra of Mg-Ca alloy which revealed qualitative analysis of phase formation. Based on figure 2, some Mg characteristic peaks are shown. It was presumed that Mg_2Ca phase characteristic peaks would be shown after milling process before sintering process due to mechanical alloying between Mg and Ca, yet no clear evidence is shown for Mg_2Ca phase. Although different milling times were used in processing Mg-7Ca alloy, the XRD pattern of each parameter is not showing any broader or sharper peak difference. But there is a peak shifting slightly to the left by increasing milling duration which is related to the tendency of Mg_2Ca formation when milling process would be prolonged. Moreover the resulted XRD spectra shows MgO phase formation at 63.3° during milling process. From figure 2 it can also be seen a peak shifting from 36.03° (Mg phase) before sintering to 34.36° (Mg_2Ca phase) after sintering process. This phenomenon is more likely caused by the existence of Mg_2Ca phase after sintering process. It had shown that sintering temperature affected phase formation in Mg-Ca alloy and also increasing of MgO phase characteristic peak. It means sintering process tends to create MgO phase which is unfavorable in Mg- 7Ca alloys. It was presumed that MgO phases could be avoided by the addition of calcium. Unfortunately, calcium addition can no longer inhibit MgO phase formation especially after sintering process at 450° and 550°C although the reaction occurred under argon atmosphere. Moreover, figure 2 shows that higher sintering temperature will increase the tendency of MgO phase formation. Indeed, it is generally accepted that chemical reactions occur more rapidly at higher temperatures. However, MgO is accepted as a a biologically non toxic oxide for biomedical devices [1].

Figure 3 depicts SEM images of Mg-7Ca alloy after sintering process, while table 2 gives an EDS quantitative analysis from figure 3. From figure 3, it can be seen that some of micro pores do exist in the Mg-7 Ca alloy. Some bubbling spot were shown in figure 3a and 3b, on the other hand figure 3c shows a better microstructure with smaller pores. Bubbling spot and porosity will affect properties of Mg-Ca alloy. Increasing milling duration tends to reduce surface roughness of Mg-Ca alloy. When sintering temperature is higher, given by figure 3d-f, it will also reduce surface roughness of Mg-Ca alloy. Specimen with 550° C sintering temperature has smoother surface, smaller pores, shows densification process occurred better than 450° C. Furthermore, based on EDS result, value of oxygen content from each Mg-7Ca alloy was high (around 19-33 wt.%). EDS result can not exactly measure the oxygen content from each specimen, but it can give an illustration of each specimen tendency on MgO phase formation. Hereby, the SEM images confirms that increasing milling time would increase MgO formation as shown by the XRD result.

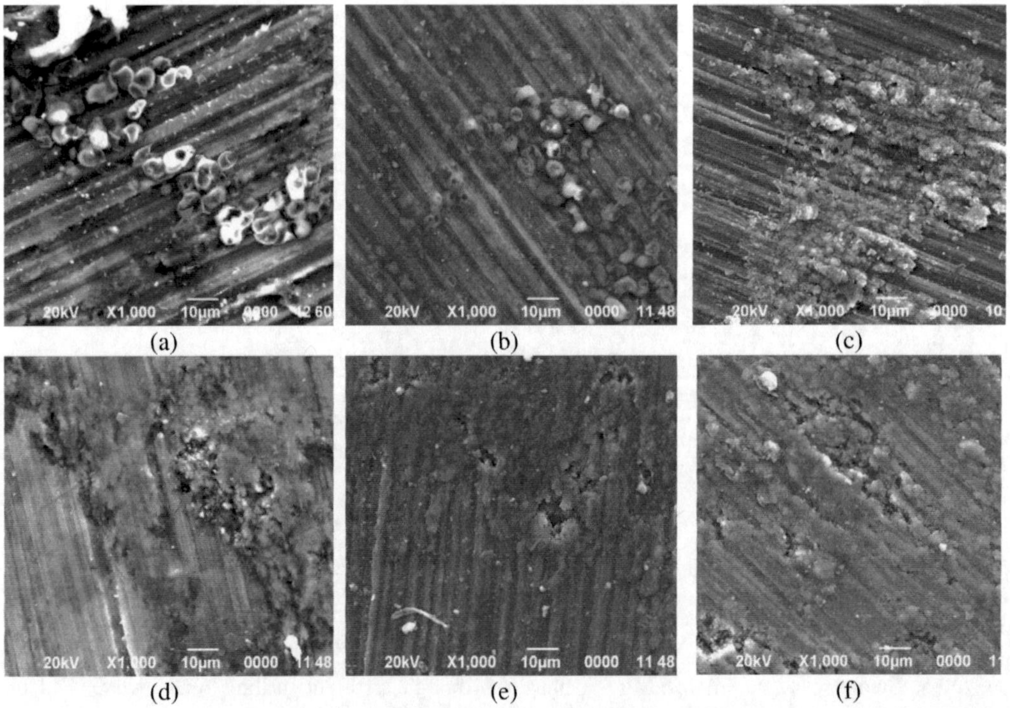

Figure 3. SEM images of as-sintered Mg-Ca alloy in different heat treatment profile. (a). 3M-450°C, (b) 5M-450°C, (c) 8M-450°C, (d) 3M-550°C, (e) 5M-550°C, (f) 8M-550°C.

Table 2. Composition analysis obtained by EDS.

Element	Specimen codes					
	3M-450°C	5M-450°C	8M-450°C	3M-550°C	5M-550°C	8M-550°C
O	20.85	19.23	33.53	32.62	27.11	28.73
Mg	79.00	80.31	66.02	67.50	72.58	70.54
Ca	0.14	0.26	0.42	0.08	0.31	0.73

To see mechanical properties trend on Mg-Ca alloys with different powder metallurgy parameter, hardness testing was done. Table 3 shows hardness number of each specimen according to its milling time. It can be seen that hardness property will increase by increasing milling time. Not only milling time, sintering temperature will also increase the hardness number of Mg-Ca alloy. Increasing mechanical properties is in line with the formation of Mg_2Ca phase which has higher mechanical properties. Further, mechanical properties are also related to total porosity. Along with SEM images where smaller pores were created in higher sintering temperature (550° C), mechanical properties also increase. Depends on the application of the end product, phase formation in Mg-Ca alloy must be controlled to avoid excessive brittleness which may lead to crack due to the precipitation of Mg_2Ca. Although mechanical properties will increase by milling time addition and sintering temperature, the corrosion potential might not give a similar trend.

Table 4 shows corrosion properties of the resulted Mg-Ca alloy which represent nobler potential corrosion in comparison with Mg in some reference. Zhang et al. [14] reported electrochemical testing of Mg and Mg-Zn alloy in SBF solution, in which Mg had -1.62 V, while Gu et al. [15] mentioned potential corrosion of Mg was -1.7 V. Here it can be concluded that better potential corrosion is achieved with the addition of Ca. From the electrochemical testing, it can be seen that corrosion rate will increase by milling time increasing. It may also relate to the formation of Mg_2Ca phase. Mg_2Ca has higher electrochemical activity than Mg phase so that galvanic corrosion may occur more [16]. Since corrosion rate is an important consideration for biodegradable implant, lower corrosion rate is preferable.

Table 3. Milling time vs average hardness property (HVN).

Specimen Codes	3M-450°C	5M-450°C	8M-450°C	3M-550°C	5M-550°C	8M-550°C
Hardness Number HVN)	42.18	45.85	49.03	44.86	47.09	51.33

Table 4. Corrosion properties of Mg-Ca alloys in Hank's solution obtained from the polarization test.

Specimen Codes	E Corr	Corr Rate (MPY)
3M-450°C	-1.54	0.138
5M-450°C	-1.58	1.962
8M-450°C	-1.55	1.418
3M-550°C	-1.58	0.311
5M-550°C	-1.57	0.424
8M-550°C	-1.59	1.640

4. Conclusion

In this study, milling time and temperature sintering were varied. It was shown that milling process for 3, 5 and 8 hours could be used to form intermetallic phase Mg_2Ca. Not only Mg_2Ca phase, after sintering process MgO phase was also existed. Increasing milling time tends to form Mg_2Ca phase which can increase the mechanical properties of Mg-Ca alloy. Based on this study, variation on sintering process affects densification process. Sintering temperature also affects mechanical properties whereas higher sintering temperature would give higher mechanical properties. Moreover, from corrosion testing, it can be seen that the longer duration of milling can also reduce corrosion resistance. This study showed that Mg-Ca alloy with different powder metallurgy process would have different corrosion rate although there were no difference of Ca content in the alloy. To be used as biodegradable implant, specimen with lower corrosion rate may be choosen. This study shows that optimization process should be continue with 550° C sintering temperature.

Acknowledgement
This work was financially supported by the 2015 Competitive Program of Indonesian Institute of Sciences.

References
[1] Seyedraoufi Z S and Mirdamadi S 2013 *J. Mech. Behav. Biomed. Mater.* **21** 1
[2] Li Z, Gu X, Lou S and Zheng Y 2008 *Biomaterials* **29** 1329
[3] Lietaert K, Weber L, Van Humbeeck J, Mortensen A, Luyten J and Schrooten J 2013 *J. Magnesium Alloys* **1** 303
[4] Rad H R B, Idris M H, Kadir M R A and Farahany S 2012 *Mater. Design* **33** 88
[5] Bamberger M, Levi G and Vander Sande J B 2006 *Metal Mat. Trans. A* **37** 481
[6] Zhang E, Yin D, Xu L, Yang L and Yang K 2009 *Mat. Sci. Eng. C* **29** 987
[7] Kartika I, Thaha Y N, Lestari F P and Sriyono B 2014 *Adv. Mat. Res.* **896** 267
[8] Staiger M P, Pietak A M, Huadmai J and Dias G 2006 *Biomaterials* **27** 1728
[9] Wan Y, Xiong G, Luo H, He F, Huang Y and Zhou X 2008 *Mater. Design* **29** 2034
[10] Kim W C, Kim J G, Lee J Y and Seok H K 2008 *Mater. Lett.* **62** 4146
[11] Wolff M, Ebel T and Dahms M 2010 *Adv. Eng. Mat.* **12** 829
[12] Kuwahara H, Al-Abdullat Y, Mazaki N, Tsutsumi S and Aizawa T 2001 *Mater. Trans.* **42** 1317
[13] Suryanarayana C 2001 *Prog. Mater. Sci.* **46** 1
[14] Zhang S, Zhang X, Zhao C, Li J, Song Y and Xie C 2010 *Acta Biomat.* **6** 626
[15] Gu X, Zheng Y, Zhong S, Xi T, Wang J and Wang W 2010 *Biomaterials* **31** 1093
[16] Bakhsheshi-Rad H R, Hamzah E, Fereidouni-Lotfabadi A, Daroonparvar M, Yajid M A M and Mezbahul-Islam M 2014 *Mater. Corros.* **65** 1178

Reduction mechanism of indonesian limonite ore by solid reducing agents

W Astuti[1], S Novita[2] and F Nurjaman[1]

[1]Research and Development Division for Mineral Technology, Indonesian Institute of Sciences (LIPI), Jl. Ir. Sutami Km. 15 Tanjung Bintang, Lampung Selatan, Lampung, Indonesia
[2]Department of Physics, Faculty of Mathematics and Natural Sciences, Lampung University, Jl. Sumantri Brojonegoro No. 01, Bandar Lampung, Lampung

E-mail: widi.mineral@gmail.com

Abstract. Limonite ore from Sulawesi Island (Indonesia) was reduced isothermally at 900 – 1200°C by pulverized coal and charcoal to produce ferronickel. XRD and XRF analyses were used to characterize the samples. The influence of reductive conditions on the reduction and metallization degree of the samples, including reduction temperature, reduction time and reducing agent types were studied. Results have shown that both the reduction and metallization degrees increase as temperature and time increase. The reduction reactions will be more completed at higher reduction temperature. It could be seen from XRD results that magnetite and Fe-Ni peaks increased at higher temperature reduction. This could be mean that the rate of reduction reaction is faster at higher temperature. The losses of samples weight also happen faster at higher temperature. It was also found that the reduction behaviors using coal differed from those using wood charcoal due to the differences in proximate content. From this study, it could be concluded that fixed carbon content in the reducing agent did not always be the most important aspect of choosing a reducing agent. The sample which has been reduced with coal and charcoal showed an almost exactly same pattern even they both had different characteristics. The relatively higher volatile matter content in charcoal gave a significant contribution to the reduction process.

1. Introduction

Nickel is one of the most important metal in the world. Mostly, nickel is used as a raw material in the stainless steel making process which accounted for 62% of nickel consumption [1]. One of the sources of nickel is nickel laterite ore. Nickel laterite ore is the primary raw material in the process of making nickel pig iron (NPI) and ferronickel. NPI or ferronickel is usually produced in the blast furnace using coal or coke as a fuel. In recent years, the rapid rise in stainless steel demand has dramatically increased the production of nickel metal by approximately 4.7% per year [2]. Therefore, the recovery of nickel from laterite ores using pyrometallurgy has been extensively investigated [2].

Indonesia is one of the largest producers of nickel ore in the world and has approximately 16% of total global nickel resources. These are in the form of lateritic ores; they are widely distributed and spread among the islands of Kalimantan, Sulawesi, Halmahera, Gag, and Papua [3]. Nickel pig iron (NPI) production is a new promising process, although it was firstly developed about 50 years ago but it has not been commercially used until some Chinese pig iron producers changed their production methods into NPI production. Chinese NPI production was firstly started in blast furnaces using low-grade laterite ores imported from Indonesia, Philippines and New Guinea [2]. The process is almost same as pig iron production processes. The only difference is that the ore contains more nickel. Not only as a fuel, coal or coke also act as a solid reducing agent in the nickel laterite reduction. The use of coal or coke as solid reducing agent can be substituted by other carbon sources such as charcoal and other biomass. However, the use of charcoal for reducing laterite ores has not been studied more particularly in term of reduction mechanism.

In this study, therefore, the reduction of nickel from a low-grade limonite ore was performed using different solid reducing agents i.e. coal and charcoal in order to compare the reduction mechanism of those reducing agents. The effects of reduction temperature and time on the reduction process were also investigated. Moreover, the reduction mechanisms of nickel were analyzed based on the change of mineral phase during the reduction process. The present study aims to provide a basis for the utilization of low-grade limonite ore from Indonesia especially using pyrometallurgical method.

2. Materials and methods

2.1. Materials

A low-grade limonite ore from Indonesia (Sulawesi Island) was used in this experiment. The samples were mineralogically and chemically characterized. X-ray fluorescence (XRF) was used to determine the chemical compositions of the sample. The mineral phases of the raw ore sample were identified by X-ray diffraction (XRD), using CuKα radiation, in the 2θ range 5° to 80°. The accelerating voltage and applied current were 40 kV and 40 mA, respectively, with a scanning speed of 2°/min and a scanning step of 0.02°. Thermogravimetric-differential thermal analysis (TG-DTA; 2000SA, Bruker, Billerica, MA, USA) of the ores samples was performed from room temperature to 1250 °C at a heating rate of 10°C/min to clarify the mineral contents of each sample. Scanning electron microscopy/energy-dispersive X-ray spectroscopy (SEM/EDS; VE-9800 SEM, Keyence, Osaka, Japan, and EDAX Genesis, Ametek, NJ, USA) was used to determine the nickel contents of the different minerals by metal mapping. The selected reducing agents are coal and charcoal. The proximate analysis of the reducing agents is listed in table 1. Bentonite was also used as a binder in the pellet making. In addition, limestone was also applied in the pellet composition as a flux in the pellet when reduction process conducted. All raw materials were ground and sieved to get the particle size of -100 mesh. The mixtures of all raw materials in certain composition i.e. limonite ore (80 wt%), reductant (15 wt%), limestone (3 wt%), bentonite (2 wt%) and some water were pelletized in order to obtain the composite pellets with a diameter of pellet approximately 1 cm.

Table 1. Proximate analysis of reducing agents (pulverized coal and charcoal) (mass fraction, %).

Reducing Agents	Total Moisture (%)	Volatile Matter (%)	Ash (%)	Fixed Carbon (%)
Coal	7,01	29,50	2,79	60,70
Charcoal	7,69	51,74	7,24	30,66

2.2. Reduction methods

Reduction process was carried out in the laboratory using muffle furnace. The dried composite pellet was placed in a ceramic crucible, which was then placed in the muffle furnace when the temperature of furnace reached the reduction temperature. Once the reduction experiments were finished, the samples were taken out of the furnace and cooled to room temperature. The reduced product was dried for weighing and XRD analysis. Thermodynamic analysis using Ellingham diagram shows that the lowest temperature for a complete iron reduction was about 720°C [4]. Based on that analysis, the temperature used in this research was 800 - 1200°C. The reduction time used in this research was 15, 30, 60, 120, and 240 minutes.

IOP Conf. Series: Journal of Physics: Conf. Series **817** (2017) 012063 doi:10.1088/1742-6596/817/1/012063

3. Results and discussion

3.1. Ore characteristics

The chemical composition of the limonite ore by X-ray fluorescence (XRF) is given in table 2. X-ray diffraction analysis (XRD) for limonite ore shows that goethite (FeOOH), olivine ($MgNiO_4Si$), and quartz (SiO_2) is the essential mineral in the limonite sample. The XRD pattern shows no independent peaks for nickel compounds. The SEM/EDS analysis is a simple method for element mapping and the data can be used to predict the minerals associated with nickel and other metals. Figure 1 show the metal mapping for each sample, obtained using SEM/EDS analysis. The figure shows that nickel is associated with both magnesium- and silica-containing minerals such as olivine, and also iron-containing mineral i.e. goethite.

The result from DTA shows the temperatures at which thermal reactions such as phase transformation and thermal decomposition take place in a material when it is heated continuously to high temperature, and also the intensity and general characteristics (endothermic or exothermic) of such reactions [5]. The result from TGA shows the weight gain or loss of a material (e.g., a mineral) as a result of absorption or gas release as a function of temperature. The TG-DTA patterns of the sample figure 2 shows that there are three endothermic peaks, at 50, 250, and 1150 °C and one exothermic peak at 838 °C. The endothermic peaks at 50 and 250 °C are associated with moisture evaporation and goethite decomposition, respectively. The endothermic peak at 567 °C arises from dehydroxylation of olivine, whereas the exothermic peak at 805 °C is associated with recrystallization of forsterite (Mg_2SiO_4); these observations confirm the presence of olivine minerals [6, 7, 8]. The endothermic peak at 1165 °C is assigned to the dehydroxylation of talc sheets, accompanied by the formation of enstatite ($Mg_2Si_2O_6$) and silica.

Table 2. Chemical composition of nickel limonite ore (mass fraction, %).

Ore samples	Element (wt%)							
	SiO_2	Fe	Ni	Co	Mg	Mn	Cr	Al
Limonite ore	34.00	19.24	0.94	0.05	11.18	0.34	1.18	1.31

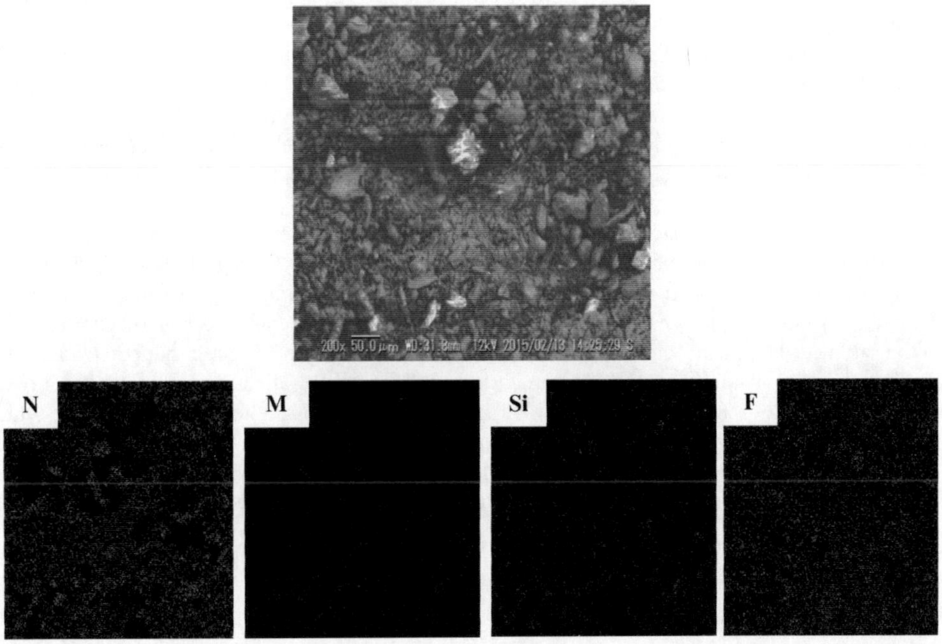

Figure 1. SEM/EDS metal mapping of limonite ore.

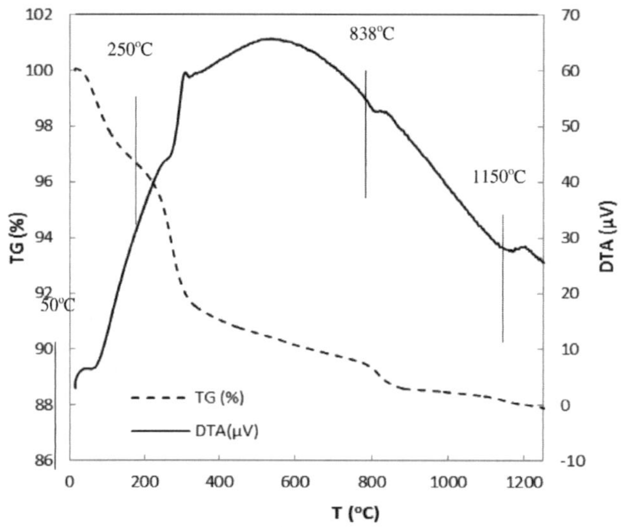

Figure 2. TG-DTA pattern of limonite ore

3.2. Reduction mechanism analysis

The reaction mechanism differences happened because the reactions which took place consist of various step such as carbon oxidation, reducing gas diffusion, and reduction reaction hence the reaction was very complex. From the analysis of the XRD results figure 3, the sample which has been reduced with coal and charcoal show an almost exactly same pattern with both had strong magnetite peaks although on the sample which has been reduced with charcoal had a slightly lower magnetite peak. However, Fe-Ni peaks on the sample which has been reduced with charcoal were higher compared to the sample which has been reduced with coal. It means that reduction reaction of the sample with charcoal as reducing agents proceed more complete because there was more magnetite which has been reduced to Fe. Based on that, it could be concluded that charcoal could be a promising substitute for coal in the reduction process of nickel laterite. From this study, it could be concluded that fixed carbon content in the reducing agents did not always be the most important aspect of choosing a reducing agent. The sample which has been reduced with coal and charcoal show an almost exactly same pattern even they both had different characteristics. The relatively higher volatile matter content in charcoal gave a significant contribution to the reduction process. This was also supported by XRD results of samples on 15 minutes reduction duration which shown that sample which has been reduced with charcoal had stronger magnetite peaks compared to the sample which has been reduced with another reducing agent. It could happen because the volatile matter on charcoal would evaporate on the early reduction stage. The changes of the fraction of reduction reaction also shows that in the early stage of the reduction process. The change of sample which has been reduced with charcoal was higher because of the volatile matter evaporation. Physical properties of the reductant also had an important role in the reduction process. The density of coal was 1.48 g/cm^3 and charcoal was 0.72 g/cm^3. The smaller density of charcoal than that of coal was indicated that samples which used charcoal as a reducing agent will be more porous.

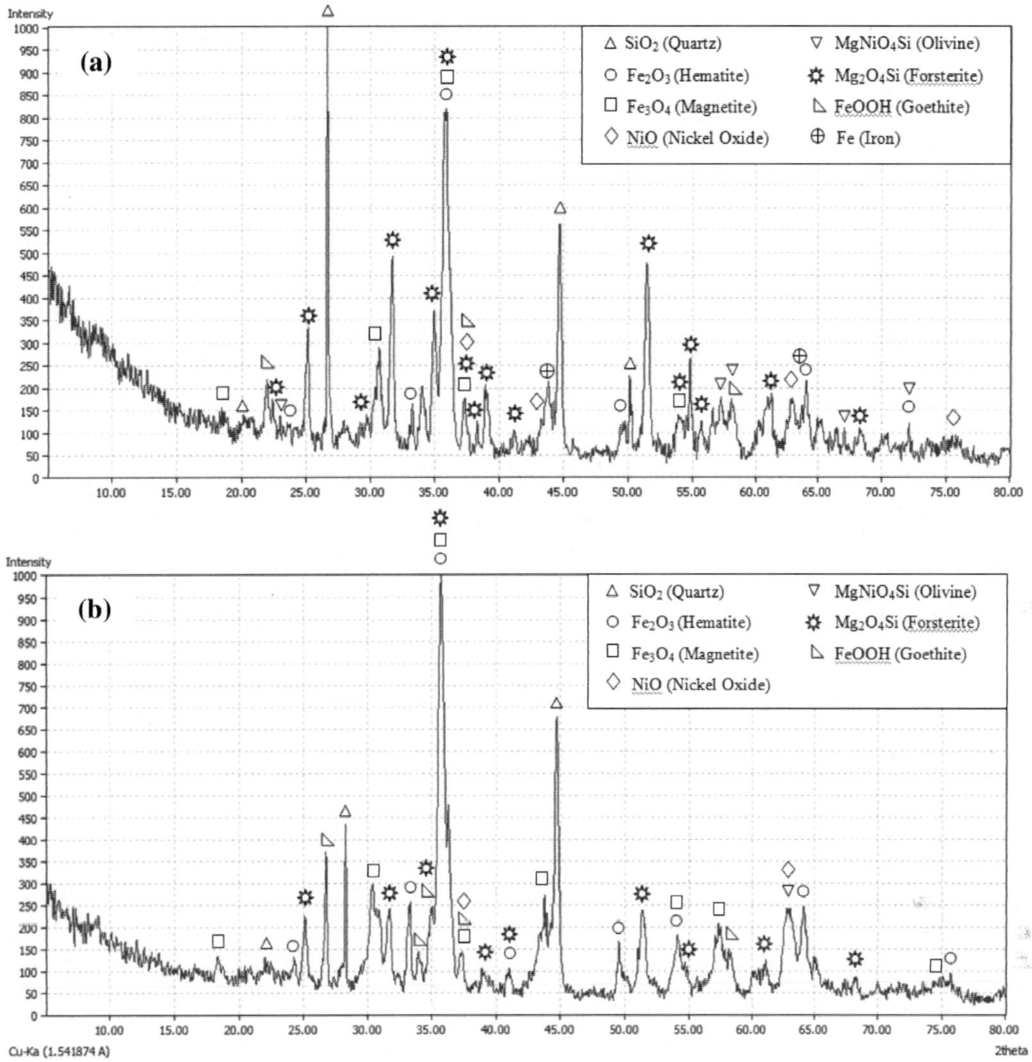

Figure 3. XRD analysis of reduced limonite, (a) coal as reducing agent; (b) charcoal as reducing agent, at 1100°C and 15 minutes of reduction time.

3.3. The effect of reduction temperature and time on the reduction process

The reaction of iron and nickel reduction will be more completed on longer duration of the reduction process. The reduction reactions were also more completed at higher reduction temperature. It could be shown by the color changes of pellet composite during reduction process as depicted in figure 4. The color of the raw material is red and the color becomes black and grey after reduction process occurs. From figure 4, it can be seen that the color of pellet composite during reduction at temperature of 800°C and 900°C did not change. However, the color of pellet become black and gray at the reduction temperature of 1100°C. The changes of color at 1100°C was started on the 15 minutes of reduction time. This could be mean that the rate of reduction reaction is faster at higher temperature. The losses of samples weight also happen faster at the higher temperature.

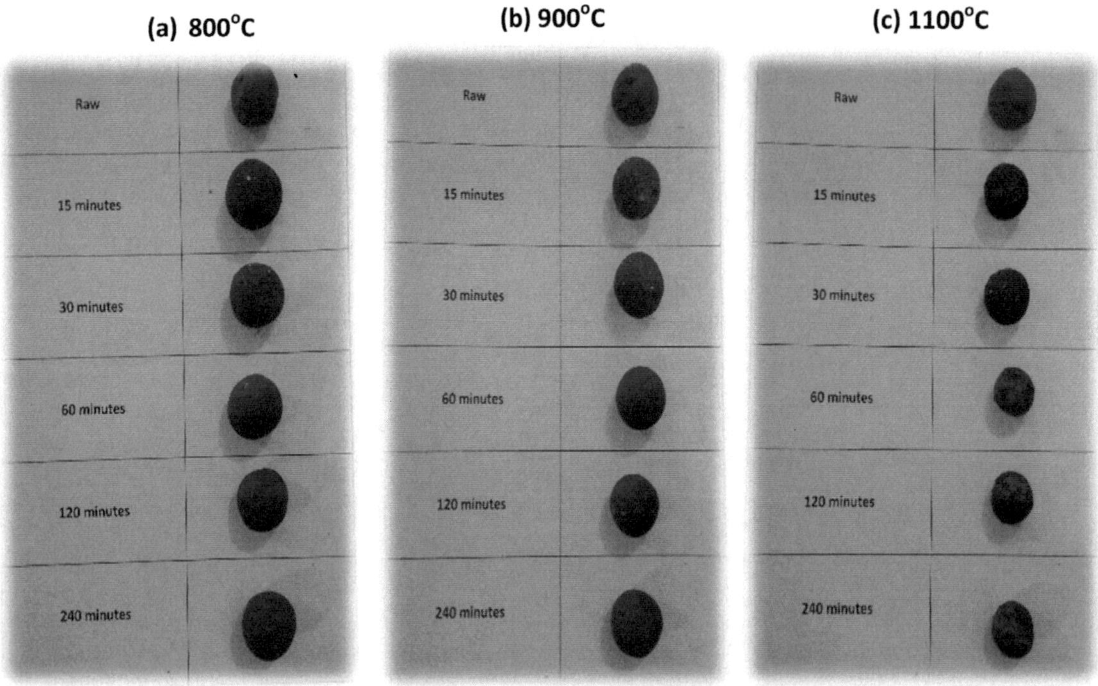

Figure 4. The color changes of pellet composite during reduction process.

4. Conclusion

Charcoal performance as a reducing agent in the nickel laterite reduction process is well enough compared to coal. Reduction temperature and reduction time significantly influence the growth behavior of Fe-Ni metallic particles. The reduction degree of limonite ore increases with increasing reaction temperature and prolonging reaction time.

Acknowledgement

The authors gratefully acknowledge Indonesian Institute of Sciences for financial and other supports that were provided within the program of Development Material, Energy, and Manufacturing Engineering.

References

[1] Barkas J 2010 *Proc. 7th China Nickel Conf.* (Shanghai) p 54 - 59
[2] Li Y, Sun Y, Han Y and Gao P 2013 *Trans. Nonferrous. Met. Soc.* **23** 3428-433
[3] Supriyadi A 2008 *Prosiding Kolokium Pertambangan 2008* (Bandung: Pusat Penelitan dan Pengembangan Tekmira) p 36 - 41
[4] Ellingham H J T 1944 *J. Soc. Chem. Ind.* p 143
[5] Speyer R F 1994 *Thermal Analysis of Materials* (New York: Marcel Dekker, Inc)
[6] Brindley G W and De Souza J V 1975 *Mineral Mag.* **40** 141 – 52
[7] Fan R and Gerson A R 2011 *Geochim. Cosmochim. Acta* **75** 6400 – 15
[8] Tartaj P, Cerpa A, Garcia-Gonzalez M T and Serna C J 2000 *J. Colloid Interf. Sci.* **231** 176 – 81

IOP Conf. Series: Journal of Physics: Conf. Series **817** (2017) 012064 doi:10.1088/1742-6596/817/1/012064

The effect of pretreatments on the dissolutions of impurities from Indonesian cassiterite mineral

L H Lalasari, A Suharyanto and F Firdiyono

Research Centre for Metallurgy and Material, Indonesian Institute of Sciences (LIPI), Kawasan Puspiptek Serpong, Tangerang Selatan 15314, Indonesia

E-mail: ifa_sari@yahoo.com; lati003@lipi.go.id

Abstract. The purpose of this research was to purify cassiterite from associated minerals (impurities) and to investigate the effects of pretreatment on the dissolution of impurities from Indonesian cassiterite mineral. The steps of this research are as follows: characterization, pretreatment and leaching processes by dilute hydrochloric acid (HCl) solution. Firstly, cassiterite was characterized by X-Ray Diffraction (XRD) and X-Ray Fluorescence (XRF) techniques, and then pretreatments were carried out including washing, roasting at 700 °C and both combinations of washing and roasting at 700°C. Cassiterite obtained from the processes of pretreatment and without pretreatment was then leached using 10% HCl solution with temperature variations of 35 °C – 150 °C for 2 hours in atmospheric condition. The results showed that the chemistry composition of cassiterite was as follows: 57.82% Sn, 0.45% Si, 1.45% Ti, 1.79% Fe, 3.43% La, 3.37% Ce and 31.69 % others. The best condition for purification of cassiterite was leaching without pretreatment that has the highest dissolution of associated mineral from cassiterite than other processes (pretreatment) for leaching at 110 °C for 2 hours. This research was able to purify cassiterite with a chemical composition as follows: 66.14% Sn, 0.27% Si, 1.38% Ti, 1.18% Fe, 2.12% La, 2.03% Ce and 26.88% others. These results are quite satisfactory because this technology is quite simple and economical.

1. Introduction

One type of mineral that is widely available in Indonesia, particularly in Bangka Belitung island is cassiterite. Cassiterite or tin oxide which has the molecular formula (SnO_2) is the primary mineral producing tin and is quite abundant in the earth's crust [1][2][3]. The cassiterite found in hard-rock deposits is associated with granitic host rock that needs further processes for separating associated minerals from SnO_2. Numerous techniques were developed such as classification, gravity concentration, and flotation techniques; however the minerals associated with such types of cassiterite deposits are harder to be seperated because of its fine liberation size [3]. The hydrometallurgical processes were also applied to obtain SnO_2 high grade of cassiterite. Some researchers used a hydrometallurgical route were as follows: Gerald et al. [4] tried to recover SnO_2 from cassiterite ore by microwave pretreatment, leaching and precipitation. Microwave treatment was selected due to the presence of associated minerals which behave differently and could be separated by microwave energy. The leaching processes conducted using -212 □m particle size of cassiterite, 1 – 8 M potassium hydroxide (KOH) solution and a temperature at 35 °C, 55 °C, 85 °C and 103 °C. The result showed the SnO_2 recovery by the microwave assisted process was 53%. It was higher than SnO_2 obtained by the untreated conventional method that was about 28%.

Published under licence by IOP Publishing Ltd

Baba et al. [5] reported that increase in the HCl concentration was accompanied with increase in the amount of cassiterite ore dissolved at various time. The rate of cassiterite ore dissolution is controlled by the hydrogen ion concentration $[H^+]$ that the reaction order is first order (for HCl concentration ≤ 4 M), with a correlation coefficient of 0.9995. The residual products consisted mainly of TiO_2, $FeTiO_3$, SiO_2 and Ta_2O_5 that was known as the unreacted materials by HCl solution. Yang et al. [6] has successful separated 99.4% iron from a low-grade tin concentrate by 150 g/L hydrochloric acid concentration at a temperature of 35°C for 2 hours. The samples consisted of more than 50% cassiterite enwrapped by pyrite and limestone. It makes the iron components and other impurities could be dissolved by HCl solution. Lalasari et al. [7] in the previous research also tried to dissolve cassiterite-Indonesia by hydrochloric acid solution in various concentrations. The dissolution of cassiterite was not rising significantly with the increase of hydrochloric acid concentrations. Cassiterite in Indonesia contained mainly associated minerals such as quartz (SiO_2), ilmenite ($FeTiO_3$), xenotime and monazite that were difficult to be dissolved with HCl solution.

Cassiterite with high purity has now been developed into several functions which are as follows: nanomaterial SnO_2 as gas sensors [8][9], SnO_2–TiO_2 nanocomposites as adsorbent [10], antibacterial [11]. Cassiterite is also utilized as raw materials to make sodium stannate (Na_2SnO_3) which are widely used as electroplating, metal alloys of tin, and catalysts for organic synthesis [12]. Before cassiterite was utilized as raw materials for synthesizing tin chemical materials, it requires further processes for separating associated minerals from SnO_2. In this study, we would investigate the effect of pretreatments prior to leaching on the dissolutions of impurities from Indonesian cassiterite mineral in variation temperatures. The leaching solvents used in the research was hydrochloric acid (HCl) that is a strong acid and more reactive than other acid solvents. The HCl solution is expected to separate associated minerals of cassiterite maximally.

2. Experimental

2.1. Material

Cassiterite was originally from Bangka Island in South Sumatra, Indonesia that has a chemical composition shown in table 1. The leaching reagent was purchased from Merck with an analytical grade of 37% hydrochloric acid (HCl).

Table 1. The chemical composition of cassiterite from Bangka Island in Indonesia

Element	Sn	O	La	Ce	Nd	Fe	P	Ti	Ca	Al	Si
Percentage (%wt)	57.82	23.39	3.43	3.37	2.18	1.79	1.68	1.45	0.63	0.52	0.45

2.2. Methods

The procedures were similar to the previous researches carried out by Lalasari et al. (2015) [7]. Firstly, cassiterite was grinded to -100 mesh of particle size and characterized by X-Ray Fluorescence (XRF) apparatus and X-Ray Diffraction (XRD). Secondly, samples was carried out pretreatment processes including washing, roasting at 700 °C and both combinations of washing and roasting at 700 °C. The samples after pretreatments were characterized by XRD analysis, respectively. In this step was also prepared a leaching reagent that 37% HCl was diluted with aquadest to make a solution of 10% HCl. The 10% HCl solution was reagent used in every leaching process. Thirdly, 10 gram of cassiterite (without pretreatment and with pretreatment) was leached in 100 ml solution of 10% HCl at variation temperatures of 35 - 150 °C for 2 hours in atmospheric conditions. The products were then filtered to separate filtrate and residue of cassiterite. Finally, filtrate was characterized by Induced Coupled Plasma (ICP) analysis and residue was characterized by XRF analysis.

3. Result and Discussion

According to Table 1 looked the chemical composition of cassiterite was follows as: 57.82% Sn, 23.39% O, 3.43% La, 3.37% Ce, 2.18% Nd, 1.79%Fe, 1.68% P, 1.45% Ti, 0.63% Ca, 0.52% Al, 0.45% Si and 2.67% others that was different with the cassiterite of previous researches by Lalasari (2015). The cassiterite used in this research has a lower tin percentage than previous samples (66.36% Sn). It probably due to the cassiterite that was not yet completely processed or pretreated furthermore such as comminution, classification, gravity concentration and magnetic separation. In this opportunity, we would explore the effect of pretreatment on the dissolution of impurities from Indonesian cassiterite mineral.

3.1. The effect of pretreatment on the formed phase

To study the pretreatments effects on the formed phase of cassiterite, samples prior to leaching with 10% HCl solution was applied some processes such as washing, roasting at 700 °C and both combinations of washing and roasting at 700 °C. The samples after pretreatments were characterized by XRD analysis that showed in figure 2.

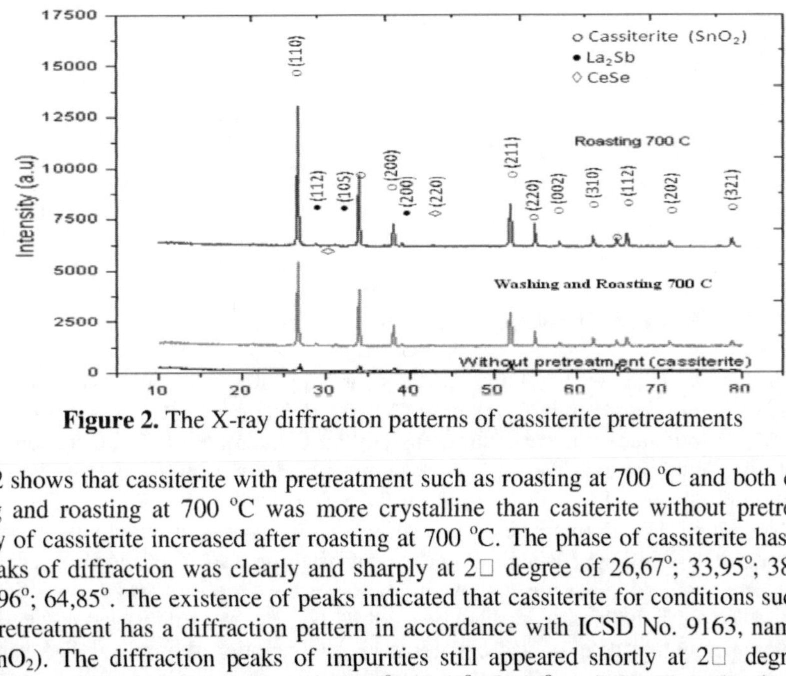

Figure 2. The X-ray diffraction patterns of cassiterite pretreatments

Figure 2 shows that cassiterite with pretreatment such as roasting at 700 °C and both combinations of washing and roasting at 700 °C was more crystalline than casiterite without pretreatment. The crystallinity of cassiterite increased after roasting at 700 °C. The phase of cassiterite has not changed that the peaks of diffraction was clearly and sharply at 2□ degree of 26,67°; 33,95°; 38,02°; 51,85°; 54,84°; 61,96°; 64,85°. The existence of peaks indicated that cassiterite for conditions such as without and with pretreatment has a diffraction pattern in accordance with ICSD No. 9163, namely stannum dioxide (SnO_2). The diffraction peaks of impurities still appeared shortly at 2□ degree of 28,89°; 31,21°; 39,05° as La_2Sb and 2□ degree of 29,87°; 42,67°; 61,96° as CeSe. The other impurities were not found such as Nd, Fe, Ti, Ca, Al, Si, P, etc because they have lower intensities of diffraction peaks.

The phase of cassiterite after pretreatment has not changed because the processes such as roasting at 700 °C and leaching with water has not caused formation of new phase of cassiterite. The washing process only dissolved organic compounds and associated minerals of cassiterite by gravity concentration and also cassiterite would keep stable although roasted at 700 °C. This pretreatment was almost suitable with experiment conducted Zhang et al [12]. He explained that cassiterite (SnO_2) seldom reacted with air during the roasting at 900 °C for 120 min.

3.2. The effect of pretreatment on the dissolved cassiterite

Based on the result of XRD analysis, pretreatment effects such as roasting and both combinations of washing and roasting have not changed the phase of cassiterite and only increasing crystallinitity of cassiterite. The effect of crystallinity on the dissolution of cassiterite would be investigated in this experiment. The cassiterite samples resulted from pretreatment were leached with 10 % HCl solution for 2 hours in variatious temperatures and pretreatments. The results are shown in Figure 3.

The dissolution of cassiterite rose gradually with the increase of leaching temperature at all conditions. The highest dissolution was showed on samples without pretreatment that the dissolved cassiterite was approximately 4.1 %. This result was enough interesting and could become a basic to purify cassiterite which was simple and economical. Cassiterite without pretreatment was easier dissolved than cassiterite with pretreatment. The increase of crystallinity would make cassiterite was difficult to be dissolved by 10% HCl solution. Cassiterite and other associated minerals become more stable after roasted. According to Figure 3, the grade of cassiterite dissolution prior to be leached by 10% HCl solution was cassiterite without pretreatment, cassiterite with washing pretreatment, cassiterite with roasting pretreatment at 700 °C and cassiterite with both pretreatment combinations of washing and roasting at 700 °C.

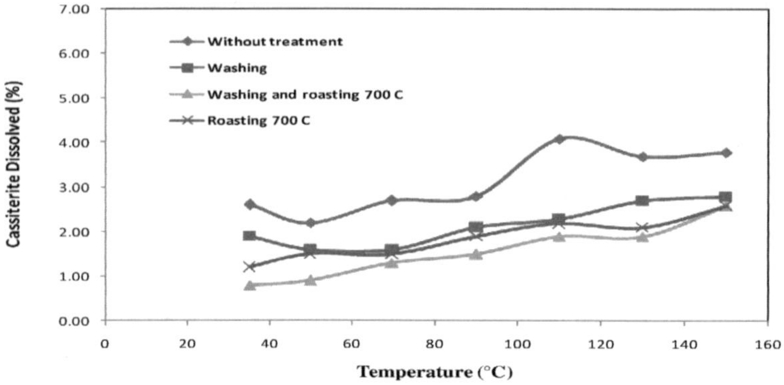

Figure 3. The dissolved cassiterite in variation conditions

Figure 3 also shows that the dissolution of cassiterite by 10% HCl solution was insignificant compared to experiment carried out by Yang et al (2014) [6]. Yang et al [6] has successful separated 99.4% iron from a low-grade tin concentrate by 150 g/L hydrochloric acid concentration at a temperature of 35°C for 2 hours. The cassiterite samples used by Yang et al were originally from China that consisted of more than 50% enwrapped by pyrite (FeS) and limestone. They were more soluble in HCl solution. The characteristic of cassiterite from China was different with cassiterite from Indonesia. Cassiterite from Bangka Island, Indonesia consisted of associated minerals such as quartz, ilmenite, rutil, xenotime and monazite that were harder to be dissolved by HCl solution [7]. The dissolution of associated minerals was proportional to the percentage of dissolved cassiterite by HCl solution. The low percentage of dissolved cassiterite by HCl solution showed the less percentage of associated mineral separated from cassiterite.

3.3. The effect of pretreatments on the dissolution of Fe, Ti, Si and Sn elements

To explore the dissolution of impurities from cassiterite by dilute HCl solution, filtrate from the dissolved cassiterite was analyzed by ICP-OES. The elements of filtrate measured in this experiment were few such as iron (Fe), titanium (Ti), silica (Si) and tin (Sn). The results are shown in figure 4.

Figure 4 describes the dissolution of Fe, Si, Ti, and Sn elements by dilution in HCl solution. As shown in figure 4, the increase of leaching temperature was accompanied with the rise in the amount of dissolved elements from cassiterite. However, the dissolution of Fe, Si, Ti and Sn elements was insignificant in all conditions. It was found the dissolution of Fe element higher than Ti, Si and Sn elements that was about 47.92 % for leaching cassiterite without pretreatment. The dissolution of Fe element was still less than the experiment carried by Yang et al [6] that could separate 99.4% iron from a low-grade tin concentrate (cassiterite) by HCl solution. These were probably caused by the fact that cassiterite in Indonesia consists of $FeTiO_3$ mineral and was more dominant than FeO and FeS_2 minerals. Therefore, the Fe element of cassiterite was difficult to be dissolved in HCl solution.

Figure 4 also shows that the dissolution of Si and Ti elements were low enough at all conditions. The highest dissolution of Si and Ti elements were approximately 17.87% and 0.49%, respectively. The Si associated mineral of cassiterite was indicated as quartz (SiO_2), whereas the Ti associated mineral was predicted as ilmenite ($FeTiO_3$) and rutil (TiO_2). These minerals have a slightly higher solubility in HCl solution. Baba et al [5] also explained the residual products of cassiterite from Nigeria consisted mainly of TiO_2, $FeTiO_3$, SiO_2 and Ta_2O_5 that was known as the unreacted materials by HCl solution. In this experiment, it was looked that the Sn element was almost insoluble in HCl solution that was only about 0.06%. These was really expected to purify cassiterite from associated minerals.

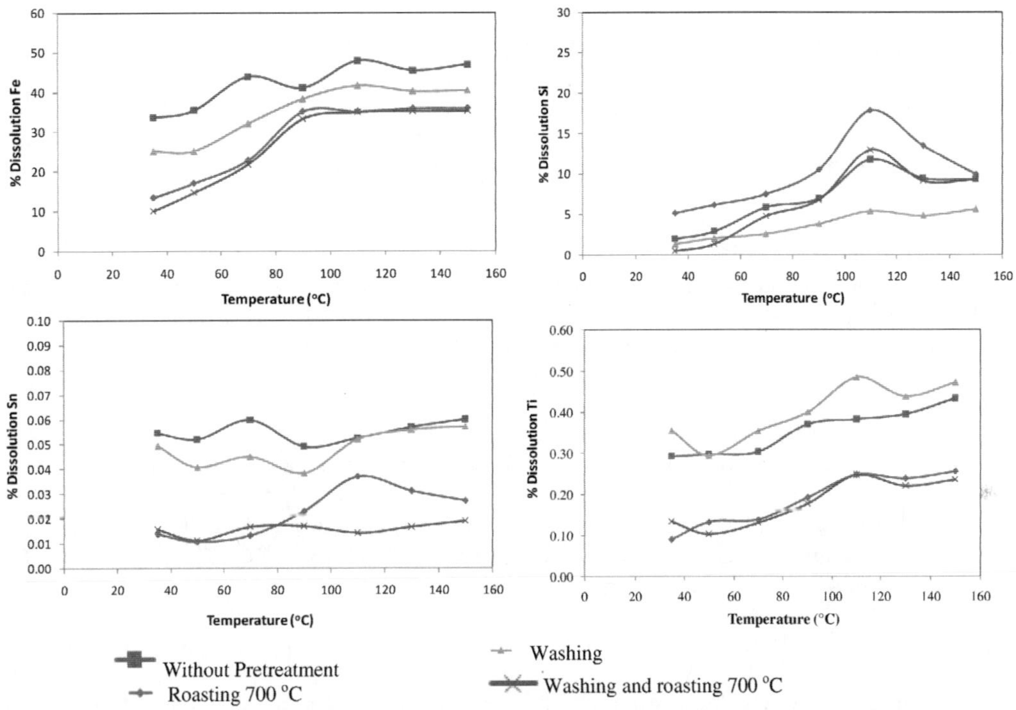

Figure 4. The dissolution of Fe, Si, Ti and Sn in variation pretreatments and temperatures

3.4. The effect of pretreatment on the chemistry composition of cassiterite residue

To know the chemical composition of cassiterite leached by dilute HCl solution, residue of cassiterite was characterized by XRF analysis. The result is shown in table 2.

Table 2. The chemical composition of cassiterite residue

Element	Sn	O	La	Ce	Nd	Fe	P	Ti	Ca	Al	Si
A (%wt)	57.82	23.39	3.43	3.37	2.18	1.79	1.68	1.45	0.63	0.52	0.45
B (%wt)	66.14	22.50	2.12	2.03	0.96	1.18	1.00	1.38	0.00	0.00	0.27
C (%wt)	59.98	22.58	3.15	3.20	2.07	1.56	1.24	1.40	0.00	0.00	0.25

Note: A is cassiterite samples; B: is cassiterite residue without pretreatment; C is cassiterite residue with pretreatment (roasting at 700 °C)

As shown in table 2 that cassiterite without pretreatment has a purity higher compared to cassiterite with pretreatment (roasting at 700 °C). The purification of cassiterite reached in this experiment was about 66.14% Sn from 57.82% Sn (cassiterite samples). The roasting pretreatment to cassiterite only increased crystallinity and caused associated minerals of cassiterite become more stable so that it was more difficult to be dissolved in HCl solution. The result was similar to the experiment by Zhang et al [12]. He tried to upgrade cassiterite using roasting at 900 °C for 120 min and leaching with 25% HCl solution. It was found that that cassiterite was insoluble in hydrochloric acid. The purity of cassiterite only increased about 62.93% Sn from 42.92% Sn.

4. Conclusion

The chemistry composition of cassiterite from Bangka in Indonesia is as follows: 57.82% Sn, 0.45% Si, 1.45% Ti, 1.79% Fe, 3.43% La, 3.37% Ce and 31.69 % others. The phase of cassiterite has not changed after carried out by pretreatment (washing; roasting and both combinations of washing and roasting. Cassiterite from Bangka in Indonesia consisted of quartz, ilmenite, rutil, xenotime and monazite mineral that were difficult dissolved by HCl solution. The best condition for purification of cassiterite was process of leaching without pretreatment that has the highest dissolution of associated mineral than other processes. This research was able to purify cassiterite with a chemical composition as follows: 66.14% Sn, 0.27% Si, 1.38% Ti, 1.18% Fe, 2.12% La, 2.03% Ce and 26.88% others.

Acknowledgments

The authors are grateful for the main financial supports from the thematic program of Research Center for Metallurgy and Material, Indonesian Institute of Science (LIPI) and thanks for contributing from Dea Sidawaty Bharani, Bogor Agricultural University (IPB) in this experiment.

References

[1] Falcon L M 1985 *J. South African Inst. Min. Metall.* **85**(10) 333–45
[2] Wills B A and Napier-Munn T J 2006 *Will's Mineral Processing Technology* (UK: Butterworth-Heinemann) p 4
[3] Angadi S I, Sreenivas T, Jeon H, Baek S and Mishra B K 2015 *Miner. Eng.* **70** 178–200
[4] Gerald O, Martin O, Emmanuel E, Asuwaji, Ayuk and Nwoye 2013 *Charact. Miner. Met. Mater.* 407–14
[5] Baba A A, Adekola F and Bale R 2009 *Sci. Focus* **2**(14) 198–207
[6] Yang J, Wu Y and Zhang X 2014 *Miner. Process. Extr. Metall. (Trans. Ins. Met. C)* **123**(4) 228-33
[7] Lalasari L H, Dewiani Y, Suharyanto A and Firdiyono F 2015 *Proc. Int. Conf. on Materials and Metallurgical Engineering* (Surabaya: Sepuluh November Institute of Technology, Indonesia) pp 1–5
[8] Miller T A, Bakrania S D, Perez C and Wooldridge M S 2006 *Functional Nanomaterials*, ed K E Geckeler and E Rosenberg (US: Ed. American Scientific Publishers) chapter 30 pp 1–24
[9] Edelman F, Hahn H, Seifried S, Alof C, Hoche H, Balogh A, Werner P, Radecka M, Pasierb P, Chack A, Mikhelashvili V and Eisenstein G 2000 *Mater. Sci. Eng.* **B69–70** 386–91
[10] Rahman M M, Khan S B, Marwadi H M and Asiri A M 2014 *J. Taiwan Inst. Chem. Eng.* **45** 1964–74
[11] Talebian and Zavvare S H 2014 *J. Photochem. Photobiol. B Biol.***130** 132–9
[12] Zhang Y, Su Z, Liu B, You Z, Yang G, Li G and Jiang T 2014 *Hydrometallurgy* **146** 82–8

IOP Conf. Series: Journal of Physics: Conf. Series **817** (2017) 012065 doi:10.1088/1742-6596/817/1/012065

Lamtoro charcoal (*l. leucocephala*) as bioreductor in nickel laterite reduction: performance and kinetics study

H T B M Petrus[1], A Diga[1], A R Rhamdani[1], I W Warmada[2], A T Yuliansyah[1] and I Perdana[1]

[1]Center of Advanced Material and Mineral Processing, Department of Chemical Engineering, Faculty of Engineering
[2]Department of Geological Engineering, Faculty of Engineering
Universitas Gadjah Mada, Jl. Grafika 2 Yogyakarta 55281, Indonesia

E-mail: bayupetrus@ugm.ac.id

Abstract. The performance and kinetic of nickel laterite reduction were studied. In this work, the reduction of nickel laterite ores by anthracite coal, representing the high-grade carbon content matter, and lamtoro charcoal, representing the bioreductor, were conducted in air and CO_2 atmosphere, within the temperature ranged from 800°C and 1000°C. XRD analysis was applied to observe the performance of anthracite and lamtoro as a reductor. Two models were applied, sphere particle geometry model and Ginstling-Brounhstein diffusion model, to study the kinetic parameters. The results indicated that the type of reductant and the reduction atmosphere used greatly influence the kinetic parameters. The obtained values of activation energy vary in the range of 13.42-18.12 kcal/mol.

1. Introduction

Nickel is one of the most important materials in the world. 66% of nickel consumption used for stainless steel making, while the others 18% used in other alloys making, and the rest used for other purposes. There are two main nickel sources, sulfide ore and laterite ore. At least, 60% of nickel demand is supplied by sulfide ore besides 80% of nickel deposit lies on laterite ore [4]. Laterite is a highly weathered material rich in secondary oxides of iron, aluminium, or both. In addition, laterite often contains minor amounts of nickel, cobalt, and chromium [5]. One of the reasons that laterite ore is not used as the main nickel source is because of its low nickel content but recently China as the biggest stainless steel producer had found a way of utilizing laterite ore. They used it to produce nickel pig iron (NPI), as a substitute for pig iron in the stainless steel making process, using blast furnace or rotary kiln – electric furnace (RKEF). One of the important processes in NPI making is the reduction process. In the conventional NPI production, coal was used as a reductor. There have been several studies of nickel laterite reduction. But to the best of our knowledge, very few works have been presented on studying the kinetic of nickel laterite reduction. The kinetic study of nickel laterite reduction using gaseous reductant shows that the reduction mechanism was diffusion controlled [7] but the kinetic study of nickel laterite reduction using solid reductant has not been studied yet. Therefore, this present work investigated the kinetic of nickel laterite reduction using lamtoro charcoal as bioreductor. The performance and the kinetic parameters were studied.

2. Materials and method

Anthracite and lamtoro charcoal were used as reducing agent. Laterite and anthracite coal were provided by Indonesian Institute of Sciences and lamtoro charcoal was obtained from the traditional charcoal maker. All of the materials above were crushed using a mortar and then screened to get the desired size (-100+120 mesh). Laterite was characterized by XRD (Shimadzu 7000), XRF (Thermo Scientific Niton XL3t). The proximate analysis of reductants was done by PAU UGM as shown in table 1 and 2.

Table 1. Main Components of Laterite from Pomalaa, Southeast Sulawesi.

Components	**SiO_2**	**$Fe_2O_3/Fe^{*)}$**	**$NiO/Ni^{*)}$**	**MgO**	**Al_2O_3**	**Na_2O**	**Cr_2O_3**
(wt.%)	26,94	43,83/30,60	3,19/1,73	12,67	5,13	2,73	1,63

*) Analyzed using XRF Metal

Table 2. Proximate Analysis of Reductors (%wt).

Reductor Type	**Moisture**	**Volatile matter**	**Ash**	**Fixed Carbon**
Anthracite coal	2,3	7,4	2,5	87,9
Lamtoro charcoal	7,0	17,8	4,4	70,8

1.1. Pelletization

Laterite, starch, and anthracite charcoal were mixed together with laterite/reductant/binding agent ratio of 4:1:0.3. Then 5 mL of distilled water was added to the mixed sample and shaped into a ball pellet by hand. The pellet was dried at 110°C for 5 h. By using the same procedure, other pellets were made with lamtoro charcoal and carbon raiser as a reductor.

1.2. Reduction of the pellets

The pellets were roasted using two different atmospheric conditions, air and CO_2. Muffle furnace was used for air atmospheric roasting and tube furnace was used for CO_2 atmospheric roasting with CO_2 flow rate of 3 L/min. The pellets were roasted at temperature of 800, 900, and 1000°C with roasting time varied from 15 to 240 minutes. After the roasting process, the pellets were cooled and placed in a desiccator. Thereafter, the pellets were crushed using a mortar and were characterized using XRD (Shimadzu 7000).

1.3. Fraction of reduction reaction

To evaluate the kinetics parameters, the term fraction of reaction (*f*) was used. Eq (1) represents the formula for fraction of reaction. The equation was modified from the term used in iron ore reduction process [1]. The maximum weight loss of a pellet in Eq (1) was obtained from the maximum weight loss of laterite and reductant. The maximum weight loss of laterite was obtained from the total of removable oxygen from laterite. The total of removable oxygen from laterite is the oxygen binds to the iron and nickel.

$$f(t) = 100\frac{\Delta W_t}{\Delta W_{max}} \quad (1)$$

where:
f = fraction of reaction.
ΔW_t = weight loss of pellet (g).
ΔW_{max} = maximum weight loss of pellet (g).

1.4. Kinetic models

The experimental data (*f*) were applied to the following mathematical models:

Sphere Particle Geometry Model

$$1-(1-f)^{1/3} = kt \tag{2}$$

where:
t = time (s).
k = kinetic constant (1/s).

$$k = k_0 exp\left(-\frac{E_a}{RT}\right)$$

where:
k_0 = pre-exponential factor (1/s).
E_a = activation energy (kcal/mol).
T = reduction temperature (K).
R = universal gas constant (kcal/mol/K).

Ginstling-Brounhstein Diffusion Model

$$1-\frac{2}{3}f-(1-f)^{2/3} = kt \tag{3}$$

3. Results and discussions

3.1. Materials Characterization

The chemical compositions of nickel laterite used in this research were showed in table 1. The laterite was dominated by Fe and Si element and based on the analysis of the XRD pattern using Match 2 software figure.1, the main mineral composition of the laterite was goethite (FeOOH), lizardite ($Mg_3(Si_2O_5)(OH)_4$), clinochlorite ($(Mg,Fe^{2+})_5Al(AlSi_3O_{10})(OH)_8$), and quartz ($SiO_2$). The proximate analysis of the reductants was done by PAU UGM and the results were showed in table 2. The results showed that all of the reductants used had different characteristics. Anthracite coal had the highest fixed carbon content and the lowest ash content, lamtoro charcoal had the highest moisture and volatile matter content while carbon riser had the lowest fixed carbon and volatile matter content.

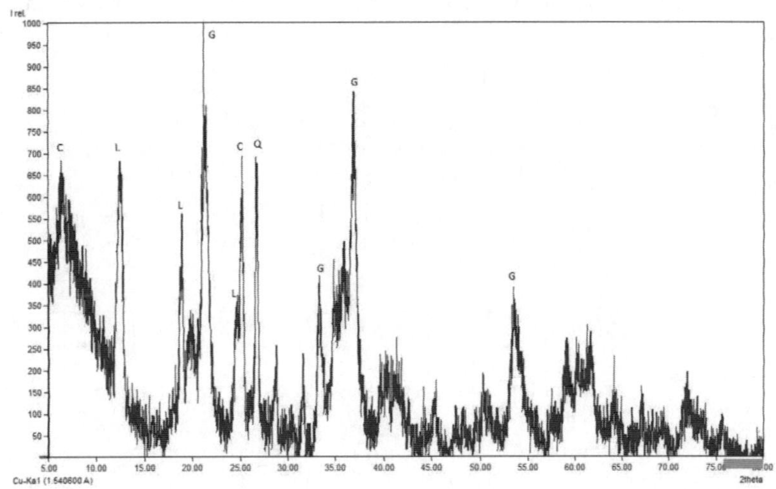

Figure 1. XRD pattern of initial laterite ore analyzed with Match! 2 Software.

3.2. Fraction of reaction

Experimental data which were the sample mass from time to time converted into a fraction of reduction reaction (*f*). The maximum weight loss from a pellet made with one type of reductant is different from a pellet made with the other type of reductant because the ash content of a reductant is different from the other reductant. The maximum weight loss of the pellet is 80.4% for the sample made with anthracite coal, 81.9% for the sample made with lamtoro charcoal, and 73.5% for the sample made with carbon raiser. Figure 2 shows the graphical representation of the fraction of reduction vs time.

From the experimental data, the fraction of reaction value was already reached a constant value before reached its maximum value at certain time. The maximum value of the fraction of reaction was 48%. It could happen because the reduction reaction has stopped. The reason of it could be two things. The first reason is the ratio between laterite and reductant was not high enough which makes the reductant has already used up before could reduce the minerals completely. The second reason is the existence of reaction between iron oxide and quartz which form fayalite ($FeSiO_4$) [2]. Fayalite has relatively low melting point (1100°C) [6] and already melt during the reduction process which makes the porosity of the pellet decreased. The decreasing of porosity of pellet could prevent reducing gas diffusion. On the other hand, fayalite could also act as a nucleating agent on iron metal forming process. During the reduction process, the iron metal layer formed on the surface of fayalite and prevent the contact of reducing gas. The fraction of reaction reached a constant value faster at a higher temperature. It could happen because the reaction rate is faster at a higher temperature. The reaction has stopped even when CO gas is used as a reductant instead of using solid reductant like in this study [7]. It concluded that the second reason was better at explaining why the reaction has stopped. The reduction reaction was fast at the early stage of the reduction process because, at the early stage, the devolatilization of the reductant occurs alongside the dehydroxylation of the minerals in the laterite which makes the weight loss of the sample is very high.

There was a slope change on the fraction of reaction vs time data on 0-5 minutes and 5-60 minutes reduction duration. This could happen because there were changes of reduction mechanism between these two periods. In the early stage of reduction, the process will follow the reaction controlled mechanism and after that, the process will follow diffusion controlled mechanism [7]. The changes of slope in the previous study happened at 20 minutes reduction duration [7], but in this study, the changes happened at 5 minutes duration time. It could happen because the reaction happened in this two study is different. The changes of slope happened in the very early stage of the reduction. These early changes made the evaluation of kinetic parameters could only be done starting at 5 minutes reduction time.

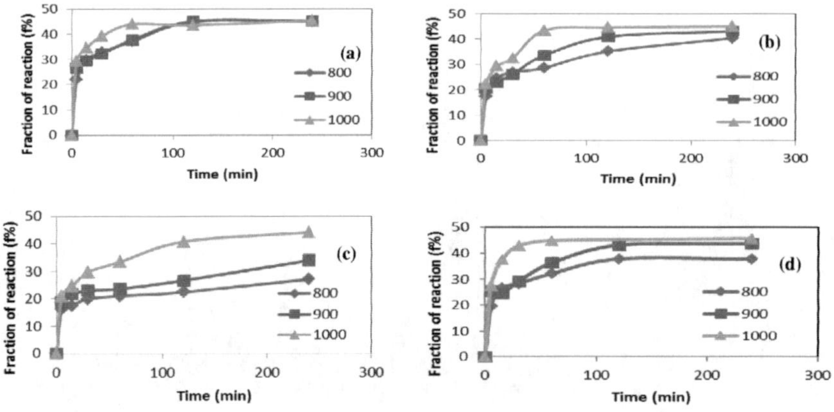

Figure 2. Fraction of reaction as a function of time of the laterite with anthracite (a), and lamtoro charcoal (b) in air atmospheric condition and with anthracite (c) and lamtoro charcoal (d) in CO_2 atmospheric condition.

IOP Conf. Series: Journal of Physics: Conf. Series **817** (2017) 012065 doi:10.1088/1742-6596/817/1/012065

The maximum value of the fraction of reduction was almost the same (around 47%) for both reductions happened in the air and CO_2 atmosphere, but based on the XRD analysis of the reduced sample, the mineral phase of both samples was different. The comparison of XRD pattern of the reduced samples was shown in figure. 3. The reduction in CO_2 atmosphere proceeds completely, compared with the reduction in air atmosphere based on the loss of hematite and magnetite peaks. It could happen because there is oxygen in the air atmosphere reduction which could oxidized the reduced mineral in the sample.

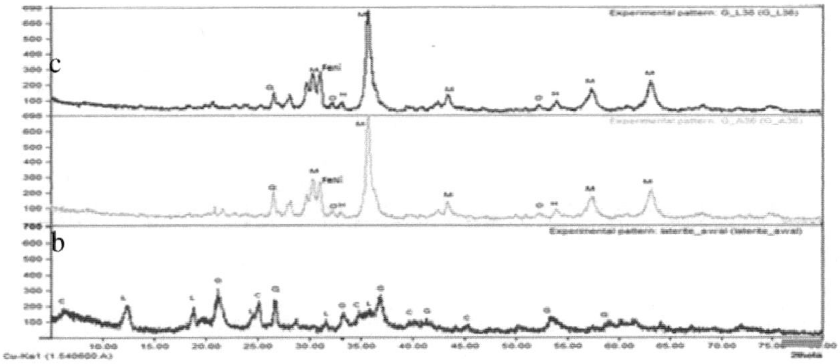

Figure 3. XRD pattern coal-laterite roasted in air atmospheric condition using lamthoro chalcoal (c) anthrasite (b) for 240 minutes at 1000°C and original nickel laterite (a) [Q: quartz; O: olivine; H: hematite; M: magnetite].

As it can be seen in Figure 3 that the reduction of nickel laterite using lamthoro charcoal is similar to that of coal represented by anthracite. The formation of ferronickel has been reached in similar operating condition at 1000°C for 240 minutes.

3.3. Kinetic Analysis

For the kinetic analysis of the experimental results, the kinetic model equations (2) – (3) have been applied to the experimental data for the temperatures of 800, 900, and 1000°C. The graphical representation of the fitting of experimental data to the kinetic model was shown in figure 4 and figure 5. Based on the simulation, both models could represent the reduction process well enough because the error in the data fitting process was small enough for both models applied. However, based on the literature study, the diffusion model was considered as the model that could represent the process. It also fitted with the reason why the reaction stopped at a certain time. The kinetics parameters obtained from the evaluation of the data using diffusion model was listed in table 3. The variation of activation energy means that the reaction mechanism happened is different for each reductant and atmosphere condition. The reaction mechanism differences happened because the reactions which took place consist of various step such as carbon oxidation, reducing gas diffusion, and reduction reaction hence the reaction was very complex.

IOP Conf. Series: Journal of Physics: Conf. Series **817** (2017) 012065 doi:10.1088/1742-6596/817/1/012065

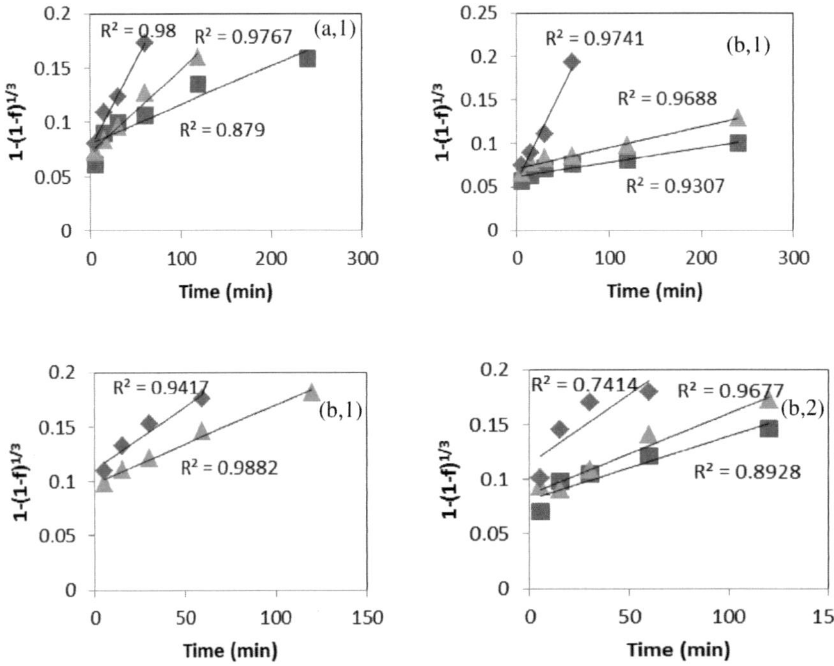

Figure 4. Applications of the sphere particle geometry model to the experimental data of sample reduced with anthracite coal (a) and lamtoro charcoal (b) in air atmosphere (1) and CO_2 atmosphere (2) at temperature of reduction of 800 (■), 900 (▲), and 1000°C (◆).

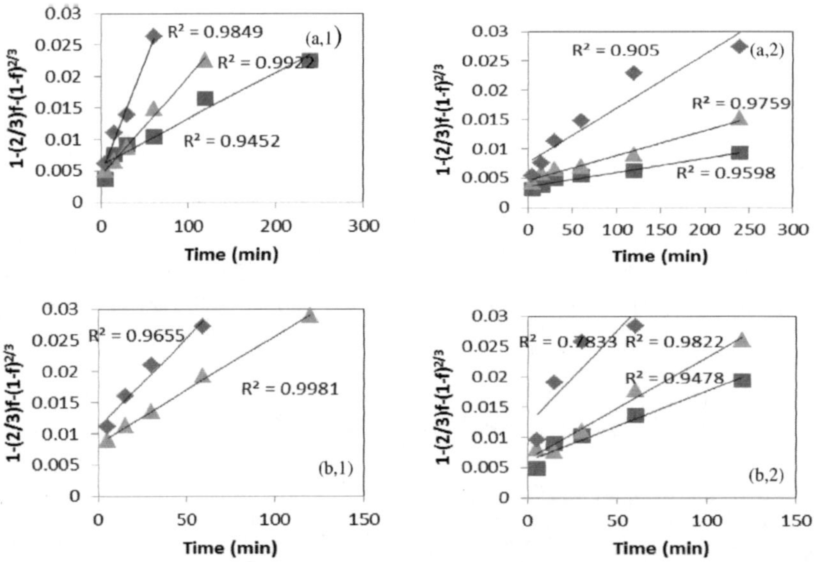

Figure 5. Applications of the Ginstling-Brounhstein diffusion model to the experimental data of sample reduced with anthracite coal (a) and lamtoro charcoal (b) in air atmosphere (1) and CO_2 atmosphere (2) at temperature of reduction of 800 (■), 900 (▲), and 1000°C (◆).

Table 3. The value of Activation Energy (E_a) dan k_0 from Simulation of Ginstling – Brounhstein Diffusion.

Reductant Type	Atmosphere			
	Air		CO2	
	k_0 (1/min)	Ea (kcal/mol)	k_0 (1/ min)	Ea (kcal/mol)
Anthracite coal	0.3182	17.4445	0.1083	18.1217
Lamtoro charcoal	0.0930	14.6658	0.0182	13.4162

4. Conclusions

From the results the following conclusions were drawn the maximum value of the fraction of reaction is about 47% and lamthoro biochar could perform as good as the coal as a reductor. Both models applied were good at representing the experimental data, but Ginstling-Brounhstein diffusion model was considered better at representing the process because it has a relatively smaller error and was supported by literature study. From applying Ginstling-Brounhstein diffusion model, the obtained value of activation energy was varied around 13.42 kcal/mol to 18.12 kcal/mol depends on the type of reductant and the reduction atmosphere used.

Acknowledgments

The authors acknowledge the financial support provided by Directorate General of Higher Education (DIKTI), Ministry of Technology Research and Higher Education. The authors also acknowledge materials and analysis equipment support from Indonesian Institute of Sciences (LIPI).

References

[1] Agrawal B B, Prasad K K and Ray H S 1990 *ISIJ Int.* **30**(8) 997-99
[2] Jiangan C, Tichang S and Qi L 2014 *J. Chem. Pharm. Res.* **6**(3) 671-78
[3] Khawam A and Flanagan D R 2006 *J. Phys. Chem. B* **110** 17315-28
[4] Pickles C A 2004 *Miner. Eng.* **17** 775-84
[5] Olapinekun E O 2000 *Int. J. Miner. Process.* **60** 9-14.
[6] Vogt J H L 1922 *J. Geol.* **7** 611-30
[7] Zevgolis E, Zogradifis C, Halikia I and Perrarki M 2009 *Proc. Int. 3rd AMIREG International Conference: Assessing the Footprint of Resource Utilization and Hazardous Waste Management* (Greece: Heliotopos Conferences) pp 459-465

IOP Conf. Series: Journal of Physics: Conf. Series **817** (2017) 012066 doi:10.1088/1742-6596/817/1/012066

The effect of solid solution treatment on the hardness and microstructure of 0.6%wt C-10.8%wt Mn-1.44%wt Cr austenitic manganese steel

F Nurjaman, F Bahfie, W Astuti and A Shofi

Research and Development Division for Mineral Technology
Indonesian Institute of Sciences, Jl. Ir. Sutami Km. 15, Lampung 35361, Indonesia

E-mail: fajar.nurjaman@lipi.go.id

Abstract. Austenitic manganese steel is steel alloy that has high manganese content (10-14%wt Mn). The characteristics of austenitic manganese steel are good in toughness, ductility, and wear resistance. Effect of solid solution treatment on the hardness and microstructure of austenitic manganese steel was studied in this experiment. The solid solution treatment process of austenitic manganese steel, 0.6%wt C-10.8%wt Mn-1.44%wt Cr, was conducted by heating the material at varied temperatures (950°C, 1000°C, 1050°C) for an hour and then quenching it in two different quenching media, i.e. oil and water. Further, the samples were tempered at three different temperatures (300°C, 400°C, and 500°C) for 2 hours. The treated materials were analyzed by Rockwell Hardness Tester to obtain the information of materials hardness and by an optical microscope and XRD to investigate the microstructure phase of the treated materials. Heating the austenitic manganese steel at 950°C for an hour followed by water quenching dissolved all carbide in as-cast condition and resulted the fully austenitic on its microstructure. Carbide precipitation occurred due to the prolongation of soaking time in solid solution treatment and tempering process. The optimum hardness of sample was 53.3 HRC, which was resulted by heating this material until 1000°C for an hour, followed by water quenching and tempering at 400°C for 2 hours.

1. Introduction

Austenitic Manganese Steel, also known as Hadfield Steel was found by Robert A. Hadfield in 1882. It has chemical composition 1.0-1.4%wt C and 11-14%wt Mn or the C/Mn ratio is 1:10 [1]. Manganese steel is a wear resistant material with very good toughness and hardness after it was heat-treated and work-hardened. As a wear resistant material, austenitic manganese steel is commonly used in crushing machine components, railways, and heavy equipment components [2]. Austenitic manganese steel has 379 MPa of yield strength, 965 MPa of ultimate tensile strength, 190 HBN of hardness, and 169 J of Charpy V-Notch impact [3]. The microstructure of this material after heat treated is fully austenitic with a crystal lattice face centered cubic where the structure of this lattice consists of interstitials of carbon atoms and substitution of manganese atoms. Although it has the same structure as stainless steel, this material has very low corrosion resistance [4]. In as-cast condition, this austenitic manganese steel is brittle (having low toughness and elongation) [5]. It is caused by the formation of carbide (Mn_3C) along with the grain boundaries on its microstructure due to the slow cooling in solidification process. The experiment to improve its mechanical properties had been done by addition of alloying elements, such as Nb and Ti for obtaining fine austenitic structure and a lot of hard dispersed carbide [6]. Heat treatment is another method to improve the mechanical properties of the austenitic manganese steel. Solid solution treatment is one kind of heat treatment process that is commonly applied in austenitic manganese steel to obtain homogeneous austenitic phase on its microstructure. This heat treatment process includes heating this material at 1000-1100°C and then it is quenched rapidly. This fully austenitic structure will transform into martensite after work-hardening and the hardness will increase. It is called strain induced martensite [7]. The other heat treatment process to improve the hardness of austenitic manganese steel is by precipitation hardening. It was reported by Aribo, S. et al, that the artificial aging at 700°C for 2 hours increased the hardness of austenitic manganese steel which is caused by carbides particles finely dispersed in the austenite matrix [8].

However, there is still less information about the evolution of austenitic manganese steel microstructure after being heat treated at various solid solution and tempering temperature. The objective of this work is to obtain the optimum hardness of the austenitic manganese steel by solid solution treatment and tempering process. In this experiment, the effect of solid solution and tempering temperature, also two different kinds of quenching media on the hardness and microstructure of the austenitic manganese steel was clearly investigated.

2. Material and Method

The solid solution treatment was applied to the austenitic manganese steel. The chemical composition of the sample is presented in table 1. Austenitic manganese steel was heated at various austenitization temperature, i.e. 950°C, 1000°C, 1050°C for an hour and then it was quenched rapidly by using two different quenching media, i.e. oil and water. The effects of the austenitization temperature and quenching media on the hardness and microstructure of austenitic manganese steel were studied. The measurement of hardness was conducted by using Rockwell hardness testing machine, while the microstructure was analyzed by an optical microscope (Nikon Eclipse MA 100, Japan) and X-Ray Diffraction (XRD). The sample with the optimum hardness resulted from this austenitization and quenching process was obtained. The experiment was continued by investigating the soaking time in the austenitization process that the sample was heated at optimum austenitization temperature, soaked for 1, 2, and 3 hours, and quenched rapidly. The tempering temperature was also investigated by heating the sample at 300°C, 400°C, and 500°C for an hour.

Table 1. Chemical composistion of austenitic manganese steel

Element	C	Mn	Cr	Si	Ni
% wt	0.59	10.81	1.44	0.37	0.04

3. Result and Discussion

3.1. Characterization of as-cast austenitic manganese steel

The microstructure of austenitic manganese steel 0.6%wt C - 10.8%wt Mn - 1.44%wt Cr in as-cast condition comprised austenitic and some of the dendritic carbides $(Fe,Mn)_3C$, as shown in figure 1. The composition of carbides was analyzed by Energy Dispersive X-Ray Spectroscopy (EDS). The carbides were also found in XRD analysis, as shown in figure 2. The carbides were formed and precipitated along austenite grain boundaries due to the slow cooling in solidification process. This carbide network resulted a negative effect on toughness, therefore the solid solution treatment must be applied on this austenitic manganese steel to improve its mechanical properties. From figure 1 and 2, the matrix was austenite. The hardness of this austenitic manganese steel in as-cast condition was only 12 HRC due to the formation of soft austenite in matrix.

IOP Conf. Series: Journal of Physics: Conf. Series **817** (2017) 012066 doi:10.1088/1742-6596/817/1/012066

The influence of solid solution temperature, soaking time, quenching media, and tempering temperature on the hardness and microstructure of austenitic manganese steel 0.6%wt C-10.8%wt Mn-1.44%wt Cr is discussed briefly in this paper.

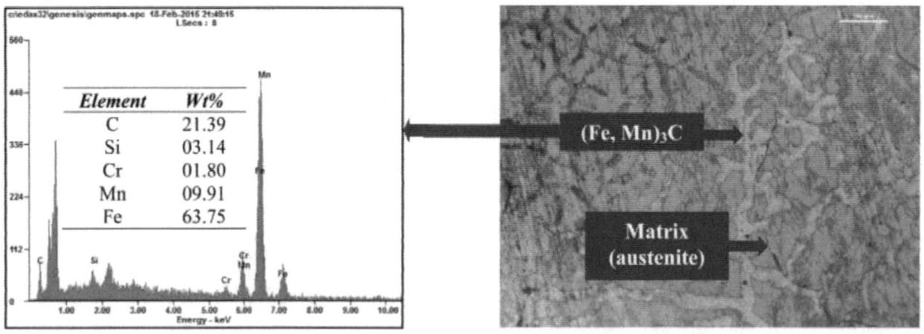

Figure 1. Microstructure of austenitic manganese steel in as-cast condition

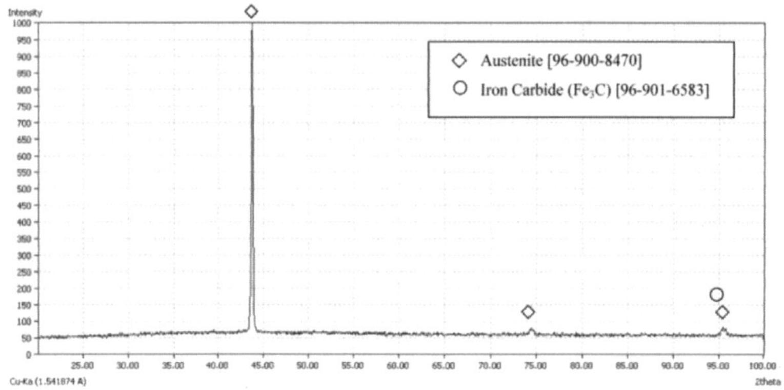

Figure 2. XRD analysis of austenitic manganese steel in as-cast condition

3.2. The effect of austenitization temperature and quenching media

Solid solution treatment was conducted to dissolve all carbides that were found in as-cast condition. XRD pattern in figure 3(a) shows the microstructure of austenitic manganese steel after solid solution treatment at 950°C for an hour and quenched by oil. It was found that ferrite and carbide appeared on its structure. While from figure 3(b) when the water was used as quenching media, there was no-carbide structure and the austenite was replaced the ferrite structure. From figure 5, it can be seen that although the trend of hardness in quenching by oil and water was similar, the hardness of quenched material by water was higher than that of quenched material by oil. Water has higher cooling rate than oil, thus the matrix will be dominated by austenite and no-ferrite structure. The formation of ferrite occurred because of the slow cooling after solid solution treatment. This ferritic matrix resulted the low hardness of this austenitic manganese steel. Ferrite is softer than austenite, therefore the hardness resulted from oil quenching was lower than that resulted from water quenching. figure 3(c-d) show that there was no-carbide structure. All carbides were dissolved into the matrix by heating this material at 1000°C and 1050°C for an hour and quenching rapidly by water. Carbon will be dissolved in the austenitic matrix at a critical temperature of the austenitic formation. It was called a solid solution temperature. Generally, it was 1050°C for manganese steel which has 0.9-1.4%wt C and 1-15%wt Mn [7]. The solid solution treatment temperature in this experiment was lower i.e. 950°C, due to the lower carbon content in this 0.6%wt C-10.8%wt Mn-1.44%wt Cr austenitic manganese steel. The higher carbon content will be resulted in more stabilized carbide structure, consequently it takes a higher temperature to dissolve carbon into the matrix. The hardness of this austenitic manganese steel increased by heating this material from 950°C until 1000°C, however it decreased after heating at 1050°C. The highest hardness was obtained at austenization temperature of 1000°C due to the appearance of martensite phase, as shown by XRD pattern in the figure 3(c).

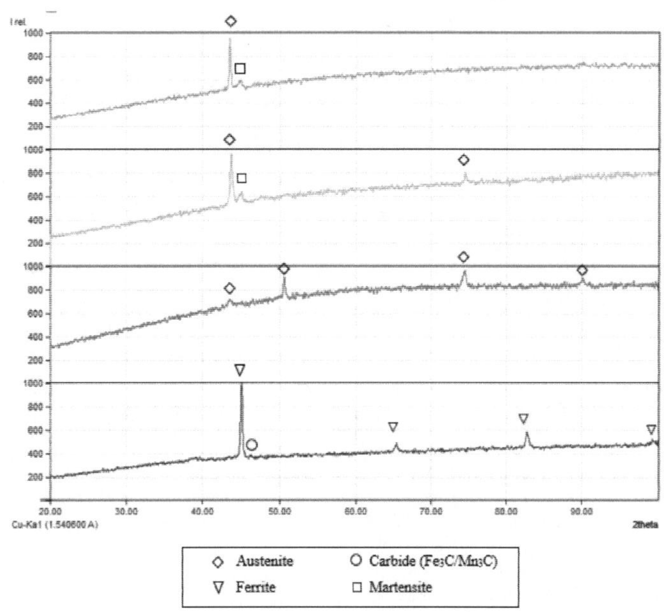

Figure 3. XRD patterns of austenitic manganese steel: (a) 950°C, oil quenched; (b) 950°C, water quenched; (c) 1000°C, water quenched; (d) 1050°C, water quenched

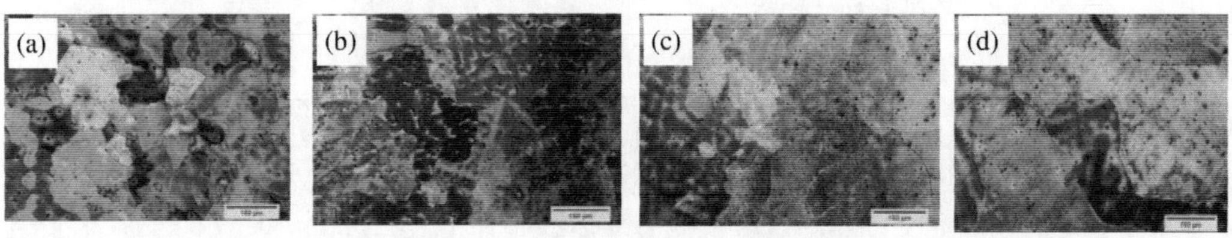

Figure 4. Solid solution treatment of austenitic manganese steel: (a) 1000°C, oil quenched; (b) 1000° C, water quenched; (c) 1050°C, oil quenched; (d) 1050°C, water quenched.

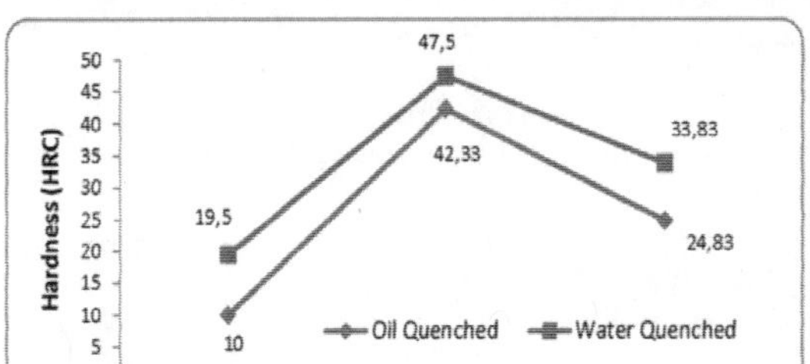

Figure 5. Hardness of manganese steel at different austenitization temperature and quenching media

The hardness of martensite is higher than that of austenite and ferrite, hence the hardness of this austenitic manganese steel increased significantly by heating at 1000°C. As reported by Schilke, et al., that the martensite formation in this manganese steel occurred because the carbon content in this material was not sufficient to stabilize the austenite structure [9]. Manganese and carbon play an important role to stabilize the austenite structure in the manganese steel. High carbon manganese steel commonly has fully austenite structure after being heated at 1050°C and quenched rapidly by water. However, in this current study, the martensite structure still existed although this medium manganese steel was heated at 1050°C, as shown by XRD pattern in figure 3(d).

The manganese steel heated at 1050°C had lower hardness than that heated at 1000°C due to the growth and coarsening of grain size. Figure 4 shows that the grain size of manganese steel heated at 1050°C was larger than that heated at 1000°C, either quenched by oil or water.

3.3. The effect of soaking time at austenitization temperature

The manganese steel was heated at 1000°C for 1 to 3 hours and then water quenched to investigate the effect of soaking time on this solid solution treatment. Figure 6 shows that the hardness of manganese steel increased due to the prolongation of soaking time in solid solution treatment. It was also confirmed by microstructure analysis that there was no carbide structure after the solid solution treatment at 1000°C for an hour, as shown previously in figure 3(c). The carbide structure was formed on its grain boundaries after heating this material for 2 and 3 hours, as shown in figure 7(a-b). The composition of this carbide shows in figure 7(c). From figure 6 and 7, it can be found that the longer soaking time of solid solution treatment in this austenitic manganese steel promoted the reaction of carbon and manganese to form carbide. This condition also led the increase of hardness because the grain size of carbide on its microstructure will reduce at the longer soaking time. The higher hardness will be obtained at the smaller grain size.

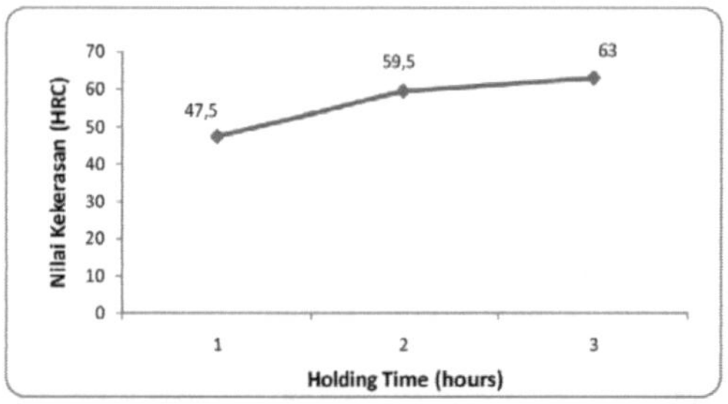

Figure 6. Hardness of manganese steel with water quenching after solid solution treatment with various holding time at 1000°C

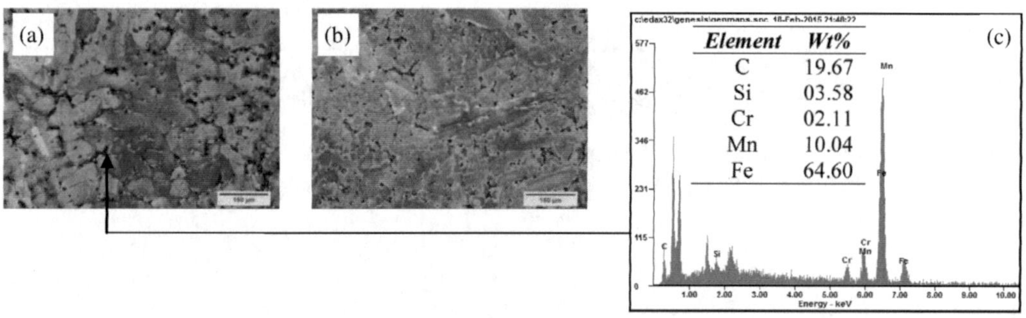

Figure 7. Microstructure of solid solution treatment on manganese steel with various holding times; (a) 2 hours; (b) 3 hours, (c) EDS of carbide on grain boundaries.

3.4. The effect of tempering at austenitic manganese steel

The manganese steel was heated at 1000°C for 3 hours continued by water quenching and tempering in various temperatures to investigate the effect of tempering on the solid solution treatment process.

Figure 6(b) shows that the tempering process reduced the hardness of austenitic manganese steel after solid solution treatment. The carbide was dissolved after tempering at 300°C. There was no-carbide structure on its grain boundaries (figure 8(a)), but the small amount of secondary carbide was formed inside the matrix. After tempering at 400°C, there were more carbide dispersed on the matrix that caused the increase of its hardness (figure 8(b)). Further, the hardness decreased after tempering at 500°C because the grain size was too large as shown in figure 8(c). The grain growth will reduce the hardness of the austenitic manganese steel. It was in accordance with Aribo, et al, that the grain size will increase due to the increase of tempering or aging temperature [8]. The large grain size will promote more dislocation in grain boundaries, consequently the finely dispersed carbide will separate from the matrix and move to fill the dislocation along the grain boundaries, as shown in figure 8(c). Figure 9 shows that the hardness of this material increased after being heated into 400°C. However, the hardness wasdecreased after being heated at 500°C. This secondary carbide will give positive effect on its mechanical properties, such as both of toughness and hardness, while the large grain size growth will reduce the hardness.

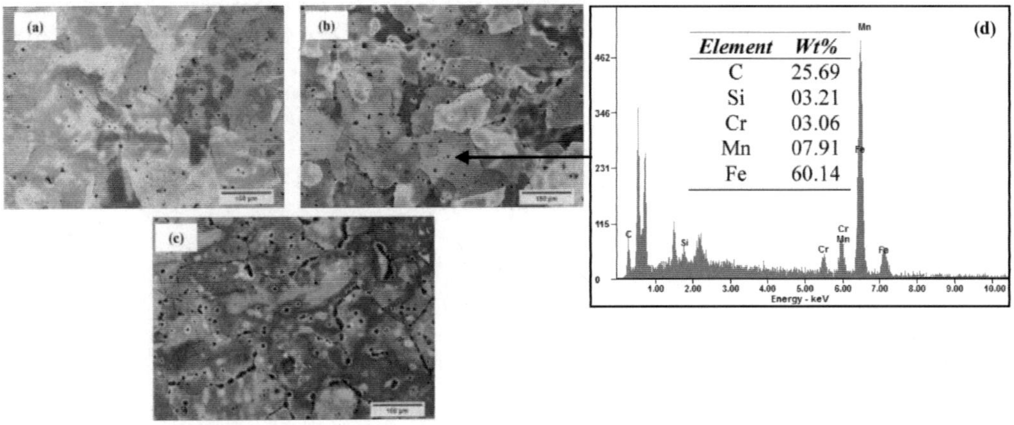

Figure 8. Microstructure of austenitic manganese steel with solid solution treatment at 1000°C for 3 hours, water quenching, and tempering for 2 hours at: (a) 300°C, (b) 400°C, (c) 500°C, (d) EDS analysis of carbide on matrix

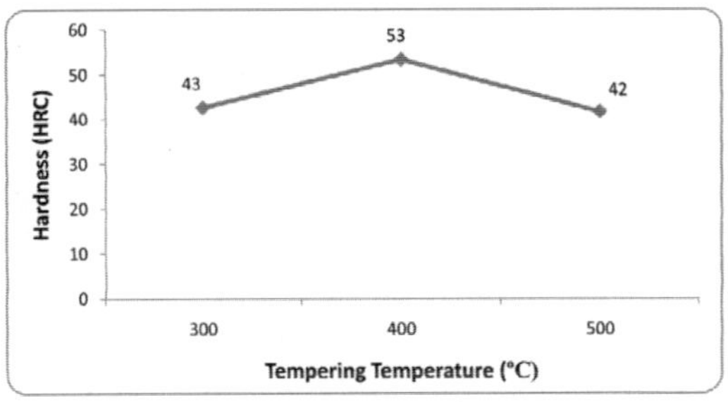

Figure 9. Hardness of manganese steel at various tempering temperature for 2 hours.

4. Conclusion

The microstructure of austenitic manganese steel with 0.6%wt C-10.8%wt Mn-1.44%wt Cr consisted of the austenitic matrix with carbide along its grain boundaries. Solid solution treatment of this austenitic manganese steel was only 950°C due to its low carbon content. Martensite was found at higher temperature solid solution treatment. Fine carbide dispersed within the matrix was obtained after tempering this manganese steel at 400°C, thus it resulted the optimum hardness. At higher tempering temperature, larger grain size and lower hardness were obtained.

IOP Conf. Series: Journal of Physics: Conf. Series **817** (2017) 012066 doi:10.1088/1742-6596/817/1/012066

Acknowledgments
We would like to thank the Indonesian Institute of Sciences for the financial support that was provided within the program of the Development of Material, Energy, and Manufacturing Engineering.

References
[1] Balogun S A, Esezobor D E, Agunsoye J O 2008 *Journal of Minerals & Materials Characterization & Engineering* **7** 277
[2] Bhero S W, Nyembe B, Lentsoana K 2013 *Proc. Int. Conf. on Mining, Mineral Processing and Metallurgical Engineering* (Johannesburg) p 174-176
[3] Dastur Y N and Leslie W C 1981 *Metall. Trans. A* **12A** 749
[4] Grajkar A, Kolodziej S, Krukiewicz W 2010 *Archives of Materials and Engineering* **41**(2) 77
[5] Khan A F, Rana A M, Islam M, Masood I, and Abbas T 2001 *Pakistan Journal of Applied Sciences* **1**(3) 317
[6] Shi Z and Gu M 1995 *J. Mater. Sci. Technol.* **11** 102
[7] Ham Y S, Kim J T, Kwak S Y, Choi J K And Yoon W Y 2010 *China Foundry* **7** 178
[8] Aribo S, Alaneme K K, Folorunso D O, Aramide F O 2010 *Journal of Minerals & Materials Characterization & Engineering* **9** 157
[9] Schilke M, Ahlstrom J, and Karlsson B 2010 *Proceedia Engineering* Vol 2 No 1 (Cambridge: Elsevier Ltd.) p 623

IOP Conf. Series: Journal of Physics: Conf. Series **817** (2017) 012067 doi:10.1088/1742-6596/817/1/012067

Influence of oxidation temperature on the oxide scale formation of NiCoCrAl coatings

E Sugiarti[1], K A Zaini[1], R Sundawa[1], Y Wang[2], S Ohnuki[3] and S Hayashi[4]

[1]Laboratory of High Temperature Materials and Coatings, Research Centre for Physics, Indonesian Institute of Sciences (LIPI), Serpong 15314, Indonesia
[2]Creative Research Institutions, Hokkaido University, Kita 21 Nishi 10, Sapporo 001-0021, Japan
[3]Laboratory of Advanced Materials, Division of Material Science and Engineering, Faculty of Engineering, Hokkaido University, Kita13 Nishi 8, Sapporo 060-8628, Japan
[4]Laboratory of Advanced High-Temperature Materials, Department of Metallurgy and Ceramics Science, Graduate School of Science and Engineering, Tokyo institute of Technology, 2-12-1 O-kayama, Meguro-ku, Tokyo 152-8550, Japan

E-mail: enis002@lipi.go.id

Abstract. Intermetalic coatings of NiCoCrAl have been successfully developed on low carbon steel substrate to improve oxidation resistance in extreme environments. The influence of oxidation temperature on the oxide scale formation was studied in the temperature range of 600-1000 °C. The measurements were made in air under isothermal oxidation test for 100 h. The surface morphology showed that a cauliflower like structure developed entire the oxide scale of sample oxidized at 800 °C and 1000 °C, while partly distributed on the surface of sample oxidized at 600 °C. The XRD analysis identified Cr_2O_3 phase predominantly formed on the oxidized sample at 600 °C and meta-stable Al_2O_3 with several polymorphs crystalline structures: η, δ, θ, κ, and α-Al_2O_3 at relatively high temperatures, i.e. 800 °C and 1000 °C. A Cross-sectional microstructure showed that complex and porous structures formed on the top surface of 600 °C and 1000 °C samples. In contrast, a very thin oxide scale formed on 800 °C oxidized samples and it appeared to act as a diffusion barrier of oxygen to diffuse inward, hence could increase in the service life of carbon steel substrate.

1. Introduction

Coatings are the most commonly used material to protect substrate against aggressive environment. NiCoCrAl coatings are often used to protect the substrate from oxidation at high temperatures [1-3]. Further, oxidation resistance of several materials is very important for some high temperature applications. Coatings with sufficient amounts of aluminium or chromium can form an Al_2O_3 or Cr_2O_3 layer on their surface by reacting with oxygen, in order to protect the base material from interaction with the high temperature environment [4]. A dense, continuous, stable, and adherent either α-Al_2O_3 or Cr_2O_3 scale is always desirable because it offers excellent protection for the substrate due to its slow growth rate and thermodynamic stability at high temperatures [5].

It has been reported that the scale formed during oxidation at a temperature between 800 - 1000 °C is often complex, typically consisting of transient aluminum oxides of the γ, δ, and θ types to the stable α-Al_2O_3 phase as well as spinel $NiAl_2O_4$ [6-10]. Many researchers have reported that at a temperature below 800 °C, Cr_2O_3 are preferentially formed and generally considered to be more protective than Al_2O_3 [11]. Moreover, previous studies showed that the pack cementation temperature can affect the oxidation behavior and the scale structure of NiCoCrAl coatings at a temperature of 800 °C [12]. However, a detailed investigation from low to high temperature oxidation behavior of these coatings has not been extensively explored.

In the present study, the influence of oxidation temperature on the oxide scale formation of NiCoCrAl coatings during 100 h oxidation in air at elevated temperatures of 600, 800, and 1000 °C were investigated. The oxidation resistance of these coatings in the corresponding oxide scale growth was compared.

2. Experimental Procedures

Carbon steel substrates measuring 15 x10 x 1.5 mm were machined out of the plate and a hole of 3.5 mm diameter was drilled near one end of each substrate. The slices were ground with SiC polishing paper up to a mesh size of 1000 and ultrasonically cleaned. NiCoCrAl multilayer coating was used to coat the substrate by electroplating and pack cementation processes. The coating parameters have been described elsewhere [13]. Isothermal oxidation test was performed at 600, 800, and 1000 °C in ambient atmosphere. The oxidation period was determined from 1 h, 5 h, 15 h, 31 h, 46 h, 66 h, 90 h, to 100 h. In each period, more than one coated sample was oxidized to check the reproducibility of the oxidation kinetic. The weight gain was measured after the coated samples were cooled using a balance with a sensitivity of 10^{-5} g. The oxide morphology and oxide composition were determined by SEM (JEOL, JSM6380LA). The compounds of the oxides were identified by XRD using X'Pert Pro with Cu-Kα radiation generated at 40 kV and 40 mA (Philips, PANalytical). A cross-sectional microstructure and elemental distribution of the samples were characterized by FE-EPMA (JEOL, JXA-8530F).

3. Results and Discussions

3.1. Isothermal oxidation

Figure 1 shows the oxidation kinetics curve of NiCoCrAl sample oxidized at 600, 800, and 1000 °C for 100 h. The weight gain of sample oxidized at 600 °C was slower than that of sample oxidized at 800 and 1000 °C. The oxidation kinetic of sample oxidized at 600 °C exhibits a linear increase in mass gain with increasing oxidation time. It revealed that the oxide growth controlled by aluminium or chromium diffusion. On the contrary, the oxidation kinetic of sample oxidized at 800 and 1000 °C follows parabolic rate law. The oxide growth occurs by diffusion of oxygen to the coated layer at initial rapid growth rate of NiCoCrAl oxidation followed by a steady growth of alumina oxide layer.

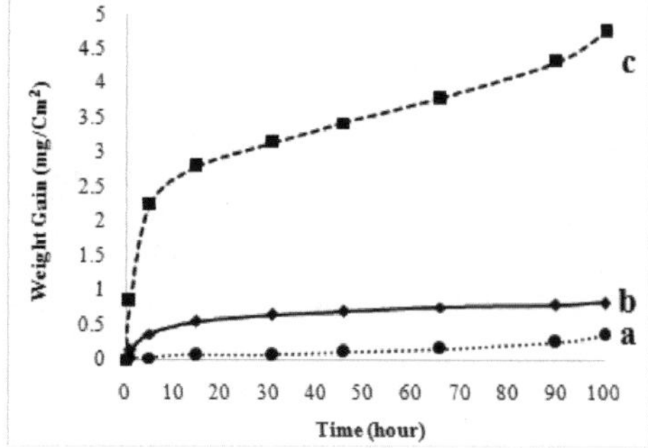

Figure 1. Isothermal oxidation at (a) 600, (b) 800, and (c) 1000 °C.

IOP Conf. Series: Journal of Physics: Conf. Series **817** (2017) 012067 doi:10.1088/1742-6596/817/1/012067

The results exhibited the sample oxidized at 800 °C has better oxidation behavior than the sample oxidized at 1000 °C due to its very low oxidation rate. For the sample oxidized at 800 °C, the weight gain increased to value between 0.5 and 0.8 mg/cm^2 after 15 h. The kinetic constant of sample oxidized at 1000 °C is one order larger than that of sample oxidized at 800 °C with $k_p = 3.01 \times 10^{-8}$ g^2 cm^{-4} s^{-1} and $k_p = 1.06 \times 10^{-9}$ g^2 cm^{-4} s^{-1}, respectively.

3.2. Surface morphology of oxide scale

Figure 2 shows SEM characterization on the surface of oxidized samples. The investigation at 600 °C implied the roughness of oxide scale entire the surface of the sample (figure 2a). It indicates the boundary structure on the surface should be related to the different phases formed in the oxide scale which has irregular dark area and diverse *cauliflower* like structure. In contrast, a very rough *cauliflower* like structure homogenously observed on the surface of the sample oxidized at 800 °C (Figure 2b), while similar structure with much less fine was also appeared on the surface of coated sample exposed to a higher temperature of 1000 °C (figure 2c). With increasing oxidation temperature, the grain structure of oxide scale was finer and uniformly deposited on the surface of coated sample.

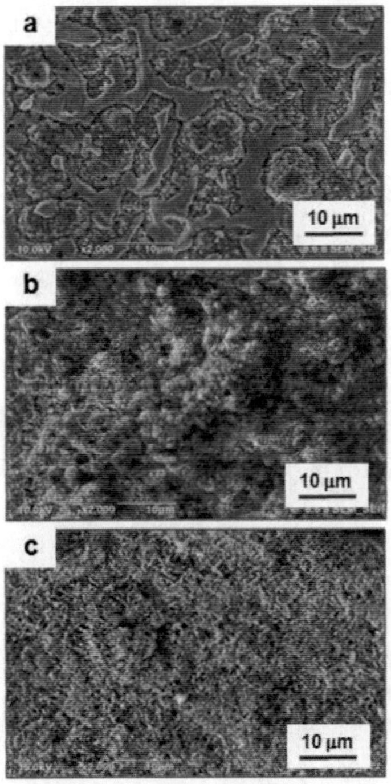

Figure 2. Surface morphology of coating samples oxidized for 100 h at (a) 600 °C, (b) 800 °C, and (c) 1000 °C.

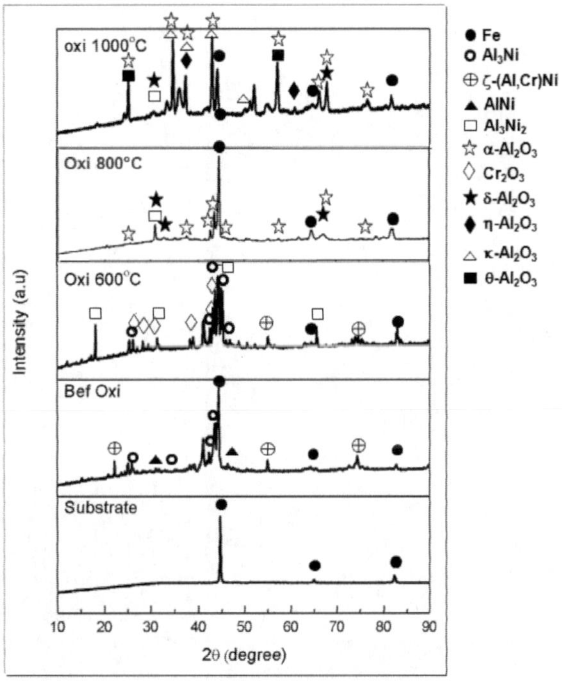

Figure 3. XRD analysis of the substrate, NiCoCrAl coating samples before and after oxidation for 100 h at 600, 800, and 1000 °C.

3.3. Phase identification of oxide scale

Figure 3 shows the phase identification of NiCoCrAl coated samples before and after oxidation. The XRD pattern shows the phases to be a mixture of predominant Al_3Ni, ζ-(Al,Cr)Ni accompanied with AlNi phase identified in the as-coated sample before oxidation which has good agreement with previous results [11]. The oxide scale was identified mainly consists of Cr_2O_3 and small amount of intermetallic phase of Al_3Ni and Al_3Ni_2 after oxidation at a low temperature of 600 °C. This result is correlated with the surface structure previously described with corundum structure of Cr_2O_3 as shown *cauliflower* like structure and intermetallic phases as shown irregular dark area. Moreover, the oxide scale was transformed from Cr_2O_3 to δ-Al_2O_3 and α-Al_2O_3 when oxidation slightly increased to 800 °C. It reveals the evolution of oxide scale structure from diverse to be more uniform for *cauliflower* like structure. Several polymorphs of alumina, i. e, η-Al_2O_3, δ- Al_2O_3, θ- Al_2O_3, κ-Al_2O_3, and α-Al_2O_3 phases were formed in the sample oxidized at 1000 °C. It indicates the formation of transient alumina at relatively high temperature of 1000 °C which has a very fine cauliflower like structure.

3.4. Microstructure of oxide scale

FE-EPMA characterization was performed in order to understand the oxide scale formation after oxidation test. Figure 4 shows a cross sectional FE-EPMA images together with composition map of oxidized sample after 100 h exposure time. The oxide scale was established with an approximate thickness of 15 μm, 6 μm, and 25 μm for the sample exposed at temperatures of 600, 800, and 1000 °C, respectively.

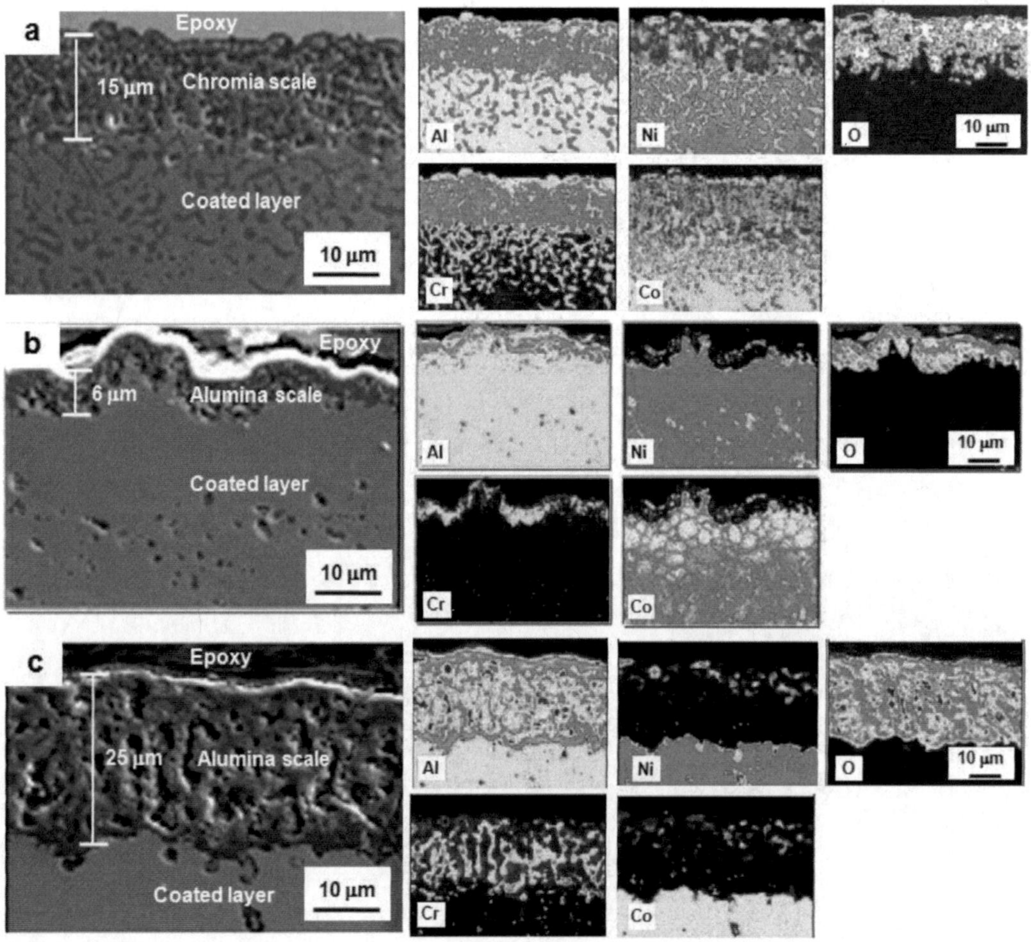

Figure 4. BSE images and elemental maps of NiCoCrAl coating sample oxidized for 100 h at (a) 600 °C, (b) 800 °C, and (c) 1000 °C.

A complex and porous oxide scale was developed at sample oxidized at 600 °C. Internal oxidation occurred between metal/oxide interfaces which are indicated as a dark precipitate beneath the oxide scale (figure 4(a)). It might cause by the high activity of oxygen to diffuse inward and reacted with the coating elements. It can be readily recognized from the intensity of oxygen either on the surface or metal layer of sample in figure 4(a). From the composition map, it clearly understands that aluminum reacted with nickel to form intermetallic phases such Al_3Ni, Al_3Ni_2 and β-NiAl as identified by XRD analysis. Furthermore, high intensity chromium content accompanied with oxygen might form chromia (Cr_2O_3) protective scale on the top layer.

In contrast, a relatively thin with localized porous was developed at the sample oxidized at 800 °C as shown in figure 4(b). The existence of high Al and O content on the top layer of sample oxidized at 800 °C producing the simple alumina (Al_2O_3) scale. Ni, Co and Cr are hardly to diffuse outward to the outermost layer because the alumina scale might play a role as a diffusion barrier for this temperature oxidation. Similar with the result of the sample oxidized at 600 °C, a thick and porous structure with large voids was developed at sample oxidized at 1000 °C as shown in figure 4(c). Low intensity chromium was diffused outward to the top layer reacting with aluminum and oxygen. Therefore, it might cause the formation of transient alumina phases with a complex structure from the initial oxidation state at 1000 °C. The intensities of nickel and cobalt were not observed on the outermost layer, thus other phases were also not identified by XRD for this sample.

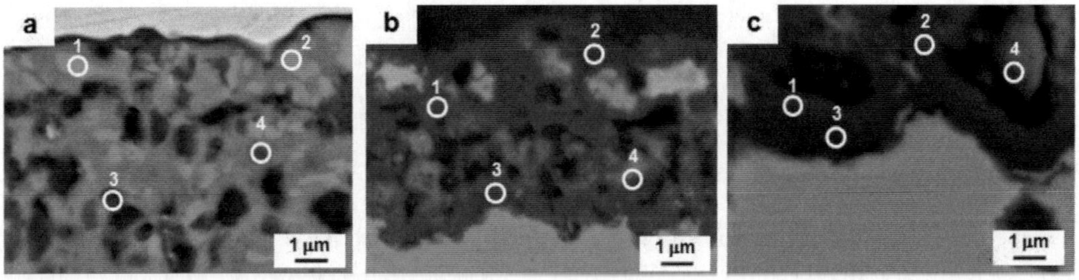

Figure 5. Magnified FE-EPMA images of oxide scale on NiCoCrAl samples oxidized at a) 600 °C, (b) 800 °C, and (c) 1000 °C for 100 h.

Table 1. Composition analysis obtained in the oxide scale of sample oxidized for 100 h at 600, 800, and 1000 °C

Sample Code	Point Analysis	Composition (at%) Ni	Co	Cr	Al	O	Phase
5a	1	7	2	9	80	2	ζ-(Al,Cr)Ni
	2	22	1	3	74		Al_3Ni
	3			47	2	51	Cr_2O_3
	4			41	3	56	Cr_2O_3
5b	1	4		5	49	42	α-Al_2O_3
	2			1	51	48	α-Al_2O_3
	3	2		6	72	20	δ-Al_2O_3
	4	2		4	74	20	δ-Al_2O_3
5c	1			1	41	58	α-Al_2O_3
	2			4	46	50	α-Al_2O_3
	3	2		15	48	35	θ-Al_2O_3
	4	1		6	74	19	δ-Al_2O_3

The morphologies of the oxide scale formed on the top of coated sample were magnified as shown in figure 5. It is clearly understood how the difference of porosity on oxide scale structure. The porosity size is about 0.6 μm, 0.2 μm, and 2 μm for sample oxidized at 600, 800, and 1000 °C respectively. It revealed the porosity of alumina scale in sample oxidized at 800 °C was denser than the other two samples. In addition, point analysis as indicated by the circle marks in fig. 5a-c was carried out in order to understand the composition of oxide scale related to the phase formation. The results are shown in table 1. It determined that the phases formation are correspond to the phases identification which previously identified by XRD analysis.

4. Conclusions

The influence of oxidation temperature on oxide scale formation of NiCoCrAl coating was investigated. At a temperature of 600 °C, the oxidation rate was determined by Cr_2O_3 formation. Generally, Cr_2O_3 scale is ideal against the oxidation below 800 °C, however Cr_2O_3 formed in this oxidized sample would not act as a protective layer due to the formation of complex with porous oxide structure and the oxidation kinetic with almost linier growth. At temperature 800 °C, δ and α- Al_2O_3 scales were formed with a thin and localized porous structure. The oxidation kinetic obeys the parabolic rate law and its K_p value was about 1.06×10^{-9} g^2 cm^{-4} s^{-1}. At temperature 1000 °C, a transformation from η-Al_2O_3, δ- Al_2O_3, θ- Al_2O_3, κ-Al_2O_3, to α-Al_2O_3 occurred from the initial to final stage of 100 h oxidation time. Transition of polymorph alumina phases might cause the formation of complex structure with thick and porous oxide scale. Consequently, the kinetic constant of this sample is one order larger than the sample oxidized at 800 °C and its k_p value was about = 3.01×10^{-8} g^2 cm^{-4} s^{-1}. Therefore, corresponding to the above described by XRD analysis and FE-EMPA results as well as the kinetic constant of oxidation, it concluded that sample oxidized at 800 °C had better oxidation resistance than the other two oxidized samples.

Acknowledgments
The authors would like to thanks to LIPI Competitive Program for financial support of this work. A part of this work was conducted at Hokkaido University, supported by "Nanotechnology Platform" Program of the Ministry of Education, Culture, Sports, Science and Technology (MEXT), Japan.

References

[1] Pace M T, Thomson R C and Wells J 2008 *Journal of The Minerals, Metals and Materials Society* 651
[2] Girolamo G D, Brentari A, Blasi C, Pilloni L and Serra E 2014 *Metal. Mater. Trans. A* **454** 5362
[3] Zhu C, Wu X, Wu Y and Liang G 2010 *Engineering* **2** 602
[4] Girolamo G D, Brentari A and Serra E, 2014 *Coatings* **4** 701
[5] Wang B, Gong J, Sun C, Huang R F and Wen L S 2003 *Mats. Sci. Eng. A* **357** 39
[6] Evans A G, Mumm D R, Hutchinson J W, Meier G H and Pettit F S 2001 *Prog. Mater. Sci.* **46** 505
[7] Tolpygo V K and Clarke D R 2000 *Mater. High Temp.* **17** 59
[8] Eni S, Zaini K A, Wang Y, Hashimoto N, Ohnuki S and Hayashi S 2015 *Oxid. Met.* **83** 453
[9] Seraffon M, Simms N J, Sumner J and Nicholls J R, 2011 *Surf. Coat. Tech.* **206** 1529
[10] Seraffon M, Simms N J, Sumner J and Nicholls J R 2014 *Oxid. Met.* **81** 203
[11] Foss B J, Hardy M C, Child D J, Mcphail D S and Shollock B *Journal of The Minerals, Metals and Materials Society* **66**(12) 2516
[12] Eni S, Zaini K A, Wang Y, Hashimoto N, Ohnuki S and Hayashi S 2015 *Adv. Mater. Res.* **1112** 353
[13] Eni S, Destyorini F, Zaini K A, Wang Y, Hashimoto N, Ohnuki S and Hayashi S 2015 *Surf. Coat. Tech.* **265** 68

The corrosivity and performance evaluation of antifouling paint exposed in seawater Muara Baru port, Jakarta

L Nuraini[1], S Prifiharni[1], G Priyotomo[1] and Sundjono[1]
[1]Research Center for Metallurgy and Material, Indonesian Institute of Sciences, Gd 470, Kawasan Puspiptek, Tangerang Selatan, Banten, 15314, Indonesia

E-mail: lutv002@lipi.go.id

Abstract. Antifouling paints were applied on a wide range of the under seawater structures in order to protect them from the growth of fouling organisms. The performance investigation of two commercial anticorrosion and antifouling paints was conducted in Muara Baru port, Jakarta. The specimens were coated by anticorrosion and/or antifouling paint. Blank specimen (without exposed) were also prepared as a control. On the other hand, bare mild steel was prepared for measure the corrosion rate through weight loss method. The test panels containing specimens were exposed up to 3 months for immersion depths of 0, 1, 2, 3 meters from sea level. Sea water parameters such as temperature, pH, salinity, conductivity and dissolved oxygen (DO) were also measured. The evaluation of coating performance was carried out such as thickness, glossy, hardness and adhesion strength. The results showed that both surfaces of anticorrosion paint and bare mild steel specimen covered by fouling organisms, whereas no fouling took place on the surface of antifouling paint. The corrosion rate of bare mild steel in the depths of 0, 1, 2, 3 meters were 12.5; 11.6; 8.3; 10.4 mpy, respectively.

1. Introduction

Corrosion and biofouling have been driving force for investigating interactions between microorganisms and solid surfaces [1]. Corrosion is the deterioration of a metal as a result of chemical reactions between it and the surrounding environment. Whereas, Biological fouling, also called biofouling, is the undesirable formation of organisms on a surface immersed in water [2]. Corrosion and fouling causes severe problems on both dynamic and static structures. Fouling on many static structures may compromise safety by reducing stability and concealing structural defects [3]. The fouling process has been considered to consist of four general stages, from the attachment of microfouling organisms to macrofouling organisms [4]. The influence of various sea water characteristics such as salinity, temperature, pH, dissolved salts and oxygen concentration on the fouling of immersed solid surfaces will also play a significant role in the growing of the fouling organism [5].

Antifouling paints have been proved as the primary method for combating biofouling in the marine industry [6, 7]. Due to the restrictions on tributyltin (TBT) compounds usage in antifouling paints since 2003 and its complete ban on all vessels in 2008 (IMO 2001), copper (Cu) has been widely used as the main biocide ingredient in antifouling paint coatings. Copper exhibits antifouling activity against organisms such as barnacles, tube worms and the majority of alga fouling species [8]. Copper(I) oxide (Cu_2O) is still being used as an antifouling pigment in marine paints [9]. However, copper has harmful effects not only on the fouling community but also on non-target species [10]. Therefore it is important to find the new environment friendly antifoulants [11] due to no similar studies in Indonesia. The objectives of this preliminary study are to evaluate the performance of two commercial anticorrosion and antifouling paints in Muara Baru port, Jakarta. The effect of difference depth of exposed and evaluation on the corrosion rate of bare mild steel using weight loss method will be comprehensively discussed. The result of the present investigation can be used for further work in order to compare the performance of coating with commercial paint and friendly antifouling paint.

2. Materials and Methods

2.1. Specimen and application of coatings

A commercial metal sheet of mild steel (C=0.13%, Sil=0.11%, P=0.02%, Mn=0.29, Ni=0.01, Cr=0.30, Cu=0.01, Ti=0.01, Ti=0.01, Fe=99%) with the thickness of 0.3 cm were used in this study. Sheets of mild steel were cut into required number of specimens of sizes, 20 cm width x 25 cm height for the immersion studies. The specimens were sandblasted according to the standard of ISO 8501-1 and coated by two commercial paints (specimen code "A and B") for anticorrosion and/or antifouling. Blank specimen (without exposed) were also prepared as a control. In contrast, bare mild steel with dimension of 7.5 cm wide x 15 cm high x 0.3 cm thick was prepared to measure the corrosion rate (specimen code "b").

2.2. Location and exposure time

The performance investigation of two commercial anticorrosion and antifouling paints was conducted in Muara Baru port, Jakarta. The test panels containing specimens were exposed up to 3 months for immersion depths of 0, 1, 2, 3 meters (specimen code "A0, A1, A2, A3; B0, B1, B2, B3; b0, b1, b2, b3") from sea level.

2.3. Evaluation of coatings for anticorrosion and antifouling properties

After immersion of specimens, the evaluation of coating properties was carried out as follow:

a. Specular gloss was determined at 60° using Horiba Gloss Checker IG-331 according to ASTM D 523
b. Thickness was measured using coating thickness gauge Elcometer 456 according to ASTM B 499
c. Hardness was tested using pencil hardness tester Elcometer 501 according to ASTM D 3363
d. Adhesion measurements were carried out using automatic adhesion tester Elcometer 510 according to ASTM D 4541 standard specification, the adhesive was a 2-pack epoxy system (Araldite 2011), the dolly diameter was 20 mm. After the dollies were glued, the adhesive cured for 24 h at 30°C and 50% RH.

Sea water parameters, i.e, temperature, pH, salinity, conductivity and dissolved oxygen (DO) were also measured.

2.4. Weight loss and corrosion rate determination

In the laboratory, fouling on the specimens was gently removed followed by cleaning the specimens to remove corrosion products on the surface of the coupons according to ASTM G1. Further, the weight of the specimens after cleaning was compared to the initial weight of the specimens. The difference of weigh indicated the metal loss during the expose period.

The corrosion rate in mils per year (mpy) was calculated from the weight loss using the equation:

$$CR\ (mpy) = \frac{W \times K}{D \times As \times T} \qquad (1)$$

Where W = weight loss in grams; K = constant; D = metal density in g/cm^3; A = coupon area (cm^2); t = exposure time (days)

3. Results and discussion

3.1. Evaluation of coatings for anticorrosion and antifouling properties

3.1.1 Gloss property

Gloss is a very important property of a paint to show the capacity of surfaces for the reflection of directed light. According to ASTM D 523, the geometry of 60° was used to measure the specular gloss on paint films which do not have a very high gloss or a very low one. The range of gloss levels was calculated from 2 to 7 units as shown in table 1. The results show that none of 10 gloss units or higher and under 5, which is a typical measurement of gloss for flat paints. The analysis level of gloss specimen A and B compared each other for the condition before and after exposure. It revealed that no significant changes on gloss level in the specimen A and B in all depth at 3 months after exposure.

Table 1. Physical and mechanical properties of antifouling paint after exposure.

Specimen code	Gloss at 60° (Gloss units)	Hardness	Thickness (μm)	Adhesion (MPa)
A0	7.0	B	710.2	2.22
A1	4.5	B	767.2	1.82
A2	5.5	B	650.8	1.96
A3	4.8	B	730.0	1.93
B0	3.1	2B	506.8	2.28
B1	2.9	2B	517.8	2.35
B2	2.6	2B	502.6	2.00
B3	2.0	2B	433.0	1.99

3.1.2 Hardness

The pencil scratch hardness test showed that all of specimen A are type B and all of specimen B are type 2B. Both of them are classified in soft type of hardness test and no significant differences.

3.1.3 Thickness

The physical performance of the paint thickness after exposures shows that paints A has the higher value of thickness than paint B. The antifouling topcoat paints will took place the process of surface polish (erosion or ablation) under water flow conditions to provide a biofouling free.

3.1.4 Adhesion test

The range of average adhesion strength test values for the antifouling coatings A and B are between 1.82 MPa and 2.35 MPa. There is no trend for particular coating system. In table 1, the same values of the pull-off strength and the type of failure indicate that the systems of coating tends to 100% cohesive antifouling, where both of coating A and B occurs fractures inside antifouling coating layer.

3.2. Weight loss and corrosion rate

Figure 1 shows the visual observation of coating A and B compared with blank specimen (without exposed) in Muara Baru seawater for 3 months. The results shows that the surface of anticorrosion coating was not undergo discolouration and fading on specimens A and B in various depth, where the surface of antifouling coating occurred a discolouration or the colour change on the surface paint from red to pink. In table 2, the phenomena of fading tend to increase due to the presence of moisture from 3 month of exposure and the growing of fouling organism.

After 3 month of exposure, the growth of marine organisms (benthos) created a dense fouling ecosystem on all surface of both anticorrosion paints for type A and B at various depth of seawater. Antifouling paints of type A and B showed that antifouling paints could minimize the attachment of marine organisms for 3 month as observed that no fouling attached in the surface on paint A. In contrast, a little of fouling organism was found in antifouling side of paint B. Therefore, the anticorrosion and antifouling paint A has better performance than paint B. This indicated that the anticorrosion and antifouling paint A has the ability to work in preventing the existence of fouling than the anticorrosion paint B. It is also implied that the antifouling paints could work well in preventing the existence of fouling with the release of some biocidal compounds, which are bound or integrated in the organic polymer matrix paint.

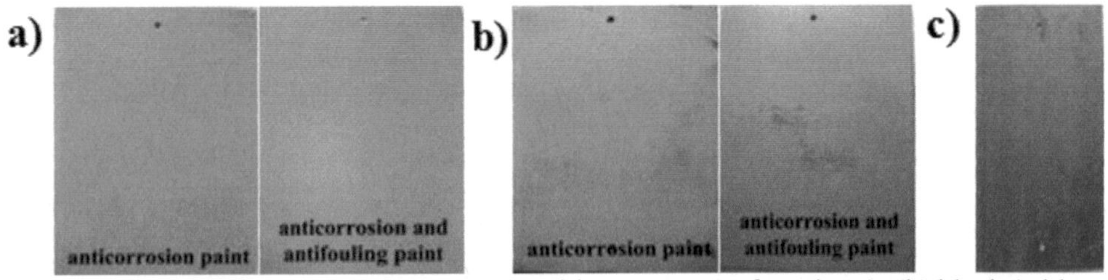

Figure 1. Photographs of specimen (a) blank (without expose) of coating A, (b) blank (without expose) of coating B and (c) without coating.

Table 2. Specimen of coating A and coating B after 3 month of exposure.

Specimen Code	Anticorrosion paint	Anticorrosion and Antifouling paint	Specimen Code	Anticorrosion paint	Anticorrosion and Antifouling paint
A0	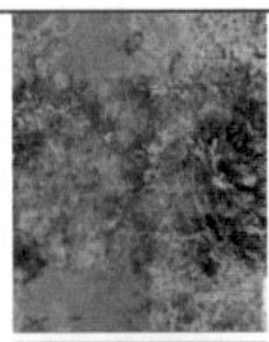	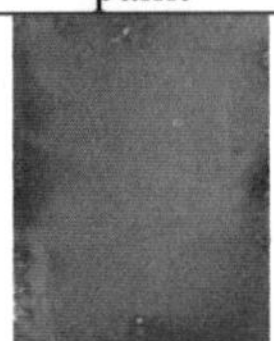	B0		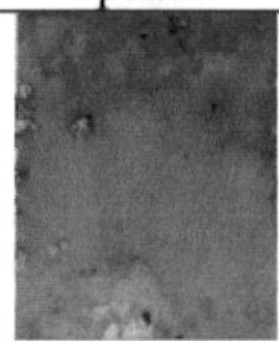
A1		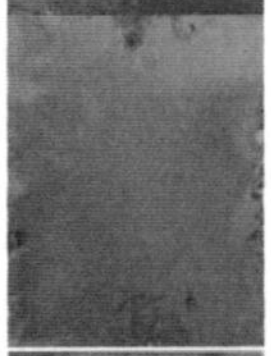	B1		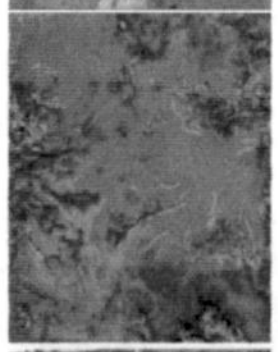
A2		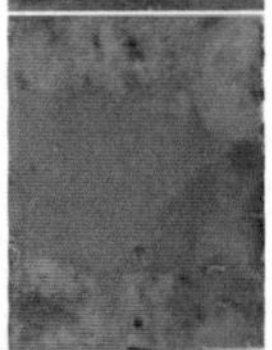	B2		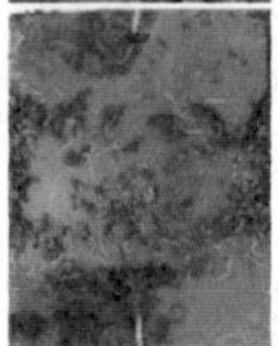
A3	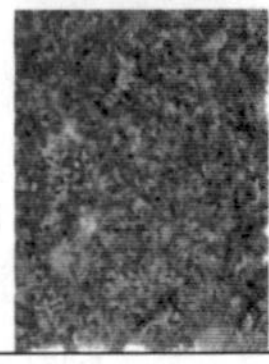	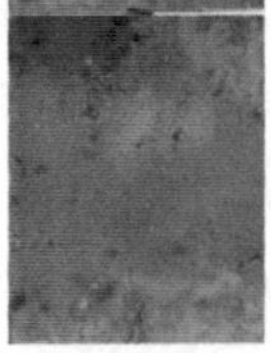	B3		

Table 3 shows the weight loss of the bare steel after exposure. The weight loss in all specimens occurred because the activity of fouling organism on the surface of specimens. The weight of the bare steels decreases due to the presence of fouling organism on the surface of the steel plates. Figure 2 shows the corrosion rate of bare specimen at various depths of 0, 1, 2, 3 meters are 12.5, 11.6, 8.3, 10.4 mpy, respectively. Furthermore, the seawater depth of 0 meter has the highest value of corrosion rate due to the location of between splash zones.

Table 2. The weight loss of specimen in various depth of seawater.

Specimen code	Weight loss (gram)
b0	13.90
b1	12.87
b2	9.18
b3	11.51

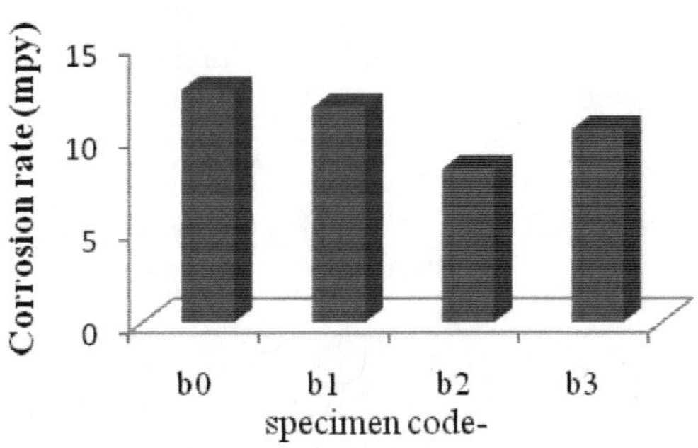

Figure 2. Corrosion rate of bare specimen at various depths of seawater.

Generally, all paints contain the same constituents of polymeric binder, pigments, solvent, and additives. The polymer or binder is the most important component in a successful hydrolyzing antifouling paint. From material safety data sheet (MSDS) of the paint, this two paints (A and B) work based on a silyl acrylate polymer, which is hydrolysed upon exposure to seawater. These copolymers blended with cuprus oxide as biocides confer a smooth surface on the copolymer and an ability of controlling/regulating biocide leaching rates [12]. However, they exhibited poor marine antibiofouling performance on static conditions. This is because the self-renewed ability of such polymers with nonreactive main chain which determined by the hydrolysis of the side groups and the shear force which generated by seawater flow [13]. Together with copper, other booster biocides were used to increase the efficiency of antifouling coatings [14]. 18 compounds are currently used as antifouling paint booster biocides with various degrees of regulation [15]. There is not specific information of MSDS paint A and B about the booster biocide used in both paint because there is no significant difference from the main component in paint A and B. The difference in result that paint A better than paint B in prevent the biofouling effect in this research might comes from the difference of booster biocide or the concentration of biocide. However, further experiment is required to clarify this hypothesis.

As explained that antifouling paint systems A and B containing copper as main fouling biocide were applied as topcoats on metal substrates. The mechanism for the release of copper compound is well known and described with the following reactions:

$$\frac{1}{2} CuO_{(s)} + H^+ + 2Cl^- \rightarrow 2CuCl_2^- + \frac{1}{2} H_2O \quad (2)$$

$$CuCl_2^- + Cl^- \leftrightarrow 2CuCl_3^{2-} \quad (3)$$

Where reaction (1) is reversible but kinetically constrained and reaction (2) is reversible and instantaneous. In the presence of dissolved oxygen, univalent copper is rapidly oxidised to the biocidal form, Cu^{2+}, in each layer. The dissolution of copper (I) oxide involved the formation of copper chloride complexes ($CuCl^{2-}$, $CuCl_3^{2-}$) in two reversible reactions [16, 17]. Variation between its species depends on pH, salinity and the presence of dissolved organic matter. It shown that Cu^{2+} ions were more toxic than Cu^+ and high concentrations of Cu^{2+} ions in seawater had deleterious effects on marine ecosystem [18].

Chen et *al.*[19] explained that the fouling layer formation begins with an attack of bacterial cells, which later developed into a small colony and subsequently increased the amount thus forming a biofilm layer. This promotes the development of complex layers of bacteria and extracellular polysaccharides.Nutrient-rich layer (biofilm) were formed followed by colonization of invertebrates and macroalgae [20]. The growth of fouling depends on many parameters, i.e, pH value, temperature, salinity and dissolve oxygen (DO) [5]. In this research, the physical condition of the sea waters of Muara Baru, Jakarta was described and shown in table 3. Based on temperature, pH, salinity and DO, there is not much different between each depth of sea weater, where the same of fouling species observed on the surface.

Table 3. Site-specific environmental conditions at Muara Baru Port, Jakarta.

Depth (meters)	Temperature(°C)	pH	Salinity (ppt)	Dissolve Oxygen (mg/L)	Conductivity
0	30.45	8.35	28.15	5.06	48.45
1	30.35	8.38	28.15	4.937	48.38
2	30.25	8.41	28.15	4.814	48.31
3	30.15	8.44	28.15	4.691	48.24

4. Conclusion

The results showed that both surfaces of anticorrosion paint A and B, and uncoated specimen were covered by dense fouling organisms. Further, there was no fouling took place on the surface of antifouling side of paint A, although the little of fouling organism was found in the antifouling side paint B. Therefore, from visual observation, we can conclude that the anticorrosion and antifouling paint A has better performance than paint B. Whereas, the corrosion rate result of uncoated specimen at the depths of 0, 1, 2, 3 meters were 12.5, 11.6, 8.3, 10.4 mpy, respectively.

Acknowledgements

The authors would like to acknowledge the Research Center for Metallurgy and Material – LIPI and Ministry of Research, Technology and Higher Education – Republic of Indonesia for financial support in this research under LIPI Priority and Advanced Research Projects 2015.

References

[1] Titah-Benbouzid H and Benbouzid M 2015 *Proc.of the 11th EWTEC Conf. (Nantes)* (EWTEC) 01199624
[2] Leggn M, Yücel M K, de Carellan G I, Kappatos V, Selcuk C and Gan T H 2015 *Ocean Eng.* **103** 237
[3] Turner A 2010 *Mar. Pollut. Bull.* **60** 159
[4] Almeidaa E, Diamantinoa T C and de Sousab O 2007 *Prog. Org. Coat.* **59** 2
[5] Yebra D M, Kiil S and Dam-Johansen K 2004 *Prog. Org. Coat.* **50** 75
[6] Yang W J, Neoh K, Kang E, Teo S L and Rittschof D 2014 *Prog. Polym. Sci.* **39** 1017
[7] Yang W J, 2012 *Biomacromolecules* **13**(9) 2769
[8] Voulvoulis N, Scrimshaw M D and Lester J N 2002 *Chemosphere* **47** 789

[9] Alyuruk H., Doner E, Karabay Z, Cavas L and Dokuz 2010 *Hacettepe J. Biol. & Chem.* **38**(4) 277
[10] Perez M, García M and Blustein, G 2015 *Mar. Environ. Res.* **109** 177
[11] Satheesh S, Baakdah M A and Al-Sofyani A A 2016 *Electron. J. of Biotechnology* **21** 26
[12] Omae I 2003 *Chem. Rev.* **103** 3431
[13] Ma C, Xu W, Pan J, Xie Q and Zhang G 2016 *Ind. Eng. Chem. Res.* **55**(44) 11495
[14] Thomas K V and Brooks S 2010 *Biofouling* **26**(1) 73
[15] Saleh A, Molaei S, Fumani N S and Abedi E 2016 *Mar. Pollut. Bull.* **105** 367
[16] Kiil S, Weinell C E, Pedersen M S and Dam-Johansen K 2002 *Trans IChemE* **80** 45
[17] Yebra D M, Kiil S, Weinell C E and Dam-Johansen K 2006 *Prog. Org. Coat.* **57** 56
[18] Konstantinou I K and Albanis T A 2004 *Environ. Int.* **30** 235
[19] Chen C, Maki J S, Rittschof D and Teo S L 2013 *Int. Biodeterior. Biodegrad.* **83** 71
[20] Jackson S M and Jones E G B 1988 *Int. Biodeterior.* **24** 277

IOP Conf. Series: Journal of Physics: Conf. Series **817** (2017) 012069 doi:10.1088/1742-6596/817/1/012069

Development of non-thermal atmospheric pressure plasma system for surface modification of polymeric materials

T P Kasih

Department of Industrial Engineering, Bina Nusantara University, Jl. KH. Syahdan, No. 9 Palmerah, Jakarta Barat, 11480, Indonesia

E-mail: tkasih@binus.edu

Abstract. Non-thermal plasma has become one of the new technologies which are highly developed now days. This happens because the cold plasma using the principle of generated reactive gases that have the ability to modify the surface properties of a material or product without changing the original characteristics of the material. The purpose of this study is to develop a cold plasma system that operates at atmospheric pressure and investigates the effect of cold plasma treatment to change the surface characteristics of the polymer material polyethylene (PE) at various time conditions. We are successfully developing a non-thermal plasma system that can operate at atmospheric pressure and can be run with Helium or Argon gas. The characteristics of plasma will be discussed from the view of its electrical property, plasma discharge regime andoperation temperature. Experiment results on plasma treatment on PE material shows the changes of surface properties of originally hydrophobic material PE becomes hydrophilic by only few seconds of plasma treatment and level of hydrophilicity become greater with increasing duration of plasma treatment. Confirmation of this is shown by the measurement of contact angle of droplets of water on the surface of PE are getting smaller.

1. Introduction

Plasma can be referred to as the fourth phase element after solid, liquid and gas phase. Unlike the normal gas phase which electrically neutral, plasma contains components of the atomic nuclei (ions) and electrons as the charged particles which have been separated because of the external energy received and have a reactive nature. Plasma can be formed naturally as happens in the sun or the elements of the stars in the sky. Plasma can also be formed by providing a high energy from proper electricity into a gas medium that makes the gas is undergoing a dissociation and ionization processes. Depending on the amount of energy transferred the transformation of the neutral gas into negatively and positively charged particles can be occurred as a part or a whole processes. The fully transformation will lead to the generation of high temperature or thermal plasma in which the temperature of all plasma components (ions, electrons, reactive species and neutrals) are the same. If the transformation occurred partially, it can allow the formation of low temperature or non-thermal plasma in which the temperature of electrons is several orders of magnitude higher than the temperature of ions or neutral species. When the thermal plasma technology has its application history for more than three decades, currently there is a much interest in the development of non-thermal plasma with their applications, specially the plasma discharge that can be generated at atmospheric pressure instead of conventional non thermal plasma at vacuum condition.

Much works have been done on the development of non-thermal plasma at atmospheric pressure plasma. Advantages both from economic and operational compared to the low pressure one have led to the variety development of non-thermal atmospheric pressure plasma as a recent trend of focus research [1]. Atmospheric pressure plasma is also generates the reactive species to modify the surface characteristics of material with the comparable result as it treats with low pressure plasma [2]. However, most of those studies were using radio frequency (RF) power supply [3-5] and microwave power supply [6, 7]. The present work deals the development of non-thermal plasma at atmospheric pressure by using a high voltage power supply (HVPS) and also discussing the application of the plasma system to modify the surface of PE film material. Development of this type of HVPS brings the advantage at which the ionization process occurred partially in the gas flow, so that the generated plasma discharge has relatively low temperature during operational. This type of plasma is suitable to be used to change the characteristic surface of particularly the thermal-sensitive material such as polymers or textile materials without any thermal damage to the irradiated surface.

Modification of the surface characteristics of the material by plasma techniques include the presence of strong interaction between the reactive species, electrons and ions formed in the plasma with the upper surface of the material. There are two things that will occur due to interaction of the material with the plasma, namely: plasma activation process, which usually involves a discharge of plasma without working gas that can form a polymer layer, such as argon gas, oxygen, nitrogen and others. The second one is the plasma polymerization process, in which the method of plasma treatment is using a working gas that can lead to polymerization process, such as gas sulfur hexafluoride (SF_6) gas tetrafluoromethane (CF_4) and others. Result of plasma polymerization is usually in the form of thin film solid formation the surface of the material. Both of two typical plasma processes mentioned above are a process that involves plasma with specific objectives to change the surface properties without changing its bulk characteristics, including the mechanical properties of the material. In this paper the development of home-made cold plasma generator and its system design at atmospheric pressure will be presented. Some characteristics of the generated plasma will be investigated and the plasma utilization in modifying the surface of polymer materials will also be discussed.

2. Experimental

2.1. Plasma generator

Atmospheric pressure plasma generator in this works was typical high voltage power supply, developed using 555 timer circuit coupled with ignition coil. Figure 1 shows the simplified block diagram of high voltage generator for generating the non-thermal plasma.

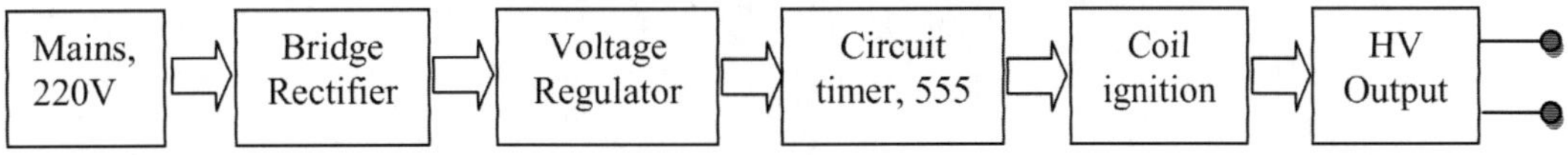

Figure 1. Block diagram of high voltage generator for atmospheric pressure plasma system.

The voltage waveform of plasma generator was measured by using Hantek DSO5072P digital oscilloscope and a high probe PD-28 with the attenuation ratio of 1000:1. Figure 2 displays the voltage waveform during plasma discharge by using Argon gas at the flow rate of 3 l/m. It displays the discharge voltage about 3 kV with a rise time 1.82 ms.

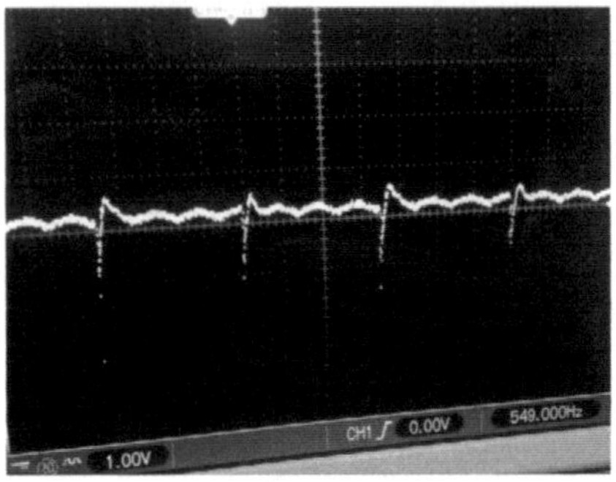

Figure 2. Voltage waveform during plasma discharge using 3 l/m Argon gas.

2.2. *Plasma treatment for surface modification*

Figure 3 shows a schematic diagram of experimental setup used for surface modification of polymeric material. High Voltage Power Supply (HVPS) connected to the SUS bar electrode which is surrounded by a glass tube. The other electrode of aluminum belt was placed at a distance of 7 mm from the edge of glass tube and connected to the ground. The stream of Argon gas from the opposite side was used for plasma generation and flowed through the glass tube at a rate of 3 liters per minute. For the purpose of exploring the application of this non thermal plasma system in modifying the surface of the polymer material, a 3 mm in thickness of polyethylene (PE) sheet was placed on the sample holder with a distance from plasma discharge nozzle as far as 2 mm. In order to verify the change of surface properties of PE sample, analysis of the water contact angle will be used in this study. A drop of water will be deposited on the surface of the material samples before and within a minute after plasma treatment. Contact angle will be measured from the water droplets by taking the image by using a digital microscope at a specific angle. Plasma treatment was conducted at applied voltage 3 kV by using argon gas at various period of time to know the effect of treatment time on PE wettability, which can be confirmed by measuring the decreasing of contact angle.

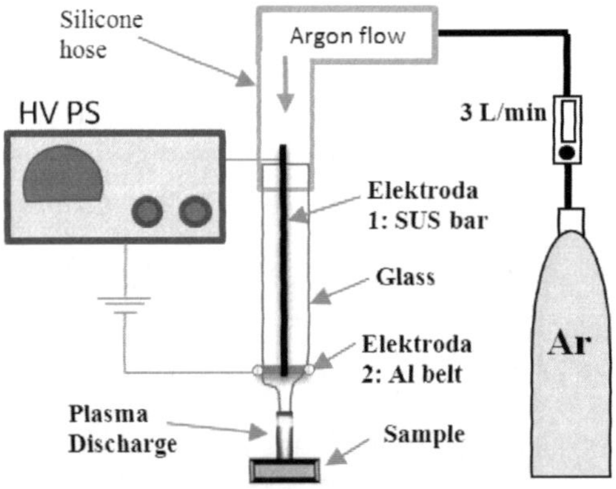

Figure 3. Schematic diagram of non thermal plasma system at atmospheric pressure for surface modification of polymeric material.

3. Result and discussion

A plasma system that operates at atmospheric pressure was successfully developed with the electrical energy source in the form of high voltage power supply is based on a vehicle ignition coil. Pictures of the plasma generated by the system in this study can be seen in figure 4 with the difference of working gas. The length of the plasma can reach as long as 6 mm with the usage of 3 l/m of working gas. The discharge appears in the uniforms phase (glow discharge) in helium based and seems to be in the form of filamentary when using argon gas. Due to more beneficial in cost and having better energy transfer efficiency than helium [8], argon gas will be used for the utilization of the plasma system in this study.

The operational temperature of the atmospheric pressure plasma can reach 35°C as measured after 120 seconds time of discharge and no significance increase observed afterward. This indicates that the characteristic of the resulting plasma is the type of non-thermal atmospheric pressure plasma. The plasma with this typical characteristic is very important for the treatment of materials with the purpose to modify the surface properties, especially for the surface treatment of the heat-sensitive material such as a polymeric material.

Application of the developed plasma discharge system above is proposed to modify the surface characteristic of polyethylene (PE). This material is preferred because PE as one of the most utilized in applications has excellent properties such as: low cost, low density, high flexibility and resistance to chemical liquids. However, the utilization of PE sometimes meets its limitation because PE also has a low surface energy properties that making PE has low adhesion properties and hydrophobic. Treatment with plasma technique is intended to improve wettability and adhesion properties of the PE surface as a potential replacement for traditional treatment that using chemical solution in a very unfriendly environment. Surface modified PE is generally known to bring the advantages into the wide area of practical application, such as: increasing its performance as a material for printing/painting in the packaging industries, enhancing the adhesion/bond ability of glue or coatings and improving the biocompatibility in the artificial components in biomedical applications [9, 10]. The surface modification of the polyethylene in this study was conducted from 10 to 60 seconds with atmospheric pressure plasma treatment. After that, a droplet of water from a pipette was dropped on the surface of the material, and the water contact angle was measured. The contact angle of plasma treated PE would be then compared with the non-treated or original one.

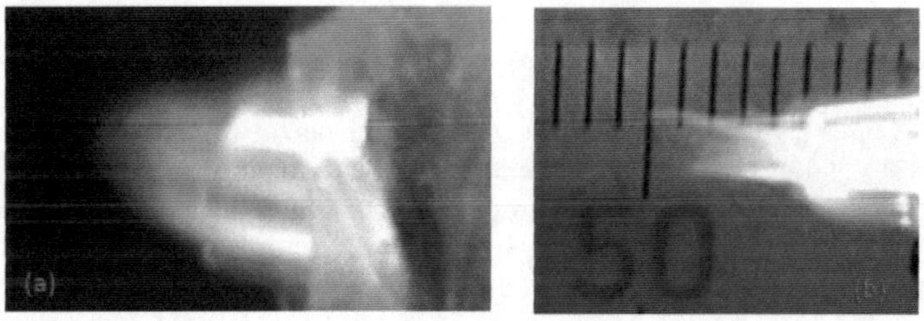

Figure 4. (a) Picture of plasma in Helium and (b) plasma discharge with Argon gas.

Figure 5 shows a picture of a drop of water beads on the surface of PE before (a) and after (b) treatment of argon plasma at atmospheric pressure. Visible differences in the contact angle as seen in that figure indicated by the decreasing of contact angle of water bead from 90° to 55° after plasma treatment for 20 seconds. This is also as a confirmation of the increase of wettability properties of the modified surface of PE.

IOP Conf. Series: Journal of Physics: Conf. Series **817** (2017) 012069 doi:10.1088/1742-6596/817/1/012069

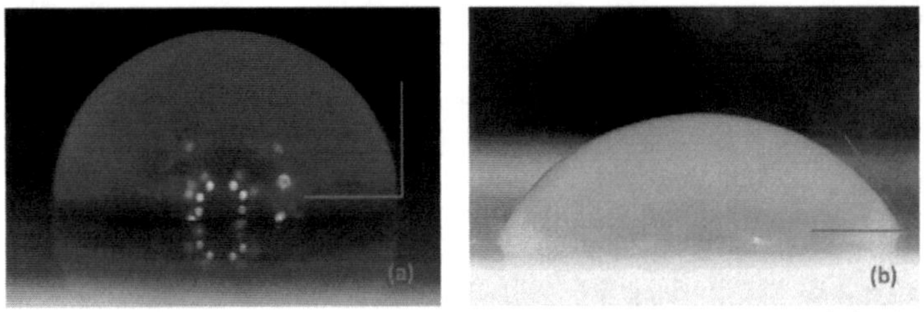

Figure 5. Contact angle of water bead on the surface of PE (a) original, and (b) after 20 seconds of plasma treatment. (Argon gas flow: 3 l/m, applied voltage 3 kV, treatment time: 20 s)

Figure 6 shows a graph of the decreasing in contact angle after various argon plasma treatment times at atmospheric pressure. From that figure, it can be seen that the plasma treatment for 60 seconds can make changes in the characteristics of the hydrophobic surface becomes hydrophilic PE (like water) to decrease the contact angle of 90° in the PE becomes 38.5°. From the graph above also shows that the first 20 seconds of plasma treatment took quite significant effect on a decreasing in contact angle, followed by a relatively stable decrease in the contact angle by addition of plasma treatment time.

The transformation in surface properties of the naturally hydrophobic PE into hydrophilic through contact angle measurements can be explained as follows: plasma discharge containing positive ions, electrons, free radicals and other reactive elements will bombard the surface of PE with each energy possessed by the plasma elements. The reaction that occurs between the plasma and the material causes the PE surface become activated. When the plasma discharge is drop off, the interaction between activated PE surface with the ambient air or atmosphere led to the creation of new polar functional groups such as: C=O, OH, COOH and other oxygen containing function groups on the surface. Those new polar groups provide stronger bond with water droplets (increasing of the wettability). The more time needed to expose the PE surface with the plasma discharge, the more oxygen functional groups will be generated. In consequence, the PE surface become more polar and make the water contact angle turns into the decreasing, indicated by the level of the water droplets on the surface of the material become wider. This is how the transformation process of PE surface has been occurred from hydrophobic into the hydrophilic. This finding is in accordance with the results of Kun et al [11] when treating the PE surface with coplanar barrier discharge. In that case, dramatic decrease of water contact angle was happened during the first 20 seconds and found the new groups of COO-, C=O and –OH through measurement by means of FTIR.

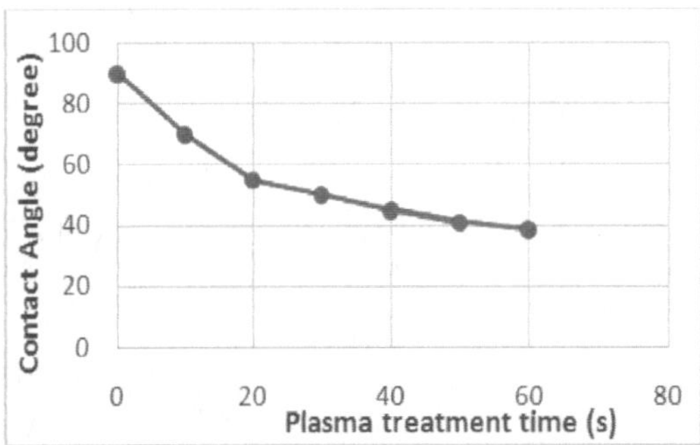

Figure 6. The graph of decreasing in contact angle by increasing plasma treatment time.

4. Conclusion

An atmospheric pressure plasma system has been developed using ignition coil-based high voltage power supply. This system can sustain the plasma discharge either by using helium and argon as working gas. It was found that the temperature of the discharge can achieve around 35°C even in the long run operation and this type of non-thermal plasma may suit for application in modifying the surface of thermal-sensitive material like PE polymer without altering its bulk properties. In attempt on exploring the utilization of the plasma system for modifying PE surface, the results showed that the PE surface can be altered into hydrophilic phase by treating it with the plasma for 60 seconds. This has been confirmed by the decreasing its contact angle from 90° to 38.5°. The present results may provide the promising application in modifying surface of thermal-sensitive material in a very short time of treatment.

References

[1] Nehra V, Kumar A and Dwivedi H K 2008 *Int. J. Eng.* **2** 53-68
[2] Noeske M, Degenhardt J, Strudhoff S and Lommattzsch U 2004 *Int. J. Adhes. Adhes.* **24** 171–7
[3] Chun H, Wei C M, Tsai C Y, Hou W T, and Juang R S 2013 *Surf. Coat. Technol.* **231** 42-6
[4] Li Y P, Zhang Z C, Shi W and Lei M K 2014 *Surf. Coat. Technol. A* **259** 77-82
[5] Igor N, Anton P, Zdeno S, Matej M, Maria O, Marian V, Jan S, Ivica J, Angela K and Miroslav S 2015 *Vacuum* **119** 88-94
[6] Mengyao Y, Baoming Z, Kunyan J, Xiaoming Q, Zhiwei X, Kunyue T, Lihuan Z, Jiajun W and Yanan J 2015 *Appl. Surf. Sci.* **327** 93-9
[7] da Maia J V, Pereira F P, Dutra J C N, Mello S A C, Becerra E A O, Massi M and Sobrinho A S S 2013 *Appl. Surf. Sci. B* **285** 918-26
[8] Kasih T P, Kuroda S and Kubota H 2007 *Plasma Processes Polym.* **4** 648-53
[9] Kazimi M RAhmed I and Faizal C K M 2011 *Proc. Int. Conf. on Chemical Innovation (Terengganu)* (Malaysia: Terengganu/ TATI University College) pp 203-6
[10] Pandiyaraj K N, Deshmukh R R, Ruzybayev I, Shah S I, Pi-Guey S, Halleluyah Jr M and Halim A S 2014 *Appl. Surf. Sci.* **307** 109-19
[11] Kun W, Jian L and Shi-qing W 2012 Proc. *2nd Int. Conf. on Electronic & Mechanical Engineering and Information Technology (Liaoning)* (France: Paris/Atlantis Press) pp 1038-42

IOP Conf. Series: Journal of Physics: Conf. Series **817** (2017) 012070 doi:10.1088/1742-6596/817/1/012070

X-ray Photoelectron Spectroscopy (XPS) analysis of low oxidation pressure behavior of FeAl –alloys at 600°C and 700°C

H Izzuddin[1, 2], **A Seemayer**[1], **G N Ankah**[1], **J Lengsfeld**[1], **P Schneider**[1] **and F U Renner** [1, 3]

[1]Max-Planck-Institut für Eisenforschung, Max-Planck-Straße 1, 40237Düsseldorf, Germany
[2]Research Center for Physics, Indonesian Insitute of Sciences, Komplek Puspiptek Serpong, Tangerang Selatan 14314, Indonesia
[3]Institute for Materials Research (IMO), Hasselt University and Associated Lab IMEC, Division IMOMEC, Wetenschapspark 1, 3590 Diepenbeek, Belgium

E-mail: hubby.izzuddin@lipi.go.id

Abstract. The low oxidation pressure behavior of binary FeAl alloys has been examined at different temperatures of 600-700°C. Polycrystalline and single-crystalline samples of Fe-10Al and Fe-25Al were oxidized for 1 h at 10-6 mbar of oxygen in an ultra-high-vacuum chamber (UHV). Prior oxidation, samples were prepared by sputter-annealing cycles in UHV. After oxidation, the samples were analyzed by X-ray Photoelectron Spectroscopy (XPS). The formed oxide layers on the surfaces consist of $FeAl_2O_4$ and Al_2O_3 showing that Al has a dominating role during the oxidation process. From XPS peaks, the composition of the Fe peaks of the oxide layer consists of Fe metal, FeAl-alloys, and FeO. Meanwhile, the composition of the Al peak is changed from Al metal, FeAl-alloys and Al_2O_3 at 600°C to Al metal, FeAl-alloys and $FeAl_2O_4$ on 700°C. From XPS sputter depth profiles, it is shown that the oxide layer was growing with increasing temperature, and the single-crystalline Fe-10Al has a thicker formed oxide layer than poly-crystalline Fe-25Al with same oxidation treatment.

1. Introduction

Iron aluminides has been applied mostly for structural applications with operational temperature between 600-800°C [1]. Oxidation on iron aluminium at 500-600°C produces oxide layers consist of α-Fe2O3, Fe_3O_4, and $FeAl_2O_4$. In this state, oxidation is dominantly controlled by diffusion of Fe ions. Above this temperature range, Fe_3O_4, and $FeAl_2O_4$ will disappear to become γ-Al_2O_3 until 900°C. Oxidation above 900°C will lead to formation of α-Al_2O_3 [2-4].

There are so many publications about oxidation study of iron aluminides in high pressures, but only a few were conducted in low oxidation pressures particularly with x-ray photoelectron spectroscopy (XPS) analysis [5-7]. Therefore, in this article, we report a study of x-ray photoelectron spectroscopy (XPS) analysis during oxidation of FeAl-alloys on 600-700°C with low base pressure system in order to understand oxidation growth mechanism.

2. Experimental Procedure

For this experiment, a poly crystalline and a single crystal of iron aluminides with compositions of 25 at.% Al (Fe-25Al) and 10 at.% Al (Fe-10Al) were prepared respectively. The polycrystalline Fe-25Al samples were produced by vacuum induction melting in Al_2O_3 crucibles, using elemental components with a purity higher than 99.9% and cold Cu moulds for solidification process in order to obtain 10 mm diameter of rods. The samples were cut to pieces with 8 mm diameter and 3 mm thickness using spark erosion technique. Then, the samples were mechanically grinded and polished up to 0.05 μm surface roughness with diamond paste and silica (SiO_2) suspension for the final polishing. Meanwhile, the single crystal sample Fe-10Al (111) was prepared by mechanically polished down to 0.03 μm roughness and oriented with an accuracy of < 0.1° (Surface Preparation Laboratory, Netherlands).

The oxidation tests were conducted on 600-700°C for 1 hour in an UHV chamber by flowing pure oxygen (O_2) gas into the chamber with base pressure 10^{-5}-10^{-6} mbar. Before oxidation, all the samples were cleaned by several cycles of sputtering using Ar+ ions (Energy E = 1.5 keV, sputter current I = 1.85 mA) and short annealing on 900°C.

The oxidized samples were characterized using X-ray photoelectron spectroscopy (XPS) (Physical Electronics Quantum 2000) in order to study the element composition and chemical state of resulted oxide layer on the sample surface. On each sample, survey spectra and sputter depth profiles were taken using a monochromatic Al Kα X-ray source, source analyzer angle of 45°, step size of 0.8 eV, pass energy of 93.90 eV and a number of sweeps in the range of 8 to 25. Survey spectra were recorded within the binding energy range of 0-1350 eV, and sputter depth profiles with 0.5-1 keV Ar^+ ions were measured of Fe 2p, Al 2p and O 1s photo lines. An oxidized silicon waver of known oxide thickness was used for sputter time calibration. Then, this calibrated sputter time with SiO2 was used for calculating the depth of sputter profile. In order to analyse the XPS data, the Casa-XPS software was used including sensitivity factors of 2.14469 for Fe 2p, 0.301653 for Al 2p and 0.733 for O 1s. The resulted XPS spectrum of each element was calibrated first with C 1s = 285 eV before data analyses begins.

3. Results and Discussion

Figure 1 shows Auger Electron (AES) spectra of sample surface, as received from the company, after some cycles of surface cleaning using sputtering-annealing method and after oxidation test. The sputtering-annealing is usually able to remove all contaminant and oxidized parts on the surface. But, during this research, even with prolonging sputter time and increasing annealing temperature, the sample surface cannot be perfectly cleaned particularly from oxygen contamination due to CO adsorption which was still present in UHV pressures (10^{-6}-10^{-10} mbar), and high reactivity of Fe and Al elements with oxygen gas.

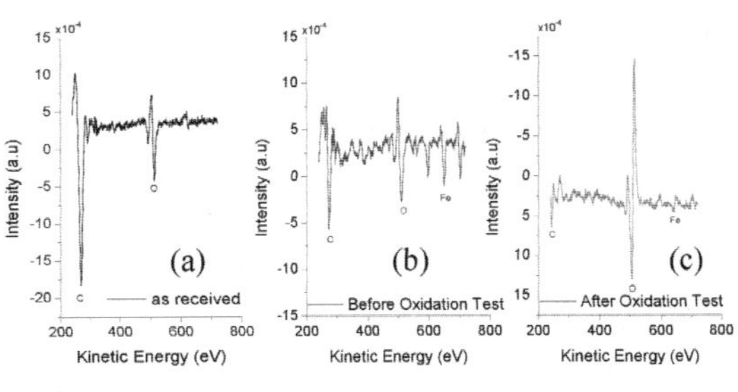

Figure 1: Auger Electron (AES) Spectra of Fe-10Al surface sample as received, after cleaning process with some cycles of prolonging sputtering-annealing, and after oxidation test.

IOP Conf. Series: Journal of Physics: Conf. Series **817** (2017) 012070 doi:10.1088/1742-6596/817/1/012070

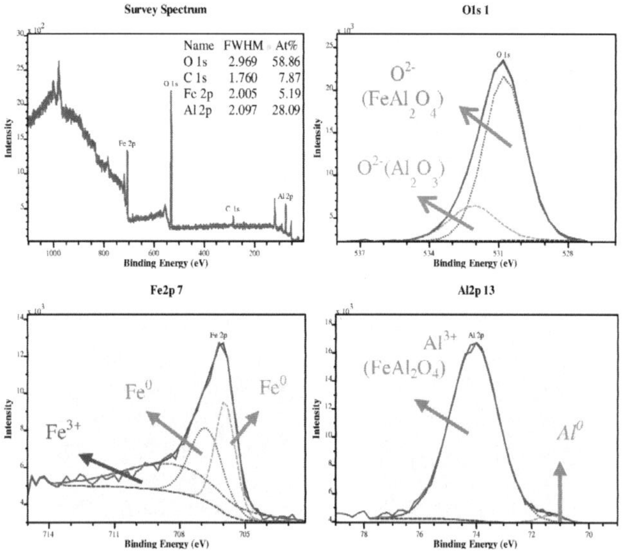

Figure 2: Surface spectrum and high resolution spectra of O, Fe and Al elements an oxidized Fe-10Al on 600°C for 1 hour.

From all survey spectrum in figure 2 and 3, with the same condition of oxidation process at 600°C and 1 hour, it shows that the Fe-10Al has the higher O content than that of Fe-25Al. With this result, the Fe-10Al has less oxidation resistance than Fe-25Al. Meanwhile, for the same composition and duration of the oxidation process, Fe-25Al produce more oxide layer after oxidation process at 600°C rather than after 700°C, as shown in figure 4.

Fe signals of oxidized Fe-10Al from oxidation on 600°C (in figure 2) show the metallic state of Fe still has higher intensity comparing with the oxide signals. But, the oxide signal has already grown enough. Meanwhile the Al spectrum has very high oxide peaks and almost no metallic Al peaks. This condition is supported from the spectrum of O peaks which consist of majority aluminium oxide compounds, $FeAl_2O_4$ and Al_2O.

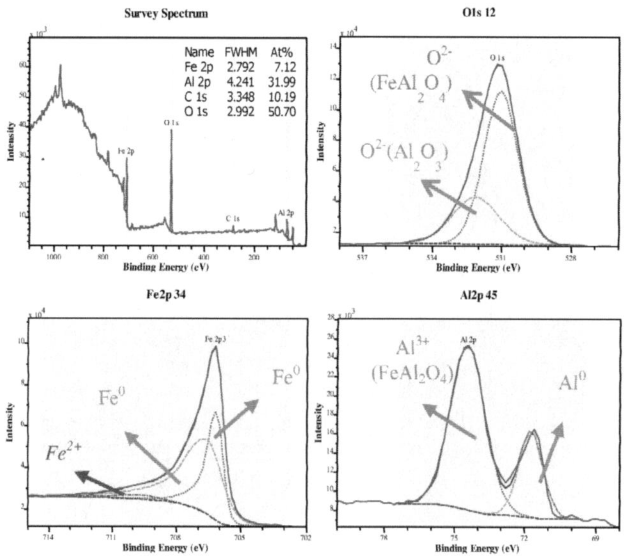

Figure 3: Surface spectrum and high resolution spectra of O, Fe and Al elements an oxidized Fe-25Al on 600°C for 1 hour.

IOP Conf. Series: Journal of Physics: Conf. Series **817** (2017) 012070 doi:10.1088/1742-6596/817/1/012070

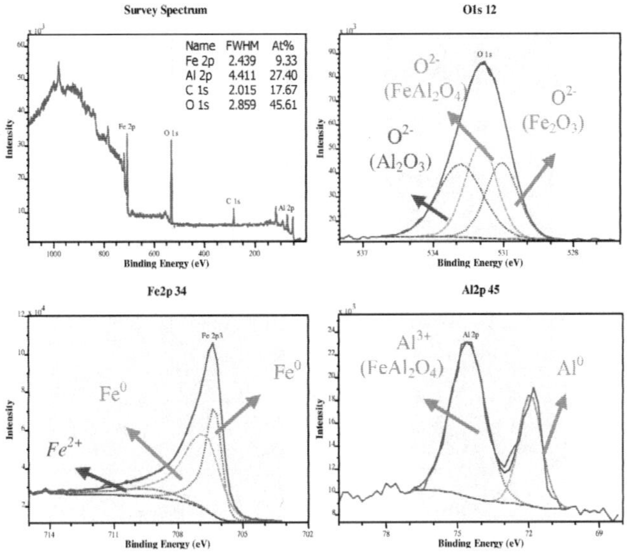

Figure 4: Surface spectrum and high resolution spectra of O, Fe and Al elements an oxidized Fe-25Al on 700°C for 1 hour.

Figure 3, Fe-25Al after oxidation on 600°C for 1 hour, it has the same characteristic of Fe, Al and O spectrum from the Fe-10Al results with the same oxidation condition, as shown in figure 2, but with lower Fe oxide peak on Fe spectrum and higher Al metallic peak on the Al spectrum. These high resolution spectra validate the survey spectra results (in figure 2 and 3) that Fe-25Al has a higher oxidation resistant than that of the Fe-10Al.

Meanwhile the high resolution spectra of the oxidized Fe-25Al on 700°C for 1 hour (in figure 4) show fewer Fe oxide signals on the Fe spectrum and higher Al metallic than oxidation results of Fe-25Al at 600°C for 1 hour. But, the interesting result comes from the O spectrum that shows trivalent oxide peaks with more iron oxide content.

The depth sputter profiles of all oxidized Fe-10Al and Fe-25Al on 600-700°C for 1 hour are shown in figure 5. The profile shown Fe-10Al with 600°C for 1 hour has the thickest oxide layers. Meanwhile, the Fe-25Al from 700°C for 1 hour has slight thicker oxide layer than from 600°C for 1 hour. This might be due to formed trivalent oxide peaks on the oxide layer.

Overall, from all XPS spectrum and depth sputter profiles of Fe-10Al and Fe25Al, the low oxidation pressures have limited the growth of iron oxide, thus aluminium become more dominant than the iron during oxidation process on higher elevated temperatures [3, 7, 8]. But, in the more oxygen pressure, such as in atmospheric pressure, the iron is more dominant during oxidation process [2-4]. These phenomena can be explained through Ellingham diagram [9]. On the same temperature, the formation of iron oxide, $FeO/Fe_2O_3/ Fe_3O_4$ requires more oxygen pressure (pO_2) than the formation of aluminum oxide. This experiment results also show no any difference of oxidation growth between the polycrystalline and single crystal samples during this experiment.

IOP Conf. Series: Journal of Physics: Conf. Series **817** (2017) 012070 doi:10.1088/1742-6596/817/1/012070

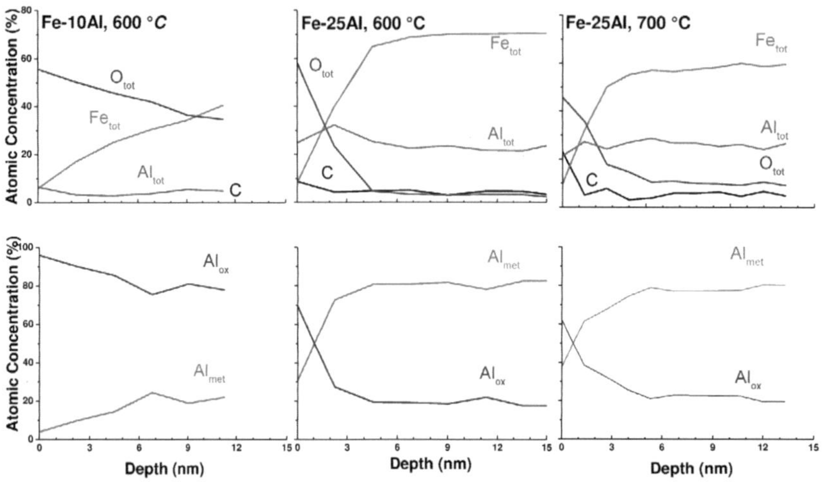

Figure 5. Sputter depth profile of oxidized Fe-25Al on 600-700°C for 1 hour respectively.

4. Conclusion

Fe-10Al has a lower oxidation resistance than Fe-25Al composition. The depth sputter profile shows that Fe-10Al produces the thickest oxide layer. Meanwhile, for the same composition, Fe-25Al from 700°C for 1 hour has thinner formed oxide layer than from 600°C for 1 hour, but with trivalent oxide formation. The most important thing from this study, the aluminium has the majority role during oxidation at 600-700°C for 1 hour with low pressure system.

References

[1] Palm M 2009 *International Journal of Materials Research (formerly Zeitschrift fuer Metallkunde)* **100** 277-87

[2] Hayashi S, Saeki I, Nishiyama Y, Doi T, Kyo S and Segawa M 2011 *Mater. Sci. Forum* **696** 63-9

[3] Kolarik V, Fietzek H, Engel W, Eltester B and Grabke H J 1997 *Mater. Sci. Forum* **251-254** 251-8

[4] Prescott R and Graham M J 1992 *Oxid. Met.* **38** 73-87

[5] Velon A and Olefjord I 2001 *Oxid. Met.* **56** 425-52

[6] Eltester B, Uebing C, Viefhaus H and Grabke H J 1997 *ECASIA '97* (Chichester: John Wiley & Sons)

[7] Ohtsu N, Nomura A, Oku M, Shishido T and Wagatsuma K 2008 *Appl. Surf. Sci.* **254** 5336-41

[8] Gao W, Li Z, Wu Z, Li S and He Y 2002 *Intermet.* **10** 263-70

[9] Ellingham H J T 1944 *Journal of the Society of Chemical Industry* **63** 125-60

IOP Conf. Series: Journal of Physics: Conf. Series **817** (2017) 012071 doi:10.1088/1742-6596/817/1/012071

Microstructure and phase composition of Fe-B-Al coatings on low carbon steel prepared by using mechanical alloying technique

R Y Sundawa, D Aryanto, A S Wismogroho and T Sudiro

Laboratory of High Temperature Materials and Coatings, Research Center for Physics, Indonesian Institute of Sciences (LIPI)
Kawasan PUSPIPTEK, Serpong, Tangerang Selatan 15314, Banten, Indonesia

Email: risma.sundawa@gmail.com

Abstract. In the present study, varying composition of FeB-Al was coated on low carbon steel by using mechanical alloying technique for 4 hours and followed by heat treatment in vacuum atmosphere of 5.9 Pa at 700 °C. The microstructure and formed phases of FeB-50 at.%Al, FeB-25 at.%Al, FeB-12.5 at.%Al and FeB coatings were intensively discussed. The cross sectional observation indicates that the coating thickness tends to increase with increasing Al content. Before heat treatment, the coatings are composed of FeB and Al phases, depending on coating composition.. After heat treatment, the intermetallic phases were identified. Interdifussion layer was also formed in the FeB-50 at.%Al coating after heat treatment.

1. Introduction

Carbon steels have been widely applied for numerous structural and constructional applications due to its good mechanical properties, easy fabrication, and low production cost. In spite of that, this material is susceptible to damage at high temperature because of its poor oxidation resistance. Various studies have been attempted to improve the oxidation resistance of carbon steel. Generally, there are two main methods to enhance the oxidation resistance at high temperature, namely alloy enrichment and coating deposition. In some cases, the second approach is more preferable than the first one because coating deposition can improves corrosion, oxidation, and wear resistance of carbon steel and maintain mechanical properties of steels [1-3].

Aluminum (Al) is one of important elements to increase the corrosion and oxidation resistance of alloys and coatings due to the fact that it can act as reservoir for the formation of a thin, compact and adherent oxide film of Al_2O_3 on the surface of materials. This oxide scale plays a significant role to retard the diffusion of oxygen to the alloy substrates [3-4]. Previous study reported that in order to improve the oxidation resistance of steel, the deposition of a thin Al layer on the steel surface and its heat treatment were carried out to form a surface layer of iron-aluminum (Fe-Al) intermetallic compounds. These have much higher melting point than Al (about 650 °C) in combination with high oxidation resistance [5]. The surface layer of Fe-Al intermetallic compounds can also be formed by applying Fe-Al coating on the surface of low carbon steel and followed by heat treatment in argon atmosphere at 600 °C [1]. Implantation of boron ions into metals and alloys results a significant increase in surface microhardness. This approach becomes an alternative method for diamond coated tools [6]. However, there are limited studies on Fe-B-Al system as a coating. The preliminary investigation before material testing at high temperature is needed to gain a better understanding on coating characteristics as microstructure and phase composition of the coating. Coating uniformity and coating-substrate compatibility can be observed through the microstructure characterization whereas the informations of formed phase are useful to predict whether the coating tends to be protective. Eventually, acquiring characteristics of coating before high temperature oxidation test is essential for coating development.

The mechanical alloying (MA) technique is a novel and potential method for surface coating on metals and alloys because this technique allows thick coating to be formed in a short period of time at room temperature and in an ambient atmosphere with minimum surface preparation before deposition [7, 8]. The fundamental principle of this process is that the collisions of ball and powder on the surface of the substrate lead to powder particles trapped on the surface and get cold welded. This continuous and repeated processes result in deposition of coating powder on the substrate [3, 7, 8]. Fe-Al, Cr-Al, TiN and Ti-Al coatings had been successfully deposited on the substrate of low carbon steel, stainless steel and Ti by using MA technique [2, 3, 7, 8]. Nevertheless, there has been no report about deposition of Fe-B-Al coatings on the substrate of low carbon steel prepared by using MA technique.

Furthermore, the previous MA studies typically used high purity powders as the sources of coating materials. Investigation on ferro-alloy lumps as deposited materials is rare. Although, these materials are more accessible and inexpensive.

In this present study, we focus on the microstructure and the phase identification of Fe-B-Al coatings with varying composition on the substrate of low carbon steel prepared by MA and followed by heat treatment. Ferro-boron (FeB) alloy lumps is used to substitute high purity powders as the source of FeB materials.

2. Experimental Methods

The samples and coatings preparation procedures used in this study were similar to that explained in the previous works [3, 9]. The commercial low carbon steel was used as substrate. Based on X-ray diffraction patern of the substrate (figure 1), it can be observed that the substrate is mainly composed by Fe. The steel plate was cut to rectangular specimens with a dimension of about 10 x 8 x 3 mm. Before coating, the surface of the specimen was polished mechanically using SiC papers for up to No. 1200 and ultrasonically cleaned in ethanol solution.

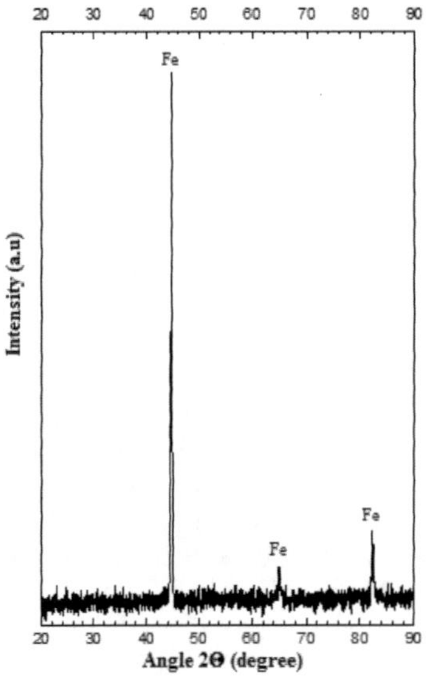

Figure 1. X-Ray Diffraction Pattern of Substrate.

IOP Conf. Series: Journal of Physics: Conf. Series **817** (2017) 012071 doi:10.1088/1742-6596/817/1/012071

In this recent work, for coating deposition, ferro boron (FeB) lumps and high purity (≥ 99%) Al powders were used as the starting materials. Chemical composition of FeB lumps is shown in table 1. FeB lumps were crushed into powders by using a high speed shaker mill for 2h. After FeB powder was prepared, FeB and Al powders with varying nominal compositions as 100FeB, 87.5FeB-12.5Al, 75FeB-25Al and 50FeB-50Al (in at%) were mixed for 1h with the ball to powder weight ratio of 5: 1. The total mass of the powder in each composition was 10 grams. Then three low carbon steel specimens were put and charged into vial. This mechanical alloying process was undertaken for 4h in ambient air atmosphere. The collisions between balls and powder on the surface of carbon steel result the deposition of coating powders on the surface of steel. The coated sample was then heat treated in vacuum of 5.9 Pa at temperature of 700 °C for 1h. Afterwards, the specimens were cooled down inside the furnace chamber to room temperature.

Table 1. Chemical composition of FeB lumps (wt.%)

C	Al	S	Si	P	B	Fe
0.5	0.2	0.01	0.5	0.05	20	Balance

The observation of deposited coating was done by using Optical Microscope. X-Ray Diffractometer (XRD SmartLab Rigaku) with Cu Kα radiation at 40 kV and 30 mA was used to indentify the phase compositions of the coatings before and after heat treatment. The coating microstructure was observed by using Scanning Electron Microscope equipped with energy dispersive X-ray spectrometer (SEM Hitachi SU3500-EDX).

3. Results and Discussions

3.1. Coating observations by optical microscope

The micrographs of cross sectional deposited coatings observed by using an optical microscope are shown in figure 2. According to the results, FeB-Al coatings are successfully deposited on the surface of low carbon steel by using MA method. The thickness of the coatings depends on the composition of deposited materials. The coating with highest Al content (FeB-50 at.%Al) results the thickest coating layer.

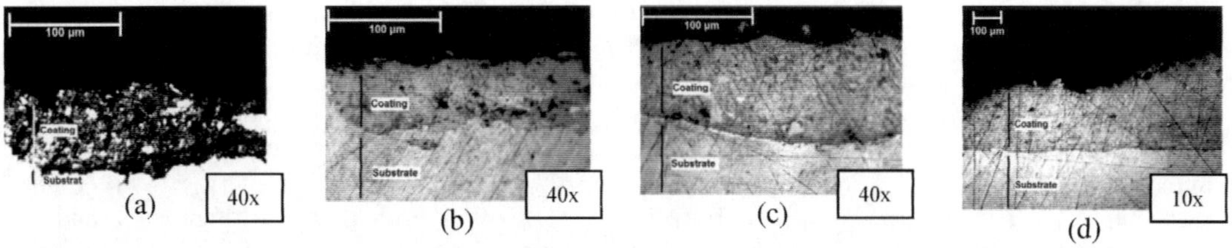

Figure 2. Optical micrographs of the cross section (a) FeB, (b) FeB-12.5 at.%Al, (c) FeB-25 at.%Al and (d) FeB-50 at.%Al coatings on low carbon steel substrate.

3.2. Phase composition of the coatings

Figures 3(a) and 3(b) show the X-Ray diffraction patterns of FeB-Al coatings with various composition before and after heat treatment at 700 °C for 1h.

As shown in figure 3(a), X-ray diffraction analysis of FeB coating shows the diffraction peaks of Fe and FeB. Meanwhile, the diffraction peaks of Fe, FeB and Al are detected in the coatings with various Al content. The diffraction peaks of FeB are detected from the substrate while FeB and Al from the coating. The peak intensity of Al diffraction increases with increasing Al content. The XRD characterization results show that the intermetallic compounds were not formed.

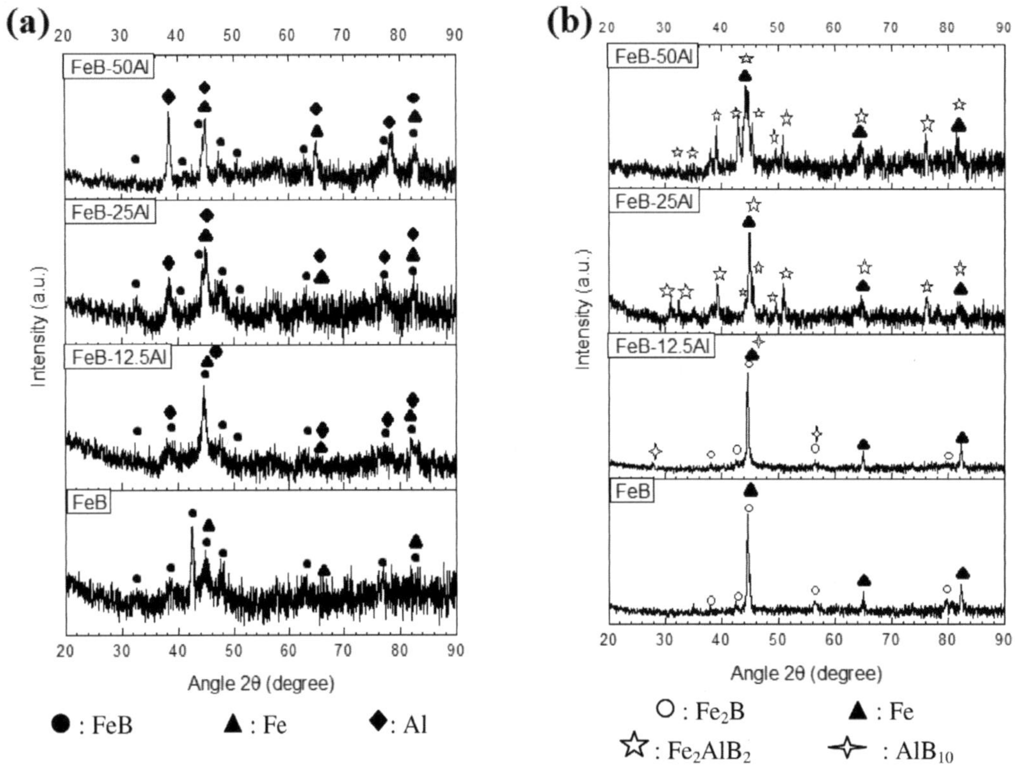

Figure 3. X-Ray diffraction patterns of FeB-Al coatings on low carbon steel substrate (a) before and (b) after heat treatment.

As showed in figure 3(b), new phases are identified for each different coating composition after heat treatment. Peaks Fe_2B diffraction are observed in the FeB-0 at.%Al and FeB-12.5 at.%Al coating. The stable compound of Fe_2B will remain stable in the presence of iron at low and elevated temperature [10]. Intermetallic phase of AlB_{10} is formed in the FeB-12.5 at.%Al coating while Fe_2AlB_2 intermetallic compound is formed in the FeB-25 at.%Al and FeB-50 at.%Al coating. The higher peak intensity indicates that crystal structure refinement was occured during heat treatment.

3.3. Cross section micrograph of the coatings

The cross-sectional observation of the coating, before and after heat treatment at 700 °C was undertaken for FeB-50 at.%Al and FeB-12.5 at.%Al coating composition. The cross-sectional micrographs of FeB-50 at.%Al and FeB-12.5 at.%Al coatings are shown in figures 4 and 5, respectively. According to the cross-sectional SEM images, the thickness of FeB-50 at.%Al and FeB-12.5 at.%Al coatings are about 162 μm and 42 μm, respectively. It can be observed that higher composition of Al results thicker coating. This phenomena could be caused by the density of Al (2.7 g/cm^3) which is lower than the density of FeB (3.3 g/cm^3) so that in the similar mass, Al has higher volume than FeB. In this experiment, the total mass of all compositions are fixed. Therefore, coating with higher composition of Al has higher and results thicker coating. It also can be clearly observed that there is no crack at coating-substrate interface.

Based on the result of EDX point analysis, the light gray particles in the figure 4(a) are composed of FeB and the dark grey matrix is composed mainly of Al. It can be seen that before heat treatment, FeB particles are dispersed in the Al matrix. This means that the particles of Al and FeB are still coexist and do not form intermetallic phases. This result has a good agreement with XRD analysis which shows that mechanical alloying does not lead to the formation of intermetallic compounds. Grey particles, that are located on the surface of substrate, have bigger size than that are located near the surface of coating because particle size of coating powder decreases with increasing milling time while the coating layer on the surface of substrate is deposited in the earlier stage of milling.

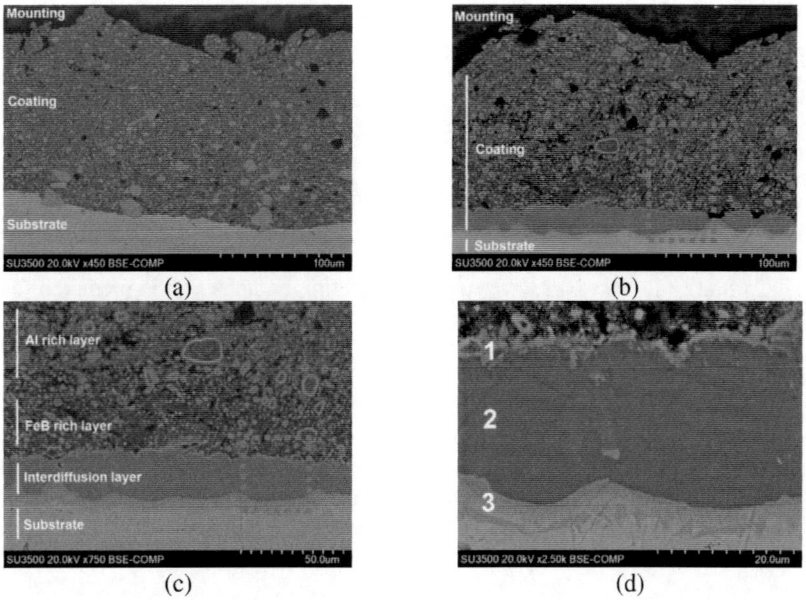

Figure 4. Cross-sectional micrographs of FeB-50 at.%Al coating on low carbon steel substrate (a) before, (b) after heat treatment, (c) and (d) higher maginification of squared area in figure 4(b) and 4(c), respectively.

Figure 5. Cross-sectional micrographs of FeB-12.5 at.%Al coating on low carbon steel substrate (a) before and (b) after heat treatment.

Meanwhile, after heat treatment, three coating layers are observed on the substrate (figure 4(b)). According to the results of EDX analysis, Al elements predominantly compose the upper layer of the coating while Fe and B elements predominantly compose a layer below it. At the interface between FeB rich layer and substrate, there is interdifussion layer which is composed mainly of Fe and Al elements. The formation of FeB rich and FeAl interdifussion layers is attributed to Al element diffusion to the steel substrate. The thickness of Al rich, FeB rich and interdifussion layers are around 106 μm, 32 μm and 24 μm, respectively.

Further detail characterization was carried out in the interdiffusion layer. At higher magnification, it can be observed that actually this layer consist of three different layers. The results of EDX point analysis at points 1, 2 and 3 shows that those three interdiffusion layers are consisted of 51.4 at.%Fe - 48.6 at.%Al; 28.6 at.%Fe - 71.4at.%Al; and 33.99at.%Fe – 66.01 at.%Al, respectively. The compositions are closed to atomic ratio of FeAl, Fe_2Al_5 and $FeAl_2$ intermetallic phases. This strongly suggests that intermetallic layers are formed in the layers number 1, 2 and 3, respectively (figure 4(d)). These intermetallic compounds cannot be identified by XRD analysis due to the fact that X-ray cannot penetrate to reach the interdiffusion area which is located around 138 μm in depth from the coating surface.

Although microstructure of FeB-12.5 at.%Al coating before heat treatmeant is quite similar with FeB-50 at.%Al coating, lower composition of Al leads to the formation of thinner coating (figure 5(a)). Meanwhile, after heat treatment, the interdiffusion layer, which is formed in the FeB-50 at.%Al coating composition, is not formed in the FeB-12.5 at.%Al coating composition. According to the EDX point analysis, the coating composition of around 39 μm in depth from surface is consisted of 7.69 at.% Al, 72.08 at.% B, and 20.23 at.% Fe. This coating composition is predicted to form AlB_{10} intermetallic compound containing Fe which is also identified from XRD characterization.

4. Conclusion

Various composition of FeB-Al coatings are successfully deposited on the substrate of low carbon steel by using mechanical alloying technique followed by heat treatment. Mechanical alloying does not lead to the formation of intermetallic compounds. The microstructure and phase composition of FeB-Al coating depended on the composition of coating. Higher composition of Al in FeB-Al coating results thicker coating. Interdiffusion layer, which is composed of intermetallic phases, is formed in FeB-50 at.%Al coating after heat treatment.

Acknowledgements

The authors are grateful to Research Center for Physics, Indonesian Institute of Sciences (LIPI) Unggulan Programme for financial support of this research.

References

[1] Seidu S O and Kutelu B J 2013 *J. Min. & Mat. Char. & Eng.* **1** 95-100
[2] Canakci A, Erdemir F, Varol T and Ozkaya S 2013 *J. Powd. Tech.* **247** 24-9
[3] Sudiro T, Sebayang P, Aryanto D, Hia A I J and Sebayang K 2015 *J. Tekn. Indonesia* **38**(3) 156-62
[4] Schweitzer P A 2007 *Fundamentals of Metallic Corrosion: Atmospheric and Media Corrosion* (Boca Raton: CRC Press) p 5
[5] Gellings P J 2005 *Introduction to Corrosion Prevention and Control* (Enschede: University of Twente Press) p 101
[6] Gogotsi Y G and Andrievski R A 1999 *Materials Science of Carbides, Nitrides and Borides* (Springer) p 270
[7] Romankov S, Komarov S V, Vdovichenko E, Hayasaka Y, Hayashi N, Kaloshkin S D and Kasai E 2009 *Int. J. Metals & Hard Mat.* **27** 492-7
[8] Romankov S, Mamaeva A, Kaloshkin S D and Komarov S V 2007 *Mat. Let.* **61** 5288-91
[9] Hia A I J, Sudiro T, Aryanto D and Sebayang K 2015 *J. Phys. Conf.: Ser.* **739** 012131
[10] Lindsay T G and Tough G S 1985 *Proc. 1st Conf. on Production Research (Nottingham)* (New York: Springer) p 340

IOP Conf. Series: Journal of Physics: Conf. Series **817** (2017) 012072 doi:10.1088/1742-6596/817/1/012072

The magnetotelluric phase tensor analysis of the Sembalun-Propok area, West Nusa Tenggara, Indonesia

F Febriani[1], D S Widarto[2], E Gaffar[3], A Nasution[4], and H Grandis[5]

[1]Research Center for Physics, Indonesian Institute of Sciences, Kompleks Puspiptek Bldg. 440-442, Serpong, Banten, Indonesia 15314

[2]Upstream Technology Center, PT Pertamina, Jakarta, Indonesia 10110

[3]Research Center for Geotechnology, Indonesian Institute of Sciences, Jl Sangkuriang, Kompleks LIPI Bldg. 70, Bandung, Indonesia 40135

[4]Department of Geology Engineering, Bandung Institute of Technology, Bandung, Indonesia 40312

[5]Department of Geophysical Engineering, Bandung Institute of Technology, Bandung, Indonesia 40132

Email: febt002@lipi.go.id

Abstract. The subsurface structure of the Sembalun-Propok area, NTB, Indonesia, has been investigated using magnetotelluric method (MT). To obtain the information of the dimensionality of the regional structure and determine the regional strike of the study area, the phase tensor analysis has been performed in this study. The results show that most of the skew angle values (β) are distributed within ± 5°. It indicates that the regional structure of the study area can be assumed as two dimensional. In addition, to determine the regional strike of the study area, we also calculated the major axes of the phase tensor. The result presents that the regional strike of the study area is about N330°E. According to the results of the phase tensor analysis, we rotated the impedance tensor to N330°E and performed 2-D inversion modeling. The result presents that the substructure model suits with the geological background of the study area.

1. Introduction

One of the geophysical methods which are useful in geothermal resources investigations is the magnetotelluric (MT) method. The MT is a passive-surface electromagnetic technique to investigate the electrical resistivity structure from the near surface to the upper mantle [1]. The MT method can investigate the lower resistivity anomalies associated with the geothermal resources since they commonly have relatively lower resistivity than initial rocks. Variation in the resistivity is related to the water abundance, temperature, and degree of mineralization [2].

The Sembalun-Propok area is located at the eastern part of Lombok Island, West Nusa Tenggara Province. Geologically, its regional structures are dominated by the normal and strike-slip faults which have the direction of NW-SE [3].

In subsurface structure modeling using MT method, information about dimensionality and regional strike of the study area are important. Without this information, we cannot choose that either 1-D, 2-D or 3-D condition for a study area as the best assumption of the MT data modeling. To obtain this information, one of the methods is the phase tensor analysis which has been proposed by Caldwell (2004). The calculation of the phase tensor does not require an assumption about the dimensionality of the underlying conductivity distribution and is applicable where both the heterogeneity and the regional conductivity structures are 3-D [4]. The method has been applied at the volcanic systems, the geothermal fields, and the fault zone [5-8].

In this study, we analyze the MT data by using the phase tensor analysis to investigate dimensionality and regional strike of the Sembalun-Propok area, West Nusa Tenggara, Indonesia.

2. Observation and Analysis of Audio-Magnetotelluric (AMT) Data

This study was carried out in the Sembalun-Propok area, East Lombok Regency, West Nusa Tenggara Province. Figures 1 (a) and (b) show the topography maps of the Lombok Island and 38 AMT observation points. In AMT exploration, time-series data of electric and magnetic fields were collected by MTU-5A manufactured by Phoenix Geophysics Ltd., in 2007. Duration of the data collection was about a few hours at each site. The data were transformed to power spectra by Fourier Transform, in a frequency range of 0.001 Hz-1000 Hz. All observed data were processed as a single site since we did not have adequate remote reference data.

In this study, we applied the MT phase tensor to obtain information about the dimensionality of the regional structure and to determine the regional strike of the study area [4]. The MT phase tensor can be written in the form of the singular value decomposition (SVD) of a square matrix [4], as follows:

$$\Phi = R^T(\alpha-\beta)\begin{bmatrix}\Phi_{max} & 0 \\ 0 & \Phi_{min}\end{bmatrix}R(\alpha+\beta) \tag{1}$$

R^T is the transposed or inverse rotation matrix. α and β can be expressed by

$$\alpha = \frac{1}{2}\arctan\left(\frac{\Phi_{12}+\Phi_{21}}{\Phi_{11}+\Phi_{22}}\right), \beta = \frac{1}{2}\arctan\left(\frac{\Phi_{12}-\Phi_{21}}{\Phi_{11}+\Phi_{22}}\right) \tag{2}$$

The angle of α expresses the tensor dependence's on the coordinate system and with the three coordinate invariants completely defines the tensor. The angle of β is called the skew angle and can be thought of as rotation and is a measure of the tensor asymmetry. The angle of β depends on the tensor's skew (Φ_{12}-Φ_{21}), which is invariant under rotation but changes sign if the coordinate system is reflected [4].

Graphically, the MT phase tensor can be represented as an ellipse where the major Φ_{max} and minor axes Φ_{min} of the ellipse depict the principal axes and values of the tensor with the orientation of the major axes specified by the angle (α-β). In general, where the regional conductivity distribution is 3-D, the skew angle β is non-zero. If the phase tensor is symmetric $\beta = 0$, this situation occurs where the regional conductivity distribution is 1-D or 2-D [4]. The information of the dimensionality and regional strike are very important as the preliminary analysis in 1-D, 2-D or 3-D MT data modeling.

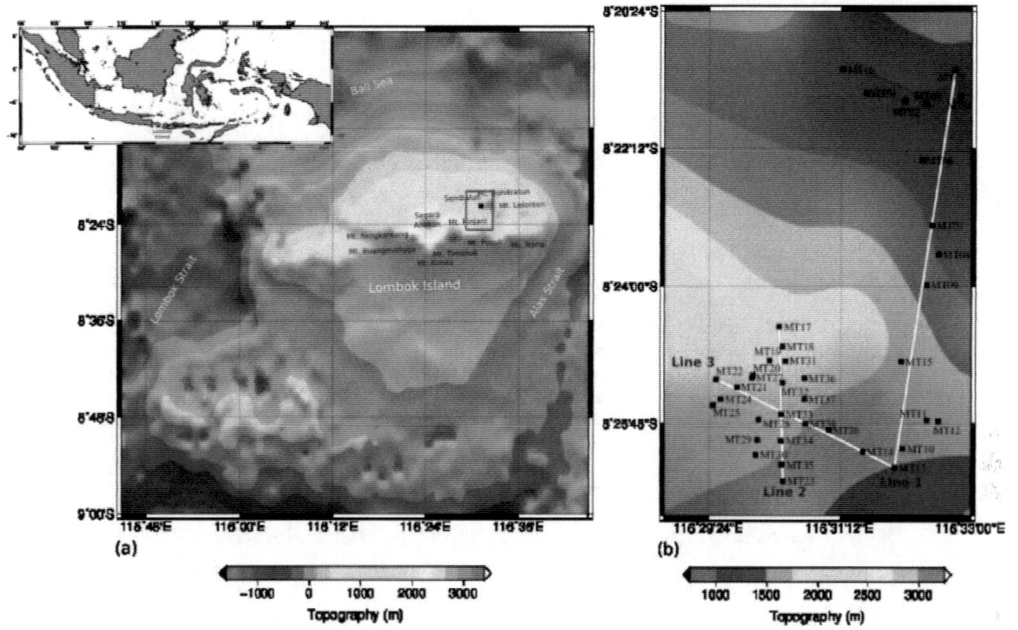

Figure 1. (a) The topography map of the Lombok Island. The red-colored rectangle in the topography map is the study area. (b) The topography map of the study area with the 38 AMT observation points. The white-colored lines are the profiles which will be analyzed in the further analysis.

3. Results and Discussion

To obtain information about the dimensionality of the regional structure, the MT phase tensor was applied on the three profiles, as presented by the white lines on the topography map in figure 1(b). The dimensionality, D, is indicated by the skew angle of the phase tensor (β value) [4]. For idealized condition, D = 1 or 2 when $\beta = 0$.

Figure 2 shows the skew angle β plot of the observation points along the Line 1, Line 2, and Line 3 for all frequency ranges. The results show that most of the β values are distributed within $\pm 5°$ for frequency higher than 0.1 Hz. For lower frequency (0.001 Hz-0.1 Hz), the β values are distributed in $-10° \leq \beta \geq 10°$. This result confirms that the resistivity structure can be assumed as 1-D or 2-D for the higher frequency. Therefore, it is safe to perform 2-D modeling for the MT data with frequency range of 0.1 Hz - 1,000 Hz.

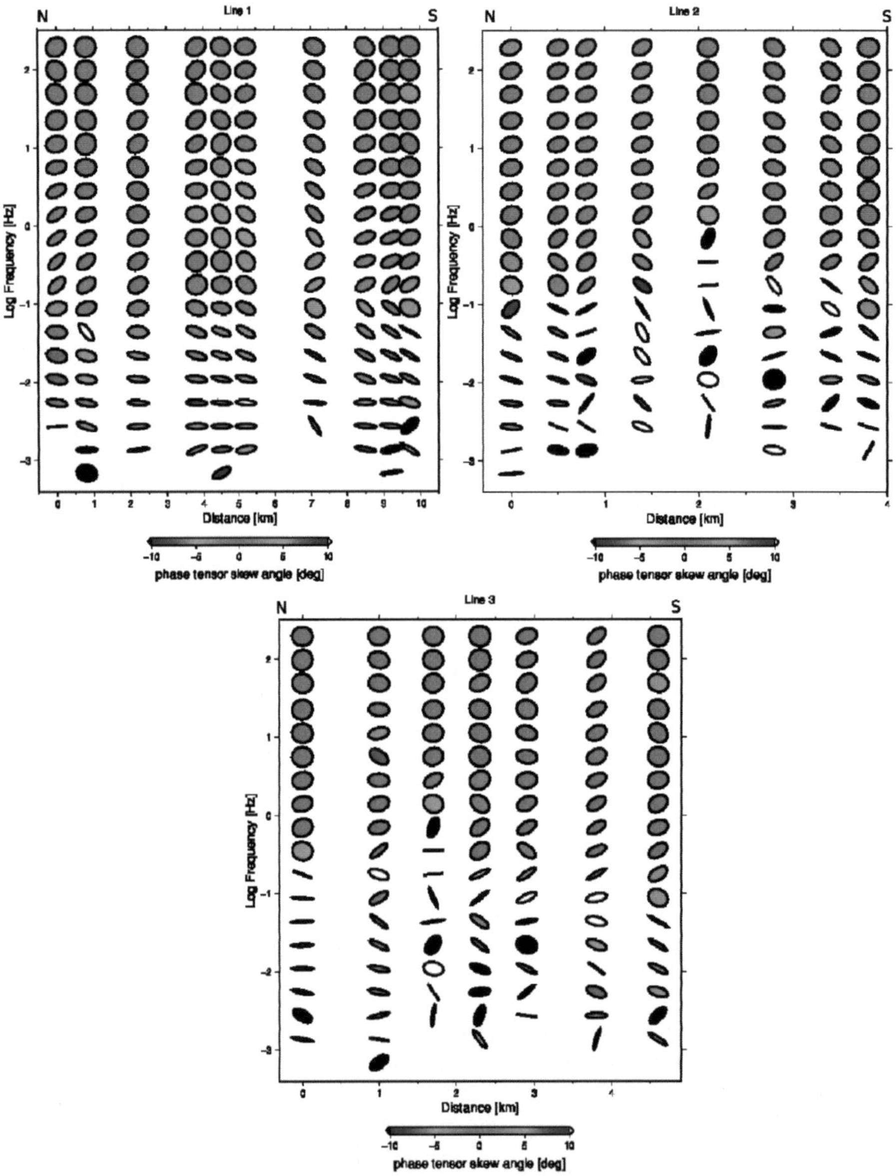

Figure 2. The skew angle β plot of the observation points along the Line 1, Line 2, and Line 3 for all frequency ranges. The horizontal axes indicate the distance between the MT observation points along the Line 1, Line 2, and Line 3. The vertical axes indicate the frequency (in log scale). The color scale indicates the skew angle β, which indicates the dimensionality.

In the next step, we estimated the regional strike of the study area by calculating the major axes of the phase tensor (α − β). Figure 3 shows the rose diagram of the (α − β) value for all frequency ranges. The estimated values have 90° ambiguity. The results of the phase tensor major axes agree well with the general regional strike. It shows that the dominated strikes are NW-SE and NE-SW directions for all frequency ranges. Using this information, we applied 2-D inversion modeling of the MT data using Ogawa and Uchida's code [9]. In this paper, we chose the Line 2 data that has frequency range of 0.1 Hz - 1,000 Hz for the modeling. Since we assumed the regional strike of the study area is NW-SE, we selected N330°E as the best value of the regional strike and rotated the impedance tensor of Line 2 to this direction.

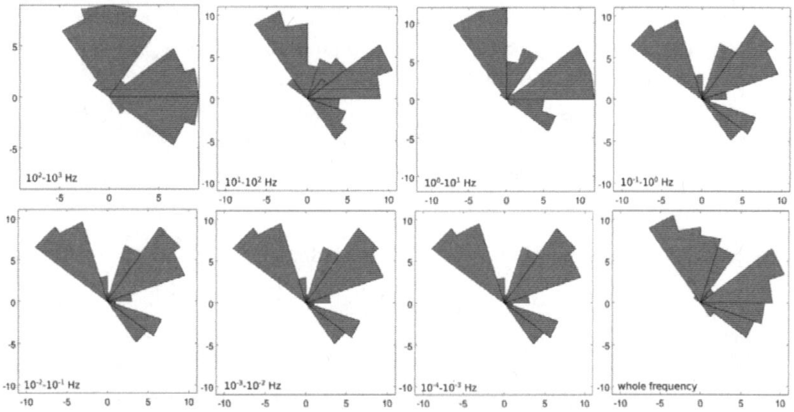

Figure 3. The rose diagram that presents the regional strike determined by the phase tensor for all sites.

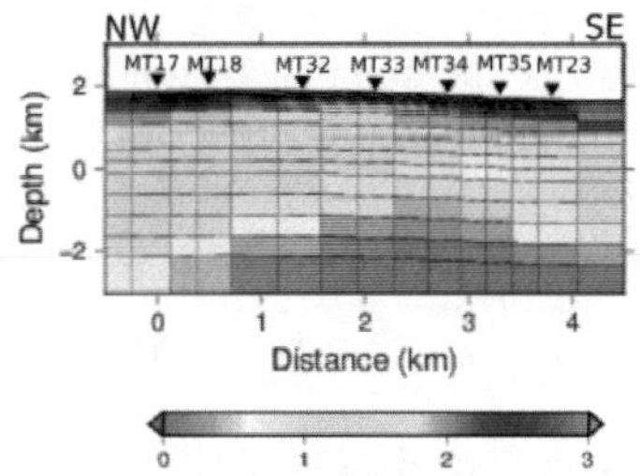

Figure 4. The result of the 2D inversion modeling of the MT observation points along the Line 2. The labels on the triangles indicate the observation sites that correspond to the Figure 1.

Figure 4 shows the result of the 2D inversion modeling of the MT data along the Line 2. The near surface along the Line 2 is dominated by high resistivity layer (> 500 Ωm). It is highly associated with the surface geology of the study area, which is characterized by the Old Quaternary volcanic of Sembalun, the permeable rocks of Sembalun-Propok and the Young Quaternary Propok-Rinjani volcanic [10]. The products mostly consist of andesitic to dacitic rocks of a calc-alkaline suite [10]. Below the resistive layer at the near surface, along the Line 2 the inversion modeling shows the layer with low-moderate resistivity layer (5 - 80 Ωm). Low - moderate resistivity zone is down to 4 km deep in the northwest and southeast regions and to 2 km deep below MT33 - MT35. It is possibly a cap rock layer of geothermal system of the Sembalun-Propok area. The result also shows that this layer is also at the near surface below the MT33 observation point. It is possibly associated with the hot spring which is located near the MT33. The third layer is the very conductive layer and possibly associated with the presence of thermal fluids.

This result confirms that the MT phase tensor can be useful as the preliminary analysis before applying 2-D inversion modeling. It helps us to obtain information about the dimensionality of the regional structure and to estimate the regional strike of the study area.

4. Conclusion

To obtain the information of the dimensionality of the regional structure and the regional strike, the MT phase tensor has been applied as the preliminary analysis in the Sembalun-Propok area, NTB, Indonesia. The result shows that the resistivity structure of the Sembalun-Propok can be assumed as 2-D for the higher frequency and the regional strikes are dominated to be NW-SE and NE-SW directions for all frequency ranges. Therefore, it is safe to rotate the impedance tensor to N330°E direction and perform the 2-D inversion modeling for the MT data with frequency range of 0.1 Hz - 1000 Hz. In the following step, we applied the 2-D inversion modeling to Line 2. The result shows that the substructure model supports the geological background of the study area.

Acknowledgments

We would like to acknowledge Mr. Chevy Iskandar who helped the data rotation of the MT data. The submission and presentation this manuscript in the ISFAP 2016 are supported by the budget of the Priority Research of the Indonesian Institute of Sciences (LIPI).

References

[1] Vozoff K 1991 *Investigation in Geophysics No. 3: Electromagnetic Methods in Applied Geophysics* vol 2, ed M N Nabighian (Colorado: Society of Exploration Geophysicists) pp 641-712

[2] Spichak V, Zakharova O and Rybin A 2007 *Proc. XXXII Workshop on Geothermal Reservoir Engineering* (Stanford: Stanford University) SGP-TR-183

[3] Soetoyo 2008 Depresi vulcano tektonik di lapangan panas bumi Sembalun, Lombok Timur, Nusa Tenggara Barat *Buletin Sumber Daya Geologi* **3**(3) (in Indonesian)

[4] Caldwell T G, Bibby H M and Brown C 2004 *Geophys. J. Int.* **158** 457-69

[5] Hill G J, Bibby H M, Ogawa Y, Wallin E L, Bennie S L, Caldwell T G, Keys H, Bertrand E A and Heise W 2015 *Earth Planet. Sc. Lett.* **432** 115-25

[6] Heise W, Caldwell T G, Bertrand E A, Hill G J, Bennie S L and Palmer N G 2015 *J. Volcanol. Geoth. Res.* **314** 39-48

[7] Bertrand E A, Caldwell T G, Hill G J, Bennie S L and Soengkono S 2013 *J. Volcanol. Geoth. Res.* **268** 36-45

[8] Tank S B 2012 *Earth Planets Space* **64** 729-42

[9] Ogawa Y and Uchida T 1996 *Geophys. J. Int.* **126** 69-76

[10] Nasution A, Takada A, Udibowo, Widarto D and Hutasoit L 2010 *Jurnal Geoaplika* **5**(1) 1-9

What did drive extreme drought event in Indonesia during boreal summer/fall 2014?

I Iskandar[1,3], W Mardiansyah[1,3], D Setiabudidaya[1], P Poerwono[1], I Yusyian[2] and Z Dahlan[2]

[1]Jurusan Fisika, Fakultas MIPA, Universitas Sriwijaya, Sumatra Selatan, Indonesia
[2]Jurusan Biologi, Fakultas MIPA, Universitas Sriwijaya, Sumatra Selatan, Indonesia
[3]Pusat Studi Geo-hazard dan Perubahan Iklim, Fakultas MIPA, Universitas Sriwijaya, Sumatra Selatan, Indonesia

Email: iskhaq@mipa.unsri.ac.id

Abstract. There are at least five climate-drivers influencing temporal variability of the Indonesian precipitation system, namely the diurnal cycle, the Madden-Julian Oscillation (MJO), the monsoon system, the Indian Ocean Dipole (IOD), and the El Niño–Southern Oscillation (ENSO). In particular, the IOD and the ENSO events will cause interannual variations in the Indonesian precipitation. During the positive IOD or the El Niño events, Indonesia experiences a severe drought event. On the other hand, during the negative IOD or the La Niña event, Indonesian will have excess precipitation. During 2014, most part of the Indonesian regions experience deficit precipitation leading to a severe drought event. This study is designed to evaluate the dynamics underlying the extreme climate event in 2014. Several climate indices, including the sea surface temperature (SST), outgoing long-wave radiation, and surface winds were evaluated to explain why Indonesia experience severe drought in 2014. The analysis indicates that the tropical Indo-Pacific climate modes (e.g. IOD and ENSO events) were in neutral conditions. It is, then, suggested that negative sea surface temperature anomalies in the Indonesian region, in particular in the eastern part suppressed convective activities leading to deficit precipitation over the maritime continent.

1. Introduction

Indonesia, so-called as the maritime continent, is located in between two oceans (Pacific and Indian Oceans) and two continents (Asian and Australian continents). As a consequence, the Indonesian climate variability is influenced significantly by the air-sea interaction in the Indo-Pacific region. On hourly and day-to-day variations, the Indonesia region is under influence of diurnal cycle. The Madden-Julian Oscillation (MJO) causes intraseasonal variability over the Indonesia region during boreal winter [1, 2]. During the active phase of the MJO event, tropical convection is developed leading to positive precipitation anomaly. On the other hand, the passive phase of the MJO event is associated with suppressed convection activity that causes negative precipitation anomaly.

On seasonal time-scale, the monsoonal wind forces seasonal variability of the Indonesian climate [3]. During the southeast monsoon from May to September, the Indonesian region will experience a dry season, while the northwest monsoon in November – February is associated with a wet season in Indonesia.

The Indonesian region is also under influence of the El Niño-Southern Oscillation (ENSO) from the Pacific Ocean as well as the Indian Ocean Dipole (IOD) from the Indian Ocean [4]. During the positive IOD and/or the El Niño events, most part of the Indonesia is experiencing a severe drought event. On the other hand, during the negative IOD and/or the La Niña events, the Indonesian region will have an excess precipitation.

A common way to highlight the IOD and ENSO events is by examining the pattern of sea surface temperature anomaly (SSTA) in the IOD and ENSO region. The IOD event is characterized by the Dipole Mode Index (DMI) indicating the SSTA different between the western tropical Indian Ocean (10°S - 10°N, 50°E - 70°E) and its eastern counterpart (10°S - 0°S, 90°E - 110°E) [5]. On the other hand, the ENSO event can be evaluated by several indices in the Niño region. However, previous study has suggested that Niño3.4 is most suitable for evaluating the ENSO impact on the Indonesian precipitation [6].

During boreal summer/fall 2014, most part of the Indonesian regions has experienced deficit precipitation leading to a severe drought event. This study is designed to evaluate the dynamics underlying the extreme climate event in 2014, focusing on the anomalously very dry condition during boreal summer/fall 2014. In order to evaluate possible driver of the event, both the IOD and ENSO events are examined. In addition, the local SSTA variability within the Indonesian region is also examined.

2. Data and Methods

2.1. Data

Monthly precipitation data from the Global Precipitation Climatology Centre (GPCC) with a horizontal resolution of 0.5° × 0.5°and covers a period of January 1990 – December 2014 are used in this study [7]. Monthly wind and outgoing longwave radiation (OLR) data are obtained from the National Center for Atmospheric Environmental Prediction–National Center for Atmospheric Research (NCEP–NCAR) Reanalysis are used in this study [8]. The data are available from January 1990 to December 2014 with a horizontal resolution of 2.5° in both longitude and latitude. Monthly extended reconstructed sea surface temperature (ERSST) Version 3 from National Oceanographic and Atmospheric Administration (NOAA) has also been used in this study. The data for a period of January 1990 to December 2014 were used in the present study.

2.2. Method

Monthly values of Dipole Mode Index (DMI) and Niño3.4 Index were used to represent climate variability in the Indo-Pacific region. The DMI is defined as the difference in SST anomaly between western tropical Indian Ocean region (50°E–70°E, 10°S–10°N) and its eastern region (90°E–110°E, 10°S–Equator) [5]. Meanwhile, the Niño3.4 Index is defined as an averaged SST anomaly in the region bounded by 5°N to 5°S, from 170°W to 120°W [9]. Note that the anomaly fields were calculated as a deviation from the monthly climatology of all fields for a period of January 1990 – December 2014.

3. Results

We first evaluate the convective activities during boreal summer/fall 2014 as an indicator for the precipitation. It is shown that the suppressed convective activities during boreal summer/fall 2014 were indicated by the positive OLR as shown in figure 1. Note that the negative OLR anomalies mean that the ocean gains heat leading to increase convective activity. On the other hand, the positive OLR anomalies indicate that the ocean losses heat from the surface heat flux, thus decrease convective activity. During boreal summer (JASO) 2014, it is clearly shown that over the western-central Indonesian region (e.g. Sumatra, Java, Kalimantan, Nusa Tenggara and Sulawesi) and eastern part of Papua, the OLR show positive anomalies. The positive anomalies leads to decrease convective activities.

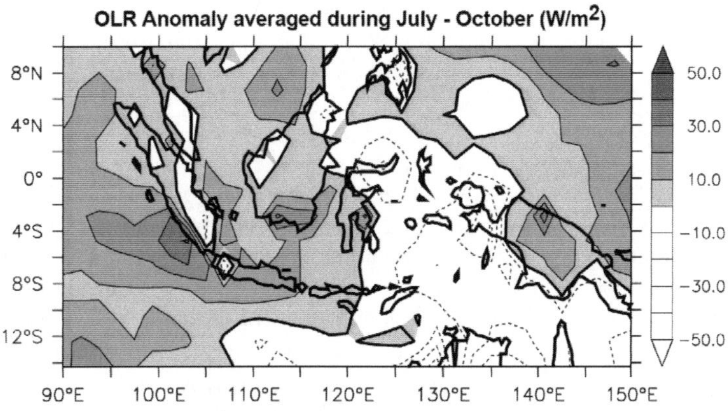

Figure 1. Map of OLR anomalies averaged during boreal summer/fall (July – October) 2014 (W/m^2).

As expected, the suppressed convective activity resulted in low precipitation (figure 2). The precipitation shows robust negative anomalies during boreal spring to boreal fall 2014. Deficit precipitation anomalies are seen over most of the Indonesian region, except over the northern part of Sumatra and over the eastern part of Papua. It is clearly seen that the precipitation patterns are associated with positive OLR anomalies as shown in figure 1.

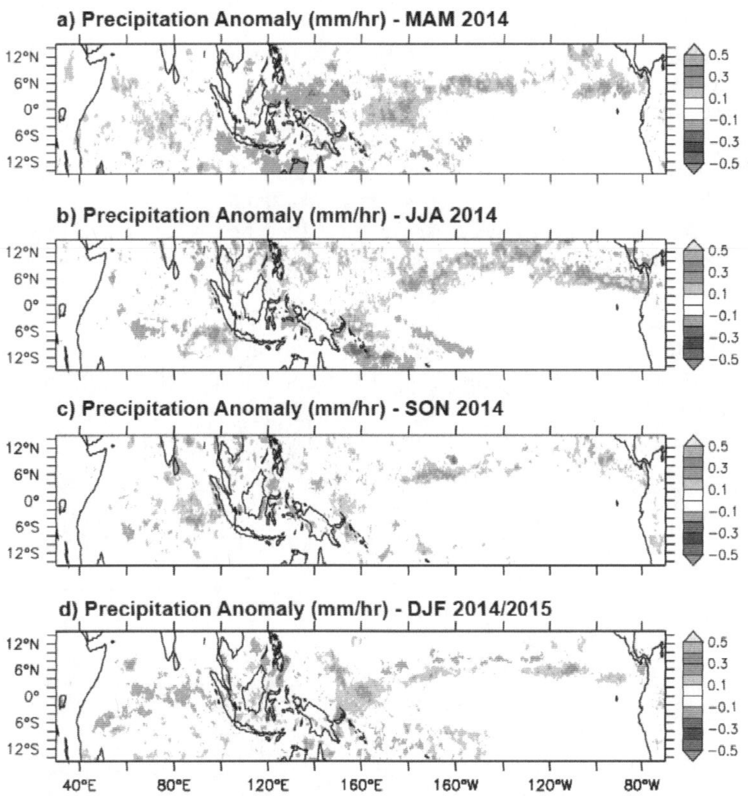

Figure 2. As in figure 1 except for the precipitation anomalies (mm/day).

In order to examine the dynamics underlying the extreme drought event during boreal summer/fall 2014, we first analyze the possible evolution IOD event by evaluating the DMI (figure 3). It was shown that the DMI has negative value during July – September 2014. The negative DMI was first observed in early June 2014. After a short warming from mid-June to early July, the index rapidly decreased until mid-August 2014. The DMI rapidly increased from early September 2014, although it remained negative until late October. A positive value was first observed in early November 2014. Note that the negative DMI indicates negative IOD event indicating positive SSTA in the eastern tropical Indian Ocean [5]. It is well known that convective activities often observed over warm SST. Therefore, the negative DMI is associated with surplus precipitation over the eastern tropical Indian Ocean and the maritime continent [5]. Since the DMI has negative value during boreal summer indicative of the negative IOD event, we may suggest that the extreme drought event during boreal summer 2014 was not correlated with the IOD event.

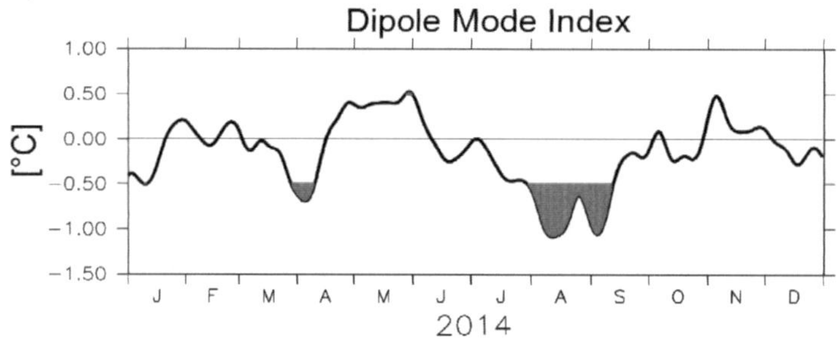

Figure 3. Time series of the Dipole Mode Index (DMI) during January – December 2014. The values exceeding one standard deviation are shaded. Positive (negative) values indicate positive (negative) Indian Ocean Dipole event.

The evolution of the ENSO event was evaluated from the time series of the Niño3.4 index (figure 4). It was shown that positive index was first observed in March 2014 and gradually increased until May 2014. The index was suddenly decreased in late June and reached a negative value in late July/early August 2014. The index again increased in September and remained positive until December 2014. Although the index had a positive value during boreal summer to winter (July – December) 2014, the maximum value of the index was still below one standard deviation. This indicated that the ENSO only has a weak signal/normal condition during boreal summer to winter 2014. Thus we may conclude that extreme drought event in boreal summer 2014 was not correlated with the ENSO signal.

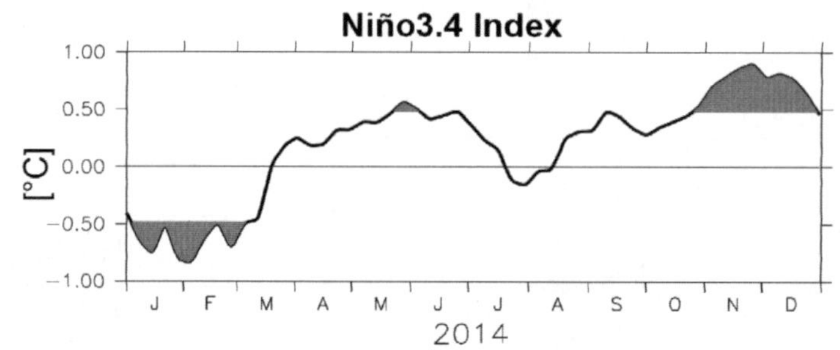

Figure 4. Time series of the Niño3.4 during January – December 2014. The values exceeding a half standard deviation are shaded. Positive (negative) values indicate El Niño (La Niña) event.

We then turn into the local SST variation associated with the monsoonal winds (figure 5). It was shown that negative SST anomaly was observed in the Indonesian seas region in July – August 2014. Negative SST anomaly was also observed in the eastern tropical Indian Ocean off south Java and east Papua as well as in the South China Sea during July 2014 (figure 5a). In August, negative SST anomaly remained in the central and eastern Indonesian seas, off south Java and off east Papua, while warming SST was observed in the South China Sea (figure 5b). In September, the SST fields have shown positive anomalies over most of the Indonesian seas, except in the Maluku Sea and off west Sumatra (figure 5c). Positive anomalies covered most of the Indonesian region in October 2014 (figure 5d). These negative SST anomalies might suppress the convective activities over the Indonesian region and thus reduced the precipitation.

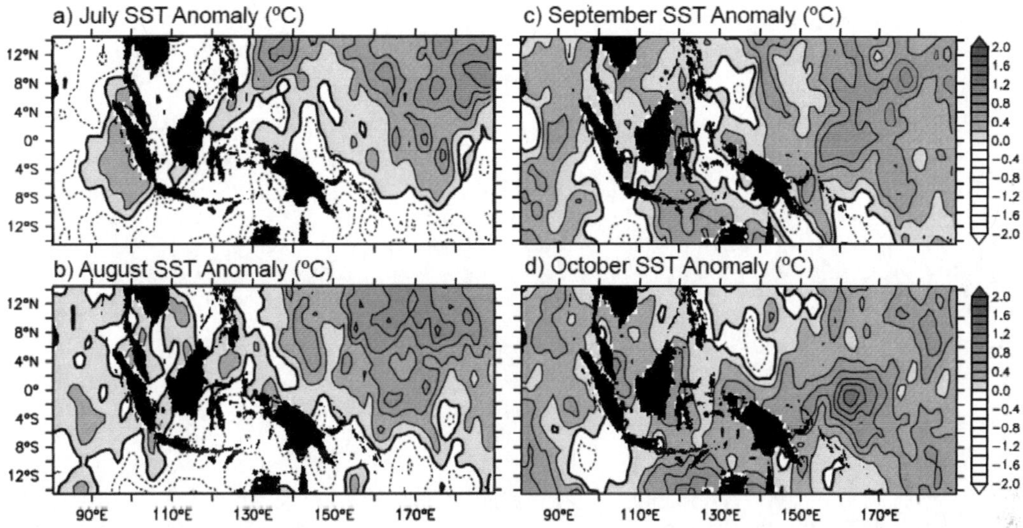

Figure 5. Maps of the sea surface temperature anomaly during boreal summer/fall (JASO) 2014.

4. Conclusion

During 2014, most part of the Indonesian regions has experienced deficit precipitation leading to a severe drought event. This study was designed to evaluate the dynamics underlying the extreme drought event occurring in boreal summer/fall 2014. The results have shown that the IOD event had a weak signal during boreal summer/fall 2014. In addition, the El Niño event has indicated a weak evolution from July – October 2014, suggesting that the El Niño might not cause a severe drought event in 2014 over the Indonesian region.

Meanwhile, examination on the sea surface temperature (SST) within the Indonesian seas region indicates that the SST was significantly below the normal condition. Thus it might suggest that this low SST, associated with the southeast monsoon season, suppressed the convective activities over the Indonesian region indicated by positive OLR anomalies. This suppressed convective activities resulted in negative precipitation anomalies leading to a severe drought in Indonesia during boreal summer/fall 2014.

Acknowledgment

This study is supported by the Ministry of Research, Technology and Higher Education through a Competency Research Grant (HIKOM) 2016.

References

[1] Rui H and Wang B 1990 *J. Atm. Sci.* **47** 357
[2] Zhang C 2005 *Rev. Geophys.* **43** RG2003
[3] Wyrtki K 1961 *Physical Oceanography of the Southeast Asian waters* (California: The University of California) p 225
[4] Yamagata T, Behera S K, Luo J -J, Masson S, Jury M and Rao S A 2004 *Geophys. Monogr.* **147** 189
[5] Saji N H, Goswami B N, Vinayachandran P N and Yamagata T 1999 *Nature* **410** 360
[6] Aldrian E and Susanto R D 2003 *J. Int. Climatology* **23** 1435
[7] Alder R F et al. 2003 *J. Hydrometeorology* **4** 1147
[8] Leibmann B and Smith C A 1996 *Bull. Amer. Meteor. Soc.* **77** 1275
[9] Trenberth K E 1997 *Bull. Amer. Meteor. Soc.* **78** 2771

IOP Conf. Series: Journal of Physics: Conf. Series **817** (2017) 012074 doi:10.1088/1742-6596/817/1/012074

Shallow dipping structure associated to the Cimandiri Fault: evidence from single station teleseismic receiver function analysis

Syuhada and T Anggono

Research Centre for Physics - Indonesian Institute of Sciences (LIPI), Tangerang, Indonesia

E-mail: hadda9@gmail.com

Abstract. The Cimandiri fault zone is located in the eastern part of the boundary between the oblique subduction of the Indo-Australian plate along Sumatera and the frontal subduction along Java. Even though some geological and geophysical investigations have been conducted and focused on this area, the geometry of the fault zone remains poorly resolved. To provide better constraints on the geometry of the Cimandiri fault zone, we analyze teleseismic receiver functions from single permanent seismic station located in the fault zone. We observe that the computed receiver functions show back azimuthal variation in the amplitude converted waves at the first 2 second of receiver functions providing clear evidence for the presence of complicated shallow structure. We implement forward modelling technique using synthetic seismograms with the same parameters as the observed receiver functions. The modelling reveals at least 2 km thick low velocity zone dipping N-E at 30º, which might be correlated to the Cimandiri fault.

1. Introduction

The seismogram from teleseismic P wave recorded at a seismic station consists of the effects of source, sub-surface structure, path propagation, and instrument response. This waveform thus can be utilized for characterizing the local crustal and mantle structure beneath seismic stations by computing a time series of receiver function. The receiver function is obtained by isolating the response of the Earth structure from the source and propagation effects. The receiver functions are calculated both in radial and transverse components. In receiver function modelling, theoretically, the isotropic laterally homogeneous media may cause zero energy on the transverse receiver function. However, the significant energy on this component may appear due to the presence of lateral heterogeneous structure such as dipping or anisotropy layers(e.g., [1]). On the radial receiver function, these factors may identify as variation in amplitude of Ps conversion waves with the back azimuth of the events (e.g., [2]). Therefore, the objective of this research is to identify the presence of dipping or anisotropic structures using receiver function method beneath a seismic station located in the area where the active fault zone may be exist.

The study area is suggested as the location of the presence of fault zone called as Cimandiri fault. The Cimandiri fault zone is located in the southern part of West Java (figure 1), in the boundary separating the Bandung zone and the southern mountains [3, 4]. This fault zone lies along the Cimandiri River, extending from Pelabuhan Ratu Bay to Padalarang. The tectonic history of Cimandiri fault zone is suggested to be related to the Meratus trends, connecting this fault zone to the Meratus mountains in south east of Kalimantan as result of northward subduction beneath Sundaland in the Cretaceous [5, 6]. Some geological investigations have been conducted to investigate the geometry of the fault with various interpretations. For instance, Dardji *et al.* in [4] suggests that the Cimandiri fault zone is a sinistral strike-slip fault; Martodjojo in [7] and Haryanto in [8] show that the Cimandiri fault is a normal fault. A recent geophysical study using audio-frequency magnetotelluric and gravity method suggests that the Cimandiri fault zone is thrust fault [9]. In this research, we follow initial receiver function work conducted in the southern part of west Java [10]. The previous study measured the crustal properties beneath this area. The study found complicated pattern on the receiver functions at station SKJI located near the Cimandiri fault zone, which suggest the presence of shallow complex structure. The new study will elaborate further about this shallow complex interface using receiver function analysis and forward modelling technique.

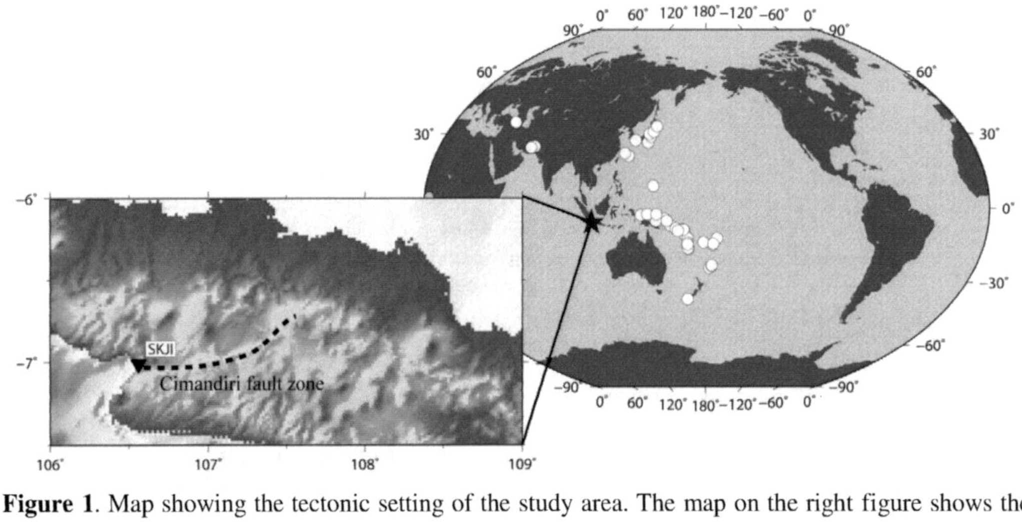

Figure 1. Map showing the tectonic setting of the study area. The map on the right figure shows the distribution of teleseismic events used in this study (white circles). A blue star in the map denotes the study area enlarged on the left figure.

2. Data and Method

In this research, we use the same seismic data recorded at SKJI station as in the previous receiver function study [10, figure 1]. All events used for the study have magnitude greater than 6 and epicentre distance between 30° and 90°. The restriction using large event magnitude is applied because this large teleseismic events might provide good signal to noise ratio, whereas this distance range is used to avoid contamination from unexpected phase arrivals from other reflections (such as from core mantel boundary [11]. To compute receiver functions, we follow the procedure as described in the previous study [10]. First, we select 60 s time windows on each seismogram (beginning at 10 s before P arrival and ending 50 S after the P arrival). We then remove the mean, the trend and the effect of instrument response. We rotate the pair horizontal components into radial and transverse components.

The iterative deconvolution technique [12] is then applied to calculate receiver functions by deconvolving the vertical component seismogram from both radial and transverse components. The stable results of receiver functions are obtained by performing a Gaussian filter with a width of 1.5 filtering frequencies above approximately 0.75 Hz. We assess the quality of the computed receiver function using a least square misfit criterion. The misfit is calculated the difference between the observed radial component and the convolution of the observed vertical component with the predicted receiver function. In this analysis we only use receiver functions with a fit of 90% or above. Figure 2 shows the receiver functions computed on radial and transverse components for station SKJI and plotted as function of back azimuth.

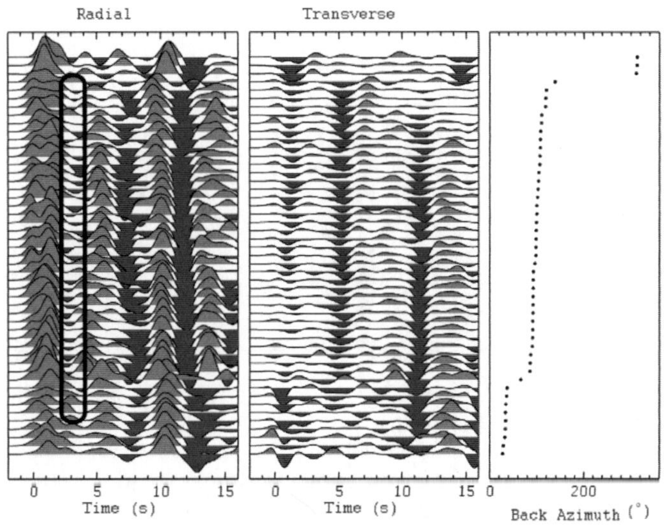

Figure 2. Receiver functions computed for station SKJI plotted as a function of backazimuth. The black ellipse on the radial component represents the feature related to the presence of dipping interface beneath the station.

3. Result and discussion

As found by previous study [10], the pattern of receiver functions on radial component is very complex in the first 5 second implying that shallow complex structure might be presence beneath the station. In this research, we only focus the ensuing discussion on that shallow complicated structure. Therefore, we search unusual pattern of phase arrival for the first 5 s of receiver function in both component such as amplitude reversal or amplitude dependency as function of backazimuth [2] due to dipping or anisotropic shallow structure. The positive pulses on figure 2 denoted by red colour represent an increase in velocity with depth. The negative pulses marked by blue colour represent a velocity decrease. Receiver functions on radial component show a possible polarity flip at 2-3 s between backazimuth of 20° and 120° (a change from positive to negative with increasing backazimuth). We also observe non-zero energy on the transverse component at zero time, which may indicate the presence of dipping structure [1].

IOP Conf. Series: Journal of Physics: Conf. Series **817** (2017) 012074 doi:10.1088/1742-6596/817/1/012074

Table 1. Model parameters used to obtain the best model of receiver functions shown in figure 3.

Depth (m)	P (g/cm^3)	Vp (m/s)	Vs (m/s)	% Anisotropy	Trend anisotropy (°)	Plunge anisotropy (°)	Strike (°)	Dip (°)
400	1730	1800	1000	-	-	-	-	-
3000	1730	1800	1000	10	270	0	90	30
30000	3200	6300	3500	-	-	-	-	-
10000	3200	6800	3800	3	0	0	0	0
	3300	8200	4700	-	-	-	120	30

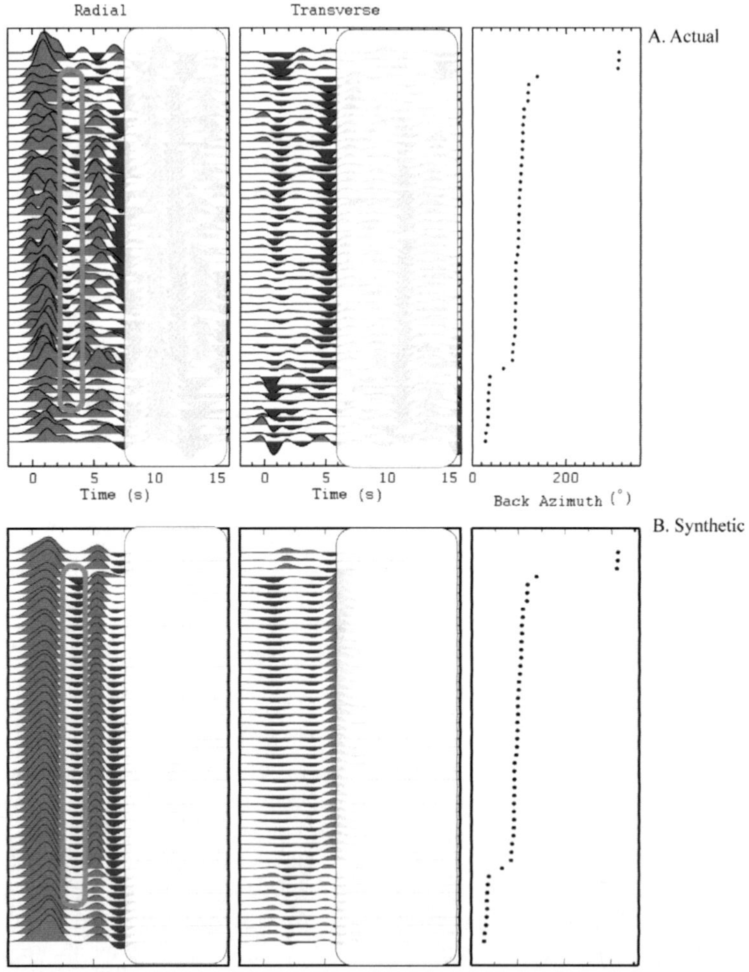

Figure 3. Comparison between receiver functions computed using actual data and syntetic data generated by the RAYSUM code and the model parameter shown in table 1. Here, we search the similarity of the features at the first 5 s (unshaded area). The green ellipse on the radial component shows the feature that fits quite well for both models.

To obtain better constraint on this possible dipping structure beneath the station, we perform forward modelling of receiver functions using RAYSUM package [13]. This code generates theoretical receiver function in the presence of dipping or anisotropic interfaces. In this forward model, we first define parameter for the input model such as the number of layers, the layer thickness, density, Vp, Vs, percent anisotropy, trend and plunge anisotropic fast axis, interface strike and dip. We make a number of trials in this forward model involving a range of structure geometries and a range of model parameters to identify model that fit the observed data. As we only focus on the first 5 second of the receiver functions, we try to match the polarity flip on the radial component receiver functions at 2-3 s. We obtain satisfactory model by adding anisotropic dipping layers structure in the near surface (table 1 and figure 3).

The best fitting model related to the shallow structure beneath SKJI station is shown by a cartoon sketch in figure 4. The model shows the presence of dipping structure with low velocity at shallow depth around 2-3 km with strike and dip around 90° and 30°, respectively. We suggest that this dipping shallow structure might be associated with the Cimandiri fault zone as inferred by other geological and geophysical studies. Dardji *et al.* in [4] suggest that the Cimandiri fault zone has the geological strike around N70-80E interpreted as a sinistral strike-slip fault zone. A recent geo-electromagnetic study also found the evidence for a dipping structure related to the Cimandiri fault zone around the station [9]. The study observed that the shallow structure of the Cimandiri fault zone is characterized by shallow conductive and low density zone at depth of 1-3 km, which is consistent with shallow low velocity zone obtained by this receiver function study.

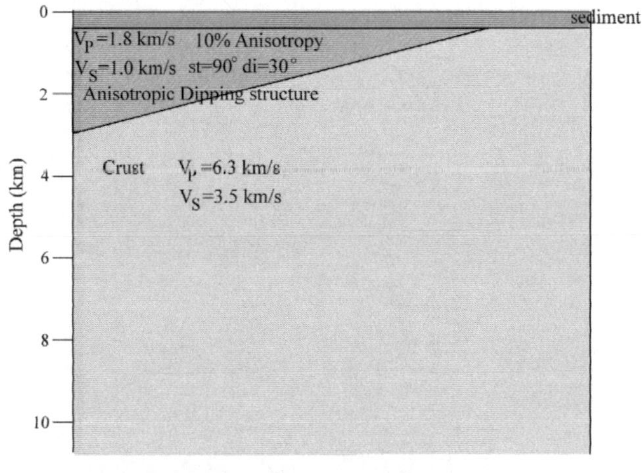

Figure 4. Illustration of the Earth structure beneath station SKJI used to generate the best model receiver functions.

4. Conclusions

We investigate the presence of dipping anisotropic structure related to the Cimandiri fault using teleseismic P wave receiver functions recorded at single permanent seismic station. We find significant backazimuthal variation in the amplitude converted waves at the first 2-3 second of receiver functions suggesting that the complicated shallow structure may be present beneath the station. We perform forward modelling of receiver functions to confirm the likely presence of this structure. The model suggests the presence of at least 2 km thick low velocity structure dipping N-E at 30° which may be related to the Cimandiri fault.

Acknowledgments
We acknowledge Indonesian Agency for Meteorology, Climatology, and Geophysics (BMKG) and GFZ Potsdam for their seismogram data.

References

[1] Jones C H and Phinney R A 1998 *J. Geophys. Res.* **103** 10065-10090
[2] Linkimer L, Beck S L, Schwartz S Y, Zandt G and Levin V 2010 *Geochem. Geophys. Geosyst.* **11** Q01S19
[3] van Bemmelen R W 1949 *The Geology of Indonesia* (The Hague: Government Printing Office) p 732
[4] Dardji N, Villemni T and Rampnoux J P 1994 *J. Southeast Asian Earth Sci.* **9** 3-11
[5] Hamilton W 1979 *Tectonics of The Indonesian Region* (US: US Geological Survey Professional Paper) p 345
[6] Martodjojo S 1984 *Evolusi Cekungan Bogor* Doctoral Thesis (Bandung: Institute Technology Bandung) p 396
[7] Martodjojo S 2003 *The evolution of Bogor, West Java basin* (Bandung: ITB Bandung Publishing) p 258
[8] Haryanto I and Sunardi E 2014 *Bulletin of Scientific Contribution* **12** 78-85
[9] Febriani F 2014 *Subsurface Structure of the Cimandiri Fault Zone, West Java, Indonesia* Doctoral Thesis (Chiba: Chiba University) p 163
[10] Syuhada and Titi Anggono 2016 *AIP Conf. Proc.* vol 1711 (New York: AIP Publishing) p 070001
[11] Macpherson K A, Hidayat D and Goh S H 2012 *J. Asian Earth Sci.* **46** 161-176
[12] Ligorría J P and Ammon C J 1999 *Bull. Seism. Soc. Am.* **89** 1395-1400
[13] Frederiksen A W and Bostock M G 2000 *Geophys. J. Int.* **141** 401-412

Crustal thickness investigation on three broadband stations in Northern Sumatra

T Anggono[1], Syuhada[1], B Soedjatmiko[1] and A Amran[2]

[1]Research Center for Physics - LIPI, Kawasan PUSPIPTEK Serpong, Indonesia
[2] Indonesian Agency for Meteorology, Climatology and Geophysics, Jl. Angkasa I, No. 2, Kemayoran, Jakarta, Indonesia

E-mail: titi.anggono@lipi.go.id

Abstract. We present a preliminary result of crustal thickness in Northern Sumatra from receiver function analysis and grid search. Total of 111 teleseismic events from three broadband stations (TPTI, KCSI, and BSI) of IA-Network were used to calculate the receiver functions. We identified direct *P* and *S* arrival from the receiver function. Converted phases *Ps* were relatively clear for all three broadband stations. *Ps-P* time was estimated about 2 - 3 s, 2 s, and 5 - 6 s for station TPTI, KCSI, and BSI, respectively. We applied *H-k* stacking method to obtain crustal thickness and *Vp/Vs* ratio beneath the three broadband stations. At station TPTI, we obtained the crustal thickness is about 19.54 ± 3.84 km, *Vp/Vs* ratio is about 1.73 ± 0.14. At station KCSI, the crustal thickness was estimated to be 37.07 ± 4.47 km, *Vp/Vs* ratio is about 1.84 ± 0.10. At station BSI, which is located to the north of these two stations, the crustal thickness was estimated to be 40.56 ± 2.26 km, *Vp/Vs* ratio is about 1.81 ± 0.05. These results show relatively large variation of crustal thickness in the Northern Sumatra.

1. Introduction

Sumatra Island is located along the Eurasian and Subducting Indo-Australian plate boundary. Along the western coast of Sumatra, the Indo-Australian plate is subducting beneath the Eurasian plate. Sumatra is dominated by volcanic arc resulting from magma generation along the underlying Wadati-Benioff zone. The incoming Indo-Australian plate is approaching the trench in oblique angle. The motion caused slip partitioning to trench-normal thrust component along the plate interface and strike-slip component accommodated by fault zone [1, 2]. The subduction process causes high seismicity activity along this region. In the last decade, two devastating megathrust earthquakes occurred in the region of northern Sumatra, which were 2004 M_W 9.2 Sumatra-Andaman earthquake and 2005 M_W 8.7 Nias earthquake.

Sumatra has been estimated to be a volcanic arc since late Permian but the intensity went up during the Tertiary period [3, 4]. It was suggested that volcanic rocks were distributed along the west coast of Sumatra during the late mid-Eocene and in Aceh area might have been resulted from back-arc volcanism [5, 6]. Cameron et al. [5] generally sequenced stratigraphy in northern Sumatra into four major volcano-sedimentary sequences. Three are pre-tertiary and the other covers Cenozoic. The oldest rocks belong to the Late Palaeozoic Tapanuli Group which is clastic and probably glaciomarine origin. In this study, we analyze receiver functions to obtain information on the crustal characteristics of its thickness and seismic properties beneath northern Sumatra.

IOP Conf. Series: Journal of Physics: Conf. Series **817** (2017) 012075 doi:10.1088/1742-6596/817/1/012075

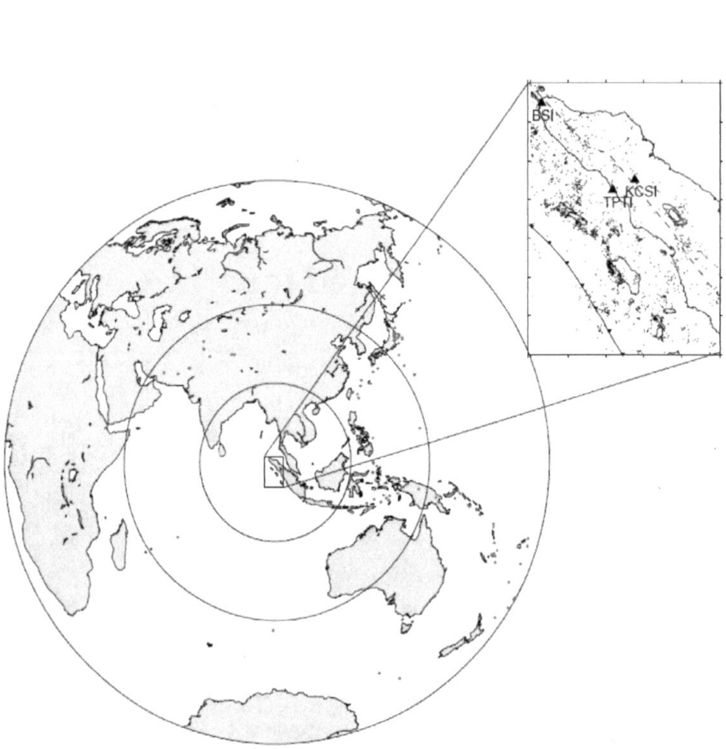

Figure 1. Study area showing the location of three broadband seismometers in Northern Sumatra (black triangles). Locations of teleseismic events used in this study are shown by red circles. Dashed line represents the Sumatra fault zone.

2. Data and method

Investigation of crustal thickness has been extensively studied by using receiver function method [e.g. 7-9]. In this method, the teleseismic waveform is used to image the crustal characteristics beneath the selected seismic stations. It is based on the use of the converted *S*-wave from the teleseismic *P*-wave (*Ps*) at the crustal-mantle discontinuity. The *Ps* phase arrives directly after the direct *P*-wave. The resulting *Ps* conversion has stronger amplitude on the horizontal components. The receiver functions are obtained from the deconvolution of vertical and horizontal components. The deconvolution process will remove the effects of source structure and propagation path.

In this study, seismogram records between 2005 and 2012 from three broadband seismometers of IA-network were used to calculate the receiver functions. Three stations are located in the northern Sumatra Island (figure 1). We computed receiver functions at these three stations from teleseismic events between distances of 30° and 95° with magnitude greater than 5.5. Total of 111 events were used to compute the receiver functions. Most of the events were from the east of the stations, which are from Western Pacific, Indonesian and Kermadec - Tonga subduction zones.

We employed time domain interactive deconvolution to calculate the receiver functions [10]. The raw seismograms were windowed between -10 and 50 s around the *P*-wave arrival and rotated into vertical, radial and transverse components. The iterative deconvolution estimates radial receiver functions using Gaussian pulses and convolves the estimation with the recorded vertical component. We filtered the receiver function with Gaussian width factor of 2.5, which corresponds to frequency band > 1.25 Hz. We selected of high quality of the receiver functions by using misfit criterion. The misfit criterion measures the similarity of receiver function to an ideal problem by calculating the difference between radial component of seismogram and the convolution of the vertical component with the obtained radial receiver function. We used receiver functions with at least 90% fit for further analysis.

We applied *H* - *k* stacking method of Zhu and Kanamori [11] to investigate the crustal properties. *H* refers to the crustal thickness and *k* is *Vp*/*Vs*. The method uses the delay time of *Ps*, *PpPs* and *PpSs+PsPs* to transform the time domain receiver function into *H-k* space for stacking. The best estimation of crustal thickness of *Vp*/*Vs* is obtained when the three phases stack is the most coherent. We applied the procedure to the three broadband seismometers using all possible receiver functions. The effect of lateral variation is reduced by stacking the computed receiver functions with many different ray parameters and back azimuths [12]. For the calculation of *H-k* stacking, we used averaged crustal velocity of 6.3 km/s and weighting parameters of 0.7 for *Ps* phase, 0.2 for *PsPs* and 0.1 for *PpSs+PsPs* multiples.

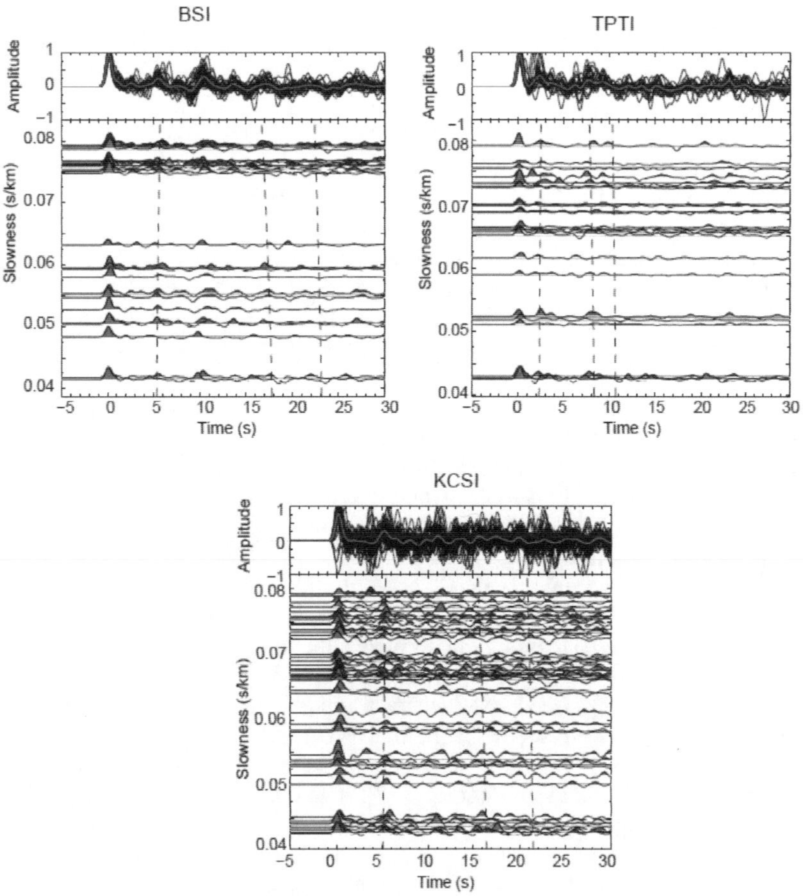

Figure 2. Radial component of receiver functions from three broadband seismometers. The theoretical arrival times of the converted *Ps*, and the *PpPs* and *PpSs+PsPs* multiple phases are also indicated by dashed lines.

3. Results and Discussion

The computed radial receiver functions organized by slowness or ray parameter are shown in figure 2. The observation at stations BSI and KCSI show relatively clear *Ps* arrival around 5.0 s, with *PsPs* and

PpSs+PsPs multiple phases are not clear enough. The mantle-crustal discontinuity or Moho layer can be indicated by the clear appearance of *Ps* phase. At station TPTI, the arrival *Ps* phase seems closer to the direct *P*-wave compared to that of stations BSI and KCSI. *PsPs* and *PpSs+PsPs* multiple phases are not clear enough to be determined manually. Station TPTI shows relatively complicated waveform of receiver functions compared to the other stations. Theoretical arrival times of converted *Ps* phase and the multiple phases are also indicated in the figure 2. It was calculated from the estimated crustal thickness obtained from the *H-k* stacking method.

The estimated crustal thickness or Moho depth and *Vp/Vs* ratio beneath three stations from *H-k* stacking analysis are shown in figure 3. The crustal thickness beneath station BSI was estimated to be 40.56 ± 2.26 km, beneath station KCSI is 37.07 ± 4.47 km and beneath station TPTI is 19.54 ± 3.84 km. *Vp/Vs* ratio beneath BSI station is estimated to be 1.81 ± 0.05, beneath KCSI is 1.84 ± 0.10, and beneath TPTI is 1.73 ± 0.14. The results of *H-k* stacking method are summarized in table 1.

Table 1. Crustal properties from H-k analysis of receiver functions.

Station	Crustal Thickness (km)	*Vp/Vs* ratio	Poisson's ratio	Number of receiver functions
BSI	40.56 ± 2.26	1.81 ± 0.05	0.281 ± 0.017	32
KCSI	37.07 ± 4.47	1.84 ± 0.10	0.291 ± 0.029	63
TPTI	19.54 ± 3.84	1.73 ± 0.14	0.250 ± 0.049	25

The obtained crustal thicknesses from -*k* stacking analysis show large variation in such small area. At stations BSI and KCSI the crustal thickness is about 38 km. Station TPTI shows the crustal thickness differs significantly of about 20 km. Station TPTI is located at the forearc basin, while both KCSI and BSI are located close to the volcanic arc [12]. Such difference in tectonic regime may cause difference in the crustal thickness. Studies by Li et al. [13] also suggested that the presence of sediment beneath a station may affect the phases between direct *P* and *Ps* phases. It was also suggested that the addition of sedimentary layer obscures the phases up to the Moho multiples, which shape the earlier part of receiver function [13, 14]. Station TPTI also has relatively few numbers of receiver functions and complicated waveforms that may affect the result of *H-k* stacking analysis. Several previous studies showed crustal thickness in Sumatra subduction region range from about 16 km at the forearc to about 30 km at back arc basin [9, 15, 16]. Condie [17] suggested that there are three crustal divisions, mainly oceanic (typically about 3-15 km thick), transitional (about 15-30 km thick), continental (about 30-70 km thick). Based on the crustal thickness and the location of the station, we suggest that Sumatra Island might be located at the transitional crust.

Vp/Vs ratio, as well as Poisson's ratio, gives an indication of crustal composition [18]. Based on *Vp/Vs* ratio, Christensen [19] divided rock characteristics into three, which are felsic (~1.7), intermediate (~1.76) and mafic rocks (~1.84). The *Vp/Vs* ratio increases with density as rock composition changes from felsic to mafic as well as to temperature. The changes might be correlated to the variation of the quartz and feldspar content [17]. It estimated that the global average of *Vp/Vs* ratio is about 1.77 by assuming average *Vp* and *Vs* of 6.45 km/s and 3.65 km/s respectively [20].

The obtained *Vp/Vs* ratio ranges between 1.73 and 1.84. Station TPTI, located in Tapaktuan, has *Vp/Vs* ratio of 1.73, which is less than global average of *Vp/Vs* continental crust. Based on the *Vp/Vs* ratio, we suggest the composition beneath this station is mainly felsic. Stations BSI and KCSI show relatively high *Vp/Vs* ratio of 1.81 and 1.84, respectively. Both stations are located or close to the Sumatra fault zone. As mentioned above that the seismic velocity may change due to rock composition, temperature and also pressure. In the fault zone, the rock changes are mostly physical or mechanical properties. The rock velocity changes might occur due to the porosity, fluid/gas content, pressure or heat. These effects generally stronger to the propagation of *P*-waves than to the *S*-waves due to the physical characteristics of the wave propagation. If the fault zone is filled by fluid than the *Vp/Vs* ratio is suggested to be high. However, if the fault zones are empty or filled by gas, the *Vp/Vs* ratio will be low [21]. Isaac et al. [22] through borehole studies at Mozumi fault, Japan estimated that the Poisson's ratio for the fault zone is about 0.263 - 0.393, which is similar to our results from *H-k* stacking analysis (table 1).

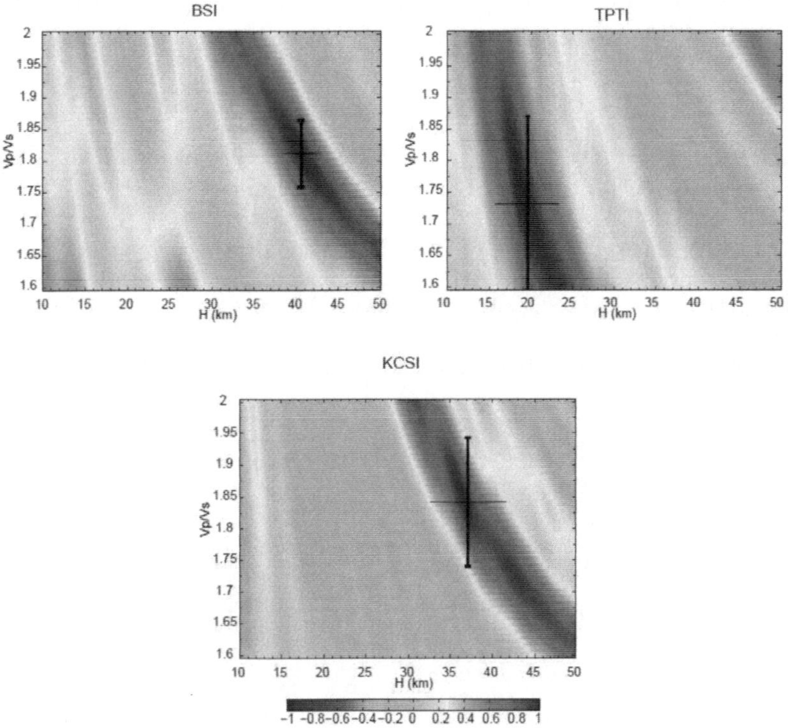

Figure 3. *H-k* stacking results for three broadband stations to estimate the crustal thickness and *Vp/Vs* ratio. Crossbar represent the maximum stacking amplitude with its uncertainty.

4. Conclusions

We have estimated the crustal characteristics beneath three broadband seismometers BSI, TPTI, and KCSI in Northern Sumatra by analysis of receiver functions. We obtained that the crustal thickness range between 20 km in the forearc basin to more than 30 km in the backarc basin. *Vp/Vs* ratio ranges between 1.73 and 1.84. We observed that *Vp/Vs* ratio changes significantly over a relatively short distance at station TPTI and KCSI. We suggest that the difference in tectonic regime may cause such differences.

Acknowledgments

This study was supported by LIPI. Some of the figures were made using Generic Mapping Tools (GMT).

References

[1] Curray J R 2005 *J. Asian Earth Sci.* **25** 187-232
[2] McCaffrey R 2009 *Annual Rev. Earth Planet Sci.* **37** 345-66
[3] Rock N M S, Syah H H, Davis A E, Hutchison D, Styles M T and Lena R 1982 *Bull. Volcanol.* **45**(2) 127-152
[4] Singh S C, Moeremans R, McArdle J and Johansen K 2013 *J. Geophys. Res. Solid Earth* **118**

208-5224
[5] Cameron N R, Clarke M C G, Aldiss D T, Aspden J A and Djunuddin 1980 *Proc. Indonesian Petroleum Assoc. the 9th Annual Convention (Jakarta)* 149-187
[6] Crow M J *2005 Tertiary volcanicity, in Sumatra Geology, Resources and Tectonic Evolution*, ed. A J Barber, M J Crow and J S Milsom (London: The Geological Society, London) p 98-119
[7] Dugda, M T, Nyblade A A and Julia J 2007 *J. Geophys. Res.* **112**(B8) B08305
[8] Park S J, Lee J M and Ryu I C 2009 *Tectonophysics* **472** 158-168
[9] Macpherson K A, Hidayat D, Feng L and Goh S H 2013 *J. Asian Earth Sci.* **64**, 245-255
[10] Ligorria J P and Ammon C 1999 *Bull. Seism. Soc. Am.* **89** 1395-1400
[11] Zhu L and Kanamori H 2000 *J. Geophys. Res.* **105** (B2), 2969-2980
[12] Macpherson K A, Hidayat D and Goh S H 2012 *J. Asian Earth Sci.* **46**, 161-176
[13] Li J, Tian B, Wang W, Zhao L and Yao Z 2007 *Bull. Seism. Soc. Am.* **97**(4), 1355-1363
[14] Syuhada S, Hananto N D, Abdullah C I, Puspito N T, Anggono T and Yudistira T 2006 *Acta Geophysica* **64**(6) 2020-2050
[15] Harmonn N, Timothy H, Frederik T, Andreas R and Penny B 2012 *Geophys. J. Int.* **189** 1306-1314
[16] Bora D K, Borah K and Goyal A 2016 *J. Asian Earth Sci.* **121** 127-138
[17] Condie K C 2005 *Earth as an Evolving Planetory System* (Elsevier Academic Press) p 13
[18] Zandt G and Ammon C J 1995 *Nature* **374** 152-154
[19] Christensen N I 1996 *J. Geophys. Res.* **101** 3139-3156
[20] Christensen N I and Moony W D 1995 *J. Geophys. Res.* **100** 9761-9788
[21] Kaypak B 2008 *J. Geophys. Res.* **113** B07307
[22] Isaacs A J, Evans J P, Kolesar P T and Nohara T 2008 *J. Geophys. Res.* **113** B12408